Progress in
Advanced Dielectrics

Progress in Advanced Dielectrics

editor

Li Jin

Xi'an Jiaotong University, China

World Scientific

NEW JERSEY • LONDON • SINGAPORE • BEIJING • SHANGHAI • HONG KONG • TAIPEI • CHENNAI • TOKYO

Published by

World Scientific Publishing Co. Pte. Ltd.
5 Toh Tuck Link, Singapore 596224
USA office: 27 Warren Street, Suite 401-402, Hackensack, NJ 07601
UK office: 57 Shelton Street, Covent Garden, London WC2H 9HE

British Library Cataloguing-in-Publication Data
A catalogue record for this book is available from the British Library.

PROGRESS IN ADVANCED DIELECTRICS

ISBN 978-981-121-042-6 (hardcover)
ISBN 978-981-121-043-3 (ebook for institutions)
ISBN 978-981-121-044-0 (ebook for individuals)

For any available supplementary material, please visit
https://www.worldscientific.com/worldscibooks/10.1142/11554#t=suppl

Desk Editor: Nur Syarfeena Binte Mohd Fauzi

Typeset by Stallion Press
Email: enquiries@stallionpress.com

Printed in Singapore

Contents

Chapter 1

The Contribution of Electrical Conductivity, Dielectric Permittivity and Domain Switching in Ferroelectric Hysteresis Loops*

Haixue Yan[†,‡,‖], Fawad Inam[†,‡], Giuseppe Viola[†,†], Huanpo Ning[†,‡], Hongtao Zhang[†,§], Qinghui Jiang[¶], Tao Zeng, Zhipeng Gao[†] and Mike J Reece[†,‡,**]

[†]*School of Engineering and Materials Science, Queen Mary University of London, Mile End Road, London, E1 4NS, UK*

[‡]*Nanoforce Technology Ltd, Mile End Road, London, E1 4NS, UK*

[§]*Department of Materials, University of Oxford, Oxford, OX1 3PH, UK*

[¶]*School of Materials Science and Engineering University of Jinan, Shandong 250022, P. R. China*

[‖]*h.x.yan@qmul.ac.uk*

[**]*m.j.reece@qmul.ac.uk*

Triangular voltage waveform was employed to distinguish the contributions of dielectric permittivity, electric conductivity and domain switching in current-electric field curves. At the same time, it is shown how those contributions can affect the shape of the electric displacement–electric field loops (D–E loops). The effects of frequency,

*This article was firstly published as "Yan, H., Inam, F., Viola, G., Ning, H., Zhang, H., Jiang, Q., Zeng, T., Gao, Z. and Reece, M. J. (2011). The contribution of electrical conductivity, dielectric permittivity and domain switching in ferroelectric hysteresis loops, *J. Adv. Dielectr.*, 1, pp. 389–406." Reprinted with the permission of World Scientific Publishing Company.

Medical AB, Järfälla, Sweden) was sintered at 850°C for 5 min under 100 MPa [15]. Doped $BaTiO_3$ powder (Syfer, UK) was SPSed at 1125°C for 5 min under 50 MPa. The bulk $Bi_{3.15}Nd_{0.85}Ti_3O_{12}$ ceramics were SPSed at 850°C and 1000°C for 3 min under a uniaxial pressure of 75 MPa. The corresponding samples are hereafter referred to as BNdT850 and BNdT1000, respectively [18]. Non-textured $La_2Ti_2O_7$ ceramic was sintered at 1350°C for 3 min under a uniaxial pressure of 100 MPa. Grain-oriented, textured $La_2Ti_2O_7$ and $Nd_2Ti_2O_7$ ceramics were prepared using a two-step SPS process [16]. In the first step, the $La_2Ti_2O_7$ and $Nd_2Ti_2O_7$ powders were sintered in a graphite die with a diameter of 20 mm for 3 min under a pressure of 80 MPa at 1250°C. After this stage, the density of the ceramics was high (≥98%), but the grain size was only slightly larger than that of the starting powders. In the second step, these sintered ceramics were placed in a graphite die with a diameter of 30 mm to sinter at 1350°C for 5 min under a pressure of 80 MPa. This caused the grains to grow in direction perpendicular to the applied pressure, resulting in a textured structure.

The bulk densities of the SPSed ceramics were measured by the Archimedes method. All the ceramics had high density (≥98%). The SPSed samples were annealed at 150°C below their sintering temperatures for 4 to 12 h to remove carbon and to reserve any reduction that took place during SPS. X-ray diffraction (XRD) patterns for the ceramics were obtained with an X-ray diffractometer (Siemens D5000) using CuK_α radiation. All the materials were single phases, except PZT according to XRD results. The microstructures were observed using a scanning electron microscope (SEM; JEOL JSM 6300).

Electrodes for electric property measurements were applied by firing silver paste (Johnson Matthey, E1100). The ferroelectric I–E (current–electric field) and D–E(electric displacement–electric field) loops were measured by a ferroelectric hysteresis measurement tester (NPL, UK) at different frequencies. This tester includes an amplifier (TREK model 610E or 609E-6), Function/Arbitrary Waveform Generator (Agilent 33120A), and a Low-noise Current Preamplifier

(Stanford Research Systems, SR570). The measurement procedure involved the application of a triangular voltage waveform. The current from electric conductivity contribution has the same triangular shape of the voltage waveform. The current from the dielectric permittivity contribution is a constant value, being independent of the applied electric field, as shown further on. The total electric displacement was calculated by integration of the total current over a cycle (a-b-c-d-e in Fig. 1(a)) in the middle of two complete cycles. The high temperature measurements were carried out in silicone oil.

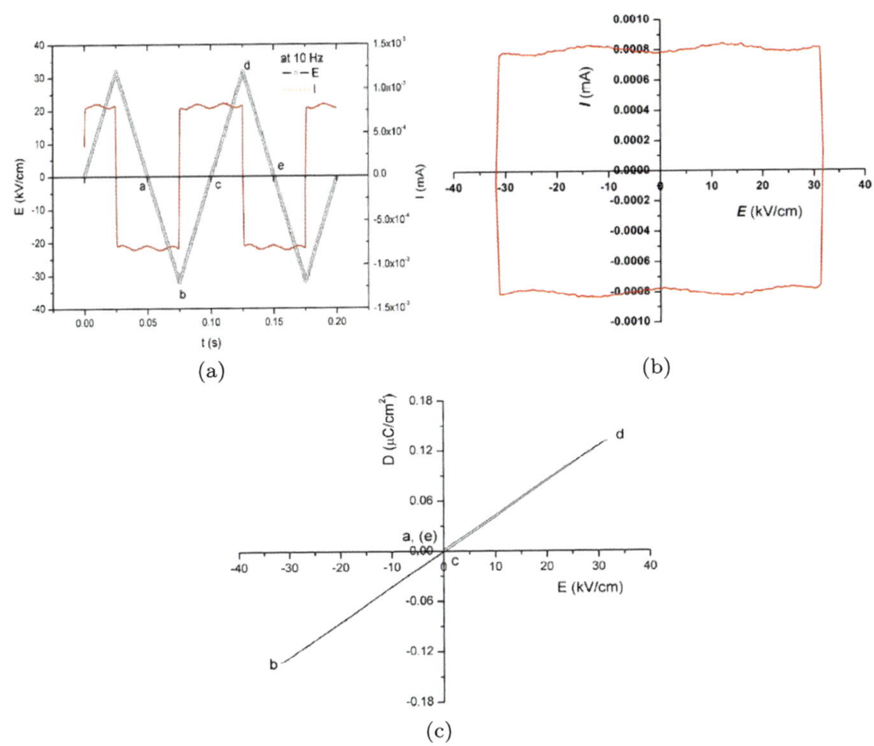

Figure 1. Electric displacement–electric field (D–E) loop measurement of Al_2O_3 ceramics at 10 Hz and 25°C. (a) Electric field versus time and current versus time, (b) current–electric field (I–E) hysteresis loop, and (c) electric displacement–electric field (D–E) loop.

3. Results and discussion

3.1. *The contribution of dielectric permittivity*

Figure 1 shows the electric displacement of Al_2O_3 ceramic, which is that of a highly linear dielectric, but not ferroelectric. The electric resistivity of the alumina is very high ($>10^{15}\,\Omega\,cm$). There is no obvious contribution to current. The current in Fig. 1(a) has only the contribution from the dielectric permittivity. A complete voltage cycle from a-b-c-d-e was selected to analyze the effect of electric field E, on current I (Fig. 1(b)). In the interval c-d (Fig. 1(a)), the electric field can be expressed as $E = \dot{E}t$, where $\dot{E}$ is the rate of the applied electric field and t is the time. The current I (Fig. 1(b)) can be expressed as follows:

$$I = \frac{dQ}{dt} = \frac{d}{dt}(CU) = \frac{d}{dt}\left(\frac{\varepsilon AEs}{s}\right) = \frac{d}{dt}(\varepsilon A\dot{E}t) = \varepsilon A\dot{E}, \tag{1}$$

where, Q is the electric charge, C the capacitance, U the voltage, E the electric field, s the sample thickness, ε the dielectric permittivity, A the sample area, and t the time. The previous equation shows that the current of Al_2O_3 depends on its dielectric permittivity, which is field independent (Fig. 1(b)). The D–E loop (Fig. 1(c)) is obtained by integrating the current with respect to time in the interval a-b-c-d-e and dividing by the area of the sample. The D–E loop of Al_2O_3 is linear and non-hysteretic having only the contribution from dielectric permittivity.

3.2. *The contribution of electric conductivity*

Figure 2 shows the D–E loop of BLFO using a triangular voltage waveform. With increasing electric field E from c to d, the current I is also increasing. The voltage and current signals have the same waveform (Fig. 2(a)) as expressed by:

$$I = \frac{U}{R} = \frac{Es}{R} = \frac{EA}{\rho}, \tag{2}$$

where R is the electric resistance, ρ the electric resistivity, A the area, and s the thickness of the sample. For BLFO ceramics, the resistivity ρ decreases with increasing voltage U. In the I–E loop (Fig. 2(b)),

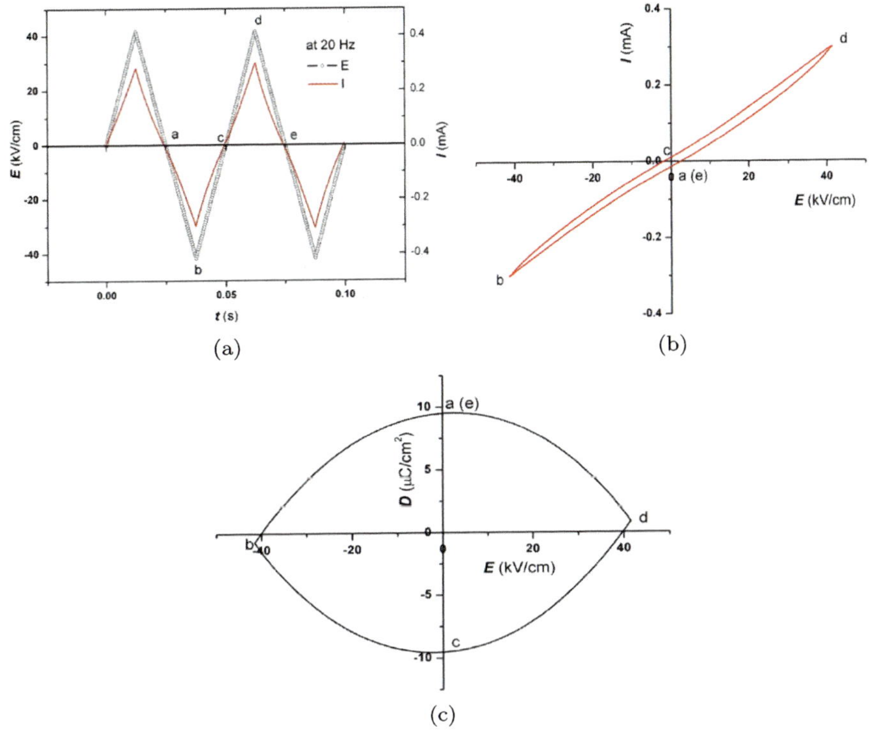

Figure 2. Electric displacement–electric field (D–E) loop measurement of BLFO ceramics at 20 Hz and 25°C. (a) Electric field versus time and current versus time, (b) current–electric field (I–E) hysteresis loop, and (c) electric displacement–electric field (D–E) loop.

the current has contributions from dielectric permittivity and electric conductivity, with the latter being the major contributor. This is because of the valence fluctuation of some of Fe ions from +3 to +2 in BLFO ceramics [19]. In the D–E loops of BLFO (Fig. 2(c)), there is no apparent contribution from domain switching.

3.3. *Domain switching current peak*

Figure 3 shows the ferroelectric hysteresis (D–E) loop for NKN sintered at 850°C. With increasing the electric field E from point c to d (Fig. 3(a)), a current peak appears before the maximum electric field was applied (at point d). This is a clear evidence of

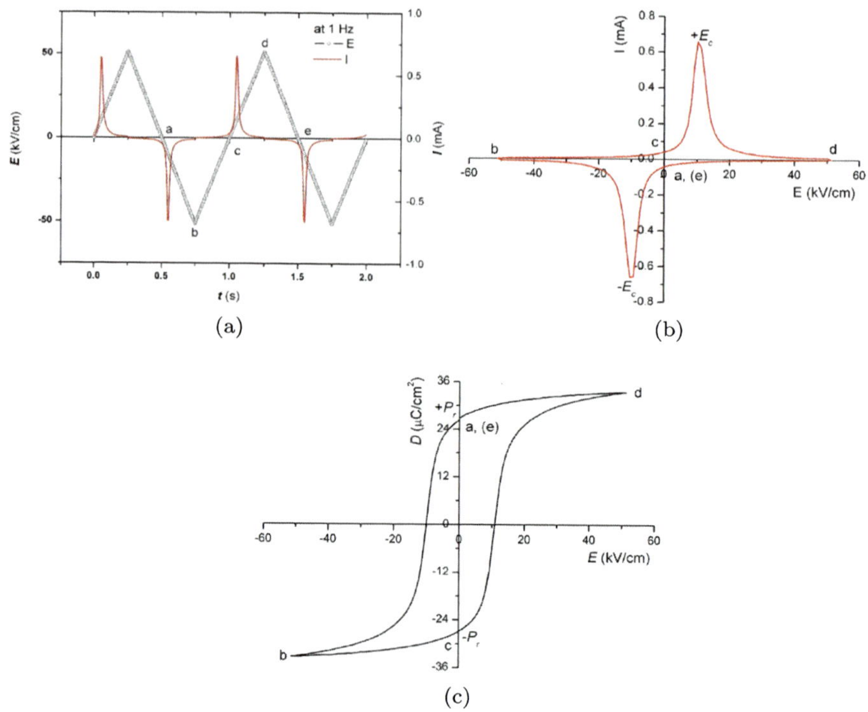

Figure 3. Electric displacement–electric field (D–E) loop measurement of NKN850 ceramics at 1 Hz and 25°C. (a) Electric field versus time and current versus time (b) current (I–E) hysteresis loop, and (c) electric displacement–electric field (D–E) loop.

ferroelectric domain switching. The current at the maximum applied field is very low compared to the peak current caused by domain switching (Fig. 3(b)). In this case, the domain switching contribution dominates the electric displacement compared to the contributions from electric conductivity and dielectric permittivity. The remnant polarization P_r is the D at zero field (Fig. 3(c)). The coercive field E_c can be identified as the electric field correspondent to the current peak (Fig. 3(b)) and in this case it coincides with the electric field correspondent at $D = 0$ in the D–E loop (Fig. 3(c)). The E_c determined from D–E loops is not an absolute threshold field because if a low electric field is applied for a long time, the polarization will eventually switch [5, 21]. This is because the switching is thermally activated and therefore rate and temperature dependent. When reporting E_c, it should be determined from the I–E current peak for a saturated loop and the electric field rate and temperature should also be reported [22].

3.4. *Saturation of D–E loop ferroelectrics*

Figure 4 shows the ferroelectric hysteresis (D–E) loop of PZT 5A. With the increasing field E in the interval c-d, the current also increases (Fig. 4(a)). In the I–E loop (Fig. 4(b)), the current peak may contain the contribution from domain switching beside the

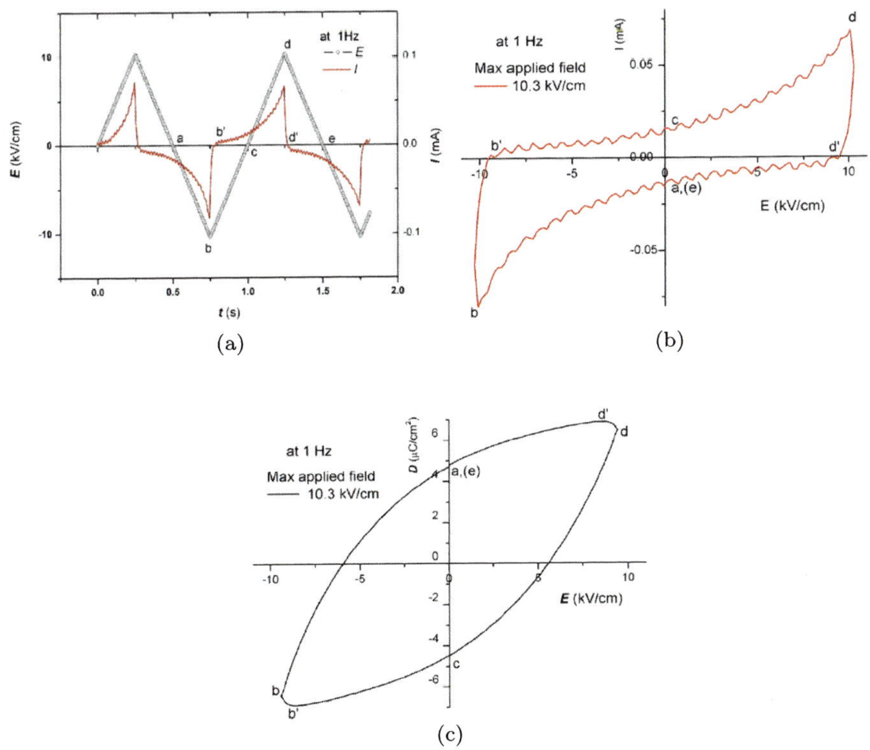

Figure 4. Electric displacement–electric field (D–E) loop measurement of PZT5A ceramics at 1 Hz and 25°C. (a) Electric field versus time and current versus time, (b) current–electric field (I–E) hysteresis loop, and (c) electric displacement–electric field (D–E) loop.

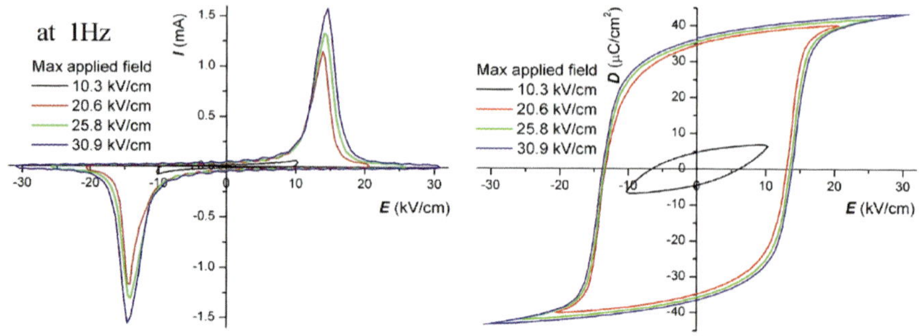

(a) (b)

Figure 5. Electric displacement–electric field (D–E) loop measurement of PZT5A at 1 Hz and 25°C. (a) Current electric field (I–E) hysteresis loop and (b) electric displacement–electric field (D–E) loop.

contribution from dielectric permittivity and electric conductivity. However, in this case, it was not possible to ascertain the occurrence of domain switching only based on low field data (Fig. 4(b)) due to the fact that the I–E loop displays a current peak in correspondence of the maximum electric field (Fig. 4(a)). Compared to the I–E loop (Fig. 5(a)) at higher field, the current at maximum field is lower than the peak current in Fig. 4(b). This indicates that the current peak in Fig. 4(b) is due to the domain switching. In Fig. 4(c), it can be noticed that the D value at point d′ where $I = 0$ (Fig. 4(b)) is higher than that at point d. This is due to the contribution of electric conductivity in the current signal, which prevents a sudden drop in the current between points d and e as occurs when the contribution to the current comes solely from the dielectric permittivity as in the case of alumina (Fig. 1(c)). In Fig. 4(c), the D–E loop obtained by integrating the current respect with time looks hysteretic, but the E field at $D = 0$ is not the E_c because the loop is pre-saturated. For the PZT 5A, the domain switching peak is clear when the field is higher than 10 kV/cm (Fig. 5(a)). The current peak positions and the remnant polarization P_r changed only slightly when the field was further increased (Figs. 5(a) and 5(b)). This suggests that the D–E loops are nearly saturated.

In many publications on ferroelectrics, P_r and E_c are wrongly reported because they are measured from pre-saturated *D–E* loops, such as that shown in Fig. 4(c). It is also impossible to confirm that a new material is ferroelectric only based on a *D–E* loop because even "bananas" can apparently also produce *D–E* loops [11].

3.5. *The effect of temperature*

It is very well known that temperature has a significant influence on the shape of the ferroelectric *D–E* loops. This is due to the fact that the spontaneous polarization, P_s , dielectric permittivity and conductivity of ferroelectric materials are temperature dependent. Figure 6 shows the *I–E* and *D–E* loops of a doped $BaTiO_3$ ceramics at different temperatures. The current peaks corresponding to domain switching are clear in all of the *I–E* curves (Fig. 6(a)). The current has contributions from the dielectric permittivity and domain switching. The absence of any tilt of the *I–E* loops (Fig. 6(a)) indicates that the contribution of the electric conductivity is negligible even at high temperature. The current peaks values decreased with increasing temperature. The *D* corresponding to the maximum applied field decreased with increasing temperature, which is consistent with the decrease of P_s at higher temperatures.

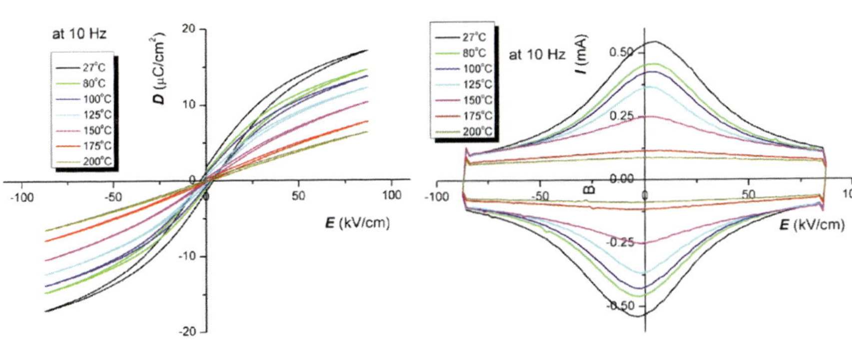

(a) (b)

Figure 6. Electric displacement–electric field (*D–E*) loop measurement of doped $BaTiO_3$ ceramics at 10 Hz and different temperatures. (a) Current–electric field (*I–E*) hysteresis loop and (b) electric displacement–electric field (*D–E*) loop.

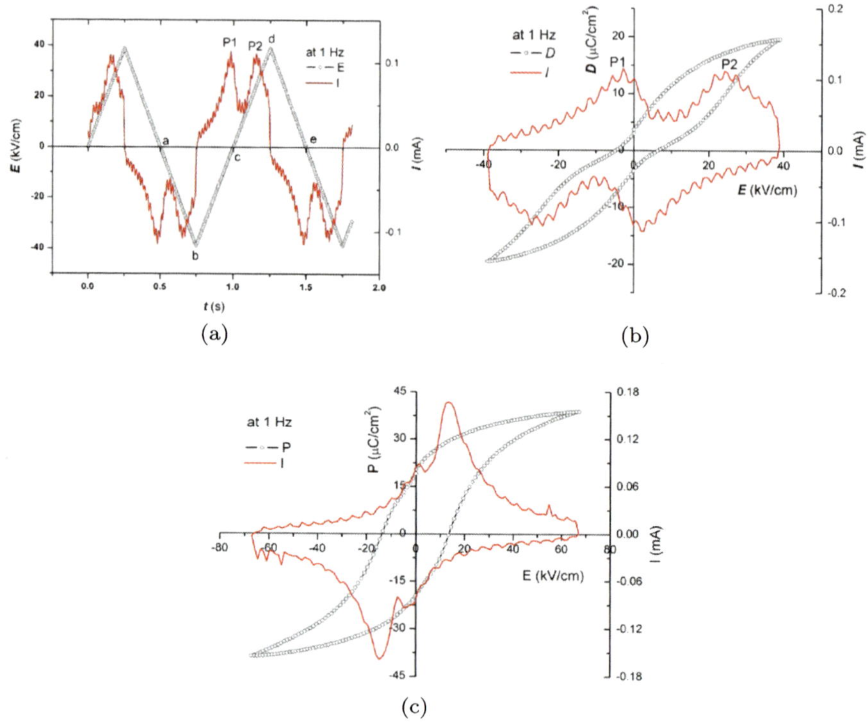

Figure 8. Electric displacement–electric field (D–E) loop measurement of PZT4D ceramics at 1 Hz and 25°C. (a) Electric field versus time and current versus time at low field, (b) I–E and field D–E hysteresis loop at low field, and (c) electric displacement–electric field (D–E) loop.

electric field correspondent to the current peaks in I–E loop (E_c) and the field correspondent at $D = 0$ in the D–E loops (Fig. 9(b)) do not coincide as in the case of NKN (Fig. 3). This further highlights why the coercive field should be defined as the field correspondent to the domain switching current peak. With increasing frequency, P_r($D = P_r$ at zero E) decreases and the coercive field increases. Increasing frequency requires a greater applied driving force to compensate for the reduced contribution from thermal activation. It also reduces the total polarization switching that is achieved for the same maximum applied field [22]. For PZT5H, two orders of magnitude change in frequency produce a 15% change in E_c [22].

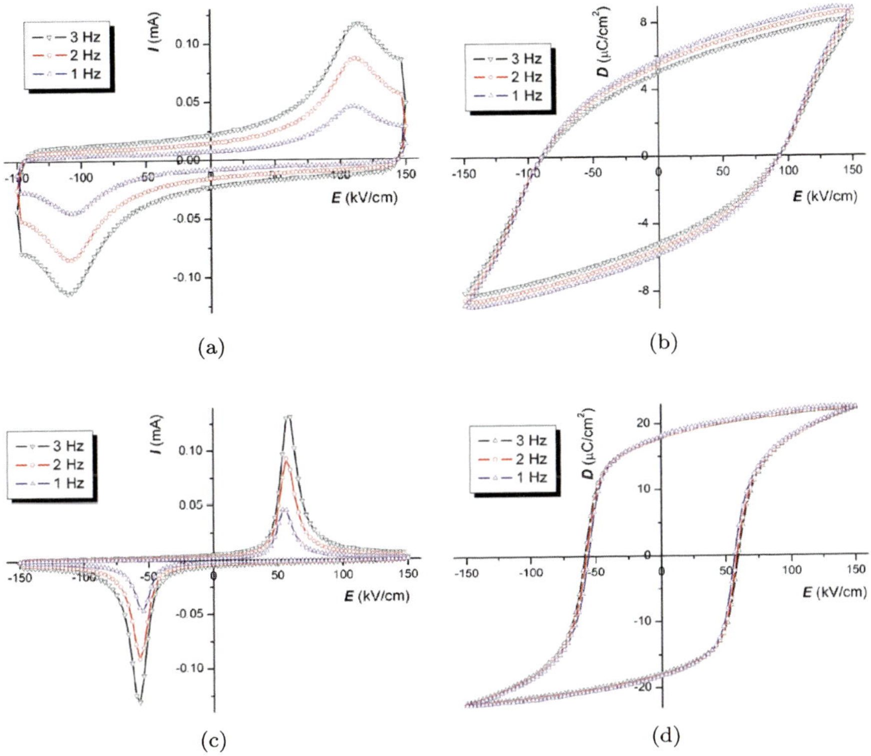

Figure 9. Electric displacement–electric field (D–E) loop measurement of BNdT ceramics at 25°C and different frequencies: (a) I–E hysteresis loop of BNdT850, (b) D–E loops of BNdT850, (c) I–E hysteresis loop of BNdT1000, and (d) D–E loops of BNdT1000.

The effect of grain size on DE loops can be observed by comparing the I–E and D–E loops relative to BNdT850 (Figs. 9(a) and 9(b)) and BNdT1000 (Figs. 9(c) and 9(d)). Based on the same field, BNdT1000 has higher P_r and lower E_c compared to BNdT850. In ferroelectric ceramics, the crystallite grains are randomly oriented and contain several ferroelectric domains. As the domains in one grain attempt to switch under the application of an electric field, they are constrained by the differently oriented neighboring grains [29]. This makes the E_c of ceramics with fine grains much higher than ceramics with coarse grains. This is consistent with the fact that the E_c of single crystals is lower than that of ceramics.

3.8. *The effect of texture*

For $La_2Ti_2O_7$ and $Nd_2Ti_2O_7$ textured ceramics, their XRD patterns from a surface perpendicular to the SPS pressing direction exhibited strong (h00) reflections [16]. The grain orientation factor, f, was 0.80 and 0.82 for the $La_2Ti_2O_7$ and $Nd_2Ti_2O_7$ textured ceramics, respectively [16]. The grains are plate-like and oriented, which is consistent with the XRD data. The thickness direction of the grains corresponds to the long a-axis of their crystallographic unit cell, which is parallel to the pressing direction. Because their ferroelectric spontaneous polarization is along the b-axis, which is perpendicular [⊥] to pressing direction, their ferroelectric switching is restricted to the plane perpendicular to pressing direction [16]. Figure 10 shows the I–E hysteresis loops of random grain oriented (non-textured) and textured $La_2Ti_2O_7$ ceramics. For the textured ceramic, the electric field was applied along a direction [⊥] to the pressing direction. The domain switching peaks were only observed in textured $La_2Ti_2O_7$

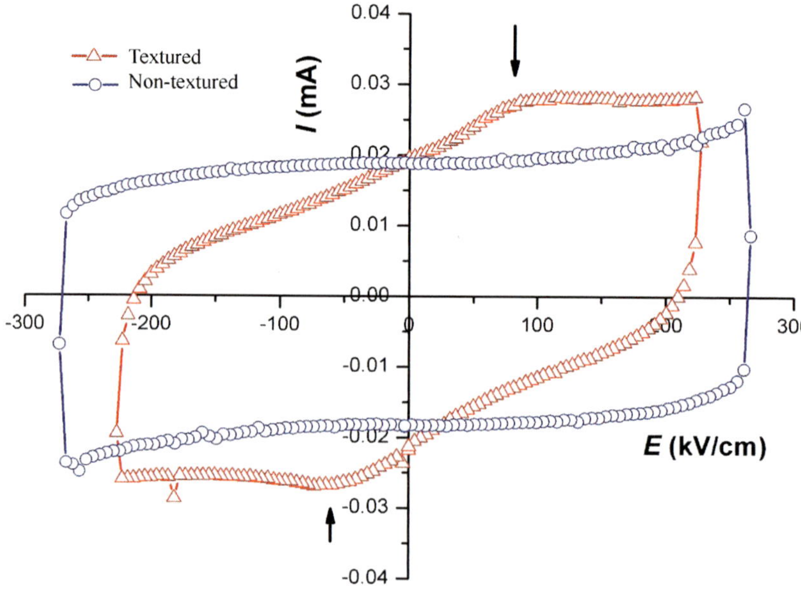

Figure 10. Current–electric field (I–E) hysteresis loop for (a) textured and (b) non-textured $La_2Ti_2O_7$ ceramics at 10 Hz and 200°C.

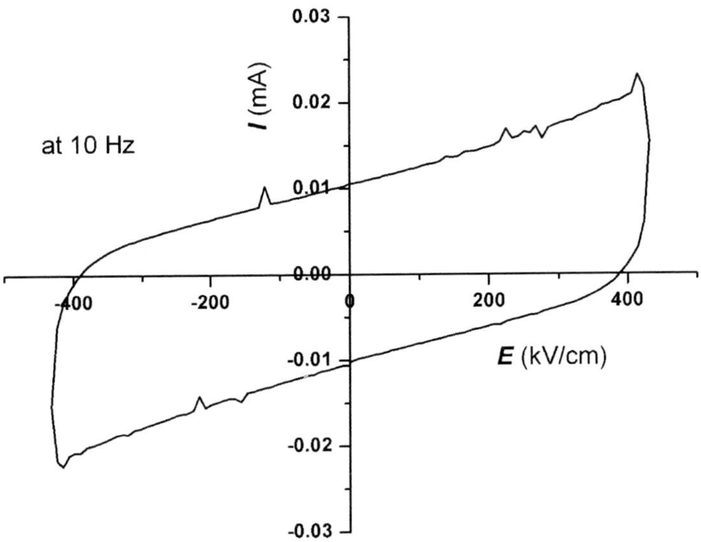

Figure 11. Current–electric field (I–E) hysteresis loop of textured $Nd_2Ti_2O_7$ ceramics at 10 Hz and 200°C.

ceramics, but not for non-textured $La_2Ti_2O_7$ (Fig. 10) [16]. This is because the E_c of textured ceramics is lower than that of ceramics with randomly oriented grains due to their pseudo-single crystal structure [30]. Figure 11 shows the I–E hysteresis loops of the textured $Nd_2Ti_2O_7$ ceramics. The electric field was applied along a direction [⊥] to the pressing direction. Although there was no domain switching current peak in Fig. 11, in fact, domain switching in $Nd_2Ti_2O_7$ was still taking place as confirmed by nonzero piezoelectric constant d_{33} after poling [16]. The domain switching current peaks in Fig. 11 was masked by conductivity and dielectric permittivity contribution.

4. Conclusions

The present Chapter provides a clear understanding of the issues involved in the measurement and correct interpretation of ferroelectric properties. Hysteretic D–E loops do not always provide solid evidence of ferroelectricity and therefore it is recommended that the presence of ferroelectricity must be confirmed by other independent methods, such as piezoelectric or pyroelectric measurements. The I–E loops should be always provided to better interpret D–E loops, and in particular the occurrence of ferroelectric switching. It is better to use a triangular voltage waveform to distinguish the effect of electric conductivity, dielectric permittivity, domain switching, and rate effects and how those contributions can affect the shape of the I–E and D–E loops. When the contribution of the electric conductivity becomes dominant, rounded D–E loops are generated, which cannot be used to estimate the coercive field and the remnant polarization. A peak in the current signal before reaching the maximum electric field indicates that domain switching is taking place. Furthermore, a clear method to define when D–E loops are fully saturated is suggested. D–E loops can be considered fully saturated when a further increase in the applied electric field does not produce a further increase in the P_r, and E_c. They become independent of the maximum field. Also, because switching is rate and temperature dependent, the field rate should also be given when reporting E_c. In addition, it was shown that oxygen vacancies produced an additional current peak in I–E loops and cause pinched D–E loops below a certain electric field threshold. The effect of temperature on D–E loops was discussed for $BaTiO_3$, indicating that P_s reduces with increasing temperature and that domain switching current peaks may still be present above the Curie temperature, due to its relaxor behavior. By reducing the grain size, the coercive field E_c increases and the remnant polarization P_r decreases. The same trend on E_c and P_r was observed by increasing the frequency of the applied electric field. Textured ferroelectric structures present lower E_c than non-textured ferroelectric materials.

Acknowledgments

This work was supported by the UK Royal Society International Joint Project (2009/R2) and Xi'an Jiaotong University international incoming short visits (2008 and 2009).

References

[1] Haertling, G. H. (1999). Ferroelectric ceramics: History and technology, *J. Am. Ceram. Soc.*, 82, pp. 797–818.

[2] Vonhippel, A., Breckenridge, R. G., Chesley, F. G. and Tisza, L. (1946). High dielectric constant ceramics, *Eng. Chem.*, 38, pp. 1097–1109.

[3] Scott, J. F. (2007). Applications of modern ferroelectrics, *Science*, 315, pp. 954–959.

[4] Shaw, T. M., Trolier-Mckinstry, S. and Mcintyre, P. C. (2000). The properties of ferroelectric films at small dimensions, *Annu. Rev. Mater. Sci.*, 30, pp. 263–298.

[5] Damjanovic, D. (1998). Ferroelectric, dielectric and piezoelectric properties of ferroelectric thin films and ceramics, *Rep. Prog. Phys.*, 61, pp. 1267–1324.

[6] Sawyer, C. B. and Tower, C. H. (1930). Rochelle salt as a dielectric, *Phys. Rev.*, 35, pp. 0269–0273.

[7] (2003). Draft 16 of a working document for a proposed standard to be entitled: IEEE standard definitions of terms associated with ferroelectric and related materials, *IEEE Trans. Ultrason. Ferroelectr. Freq. Control*, 50, pp. 1613–1646.

[8] Dunn, S. (2003). Strain behavior of thin film $PbZr_{0.3}$ $Ti_{0.7}O_3$ (30/70) examined through piezoforce microscopy, *J. Appl. Phys.*, 94, pp. 5964–5968.

[9] Dunn, S. and Whatmore, R. W. (2002). Substrate effects on domain structures of PZT 30/70 sol-gel films via PiezoAFM, *J. Euro. Ceram. Soc.*, 22, pp. 825–833.

[10] Dunn, S., De Kroon, A. P. and Whatmore, R. W. (2001). An examination of thin film lead scandium tantalum oxide (PST) using piezoAFM, *J. Mater. Sci. Lett.*, 20, pp. 179–181.

[11] Scott, J. F., Singh, M. K. and Katiyar, R. S. (2008). Critical slowing down of spin fluctuations in BiFeO3, *J. Phys.: Condens. Mater.*, 20, pp. 021001.

[12] Pintilie, L. and Alexe, M. (2005). Ferroelectric-like hysteresis loop in nonferroelectric systems, *Appl. Phys. Lett.*, 87, pp. 112903.

[13] Inam, F., Yan, H. X., Peijs, T. and Reece, M. J. (2010). The sintering and grain growth behaviour of ceramic-carbon nanotube nanocomposites, *Comp. Sci. Tech.*, 70, pp. 947–952.

[14] Rodel, J., Jo, W., Seifert, K. T. P., Anton, E. M., Granzow, T. and Damjanovic, D. (2009). Perspective on the development of lead-free piezoceramics, *J. Am. Ceram. Soc.*, 92, pp. 1153–1177.

[15] Eriksson, M., Yan, H. X., Nygren, M., Reece, M. J. and Shen, Z. J. (2010). Low temperature consolidated lead-free ferroelectric niobate ceramics with improved electrical properties, *J. Mater. Res.*, 25, pp. 240–247.

[16] Yan, H. X., Ning, H. P., Kan, Y. M., Wang, P. L. and Reece, M. J. (2009). Piezoelectric ceramics with super-high curie points, *J. Am. Ceram. Soc.*, 92, pp. 2270–2275.

[17] Ning, H. P., Yan, H. X. and Reece, M. J. (2010). Piezoelectric strontium niobate and calcium niobate ceramics with super-high curie points, *J. Am. Ceram. Soc.*, 93, pp. 1409–1413.

[18] Zhang, H. T., Yan, H. X., Ning, H. P., Reece, M. J., Eriksson, M., Shen, Z. J., Kan, Y. M. and Wang, P. L. (2009). The grain size effect on the properties of Aurivillius phase Bi3.15Nd0.85Ti3O12 ferroelectric ceramics, *Nanotechnology*, 20, pp. 385708.

[19] Jiang, Q. H., Liu, F. T., Nan, C. W., Lin, Y. H., Reece, M. J., Yan, H. X., Ning, H. P. and Shen, Z. J. (2009). High-temperature ferroelectric phase transition observed in multiferroic $Bi_{0.91}La_{0.05}Tb_{0.04}FeO_3$, *Appl. Phys. Lett.*, 95, pp. 012909.

[20] Selbach, S. M., Einarsrud, M. A., Tybell, T. and Grande, T. (2007). Synthesis of $BiFeO_3$ by wet chemical methods, *J. Am. Ceram. Soc.*, 90, pp. 3430–3434.

[21] Burfoot, J. C. and Taylor, G. W. (1979). *Polar Dielectrics and Their Applications* (Macmillan, London).

[22] Chong, K. B., Guiu, F. and Reece, M. J. (2008). Thermal activation of ferroelectric switching, *J. Appl. Phys.*, 103, pp. 014101.

[23] Dul'kin, E., Petzelt, J., Kamba, S., Mojaev, E. and Roth, M. (2010). Relaxor-like behavior of $BaTiO_3$ crystals from acoustic emission study, *Appl. Phys. Lett.*, 97, pp. 032902.

[24] Ravel, B., Stern, E. A., Vedrinskii, R. I. and Kraizman, V. (1998). Local structure and the phase transitions of $BaTiO_3$, *Ferroelectrics*, 206, pp. 407–430.

[25] Xu, B. M., Moses, P., Pal, N. G. and Cross, L. E. (1998). Charge release of lanthanum-doped lead zirconate titanate stannate antiferroelectric thin films, *Appl. Phys. Lett.*, 72, pp. 593–595.

[26] Chu, B. J., Zhou, X., Ren, K. L., Neese, B., Lin, M. R., Wang, Q., Bauer, F. and Zhang, Q. M. (2006). A dielectric polymer with high electric energy density and fast discharge speed, *Science*, 313, pp. 334–336.

[27] Merz, W. J. (1953). Double hysteresis loop of $BaTiO_3$ at the Curie point, *Phys. Rev.*, 91, pp. 513–517.

[28] Tan, Q., Li, J. X. and Viehland, D. (1999). Role of lower valent substituent-oxygen vacancy complexes in polarization pinning in potassium-modified lead zirconate titanate, *Appl. Phys. Lett.*, 75, pp. 418–420.

[29] Li, J. Y., Rogan, R. C., Ustundag, E. and Bhattacharya, K. (2005). Domain switching in polycrystalline ferroelectric ceramics, *Nature Mater.*, 4, pp. 776–781.

[30] Yan, H. X., Reece, M. J., Liu, J., Shen, Z. J., Kan, Y. M. and Wang, P. L. (2006). Effect of texture on dielectric properties and thermal depoling of $Bi_4Ti_3O_{12}$ ferroelectric ceramics, *J. Appl. Phys.*, 100, pp. 070603.

Chapter 2

Decoding the Fingerprint of Ferroelectric Loops: Comprehension of the Material Properties and Structures*

Li Jin[†,§], Fei Li[†] and Shujun Zhang[†,‡,¶]

[†]*Electronic Materials Research Laboratory,*
Key Laboratory of the Ministry of Education and
International Center for Dielectric Research,
Xi'an Jiaotong University, Xi'an 710049, China
[‡]*Materials Research Institute,*
Pennsylvania State University, University Park,
Pennsylvania 16802, USA
[§]*soz1@psu.edu*
[¶]*shujunzhang@gmail.com*

Due to the nature of domains, ferroics, including ferromagnetic, ferroelectric, and ferroelastic materials, exhibit hysteresis phenomena with respect to external driving fields (magnetic field, electric field, or stress). In principle, every ferroic material has its own hysteresis loop, like a fingerprint, which contains information related to its properties and structures. For ferroelectrics, many characteristic parameters, such as coercive field, spontaneous, and remnant polarizations can be directly extracted from the hysteresis loops. Furthermore, many impact factors, including the effect of materials (grain size and grain boundary, phase and phase boundary, doping, anisotropy, thickness), aging (with and without poling), and measurement conditions (applied field amplitude, fatigue, frequency, temperature, stress), can affect the

*This article was firstly published as "Jin, L., Li, F. and Zhang, S. (2014). Decoding the fingerprint of ferroelectric loops: Comprehension of the material properties and structures, *J. Am. Ceram. Soc.*, 97, pp. 1–27". Reprinted with the permission of The American Ceramic Society.

hysteretic behaviors of the ferroelectrics. In this Chapter, we will first give the background of the ferroic materials and multiferroics, with an emphasis on ferroelectrics. Then it is followed by an introduction of the characterizing techniques for the loops, including the polarization–electric field loops and strain–electric field curves. A caution is made to avoid misinterpretation of the loops due to the existence of conductivity. Based on their morphologic features, the hysteresis loops are categorized to four groups and the corresponding material usages are introduced. The impact factors on the hysteresis loops are discussed based on recent developments in ferroelectric and related materials. It is suggested that decoding the fingerprint of loops in ferroelectrics is feasible and the comprehension of the material properties and structures through the hysteresis loops is established.

1. Introduction

1.1. *Ferroic and multiferroic*

Ferroic materials, including ferromagnetic materials, ferroelectric materials, and ferroelastic materials etc., have been extensively studied since their discoveries, due to the fact that the subject of ferroic materials covers a large number of topics in material science, physics, and engineering aspects [1, 2]. A ferroic is a material that adopts a spontaneous and switchable internal alignment, where the alignment of spontaneous magnetization can be switched by a magnetic field in ferromagnetics, whereas the spontaneous polarization alignment can be switched by an electric field in ferroelectrics and the spontaneous strain alignment can be switched by a stress field in ferroelastics; thus, ferroic materials are those which involve at least one phase transition changing the directional symmetry of their respective prototype symmetry [3–6]. For ferroic materials, the magnetization, polarization, and spontaneous strain can be readily controlled by their conjugate magnetic-, electric-, and stress-fields, respectively, thus promising for various functional devices, such as magnetic memories, nonvolatile FRAM (ferroelectric random access memory), electromechanical devices, and shape memory alloy, to name a few [7–10]. In addition, the ferroic orderings can also be tuned by fields other than their conjugates, giving rise to "multiferroic," which combines any two or more of the primary ferroic ordering in the same phase [11, 12]. One of the most appealing aspects of multiferroics

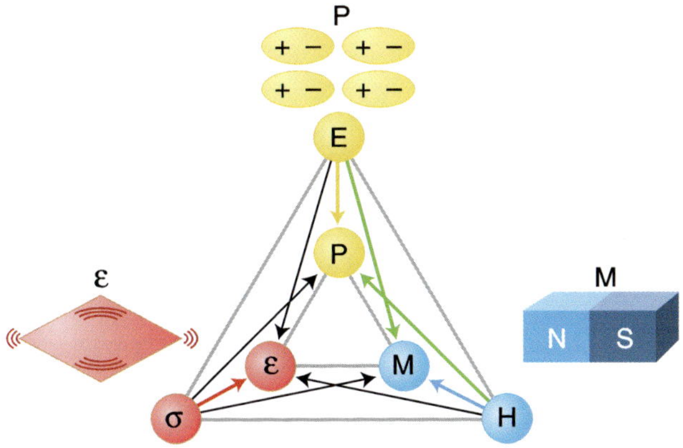

Figure 1. The electric field E, magnetic field H, and stress σ control the electric polarization P, magnetization M, and strain ϵ, respectively. In a ferroic material, P, M, or ϵ is spontaneously formed to produce ferroelectricity, ferromagnetism, or ferroelasticity, respectively. In a multiferroic, the coexistence of at least two ferroic forms of ordering leads to additional interactions. Reprinted with permission from Ref. [8]. Copyright © 2005, American Association for the Advancement of Science.

is the magnetoelectric coupling, for example, not only electric field may control magnetization but also the polarization may be tuned by magnetic field [3]. Figure 1 shows the phase control in ferroics and multiferroics. In one multiferroic material, four physical quantities affect others mutually, significantly expands the scope of functional material research and sheds light on the exploration of new functional devices design.

There are many common features among ferroic materials, such as the existence of domains, which can be switched with respect to external driving fields; the existence of phase transition from parent phase to ferroic phase, which can be regarded as derived by a small distortion of the prototype symmetry; anomalous responses near the ferroic phase transition [polymorphic phase transition (PPT)] showing a strong nonlinear behavior; and strong temperature-dependent properties in the proximity of the PPT, etc. [1, 2]. In this chapter, ferroelectric materials will be surveyed, with emphasis on the hysteresis loops. According to the principle of crystallographic symmetry, ferroelectricity can only be found in crystals with unipolar axis (10 point groups, including 1, 2, m, $mm2$, 4, $4mm$, 3, $3m$,

6, and $6mm$). Among ferroelectric family, materials with an ABO_3 perovskite structure have been extensively studied. As proposed by Slater, the ferroelectric distortion is due to the B cation "rattling" in rigid ion (oxygen) cage [13]. Later Cochran suggested that a lattice mode involving all ions could soften and lead to the displacive instability [14, 15]. Based on *ab initio* calculation including the effects of charge distortion and covalency, it was demonstrated that the hybridization between the titanium $3d$ states and the oxygen $2p$ states in Ti-O octahedra is essential for ferroelectricity [16]. Although the origin of the ferroelectricity is still controversial, these models help us understand the ferroelectricity much deeper than before.

1.2. *Ferroelectric materials*

It is almost one century since the discovery of ferroelectricity in Rochelle salt [17–19], after which, there are many developmental milestones along the way, which have been well reviewed by Kanzig [20], Cross and Newnham [21], Fousek [22], and Haertling [23]. Here, we only survey some important events in the development of ferroelectric materials. In 1921, it was Valasek [17] who first discovered the ferroelectricity in Rochelle salt, later on, ferroelectricity was reported in a new system KH_2PO_4 (KDP) in 1935 [23]. It was generally accepted that ferroelectricity was highly correlated with the hydrogen bonds in the early Rochelle salt and KDP periods [21]. However, with the discovery of the ferroelectricity in simple perovskite $BaTiO_3$ (BT) in 1944, such a hypothesis became invalid [21, 23]. The discovery of BT is very important to the development of ferroelectrics, as this finding opened a door to explore ferroelectric systems with similar crystalline structure, leading to the "perovskite" era. Meanwhile, it has challenged the scientists to model the ferroelectricity without the interference by the complex crystalline structure as encountered in Rochelle salt and KDP. In 1951, the concept of antiferroelectricity was proposed by Kittel, where the chains of ions in the crystal are spontaneously polarized, but with neighboring chains polarized in antiparallel directions [24]. In the same year, this idea was firstly identified in $PbZrO_3$ (PZ) material by Sawaguchi *et al.* [25] and Shirane *et al.* [26]. As triggered by Shirane *et al.*, the solid solution between PZ and PT [$Pb(Zr_{1-x}Ti_x)O_3$, or PZT] has been extensively studied [27–29]. In 1954, Jaffe *et al.* reported a morphotropic phase boundary (MPB) in PZT solid solution [30]. The boundary between tetragonal and rhombohedral phases is nearly vertical, demonstrating temperature-independent properties. Of particular importance is that PZTs with composition at MPB show maximum dielectric and piezoelectric properties, being much higher than those of BT. In addition, dopant strategies have been adopted in PZT ceramics, leading to a series of "hard" and "soft" materials [31, 32]. The technological importance of PZT was established in 1960s and dominated the piezoelectric material market up to now. From that time on, searching high-performance piezoelectric materials in solid solution with an MPB became a standard approach for material scientists. It should be noted that during this period, most of today's practical ferroelectric materials were discovered, including $LiNbO_3$ with corundum structure [33], $KNbO_3$/(K,Na)NbO_3 [34, 35] and $(Na_{0.5}Bi_{05})TiO_3$ [36, 37] with perovskite structure and $PbNb_2O_6$ with tungsten bronze structure [38], to name a few. In 1961, a new type perovskite ferroelectric $Pb(Mg_{1/3}Nb_{2/3})O_3$ (PMN) was firstly synthesized by Smolenskii, which was categorized to relaxor ferroelectric, manifesting itself by the diffused phase transition and strong dielectric dispersion as a function of frequency [39, 40]. Meanwhile, a series of relaxor ferroelectrics with complex perovskite structure, such as $Pb(Zn_{1/3}Nb_{2/3})O_3$ (PZN), $Pb(Yb_{0.5}Nb_{0.5})O_3$ (PYN), and $Pb(Sc_{0.5}Nb_{0.5})O_3$ (PSN), etc., have been widely studied in 1970s [41]. Analogous to PZT, relaxor end-members were reported to form solid solutions with classical ferroelectric $PbTiO_3$ (PT), with MPB compositions being located at PT ~8%–35%, exhibiting high dielectric and piezoelectric properties. Of particular significance is that some relaxor-PT ferroelectrics can be grown into single crystal forms, demonstrating superior piezoelectric and dielectric properties when compared to their polycrystalline ceramic counterparts, which have been studied in 1980s [42, 43]. However, the research fell into troughs at the beginning of 1990s due to the lack of large-size crystals. It was until 1997, Park and Shrout reported large-size relaxor-PT single crystals and their ultrahigh-field-induced strains, being on the order of 1.7%, attracted extensive attentions from both the material scientists and physicist [44, 45]. It was believed as a breakthrough in the past 50 years for the ferroelectric materials [46, 47]. Stimulated by the relaxor-PT single crystals, together with the interest from the lead-free piezoelectric materials and multiferroics, ferroelectrics receives its resurgence at the beginning of this century [10, 37].

1.3. *Hysteresis loop*

The most common feature of the ferroic materials is the occurrence of domain structure through the spontaneous breaking of the prototype symmetry, manifesting themselves as the hysteresis loops with the respective conjugate field. The appearance of the domains is to minimize the free energy when ferroic materials undergo a phase transition from high-temperature symmetric phase to low-temperature phase with a low symmetry. The symmetry of any ferroic phase of a material can be regarded as derived by a small distortion of the prototype symmetry [1]. For example, a BT crystal transfers from a cubic paraelectric phase into tetragonal ferroelectric phase at 120°C during cooling [48]. A boundary separating two adjacent domains is named as domain wall, which is determined by the crystalline symmetry. Due to the large internal stress generated during further cooling, a large domain usually splits into many small domains. The mainstream of ferroelectric research is to study the domains and their response with respect to external impacting factors, such as temperature, electric field, stress, chemical forces, and others. Hysteresis loop, as a simple and effective tool, is the most generally accepted method to understand ferroelectric materials [49].

In principle, every ferroelectric material has its own unique hysteresis loop, as a fingerprint. Through the hysteresis loops, the ferroelectricity could be identified directly. Figure 2 is a typical ferroelectric hysteresis loop, through which the characteristic parameters, such as spontaneous polarization (P_s), remnant polarization (P_r), and coercive field (E_c), can be determined. Owing to the requirement of the energy minima, the grains in polycrystalline materials are always splitting into many domains. The directions of the domains are randomly distributed in such a way to lead to zero net macroscopic polarization. When the external field exceeds

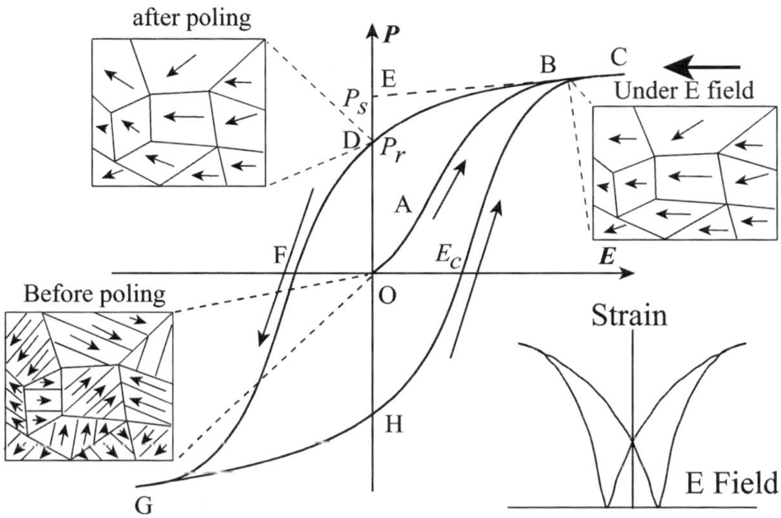

Figure 2. A typical hysteresis loop in ferroelectrics and corresponding domain reversal (polarization rotation) and strain–electric field curve.

the E_c, the polycrystalline ferroelectric ceramic may be brought into a polar state. As shown in Fig. 2, a macroscopic polarization is induced gradually by increasing the electric field strength. The drastic variation in the polarization in the vicinity of E_c is mainly attributed to the polarization reversal (domain switching), while at high field end, the polarization is saturated and the material behaves as a linear dielectric. When the electric field strength starts to decrease, some domains would back-switch, but at zero field the net polarization is nonzero, leading to the remnant polarization P_r. To obtain a zero polarization, an electric field with opposite direction is needed. Such field strength is called the coercive field (or coercivity). With increasing the opposite field strength, a similar rearrangement of the polarization is observed in the negative field part. For ferroelectric materials, the spontaneous polarization P_s may be estimated by intercepting the polarization axis with the extrapolated linear segment, as shown in Fig. 2. Since ferroelectrics usually possess ferroelastic domains (with the exception of $LiNbO_3$, which only has 180° ferroelectric domains), spontaneous strain is also induced with the external electric field simultaneously. Therefore, if the strain is monitored as well as the polarization, a strain–electric field curve, like "butterfly," can be observed.

For ideal ferroelectric system, the observed hysteresis loops should be symmetric. The positive and negative E_c and P_r are equal. In reality, the shape of the ferroelectric hysteresis loops may be affected by many factors, such as thickness of the samples, material composition, thermal treatment, presence of the charged defects, mechanical stresses, measurement conditions, and so on. Their effects on material properties could be well reflected through the loops. Therefore, by decoding the hysteresis loops, we could comprehend the material properties and structures.

There are many classical publications reviewing the hysteresis phenomena in ferroic materials [50–53]. Most of these articles emphasize the physical significance behind the experimental results. However, to researchers working on ferroelectric materials, understanding the relationship between microscopic structures and macroscopic properties through the hysteresis loops is very important, which is the purpose of this feature article. The structure of this feature article is arranged as follows. Starting from the measurement, measuring principles and techniques are introduced. The factors, especially the conductivity, which will give rise to misunderstanding of the loops, are discussed. Based on the features of the loops, four kinds of ferroelectric loops are discussed in terms of the structure and property relationship, the information which can be achieved from the loops is proposed. Following the classification, the factors affecting the hysteresis loops are introduced and discussed with an emphasis on the connection of the microscopic structures and macroscopic loop characteristics. Finally, summary and future perspectives are given.

2. Measurement of hysteresis loops

2.1. *Polarization versus electric field hysteresis loops*

Before decoding the information from the ferroelectric loop, we should make sure that the obtained loops represent the real properties of materials, instead of the potential artifacts [54, 55]. In this section, we will introduce the methods to characterize the

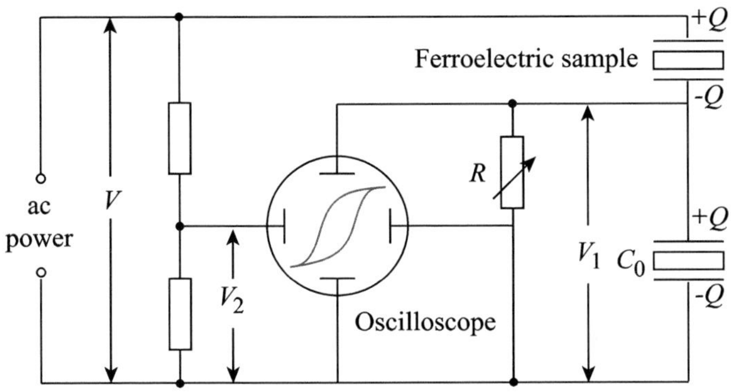

Figure 3. Schematic circuit of the Sawyer–Tower bridge for measuring the P–E characteristics of ferroelectrics.

loop in ferroelectrics. Cautions are made to avoid some obvious misinterpretation, normally due to the existence of the conductivity.

Although the first ferroelectric loop was reported in Rochelle salt by Valasek [17], the development of the ferroelectric study is rather slow. One of the reasons may be due to the difficulty in determination of the ferroelectric hysteresis loop. It was not until 1930, Sawyer and Tower developed the first electronic circuit to characterize the ferroelectric properties of Rochelle salt [56]. Figure 3 shows the schematic connection of the so-called Sawyer–Tower circuit [57]. Through it, an ac voltage is imposed on the surface of an electroded ferroelectric sample, placed on the horizontal plates of an oscilloscope; thus, the quantity plotted on the horizontal axis is proportional to the field across the crystal. A linear capacitor C_0 is connected in series with the ferroelectric sample. The voltage across C_0 is proportional to the polarization (P) of the ferroelectric sample. In fact, dielectric displacement (D) and polarization are connected by Eq. (1):

$$D = \varepsilon_0 E + P, \tag{1}$$

where, D is the charge density collected by C_0. Compared to the larger value of P, the contribution by $\varepsilon_0 E$ can be omitted. Therefore, the obtained D is considered as P in practice. With the well-developed electronic techniques, the Sawyer–Tower circuit is no longer used as its original form. Most testing systems for P–E relationship can be achieved by commercial apparatus mainly form two companies, i.e., aixACCT [58] and Radiant [59]. In the equipment, P is collected through charge or current integration technique following with compensation by a variable resistor R (see schematic circuit in Fig. 3) [57].

Although the measuring technique is not the main obstacle to study the P–E hysteresis loops, interpreting the loops is much more challenging to the neophyte researchers working on ferroelectrics. Misinterpretations of the P–E hysteresis loops are frequently made due to the strong interference by conductivity. To understand the information from the P–E loops, let us consider three types of samples with different electric features. Figure 4 shows the relationships among electric field (E), current (I), and polarization (P) with respect to ac electric field for a linear resistor, a linear capacitor, and a ferroelectric crystal, respectively. Here, the term "linear" means that the resistance or the capacitance does change with respect to electric field (both triangular and sinusoidal waveforms can be used for the measurements). For linear resistor, I keeps the same phase with E, resulting in a linear relationship between them. The corresponding P–E characteristic is shown in Fig. 4(a). Clearly this resistor exhibits a symmetric loop with respect to both horizontal and vertical axis. The case for linear capacitor is shown in Fig. 4(b). Since I of the capacitor is the differential of E with time, a constant I is obtained with respect to the constant E, but changes the sign for the reversal of E direction. As observed in the I–E curve, a square loop is illustrated with a clockwise flowing direction. Note that the corresponding P–E characteristic of such a linear capacitor shows a linear feature. In contrast, the case for a real ferroelectric sample is totally different from the above two cases. As shown in Fig. 4(c), with increasing the electric field amplitude, there are obvious current peaks through the sample. The electric field correlating to the maximum current is identical to the E_c determined by the P–E loop. It should be noted that the current maxima do not occur at the maxima of electric field, due to the fact that the current is related to the domain reversal, other than the conductivity. Only the

(a)
(b)
(c)

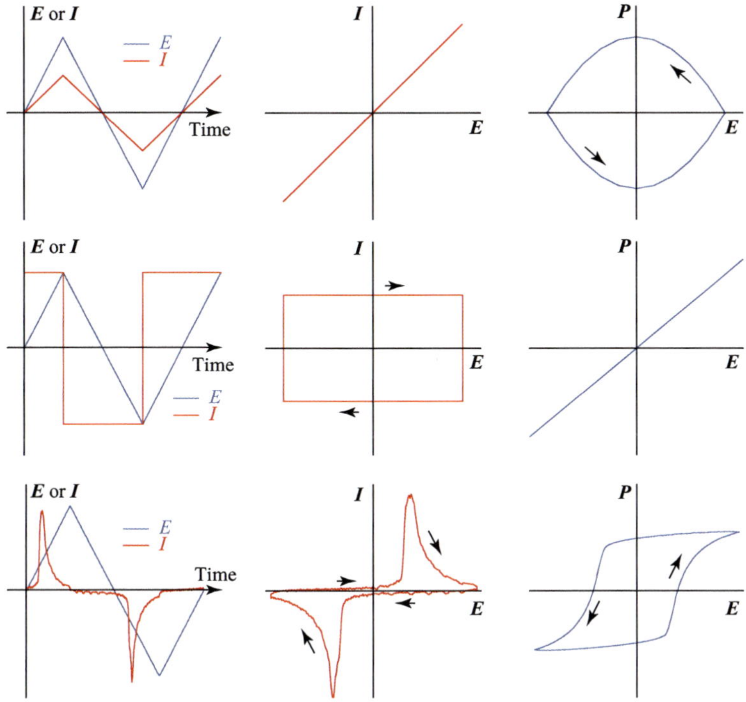

Figure 4. Schematic plots showing typical relationships among electric field (E), current (I), and polarization (P) with respect to a cycle of triangular electric field for (a) a linear resistor, (b) a linear capacitor, and (c) a ferroelectric crystal, respectively.

current maxima resulting from the domain reversal or switching is the sign of ferroelectricity. In this case, the P–E relationship exhibited real hysteresis loop, which always possess an anticlockwise feature between P and E.

Note that we only determine the charge owing to the switching of the polarization using the Sawyer–Tower circuit. In this case, we assume an ideal ferroelectric insulator [60]:

$$Q = 2P_r A. \tag{2}$$

However, the real case of ferroelectric determination is more complex, as conductivity always coexist with the capacitive ferroelectric samples. Thus [60],

$$Q = 2P_r A + \sigma E A t, \tag{3}$$

where σ is the electrical conductivity and t is the measuring time. It will deteriorate the identification of the ferroelectricity if the conductivity is large. According to the criterions given by Dawber *et al.*, "large" in this case is $\sigma > 10^{-6}$ S/cm, whereas "small" is $\sigma < 10^{-7}$ S/cm [60]. As shown in Fig. 5(a), for a real ferroelectric sample, a transition from typical ferroelectric hysteresis loop into round loop was observed with increasing the conductivity gradually. With large enough conductivity, the polarization contributed by domain switching is totally submerged. It can be considered as a combination of the P–E characteristics shown in Figs. 4(a) and 4(c). In this case, there is evidence for ferroelectricity; however, it is difficult to be extracted from the conductivity. Figure 5(b) gives

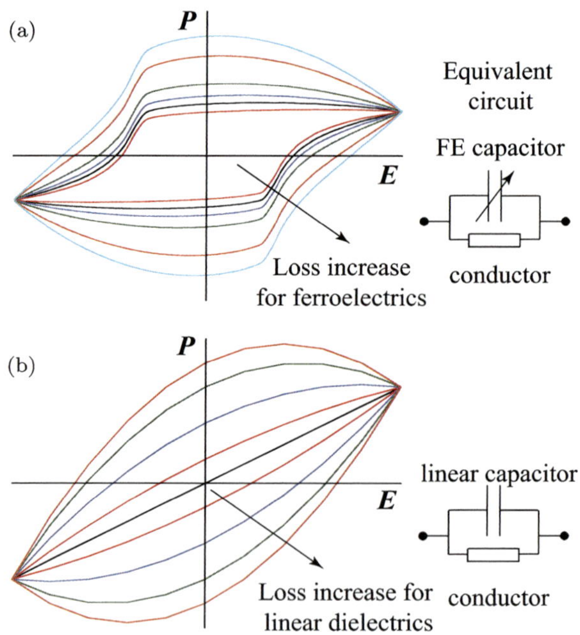

Figure 5. Schematic illustrations of the P–E characteristics for (a) a ferroelectric sample with increasing conductivity and (b) a linear capacitor with increasing conductivity, with the corresponding equivalent circuits.

another example, in which a capacitor and a resistor are combined. Though hysteresis features appear with increasing conductivity, neither saturated polarization nor switching current can be observed in this combined system, thus not relate to the ferroelectricity. In undoped $BiFeO_3$ thin film, similar P–E loops as shown in Fig. 5(a) were reported [61]. It was also found that significant improvement of the ferroelectric properties of $BiFeO_3$ thin film was achieved through control of electrical leakage by Nb doping.

It should be emphasized here that the conductivity in ferroelectrics also presents itself in hysteresis loops through another way, that is, a large gap between the starting and ending applied fields. Figure 6 gives the P–E hysteresis loops measured on $BiFeO_3$ ceramics, exhibiting a large discrepancy between the starting and ending fields, that is, so-called "gap", inherently associated with the high conductivity due to the existence of Fe^{2+}–Fe^{3+} [62].

2.2. *Strain versus electric field curves*

To measure the induced strain as a function of electric field, the key point is the determination of the displacement, which usually ranges

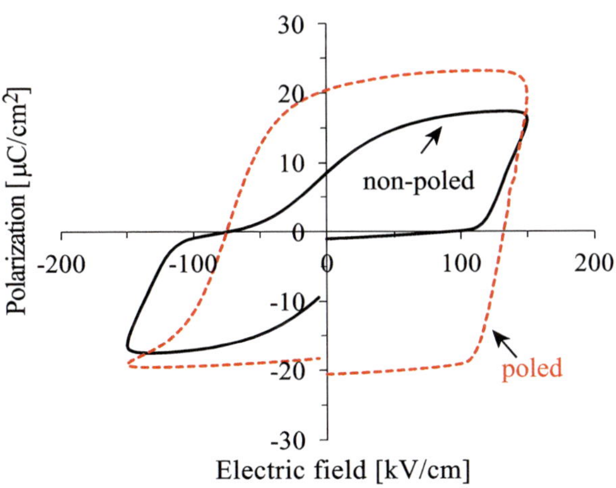

Figure 6. P–E hysteresis loops of $BiFeO_3$ ceramics sintered before (nonpoled) and after poling at 80 kV/cm of DC electric field (poled). Reprinted with permission from Ref. [62]. Copyright © 2012, AIP Publishing LLC.

from 1 nm to 100 μm, depending on the studied material and its dimension. There are two main methods (A) and (B) that have been generally used.

2.2.1. *Magnetic induction method (LVDT)*

The critical part for realization of this method is a linear variable differential transformer (LVDT), which is an electromechanical transducer that produces an electrical output proportional to the displacement of a separate, moveable, high-permittivity core [63]. Normally the transformer consists of three windings or coils, one primary and two secondary. The two secondary windings are connected in series. Application of an ac signal to the primary winding V_{p} induces a magnetic field inside the transformer. The magnetic flux couples to the secondary windings via the moveable core, which leads to an output signal V_{s}. The two secondary windings are connected in such a way that the output voltage is the difference between the induced signals in each winding. Practically, the displacement of the core could be coupled to the displacement of a sample under applied field via a stiff and nonmagnetic metal rod. For more information, readers could refer to Ref. [64].

2.2.2. *Optical method*

This method is based on the reflection of monochromatic light [65]. To this method, the photonic sensors by MTI Instruments Inc. (Albany, NY) are frequently used. The photonic sensor is a fiber-optic measurement system designed for both vibration and displacement measurement. In principle, it can measure displacement in the range from 10 nm to 5 mm statically or at frequency up to 150 kHz. The probe consists of a bundle of optical fibers, half of which emit light, while the other half receives the light reflected back from an optically reflective surface. The intensity of reflected light detected by the sensor is a function of the distance between the probe and the sample surface. When the probe is in contact with the reflective surface ($x = 0$), no light is reflected back to the senor ($I = 0$). When the probe is placed a long way from the reflective surface ($x = \infty$), no light is reflected back and again the intensity is zero. Between these two points, there is a position of maximum intensity ($I_{\max}$). To practice, there is a region of linearity between the maximum in intensity ($I = I_{\max}$) and the zero point ($I = 0$). By calibrating the position of maximum intensity for surface of varying reflectivity, and therefore different reflected intensities of light, the sensor can give an accurate measure of the distance between the surface and probe and thus the distortion of the sample.

Recently, interferometric methods including homodyne, heterodyne, and Fabry-Perot techniques are also frequently used to measure small displacements [66]. Two main optical schemes have been proposed for homodyne interferometer: the single beam Michelson interferometer, in which the displacement of only one of the two major surfaces of the sample is monitored [67], and the double-beam Mach–Zehnder interferometer in which the difference of the displacements of both major surfaces of the sample is taken into account [68].

2.2.3. *Piezoresponse force microscopy*

In addition, piezoresponse force microscopy (PFM) has been used to determine the local movement of the domain walls, with a resolution of the order of nanometer scale [69, 70]. By applying the electric field to the sample through the needle of the PFM, not only domain wall evolution with respect to the driving field can be readily recorded by this technique but also the electric field-dependent piezoelectric properties can be obtained. This method offer more information of the local piezoelectric properties than traditional methods.

2.2.4. *Strain gauge*

Strain gauge method has also been used for measuring the strain versus electric field curves of material. In this method, the strain of materials is monitored by strain gauge. This method is preferable to measure the coupling effect of strain versus electric field under mechanical stress loading [71–73].

3. Classification of hysteresis loops

There are many kinds of ferroelectric systems reported since the discovery of the ferroelectricity in Rochelle salt, and each system possesses unique characteristic hysteresis. In this section, the ferroelectric loops are categorized into four groups according to their morphological features, then the information obtained from the hysteresis loops is discussed and the potential applications related to these loops are introduced. Finally, the accompanying strain–electric field (S–E) curves are discussed based on the classification of P–E hysteresis loops.

3.1. *Classification of polarization–electric field hysteresis loops*

3.1.1. *Classic ferroelectric loops*

The most common ferroelectric hysteresis loop possesses the feature as shown in Fig. 2, which can be categorized into classic hysteresis loop, in which the polarization shows a drastic variation when the external electric field is in the vicinity of the coercive field, demonstrating that most domains are reversed or switched by the applied external field with a strength of coercive field. Strictly speaking, the P–E loop near the E_c is not absolutely perpendicular to the horizontal axis suggesting that the domain wall motion or switching yet exists above the coercive field. This phenomenon is attributed to the clamping of domain walls, which can be electrical, mechanical, or chemical driven. For different materials, the slope of this part varies significantly. At high field region, most of the domains align themselves along the electric field direction and the polarization approaches its saturated value, with a linear response between the polarization and electric field.

From the P–E loops, the spontaneous polarization P_s, remnant polarization P_r, and coercive field E_c can be easily achieved. The polarization is very important for the evaluation of piezoelectric characteristics of ferroelectric materials, following Eq. (4) [74, 75]:

$$d \propto \varepsilon P_s Q, \tag{4}$$

where ε is the dielectric permittivity and Q is the electrostrictive coefficient. Generally, the P_r values for ferroelectric materials fall in a wide range, being on the order of 0.2–0.5 C/m^2 for perovskite bulk materials, whereas the highest value of P_r was reported to be $\sim 0.7\,C/m^2$ for $LiNbO_3$ crystals [50]. Meanwhile, the coercive field, which is the indicator of the "hardness" of the domain reversal, was found to be in the range of 0.5–150 kV/cm, where the $BiInO_3$-$PbTiO_3$ perovskite ceramic with tetragonal phase was reported to possess a very high value of 125 kV/cm [76].

It should be noted even for the same material system, ferroelectric hysteresis loops of single crystalline and polycrystalline ceramic samples show a large difference. This is mainly attributed to the clamping effect of domains with respect to grain boundaries. It can be seen in Fig. 7, the shape of the P–E loop for BT crystal is rather square, whereas for BT ceramics the loop is slanted at a certain degree. By applying the same electric field, saturated polarization can be induced in BT crystal instead of ceramic, suggesting that the polarization switching in crystal is much easier than that in ceramic, owing to the absence of grain boundary. In addition, due to the crystal symmetry, domains in tetragonal BT single crystal could be switched completely with respect to the external field applied along [001] direction. In contrast, because of the random distribution of

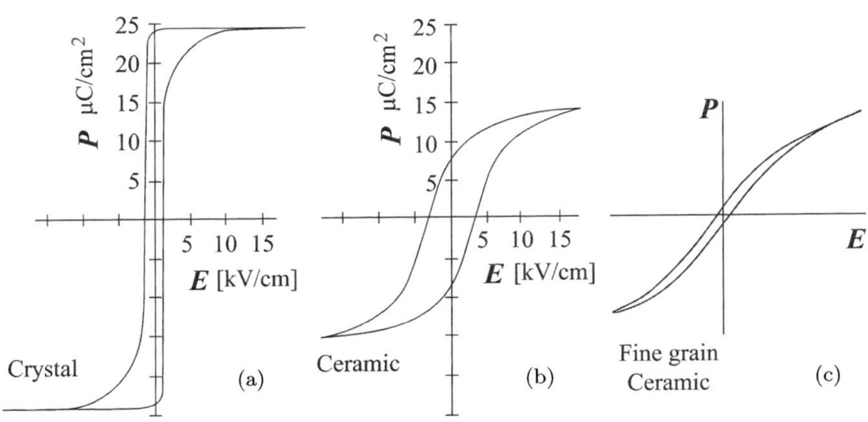

Figure 7. Room-temperature P–E hysteresis loops for BT single crystal and ceramics with different grain size. (a) BT single crystal, (b) BT ceramics with coarse grain, and (c) BT ceramics with fine grain. Reprinted with permission from Ref. [75]. Copyright © 1993, BirkhäuserVerlag Basel.

the grains, maximum 83% polarization can be switched in ceramics without considering the clamping effect by adjacent grains. This partially explained the lower spontaneous polarization observed in BT ceramic. Thus, BT single crystal was found to possess a lower E_c with a higher P_r, when compared to those of BT ceramics. Except the strong anisotropic characteristic observed in single crystals, other effects, such as phase, grain size, and density, will also contribute to the magnitude of P_r, E_c, and squareness of the P–E loop, which will be discussed in detail in Section 4.

Figure 7 shows ferroelectric hysteresis loops for BT crystal, coarse grain, and fine grain ceramics, exhibiting a transition from square loop to slim loop with decreasing grain size. Here, crystal sample can be considered as a very large grain without grain boundary. Therefore, the clamping effect due to the neighboring grains is absent in crystals. It should be noted that the P–E loop of crystal is measured along its spontaneous polarization direction, whereas the P–E along other directions behave very differently, due to the strong anisotropy of the crystals, which will be discussed in following section.

3.1.2. *Double hysteresis loops*

It is well known that in antiferroelectrics, two opposite polarizations arrange at two nearby crystalline lattice, as shown in Fig. 8. Due to such arranging characteristic, the net polarization of antiferroelectrics is zero in their virgin state (or without exposing to external electric field). With applying high enough electric field, at which macroscopic polarization can be induced, an antiferroelectric-to-ferroelectric phase transition occurs. The induced polarizations switch back to zero through another ferroelectric-to-antiferroelectric phase transition after removing the external electric field. These features of antiferroelectrics determine their unique hysteresis loops, that is, so-called double hysteresis loop. Figure 9(a) shows a typical antiferroelectric double hysteresis loop. It can be seen that before applying electric field, macroscopic net polarization is zero. With increasing field to a critical value (E_f), an antiferroelectric-to-ferroelectric phase transition is induced following with a rapid

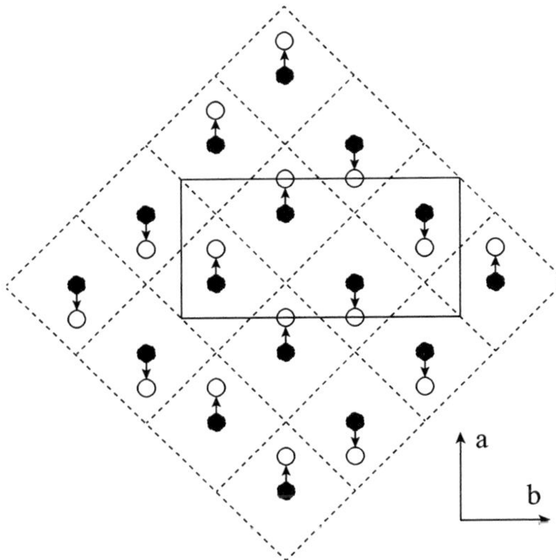

Figure 8. Two-dimensional projection of unit cell in the a–b plane showing antiparallel displacements of Pb^{2+} resulting in antiferroelectric behavior. After Lines and Glass [50].

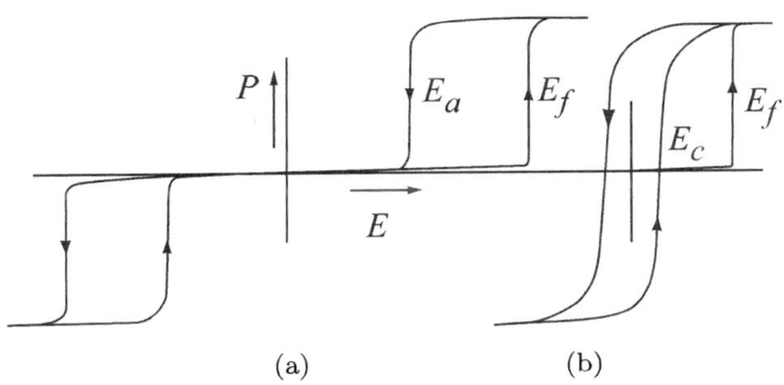

Figure 9. Schematic figure showing P–E hysteresis loops for (a) typical antiferroelectric sample and (b) field enforced metastable ferroelectricity from a sample initially antiferroelectric. Reprinted with permission from Ref. [77]. Copyright © 1964, Elsevier.

increase in the polarization, corresponding to the polarization of the ferroelectric phase at field above E_f. As the induced ferroelectric phase is metastable, a ferroelectric-to-antiferroelectric phase transition occurs with decreasing electric field, approaching the value of E_a. In the negative electric field part, a similar loop also exists.

P–E loop as observed in antiferroelectrics. The difference between the pinched and double loops from the morphology is that for pinched loop, the polarization at zero ac field not necessarily be zero. In these materials, domains/domain walls are mainly pinned by the charged defects dipoles. Even the applied electric field can induce a macroscopic net polarization, the polarization will switch back to its unpoled (virgin) state by the restoring force, which is generated by the coupling of defects dipoles and domains/local polarizations. This is totally different from the double hysteresis loop achieved in antiferroelectrics, where the zero net polarization at its virgin state is due to the opposite arrangement of the unit cell distortions. Double hysteresis loops with an open gap at original point are frequently reported in antiferroelectric materials, see Fig. 12. Strictly speaking, however, such an open loop is mainly due to the coexistence of the ferroelectric and antiferroelectric phases.

It should be noted that the debates on recently reported $(Bi_{0.5}Na_{0.5})TiO_3$ (BNT)-based lead-free system, where the double hysteresis loop was attributed to the antiferroelectric phase [82–84],

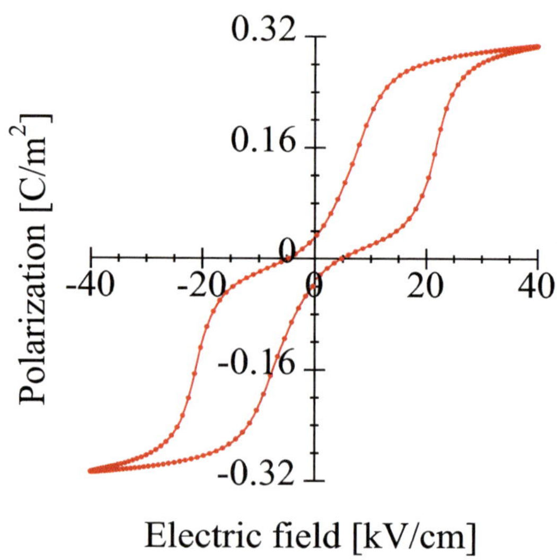

Figure 12. P–E hysteresis loop for $(Pb_{0.94}La_{0.04})[(Zr_{1-x}Sn_x)_{0.84}Ti_{0.16}]O_3$ with $x = 0.36$ ceramic after Zhang [81].

whereas others associated these phenomena to the nonpolar nanoregions [85]. Other factors giving rise to double hysteresis loop include, but not limited to, fatigued ferroelectric samples due to the space charge, which will be discussed in Section 4.

3.1.3. *Asymmetric hysteresis loops*

In above two kinds of hysteresis loops, the common feature is that both of them exhibit a symmetric shape with respect to the origin point. However, for poled hard ferroelectrics, the hysteresis loops always showing an asymmetric feature as illustrated in Fig. 13. It can be seen that both the negative E_c^+ and positive E_c^- shift to right or left with respect to the horizontal axis at a certain degree. Normally this shift is attributed to the existence of an internal bias field, which is defined as $E_{\text{int}} = (E_c^+ - E_c^+)/2$. In "hard" ferroelectrics, due to the pining effect by defect dipoles or defect clusters on domain walls, the poling and depoling processes are much more difficult than those of ferroelectrics without pining effect. A macroscopic polarization is built up during the poling process, which is stabilized through the internal bias field during aging. On the contrary, even a net

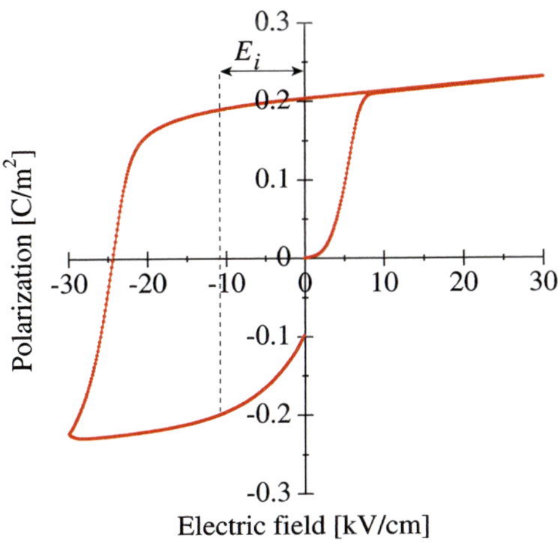

Figure 13. Schematic illustration of an asymmetric P–E hysteresis loop in a pole hard ferroelectric sample.

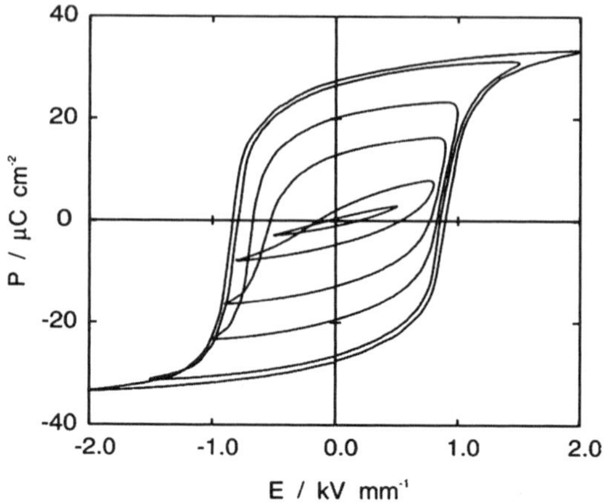

Figure 14. *P–E* hysteresis loops obtained with increasing field amplitude for a PC5H ceramic. Reprinted with permission from Ref. [86]. Copyright © 1999, Taylor & Francis.

polarization is induced by poling in soft ferroelectrics, the measured hysteresis loops will transfer from asymmetric to symmetric shape with electric field strength exceeding E_c, due to the lacking of pining defect dipoles. The evolution of this transition is illustrated in Fig. 14.

From the asymmetric loop, the P_r, P_s, E_c, and E_i can be obtained. The internal bias is closely related to the acceptor-oxygen vacancy defect dipoles, other factors, such as doped level, grain boundary, sintering condition, poling status, aging time, and measured temperature, will affect the value of the internal bias, which will be discussed in Section 4. The internal bias is an important parameter to evaluate the hard ferroelectric materials, which is associated with the clamping effect of the domain wall motion and polarization rotation, account for the decreased power dissipation (such as dielectric loss and mechanical loss) [87, 88]. Therefore, ferroelectric materials with higher internal bias are desirable for high power electromechanical applications, due to the fact that the internal bias is proportionally related to the mechanical quality factor [89]. Although the coercive field is not necessarily related to the power dissipation, higher coercive field is desired for high power applications, by offering higher field stability [90, 91].

As proposed by Carl and Härdtl [87], there are two methods to determine E_{i} with respect to the poled and unpoled states. The determination of E_{i} can be accomplished through either P–E or I–E characteristic curves. Note that the E_{i} determined from unpoled and poled states shows a large difference. This may be due to the different domain structures in unpoled and poled states.

3.1.4. *Slim hysteresis loops*

It is generally accepted that the domain wall motion accounts for the hysteresis phenomena [50, 52]. Although X-ray diffraction results suggest a pseudocubic phase structure of relaxor ferroelectrics, microdomains or polar nanoregions (PNRs) instead of the macroscopic domains do exist in a wide temperature range in the vicinity of the phase transition [92, 93]. Due to their much smaller characteristic size, microdomains respond to the external field much faster than macroscopic domains and frequently result in a slim loop, as shown in Fig. 15 [94]. Apart from the typical ferroelectric loop, the

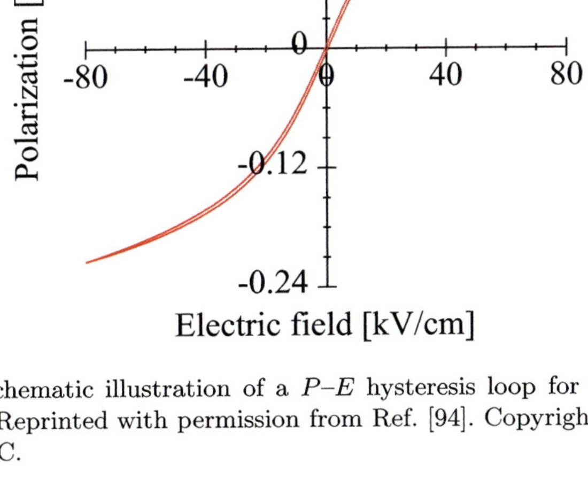

Figure 15. Schematic illustration of a P–E hysteresis loop for typical relaxor ferroelectrics. Reprinted with permission from Ref. [94]. Copyright © 2013, AIP Publishing LLC.

slim loop does not possess obvious hysteresis but with nonlinearity, revealing the existence of the microdomains, due to the fact that the macroscopic domain switching gives rise to hysteresis, while microdomains or PNRs lead to nonlinear response with respect to external field [95, 96]. The slope of the curve, which relates to the dielectric constant, decreases as electric field increases. At low field, the microdomains are switched by the external field and result in a high dielectric response, which is reflected by the steep slope of the loop, whereas at high field ends, macroscopic polarizations are induced and aligned along the direction of the field [97]. However, without contribution by the domain wall motion, the slope of the loop approaches its saturated value. Due to the high dielectric constant of relaxor ferroelectrics, being in the range of 5000–10000, low hysteresis, and the diffused dielectric maxima, relaxor ferroelectrics can be used for energy storage media [98]. From the P–E loop, the integration area between the polarization axis and the discharge curve can be calculated, which is related to the stored energy, being much higher than that of normal ferroelectrics [99].

The essence of such kind of slim hysteresis loop is the existence of microdomains, which is a transition state between the macroscopic polar state and nonpolar state. It should be noted that in normal ferroelectric and antiferroelectric systems, "2S" slim loops are frequently observed but in a relatively narrow temperature region close to T_C, in contrast to the "1S" slim loop observed in relaxor ferroelectrics over a wide temperature range [100]. Figure 16 shows a hysteresis loop of lead lanthanum zirconate titanate (PLZT)-based antiferroelectrics, exhibiting a typical "2S" shape near its Curie temperature [23].

In antiferroelectrics, both E_f and E_a shift to higher values with increasing temperature, as more energy is required to keep the induced metastable ferroelectric phase [101]. However, macroscopic ferroelectric phase cannot be induced even at higher field when temperature approaches T_C, which can be confirmed by the vanishing of double hysteresis and appearance of the "2S" slim loop near T_C [102]. Due to the existence of nonlinearity, metastable polar phase is highly expected in this temperature region.

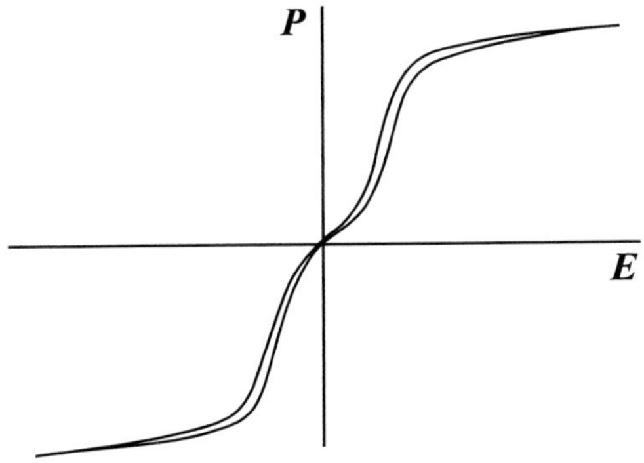

Figure 16. Schematic illustration of a P–E hysteresis loop for a PLZT 8.6/65/35 ceramic. After Haertling [23].

3.2. *Classification of strain–electric field curves*

For ferroelectric materials, application of electric field to the sample not only induces the polarization but also the strain. Generally, they are correlated through Eq. (5) [94]:

$$S = Q\varepsilon^2 E^2, \tag{5}$$

where Q is the electrostrictive coefficient, which is insensitive to temperature. It should be noted that Eq. (5) is not an accurate equation to describe the relation between strain and electric field, due to the fact that the microstructure of materials, anisotropy, and nonlinear properties are not considered. According to the classification of the hysteresis loops, the S–E curves can be categorized into four groups with bipolar and unipolar strain curves, as given in Fig. 17. (1) To ferroelectrics with classic hysteresis loops, the bipolar S–E curves show symmetric "Butterfly" curves. When the dc field is relatively small, the S–E curves obey linear relationship, corresponding to the piezoelectric effect. The large hysteresis in S–E curves and the large negative strain are mainly attributed to the ferroelastic domain wall switching. In unipolar mode, the first cycle shows a very large hysteresis, mainly due to the extrinsic contribution by domain switching. After the first cycle, a large remnant strain is maintained. Relatively small hysteresis can be achieved with further unipolar electric field cycles, as shown in Fig. 17(a). Generally,

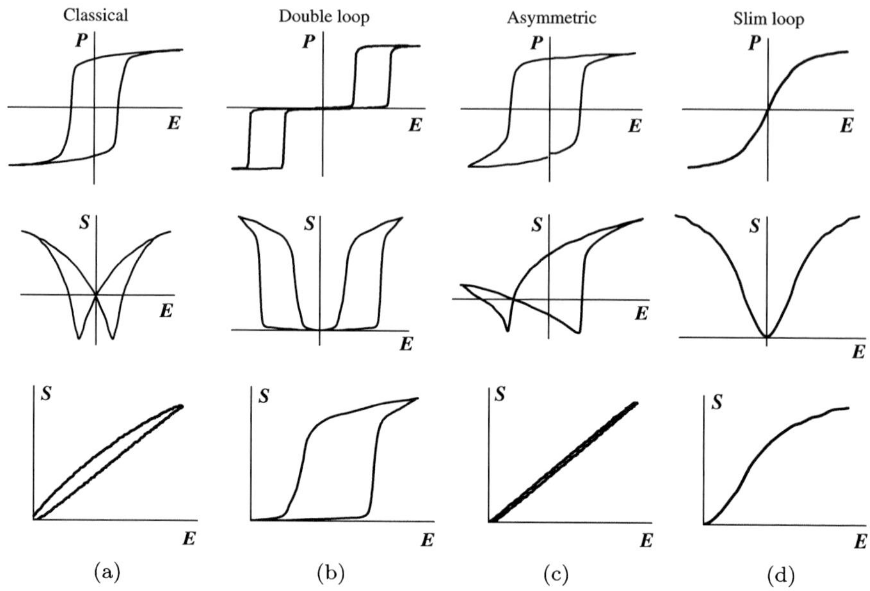

Figure 17. Four typical kinds of P–E hysteresis loops and their corresponding S–E curves for ferroelectric, antiferroelectric, and relaxor ferroelectric materials with different features (a) classical loop, (b) double loop, (c) asymmetric loop and (d) slim loop.

typical soft PZT exhibit such an S–E characteristic, from which, the high field d_{33}^* can be calculated by the slope of the S–E curve (usually at 20 kV/cm), while the strain hysteresis (related to the piezoelectric loss) can be evaluated by the fraction of the strain at half maximum field [103, 104]. (2) To antiferroelectrics, their bipolar S–E curves show a double loop feature. Once the net polarization is induced, much higher strain with large hysteresis is expected, as shown in Fig. 17(b). In unipolar mode, antiferroelectric has large strain, induced by the antiferroelectric-to-ferroelectric phase transition, accounts for the large displacement, which will benefit large displacement transducer or actuator applications. However, the large hysteresis, which inherently associates with the phase transition, will greatly impair precision actuation application. (3) To hard ferroelectrics, the bipolar S–E curves of unpoled aged state possess similar double S–E curves as that of antiferroelectrics, whereas strong asymmetric "butterfly" S–E curves are observed in poled aged state, as shown in Fig. 17(c). For hard materials, the unipolar strain is linear and hysteresis free, due to the fact that the internal bias stabilizes the domain wall motion, greatly reduces the extrinsic contribution. This is important for electromechanical applications where high strain precision or no dc bias is required. (4) Finally, to relaxor ferroelectrics, the bipolar S–E curves are shown in Fig. 17(d). Due to the absence of macroscopic domains, there is no or very low hysteresis being observed in S–E curves. Of particular interest is that the dc field-induced strain in relaxor ferroelectrics is hysteresis-free and temperature independent, thus, finding their potential applications for high precision actuation in a wide temperature range. Analogous to antiferroelectrics, the bipolar strain curves of relaxor ferroelectrics were found to possess only positive strains, whereas ferroelectric materials were found to show both positive and negative strains as a function of ac field, due to the lack of ferroelastic domains in antiferroelectrics and relaxor ferroelectrics. In addition, it should be noted that hysteresis-free unipolar strain behavior with large strain level of 1.7% has been reported for the relaxor-PT ferroelectric single crystal. This is due to the lack of domain wall motion, which is inherently associated with the engineered domain configuration [105].

Besides the P–E and S–E hysteresis loops, there is another kind of hysteresis loops, that is, piezoelectric response-electric field loops, which are frequently reported in ferroelectric materials, especially in thin films. In brief, we limit the discussion on longitudinal piezoelectric coefficient d_{33} as a function of dc bias field E_{dc} [106]. Based on Landau–Ginzburg–Devonshire (LGD) theory calculations, the intrinsic piezoelectric coefficients would initially increase with an applied negative bias field until depoling and repoling set-in. Then, the d_{33} tunes the sign and becomes smaller gradually with increasing the bias field. In some systems, the predicted response has been observed in both bulk material and thin films [106, 107]. In fact, not only the intrinsic effects control the response during this processing, extrinsic domain wall motion also contributes to a similar morphological response as intrinsic calculation by means of the back-switching of domains. In some hard PZT systems, where domain wall motion is pinned by means of defect dipoles or charge injection (fatigue), d_{33}-E_{dc} loop exhibits the features as that observed in P–E hysteresis loops. A maximum d_{33} was observed at highest field, due to the lack of back-switching of non-180° ferroelastic domain walls [107].

From viewpoint of applications, prediction of the response of ferroelectric materials subjected to external field is necessary for subsequent transducer design. Hwang *et al.* predicted the polarization and strain for an individual grain from the imposed electric field and stress through a Preisach hysteresis model. By averaging the response of the individual grains, which were considered to be statistically random in orientation, they successfully predicted the response of the bulk ceramics with respect to the external loads [108]. Recently, a two-dimensional model that consists of four energy wells, four saddle points, and one energy maximum was reported. Using the parameters determined from experimental data of $BaTiO_3$ single crystals, the responses of the polarization subjected to multiaxial in-plane electric field loading at various frequencies were calculated [109]. Then, by employing stochastic homogenization techniques, a macroscopic model suitable for nonhomogeneous, polycrystalline compounds was developed [110]. To predict the response of ferroelectric materials under different excitations without having to perform too much experimental work, several behavioral laws linking to the electrical field, temperature, and mechanical stress were proposed by Guyomar *et al.* [111, 112]. The scaling law could also predict the piezoelectric coefficient under stress using only pure electrical measurements, and the dielectric constant under an electrical field using pure mechanical measurements. Due to the limitation of this article, it is impossible to include all the important literatures on modeling within such a short paragraph. However, we believe that with the clues supplied from these references and references within them, readers could readily find a lot of information.

4. Impact factors on ferroelectric hysteresis loops

There are many factors affecting the characteristics of ferroelectric hysteresis loop. Some factors are associated with the material itself, while others come from the measurement conditions. In this section, we will discuss the ferroelectric hysteresis loop with respect to the grain size, phase transition, dopants, frequency, amplitude of electric field, temperature, frequency, stress, etc.

4.1. *Effect of materials*

4.1.1. *Grain size and grain boundary*

Based on LGD theory in the case of isolated particles, the size dependence of the crystal lattice distortion and polarization in ferroelectric phase has been calculated [113, 114]. It is suggested that there is a critical size, below which, a transition from a ferroelectric phase to a cubic paraelectric phase will occur with the particle size being on the order of a few nanometers to a few tens nanometers accompanying with the disappearance of polarization [113, 114]. For example, the critical grain size is 4–20 nm for PT [115, 116] and 10–100 nm for BT [117, 118]. In ferroelectric polycrystalline ceramics with large grains, domain patterns are formed to balance the depolarization field, which mainly comes from the charges accumulated at the grain boundary [111]. With decreasing the grain size from micrometer to nanometer, this process is not energetically favorable in fine grain ceramics, where the compensation via surface charges and/or polarization gradient is possible [119]. The grain size decrease leads to the following results. First, the lattice distortion will become less pronounced as the grain size decreases [119, 120]. This effect is similar to the hydrostatic pressure effect on polycrystalline ceramics, which will be discussed in Section 4.3. Both of them favor a high-symmetric phase, that is, promoting to a cubic paraelectric phase. As reported in BT ceramics, the tetragonal distortion ($c/a-1$, where c and a are the unit cell parameters) was found to decrease linearly with decreasing the grain size from 1200 to 50 nm, accompanying with a T_C shift from 127°C to 105°C [119]. Second, multidomain structure is prone to transfer to single domain state with respect to grain size decreasing [120]. It was suggested that domain size is proportional to square root of grain size [121, 122]. However, this law can only be hold for the grain size larger than hundreds nanometer, below which it is invalid. As reported by Arlt, there is a transition from multidomain state into single domain state with decreasing the grain size in BT ceramics [120]. In this case, both domain wall motion and domain switching become difficult due to the increased clamping effect by the neighboring grains and the absence of domain walls. Based on the above, contributions from both intrinsic lattice and extrinsic domain wall to the polarization are reduced with decreasing grain size, resulting in a smaller value of P_r. Thus, the ferroelectric hysteresis loop is expected to transfer from square shape to slanted one. This was clearly illustrated in Fig. 7, which gives the experimental results for grain size effect in BT system. In other ferroelectric systems, such as PZT [123–125], PMN-PT [126, 127], $BiScO_3$-$PbTiO_3$ [128], and $BaTi_{0.995}Mn_{0.05}O_3$ [129], the grain size effect has also been extensively studied, similar results have been achieved as observed in BT ceramics. Figure 18 shows two representative hysteresis loops observed in PMN-PT [127] and $BiScO_3$-$PbTiO_3$ [128] systems with different grain sizes. The common feature among these data is that with decreasing grain size, the loops transfer from square shape to slanted shape and finally to slim shape, accompanying by a decrease of P_r as a function of grain

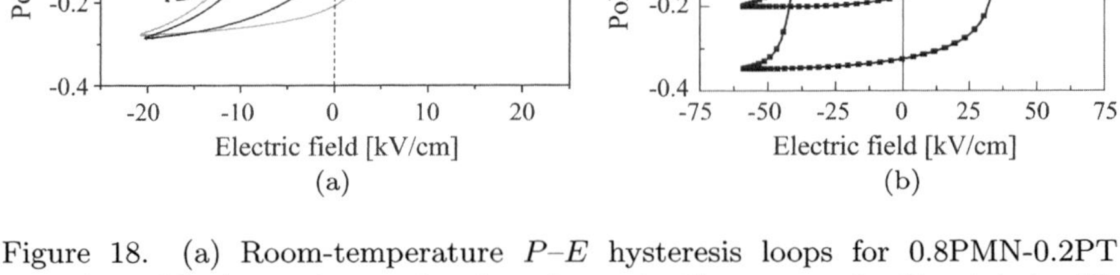

Figure 18. (a) Room-temperature P–E hysteresis loops for 0.8PMN-0.2PT ceramics with decreasing grain sizes down to the nanoscale. Reprinted with permission from Ref. [127]. Copyright © 2008, The American Physical Society. (b) P–E hysteresis loops for $0.375BiScO_3$-$0.625PbTiO_3$ ceramics with average grain sizes of (1) 28, (2) 80, (3) 375, and (4) 3000 nm. Reprinted with permission from Ref. [128]. Copyright © 2010, IOP Publishing Ltd.

size. These results could be well understood from both intrinsic and extrinsic points of views. Recently Liu *et al.* [130] proposed that the existence of low permittivity paraelectric grain boundary and its influence on the grain microstructures played a key factor to the grain size effect. A two-dimensional polycrystalline phase-field model was developed to simulate the hysteresis behaviors of the nanoscale BT ceramics, showing apparent grain size dependence of the hysteresis loops, where the noticeable vortex polarization structures were observed as the grain size reduced to tens of nanometers. According to this simulation, a similar evaluation of the hysteresis loops as observed in above-mentioned systems was confirmed [130].

Compared to remnant polarization, the scenario of coercive field was found to be more complicated. In some systems, it was found that E_c decreased gradually with decreasing grain size [127], whereas in some other systems, the E_c was reported to maintain the same values with grain size [126, 131]. Theoretically, intrinsic E_c corresponds to a field strength at which, polarization is switched completely without considering the extrinsic domain wall contribution [132]. However, in practice, the real electric field required for polarization switching is much lower than the theoretical value, being about 10% of intrinsic E_c, due to the aid of domain nucleation and domain wall motion [133]. Generally, E_c is determined from the hysteresis loops, being equal to the electric field at which the polarization is zero. However, for a very slanted or slim ferroelectric hysteresis loop, ferroelectric property is greatly reduced, which is demonstrated by the relatively small P_r. This hysteresis loop can be deemed as the combination of linear dielectric response and conductivity. In this case, even the P–E characteristic curve has an intercept with respect to the electric field axis. It is meaningless to consider the value as E_c, as shown in Fig. 16(a) for the 0.8PMN-0.2PT ceramics with grain size less than 140 nm.

It is important to note that the coercive fields of single crystals are much lower than those of ceramics with the same composition, for example, the coercive field was found to be on the order of ~2.5 and ~5 kV/cm for PMN-PT crystal and ceramic, respectively, due to the fact that no grain boundaries exist in single crystals, which clamp the domain reversal and domain wall motion.

4.1.2. *Phase and phase boundary*

Ferroelectric phase plays an important role on hysteresis loop, as the symmetry determines the spontaneous polarization directions, where the polarization may exist at certain external conditions, such as temperature, chemical composition, pressure, electric field, etc. [49]. For example, polarization can be developed along six $\langle 100 \rangle$ directions in tetragonal phase, while it can be developed along eight $\langle 111 \rangle$ directions in rhombohedral phase [63]. Since the discovery of the MPB in PZT solid solution, much attention has been devoted to the ferroelectrics with an MPB. The end-members of MPB systems have different phase structures. Taking $Pb(Zr_{1-x}Ti_x)O_3$ as an example, PZ has a rhombohedral symmetry while PT has a tetragonal symmetry at room temperature. With increasing the PT concentration, rhombohedral PZT will transfer to tetragonal phase, where the Zr/Ti ratio at MPB region is around 52/48. As shown in Fig. 19, this phase transition is an abrupt structural change as a function of composition, with the phase boundary being nearly independent of temperature, which ensures the unique performance of the MPB composition [31].

Figure 20 shows *P–E* loops for Nb-modified PZT ceramics with different phases (Zr/Ti ratio), where the P_r was found to gradually reduce from $0.4\,C/m^2$ for rhombohedral phase, through $0.28\ C/m^2$ for MPB, and finally to $0.2\,C/m^2$ for tetragonal phase [134]. This trend is consistent with the data of pure PZT ceramics reported by Jaffe *et al.* [31]. Interestingly, the highest P_r was achieved in PZT composition with 56 mol% Zr. A simple explanation of this phenomenon is that rhombohedral phase has eight equivalent direction of domain state, while it is only six for tetragonal phase, therefore higher P_s and P_r can be expected in rhombohedral phase. The ferroelastic domains (109° and 71° domains) in rhombohedral phase make the domain switching easier than that in tetragonal phase (90° ferroelastic domains), where the lattice distortion is induced by the gradually increased tetragonality (c/a ratio) as the

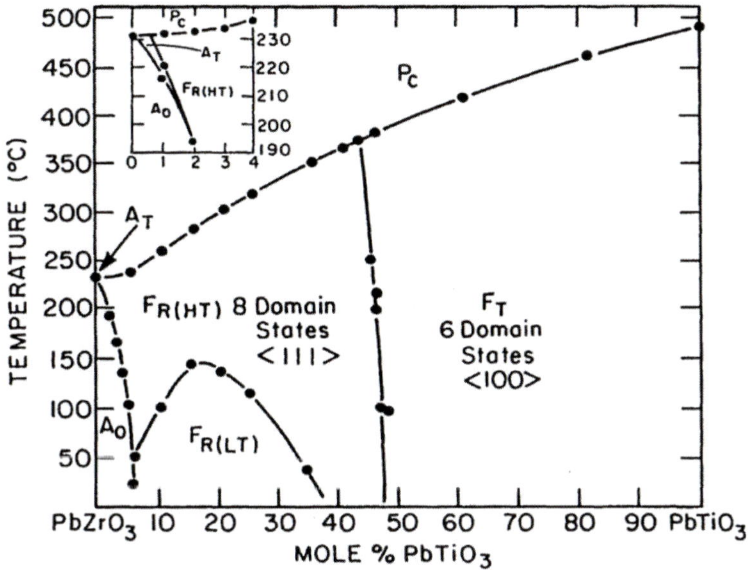

Figure 19. Phase diagram of PZT solid solution. After Jaffe *et al.* [31].

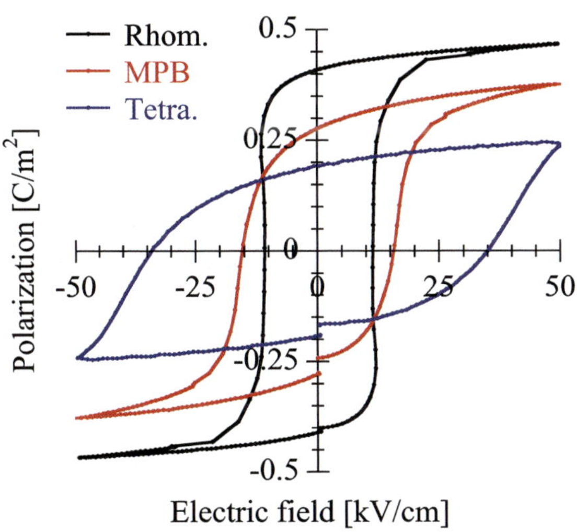

Figure 20. *P–E* hysteresis loop of Nb-doped PZT ceramics as a function of composition. After Jin [134].

PT concentration increases. In this case, due to the very large stress induced by the lattice distortion, domain walls are highly clamped, leading to higher E_c and lower P_r, as reflected by the ferroelectric hysteresis loops of the tetragonal phase [135].

However, the above discussion seems to be not well held in lead-free piezoelectric materials. For $x(Bi_{1/2}Na_{1/2})TiO_3$-$(1 - x)$ $(Bi_{1/2}K_{1/2})TiO_3$ (BNKT), the P_r at rhombohedral phase is larger than that in tetragonal phase, similar to PZT, while the E_c exhibits contrary tendency to PZT, with higher E_c in rhombohedral phase, as given in Fig. 21 [136]. The c/a ratio for tetragonal BNKT ($x = 0.7$) is about 1.015, which is smaller than that observed in tetragonal PZT. Due to the smaller lattice distortion, domain switching may be easier when compared to the tetragonal PZT, this may explain the above-mentioned phenomena. In addition, the P_r of MPB composition exhibited the highest values in BNKT ceramics, similar results have also been reported in BNT-BKT-BT lead-free system [137].

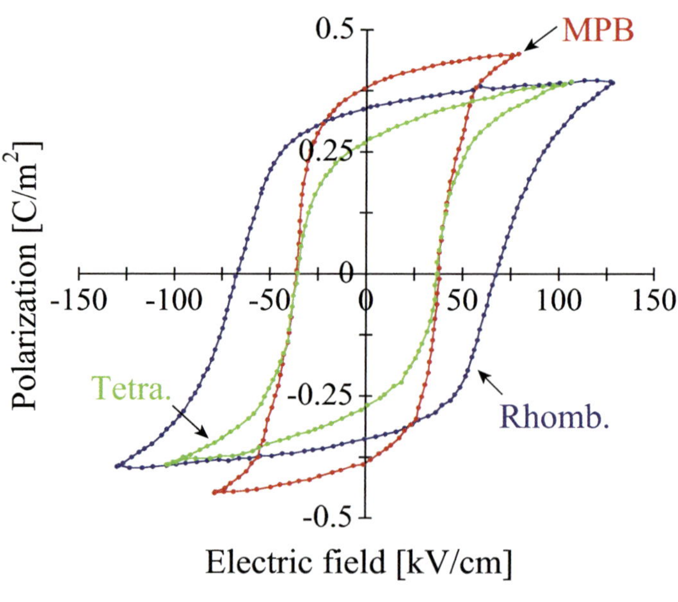

Figure 21. P–E hysteresis loops of BNKT100x ($x = 0.70$, 0.80, and 0.94) at RT and 50 Hz. Reprinted with permission from Ref. [136]. Copyright © 2006, The Japan Society of Applied Physics.

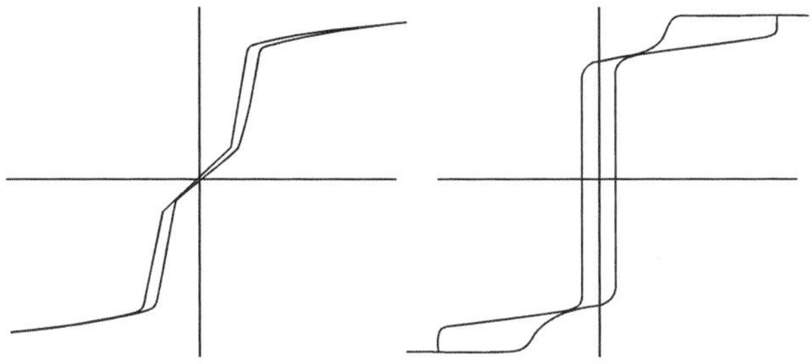

Figure 22. (a) Double loop observed in BT crystal at 111.4°C. Reprinted with permission from Ref. [80]. Copyright © 1953, The American Physical Society. (b) Triple loop observed in BT crystal. Reprinted with permission from Ref. [138]. Copyright © 1956, The American Physical Society.

The hysteresis loop of ferroelectrics around the phase transition point shows interesting characteristics, and no matter it is induced by composition or temperature. For example, in BT single crystals, a double hysteresis loop was reported at temperature approaching T_C, due to the electric field-induced ferroelectric–paraelectric phase transition, as shown in Fig. 22(a) [80]. Similarly, a triple hysteresis loop was also reported in BT crystals at temperature slightly below the orthorhombic–tetragonal phase transition temperature with the electric field applied along *c* axis [Fig. 22(b)] [138]. The middle part of the loop is induced by the domain switching of the orthorhombic phase, whereas the upper positive and lower negative parts are the result of orthorhombic–tetragonal phase transitions. Similar phenomena were also observed in PMN-PT single crystals, where a "2S" slim hysteresis loop was achieved at temperature of T_C, while triple hysteresis loop exists for PMN-PT with MPB compositions.

4.1.3. *Doping*

As mentioned above, PZTs exhibit soft and hard characteristics through adding of donor and acceptor dopants, respectively. Normally, in soft PZT, lead vacancies are generated to keep electric charge equilibrium. It suggests that domain wall mobility is increased in soft PZT, resulting in facilitated domain switching and domain wall motion. Therefore, a square ferroelectric hysteresis loop with a larger P_r and lower E_c is frequently observed. In contrast, due to the existence of oxygen vacancies in hard PZTs, which are generated through the acceptor doping, domain wall mobility is reduced. In this case, higher field is needed to induce polarization. It is believed that small amount of dopant in ceramic materials mainly affected extrinsic domain wall mobility, while having minimal effect on the intrinsic crystalline lattices [139, 140]. The corresponding ferroelectric hysteresis loop is shown in Fig. 17. Hard PZT in unpoled state could be confused as antiferroelectrics because of the similar double hysteresis loop [79].

Different from the regularity summarized from PZT ceramics, the doping effect in antiferroelectric ceramics is much complicated. $Pb(Zr,Sn,Ti)O_3$, as the most studied antiferroelectric system, both the E_a and E_f are increased but with a narrowing of the loop width $E = E_f - E_a$ when the Ti component decreases, demonstrating decreased hysteresis and energy loss, with increased energy storage density, as shown in Fig. 23 [81]. In addition, various impurities have been added in this system, for example, Sr^{2+} and Ba^{2+} for Pb^{2+} on

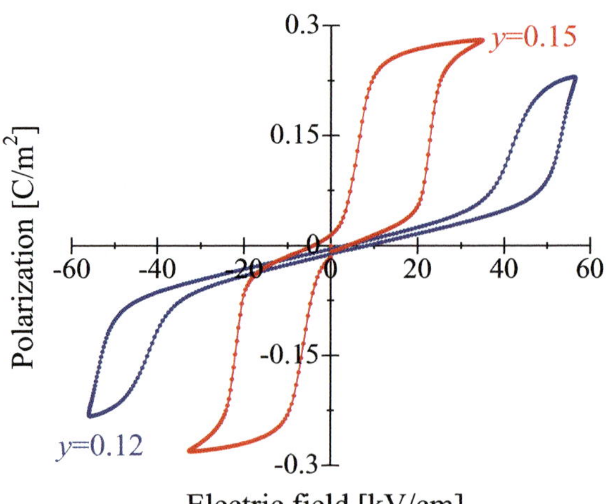

Figure 23. P–E hysteresis loops of antiferroelectric $(Pb_{0.94}La_{0.04})[(Zr_{0.70}Sn_{0.30})_{1-y}Ti_y]O_3$ ceramics with $y = 0.12$ and $y = 0.15$, respectively. After Zhang [81].

A site and Nb^{5+} for Zr^{4+}/Ti^{4+} on B site, were used for reducing the antiferroelectric–paraelectric phase transition temperature and increasing the mobility of the domain walls, respectively [141], while the doping of La^{3+} may induce transition from normal antiferroelectrics into relaxor antiferroelectrics [142].

4.1.4. *Anisotropy*

For ferroelectric ceramics, due to the random distribution of grains, the properties measured along any direction in principle are the same. However, for ferroelectric single crystals and textured ceramics, the properties are strongly dependent on crystallographic directions, which can be clearly demonstrated by the hysteresis loops. Figure 24(a) shows the polarization hysteresis for rhombohedral PMN-030PT single crystals measured along crystallographic orientations [001], [110], and [111] [143] revealing that both the remnant polarization P_r and coercive field E_c are orientation dependent. The values of P_r were found to be on the order of 0.24, 0.34, and 0.41 C/m^2 for [001], [110], and [111] orientations, whereas the values of E_c were

(a) (b)

Figure 24. *P–E* hysteresis loops for (a) PMN-30PT crystals along crystallographic orientations ⟨001⟩, ⟨110⟩, and ⟨111⟩ (Reprinted with permission from Ref. [143]. Copyright © 2009, AIP Publishing LLC) and (b) BNT single crystals along [100], [110], and [111] directions (Reprinted with permission from Ref. [146]. Copyright © 2013, The Japan Society of Applied Physics).

2.5, 2.9, and 3.2 kV/cm, respectively. It is known that there are eight possible polarization orientations along the pseudocubic $\langle 111 \rangle$ direction for rhombohedral PMN–PT single crystals (3 m symmetry). Upon applying an electric field, the dipoles reorientate as close as possible to the applied electric field direction. For [001] poled crystals, there are four equivalent polar vectors along the $\langle 111 \rangle$ direction, with an inclined angle of 54.7° from the poling field. Following "4R", "2R", "1R", and others are technical nomenclatures used in the framework of domain engineering as proposed in relaxor ferroelectric single crystals [10, 144, 145]. The four $\langle 111 \rangle$ domains are equivalent with a domain engineered configuration "4R", resulting in a macrosymmetry 4 mm [144]. For [110] poled crystals, there are two equivalent polar vectors along the $\langle 111 \rangle$ direction, which will rotate 35.5° toward the applied field direction of [110] with a designated domain engineered configuration "2R" For this case, the macroscopic symmetry is 2 mm. In contrast, there is only one polar vector along the $\langle 111 \rangle$ direction for [111] poled ferroelectric crystals, thus it will form a monodomain state, designated "1R," exhibiting macroscopic symmetry 3 m. According to the domain engineered configurations, the polarization level derived from the hysteresis loops should correspond, in theory, to the intrinsic value along the polar axis of the monodomain crystal P_s, following $P_{r\langle 001 \rangle} = \frac{P_{r\langle 110 \rangle}}{\sqrt{2}} = \frac{P_{r\langle 111 \rangle}}{\sqrt{3}}$ [44, 145]. As expected, the P_r values for the different orientations obtained from Fig. 24(a) are in good agreement with the predicted values. For lead-free ferroelectric crystals, such as BNT, the ferroelectric hysteresis loops measured along different crystallographic directions also exhibited large difference, as shown in Fig. 24(b) [146].

For ferroelectric ceramics with textured configurations, the anisotropy reflected in the hysteresis loop is obvious due to the orientated growth of the grains. Figure 25 presents three groups of ferroelectric hysteresis loops determined from randomly oriented and textured ceramics with different structures [147]. Figure 25(a) shows the P–E hysteresis loops for randomly oriented and textured $(Na_{1/2}Bi_{1/2})TiO_3$-5.5 mol% $BaTiO_3$ ceramics with perovskite structure. As these ceramics have a rhombohedral symmetry, the P_r

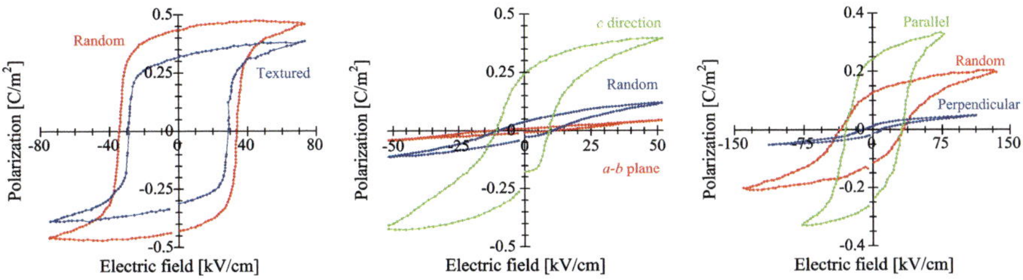

Figure 25. *P–E* hysteresis loops for (a) perovskite structure: randomly oriented and textured $(Na_{1/2}Bi_{1/2})TiO_3$-5.5 mol% $BaTiO_3$ ceramics (Rhombohedral phase and textured along ⟨001⟩ direction) (reprinted with permission from Ref. [147]. Copyright © 2003, Kluwer), (b) tungsten bronze structure: randomly oriented and textured $Sr_{0.53}Ba_{0.47}Nb_2O_6$ ceramics (reprinted with permission from Ref. [148]. Copyright © 2002, Materials Research Society), and (c) bismuth layer structure: randomly oriented and textured $Bi_4Ti_{2.96}Nb_{0.04}O_{12}$ ceramics (After Hong *et al.* [149]).

determined along ⟨001⟩ direction must be $1/\sqrt{3}$ of that determined along ⟨111⟩ direction, with an angle of 54.7°. In ceramics, however, grains are randomly distributed in three dimensions. The angles between the measurement direction and the spontaneous polarization direction (⟨111⟩) of various grains change from 0° to 54.7°. Therefore, the averaging of polarization in three-dimensional space, randomly oriented ceramics has a higher P_r than ⟨001⟩ textured materials. For $Sr_{0.53}Ba_{0.47}Nb_2O_6$ with tungsten bronze structure [148] and $Bi_4Ti_{2.96}Nb_{0.04}O_{12}$ with bismuth layer structure [149], the maxima P_r are observed along their *c* directions. Therefore, in textured ceramics, the higher P_r is observed along the *c* axis. In contrast, lower P_r is observed in the direction perpendicular to the *c* axis. For randomly oriented ceramic samples, the values of P_r are between the values of these two limitations, as illustrated in Figs. 25(b) and (c).

Analogous to the *P–E* characteristics, the bipolar *S–E* curves of ferroelectric crystals are also associated with the crystal orientation. Figure 26 shows the strain–electric field behaviors for 001- and [111]-oriented rhombohedral PMN-0.29PT crystals, where strong orientation dependence of *S–E* behaviors was observed. The positive strain of [001]-oriented rhombohedral crystal is much higher than that of

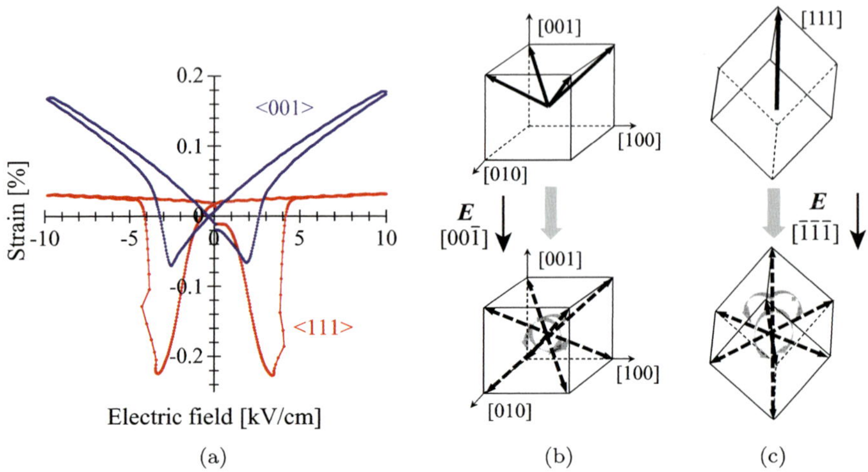

Figure 26. (a) Strain versus electric field curves for PMN-0.29PT crystals measured along [001] and [111] directions at room temperature. (b) and (c) Schematic illustration of the domain switching for rhombohedral crystal. (b) [001] oriented crystal. At the coercive, which is antiparallel to the [001] direction, [111], $[\bar{1}11]$, $[1\bar{1}1]$, and $[\bar{1}\bar{1}1]$ domains transform to the $[11\bar{1}]$, $[\bar{1}1\bar{1}]$, $[1\bar{1}\bar{1}]$, and $[\bar{1}\bar{1}\bar{1}]$ domains). (c) [111] oriented crystal. At the coercive, which is antiparallel to the [111] direction, [111] domain could transform to the $[\bar{1}11]$, $[1\bar{1}1]$, $[\bar{1}\bar{1}1]$, $[11\bar{1}]$, $[\bar{1}1\bar{1}]$, and $[1\bar{1}\bar{1}]$ domains.

[111]-oriented crystal. This is because that [001] poled rhombohedral crystal is in "4R" domain engineered configuration, whereas [111] poled rhombohedral crystal is in "1R" single domain state. The piezoelectric coefficient of domain engineered crystal is much higher than that of single domain state [150], due to the "polarization rotation process," which contributes to a high and linear strain. For the part of negative strain, however, it can be seen that the strain of [111] poled PMN-0.29PT crystal is about five times higher than that [001] poled crystal at the coercive field. As shown in Fig. 26(b), at the coercive field, the negative strain induced by domain switching was minimal for [001]-oriented rhombohedral crystal because the spontaneous strain of eight possible domains ($[11\bar{1}]$, $[\bar{1}1\bar{1}]$, $[1\bar{1}\bar{1}]$, $[\bar{1}\bar{1}\bar{1}]$, [111], $[\bar{1}11]$, $[1\bar{1}1]$, and $[\bar{1}\bar{1}1]$) are equivalent with respect to the [001] direction. Thus, the negative strain of [001]-oriented rhombohedral crystal is induced by linear piezoelectricity, being equal to $-d_{33}E_c$

approximately. On the contrary, due to the spontaneous strain of eight possible domains that are not equivalent to [111] direction, the negative strain for [111] poled rhombohedral crystals was induced by the non-180° ferroelastic domain switching from the [111] domain to $[11\bar{1}]$, $[\bar{1}1\bar{1}]$, $[1\bar{1}\bar{1}]$, $[\bar{1}11]$, $[1\bar{1}1]$ or $[\bar{1}\bar{1}1]$ domains, as shown in Fig. 26(c), resulting in high level of negative strain. At higher electric field, admittedly, every domains are transformed to $[\bar{1}\bar{1}\bar{1}]$ domain, being again in single domain state.

4.1.5. *Thickness of the sample*

For bulk ferroelectrics, including the ceramics and crystals, the studied thickness is generally higher than 200 μm, which has limited impact on their ferroelectric properties. However, the P_r was reported to decrease and E_c increase when the sample thickness further down to $< 150\,\mu$m, as shown in Figs. 27(b) and 27(d), due to the fact that the domain size is on the order of the sample scale, thus the sample boundary clamps the domain reversal and restricts the polarization rotation, accounting for the decreased P_r and increased E_c [151]. Of particular significance is that the fine grain PMN-PT ceramics and $Pb(In_{1/2}Nb_{1/2})O_3$-$Pb(Mg_{1/3}Nb_{2/3})O_3$-$PbTiO_3$ (PIN-PMN-PT) crystals exhibit minimal scaling effect, as given in Figs. 27(a) and 27(c), which can be explained by the smaller domain sizes, being on the order of $\sim 2\,\mu$m, much smaller when compared to the sample thickness [151, 152].

On the contrary, for ferroelectric thin films, an unavoidable topic needs to be addressed is the thickness effect on their properties. By decreasing the thickness of the films from micrometer to nanometer scale, such an effect becomes more pronounced. Ma *et al.* studied the hysteresis loops of $(Pb_{0.92}La_{0.08})(Zr_{0.52}Ti_{0.48})O_3$ films with the thickness varying from 3100 to 350 nm and found that the PE loops transformed from square shape to slanted shape, as shown in Fig. 28 [153]. The P_r decreased with decreasing the thickness, while E_c showed the opposite trend. This observation could be well interpreted in terms of the domain wall clamping by the tensile stress. Due to the different thermal expansion coefficients and the lattice mismatch between the substrate and ferroelectric materials,

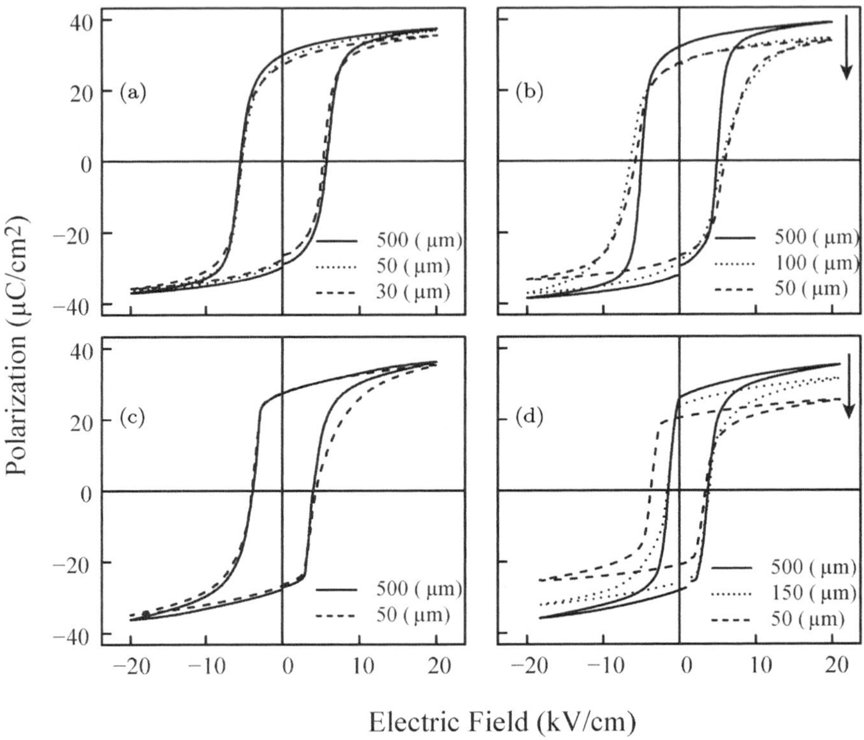

Figure 27. *P–E* hysteresis loops for relaxor-PT materials as a function of thickness. (a) Fine-grained PMN-PT, (b) coarse-rained PMN-PT, (c) PIN-PMN-PT crystals, and (d) PMN-PT crystals. Arrow indicates decreasing thickness. Reprinted with permission from Ref. [151]. Copyright © 2010, Wiley.

there exists a tensile stress at the interface between them. This tensile stress deceases gradually from the interface to the ferroelectric films. With the in-plane tensile stress, the domain wall motion and domain switching are clamped. Higher field is required to move or switching the domain wall and the switching regions are expanded compared with free standing state of the materials [154–156]. As the polarization measured by the Sawyer–Tower circuit is a collective response of the dipoles, the fraction of the clamped domain volume becomes higher with decreasing film thickness. Therefore, a decrease of P_r accompanied with an increase of E_c was observed with decreasing the film thickness.

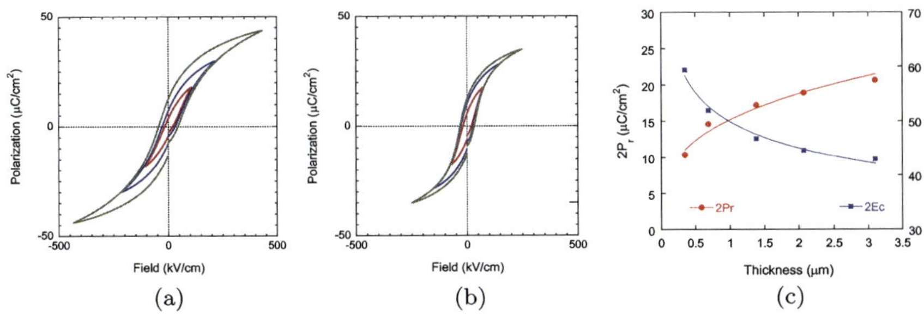

Figure 28. P–E hysteresis loops of $(Pb_{0.92}La_{0.08})(Zr_{0.52}Ti_{0.48})O_3$ films with the thickness varying from (a) 350 nm to (b) 3100 nm, and (c) the polarization and coercive field as a function of film thickness. Reprinted with permission from Ref. [153]. Copyright © 2013, IOP Publishing Ltd.

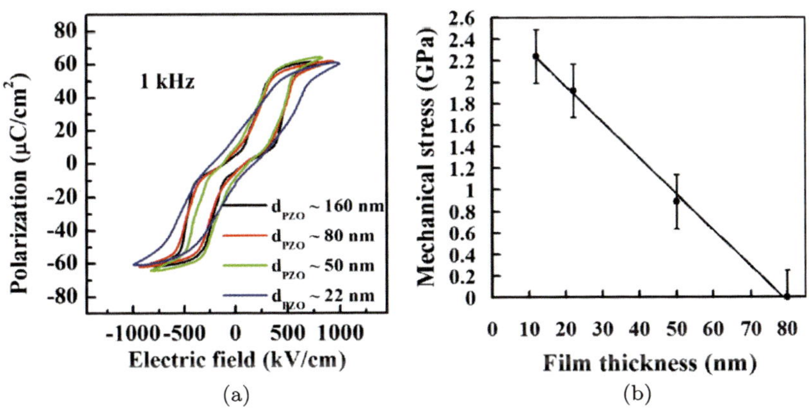

Figure 29. (a) Room-temperature P–E hysteresis loops of four representative thin films. (b) Plot of the mechanical stress values versus $PbZrO_3$ film thickness. Within the investigated range, the film thickness and stress value follow a roughly linear behavior. Reprinted with permission from Ref. [157]. Copyright © 2011, The American Physical Society.

For antiferroelectric $PbZrO_3$ thin films, smaller thickness was reported to favor the ferroelectric phase due to the large compressive stress as the thickness decreases [157]. A transformation from antiferroelectric double loop to ferroelectric single loop was found to occur in $PbZrO_3/SrRuO_3/SrTiO_3$ epitaxial heterostructures when the $PbZrO_3$ layer thickness is below 22 nm, as shown in Fig. 29(a). There is a compressive stress between the $PbZrO_3$ and $SrRuO_3$

interface, due to their different lattice parameters. Figure 29(b) gives the compressive stress as a function of the layer thickness, where it was found that the compressive stress reached 2 GPa when the layer thickness decrease to 22 nm. By such a large stress, the mechanical stress presenting in the $PbZrO_3/SrRuO_3$ interface overcomes the small free energy difference between the orthorhombic antiferroelectric phase and rhombohedral ferroelectric phase, stabilized the ferroelectric phase and prevented any further phase transition [25, 158, 159]. It should be noted that the P_r obtained from the ferroelectric loop is still low, due to the fact that even the ferroelectric phase was stabilized by the mechanical stress, the domain wall motion is yet hard due to the clamping effect at the interface.

4.2. *Aging*

4.2.1. *Aging without poling*

Aging process includes aging without poling and after poling, to release the stress induced by the high temperature or electric field. During the cooling through the T_C, domain realignment does occur slowly in small steps as time elapses. The mechanism of this aging process is mainly attributed to the presence of mobile charge species, such as defect or defect dipoles, which usually stabilize the domain pattern and decrease the domain wall contribution to the polarization response [160, 161]. This phenomenon is more obvious in accepted-doped ferroelectric materials, such as Fe^{3+}-doped PZT ceramics [79, 162], where the oxygen vacancies are present to keep the electronic equilibrium due to the lower valence replacement of the Fe^{3+} ions for (Zr^{4+}, Ti^{4+}) ions, forming the defect dipoles $(Fe'_{Zr,Ti}-V_O^{\bullet\bullet})^{\bullet}$, which clamp the domain wall motion. However, the exact mechanism dominating the stabilization of the domain walls is still under debate [87, 163–169]. Three models have been proposed to answer this question. (1) the bulk effect (charged defects align along the polarization within ferroelectric domains) [163–165], (2) the domain wall effect (charged defects diffuse to domain walls creating pinning centers) [166, 167], and (3) the interface effect (charged defects drift and accumulate at the grain boundaries and other interfaces) [168, 169]. Although the bulk and domain wall models are based on experimental observations, the contribution by the surface interface cannot be excluded.

It should be noted that classical ferroelectric hysteresis loop can be observed in samples with pinning centers, if the aging state is disturbed. For example, when the sample is quenched from high-temperature paraelectric state, normally square hysteresis loops are observed, due to the fact that the pinning centers distributes randomly with respect to the polarization and cannot form the restoring force immediately. However, as shown in Fig. 30, when time elapses, domain patterns are stabilized by the charge defects gradually and the macroscopic net polarization becomes zero [170]. A polarization is only induced when the sample subjects to an external electric field which is larger than the pinning force generated by the charged defects. However, the distribution of the charged defects cannot be disturbed immediately. Upon removing the field,

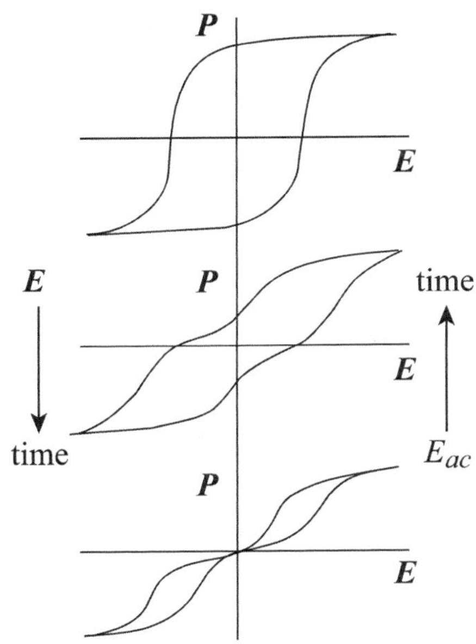

Figure 30. The evolution of P–E hysteresis loops of ferroelectric ceramics by pinning centers with respect to time and electric field. After Jonker [170].

there is a restoring force switching the domain pattern back to the original state as much as possible due to the existence of the well-distributed charged defects, resulting in a constriction effect on the loop (pinched loop) [79, 165].

Of particular interest is that similar double hysteresis loops are also observed in antiferroelectrics. The mechanisms of the double hysteresis between the aged ferroelectrics and antiferroelectrics, however, are completely different. In antiferroelectrics, there are two phases, including stable antiferroelectric phase and field-induced metastable ferroelectric phase during the dynamic hysteresis measurement, while there is only one ferroelectric phase in aged ferroelectrics. In addition, in antiferroelectrics, ac field loading cannot affect the morphology of the loop, while the ac loading can alleviate the pinning effect and open the pinched loop gradually in aged ferroelectrics, as shown in Fig. 30 [170]. Thus, this feature can be a criterion to distinguish the aged ferroelectric from real antiferroelectric.

4.2.2. *Aging after poling*

The above discussion was focused on the double hysteresis loops observed in aged ferroelectrics after sintering or electroding in an unpoled state. Here, the aging behavior for samples after poling and the corresponding *P–E* loop characteristics will be discussed.

Poling is a reorientation process of domains. As discussed in earlier section, B-site acceptor dopants in ABO_3 perovskite ferroelectrics are believed to substitute high valence B cations by low valence of cations, such as Mn^{2+}, Mn^{3+}, and Fe^{3+}, resulting in oxygen vacancies, leading to the development of acceptor-oxygen vacancy defect dipoles. The dipoles align themselves along a preferential direction for the spontaneous polarization, and/or move to the high-stressed areas of domain walls or grain boundaries by diffusion, pin the walls, and stabilize the domains. The build-up of these parallel defect dipoles to the local polarization vector leads to an offset of *P–E* behavior or internal bias (see Fig. 31 of Mn-doped PMN-PZT crystal). Compared to the obvious internal bias observed in PMN-PZN-Mn crystal, PMN-PZT crystal exhibits a symmetric hysteresis

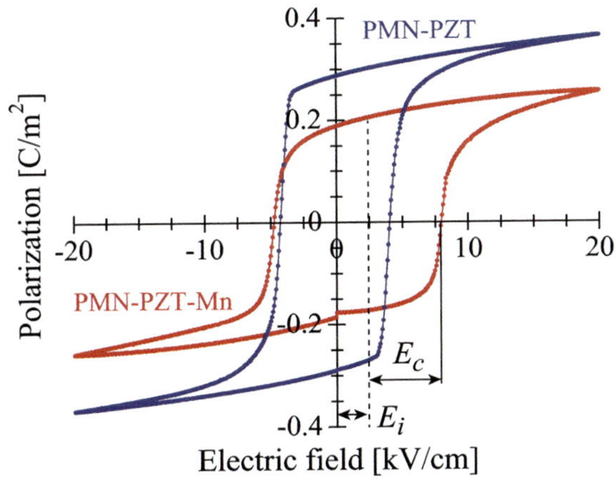

Figure 31. P–E hysteresis loops of pure and Mn modified PMN-PZT single crystals, measured at 20 kV/cm field after poling and being aged for 24 h. Reprinted with permission from Ref. [171]. Copyright © 2008, AIP Publishing LLC.

loop with respect to the ac electric field, although it has already been poled before the P–E characterization [171].

Besides the pining effect due to the defect dipoles, the space charge accumulated during aging process should also be considered to explain the evolution of the hysteresis loops during aging in poled ceramic samples. Okazaki and Nagata have proposed a model to include the space charge effect for aged piezoelectric ceramics, as shown in Fig. 32, in which the smaller rectangles represent the grain boundary, whereas the larger one represents grains [57]. Based on this model, the aging process is divided into four steps: (1) In initial state, the total polarization P_r is a sum over all domains and spontaneous polarization is P_1. (2) After poling, domains arrange with the poling field direction, and the charges are induced at the boundaries (P_3). (3) In the aging process, P_r gradually reduces, due to the fact that the space charges accumulate and lead to P_2 in the grain. Meanwhile, the induced charges at the boundaries are reduced too. (4) In the final state, the polarization of the domain is screened completely by space charges, which accumulate at the ends of the domain. In every state, the corresponding P–E hysteresis loop is shown in Fig. 32, showing an internal bias field. It should be noted that the internal bias field

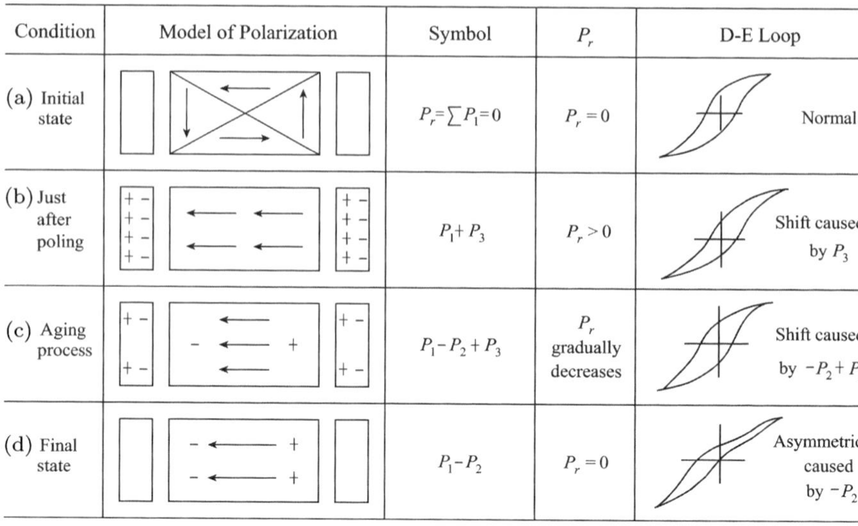

Figure 32. Space charge model of aging of poled ferroelectrics. P_1 is the spontaneous polarization, P_2 is the space charge polarization inside the domains, P_3 is the space charge polarization at the grain boundary, and P_r is the total polarization. After Xu [57].

will increase as a function of aging time after the poling process, following a time law of the form [172, 173]:

$$E_i(t) = A \log t + B, \tag{6}$$

where A depends on the temperature: with rising temperature, E_i builds up more quickly. Furthermore, A depends somewhat on the previous history of the samples. This will lead to a more asymmetric behavior of the P–E loops.

4.3. *Effect of measurement conditions*

4.3.1. *Applied electric field amplitude*

Ferroelectric materials exhibit strong nonlinear dielectric and piezoelectric responses with respect to the electric field strength. According to the field strength, the P–E behavior can be divided into three regions as shown in Fig. 33 [86]. At field region below the threshold field E_t, the dielectric permittivity ε' is almost independent

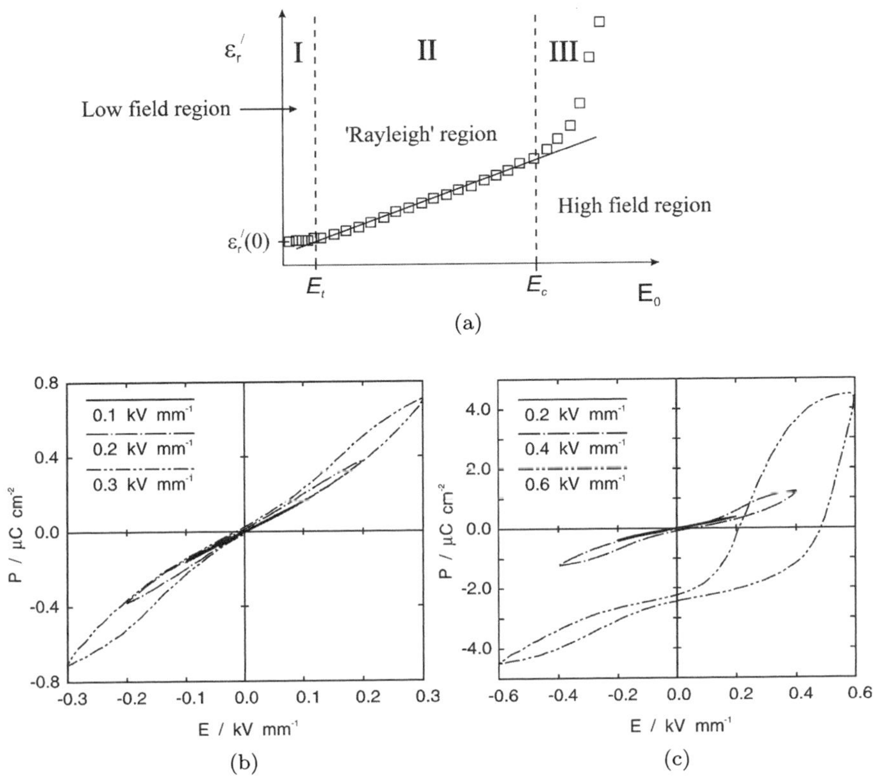

Figure 33. (a) Schematic view of the field dependence of dielectric constant in ferroelectric ceramics over a wide range of field strength. (b) and (c) Corresponding hysteresis loops measured at three regions as given (a). Reprinted with permission from Ref. [86]. Copyright © 1999, Taylor & Francis.

of the field strength. Between the E_t and E_c is the intermediate Rayleigh region, where ε' shows a linear increase with the electric field strength. It is believed that in this region some large-scale domain wall translation occurs. At high field region above the E_c, normally there is very large hysteresis in P–E behavior and an obvious increase in the dielectric permittivity apart from the linear relationship between the ε' and E_0, due to the high degree of irreversible process, that is, domain nucleation and growth. It should be noted that the boundary between the low-field region and Rayleigh region is blurry in some systems without random pinning defect dipoles. For soft

PZT, the threshold field E_t is too low to be distinguished from the Rayleigh region [174], whereas for BS-PT ceramics, there is no evidence to support the existence of E_t [175]. Furthermore, the transition from Rayleigh region to high-field region is not accurately separated by E_c.

In fact, Rayleigh region is of special interest due to the fact that the Rayleigh law proposed in 1887 has been successfully used to model the nonlinear hysteresis loop for ferroelectric materials at the intermediate field level [176]. The Rayleigh relation for dielectric response can be expressed as [176]

$$\varepsilon'(E_0) = \varepsilon'(0) + \alpha E_0, \tag{7}$$

$$P = (\varepsilon'(0) + \alpha E_0)E \pm \left(\frac{\alpha}{2}\right)(E_0^2 - E^2), \tag{8}$$

where α is the dielectric Rayleigh coefficient, and $\varepsilon'(0)$ is the initial dielectric permittivity, which is a field-independent term and represents the contribution by intrinsic lattice and contribution from reversible domain wall vibration, ε_0 is the vacuum dielectric permittivity, E_0 is the amplitude of the ac electric field. Normally the dielectric response at Rayleigh region can be well depicted by Eq. (7). The term αE_0 represents the contribution by the irreversible domain wall motion. In Eq. (8), the "+" sign corresponds to the decreasing field, whereas the "−" sign corresponds to the increasing field. In practice, the maximum E_0 used for Rayleigh modeling is limited below 0.5 E_c, because above 0.5 E_c there is an obvious deviation of P–E loops from symmetric shape into asymmetric shape [86]. Figure 34 gives an example of the Rayleigh fitting to the measured loop of a PLZT film [177]. Nowadays, the Rayleigh method has been widely used for analyzing the contribution of domain wall to dielectric and piezoelectric effects for ferroelectrics [105, 178–184]. With further increasing the applied electric field, ferroelectric phase transition will occur. The field level to induce the phase transition is closely related to the composition and temperature, lower electric field is required to induce phase transition with composition being on the proximity of MPB or at elevated temperature.

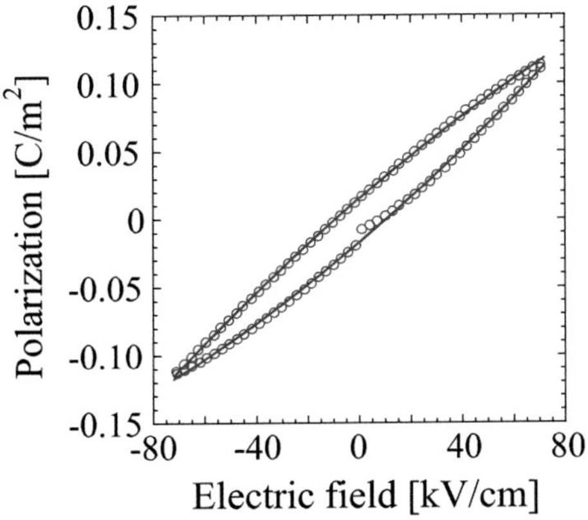

Figure 34. *P–E* hysteresis loops of a PLZT film measured at Rayleigh region and modeled by Rayleigh law. Reprinted with permission from Ref. [177]. Copyright © 2008, AIP Publishing LLC.

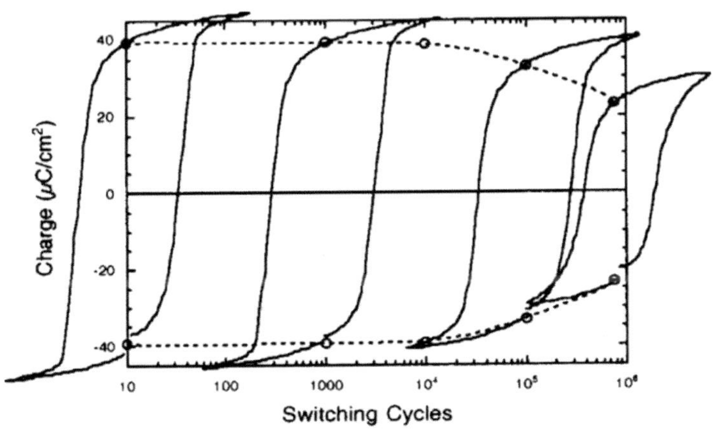

Figure 35. A schematic illustration of polarization decay as a function of the number of the switching cycles. Reprinted with permission from Ref. [190]. Copyright © 2001, AIP Publishing LLC.

4.3.2. *Cyclic field fatigue*

In ferroelectric materials, the polarization fatigue is defined as the loss of switchable polarization with respect to cyclic electric field [185–189]. As shown in Fig. 35, after the loading of bipolar cycling, the switchable polarization decreases drastically while the coercive field increases in contrast [190]. Lou has reviewed the impact factors to the fatigue phenomena for ferroelectric thin film and some bulk ferroelectric materials [189], including the influence from experimental conditions (electric field, temperature, frequency, oxygen partial pressure, optical, and thermal fatigue), and the influence of the electrodes and material modifications (conductive oxide electrodes, interface quality, interface layer, crystal microstructure, doping, processing condition, and anisotropy). Although many investigations have suggested the degradation of the polarization (including both saturated and remnant polarization) in such a fatigue process, the change in the coercive field is subtler than that observed in the polarization, as observed in Fig. 36 for (K,Na)NbO_3-based lead-free ceramics [191].

It is generally accepted that fatigue is a result of charge injection and accumulation of space charge that pins domain walls or retards the nucleation of reversion domain to permit switching. Many strategies have been developed to overcome this problem, mainly through three methods [190]. (1) Doping the ferroelectrics with donor dopants to reduce the oxygen vacancies concentration

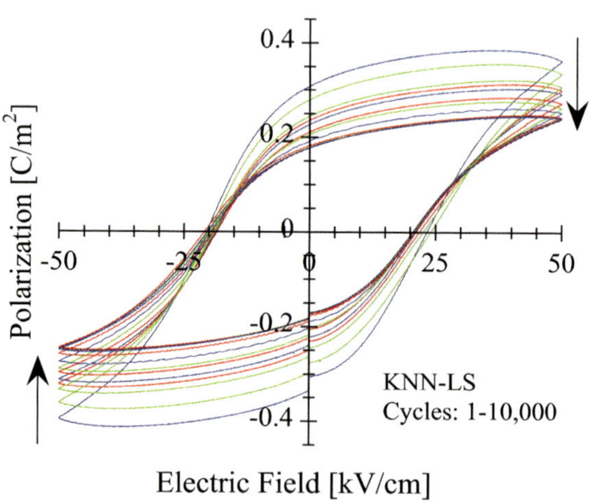

Figure 36. *P–E* hysteresis loops as a function of switching cycles for KNN-LS material. Reprinted with permission from Ref. [191]. Copyright © 2008, AIP Publishing LLC.

[192–194]. (2) Using oxide electrodes for PZT thin film ferroelectrics [195–197]. (3) Searching for ferroelectric materials with fatigue-resistant characteristics, such as $SrBiTa_2O_9$ and $SrBiNb_2O_9$, and $Bi_{3.25}Sm_{0.75}Ti_3O_{12}$ systems [190, 198, 199].

Recently, the fatigue phenomena observed in relaxor-PT single crystals offer another approach to improve the fatigue resistance by means of so-called engineered domain structures. It was reported that there was no fatigue being observed in [001]-oriented PZN-PT single crystal with rhombohedral phase, while severe fatigue occurred in [011]- and [111]-oriented crystals [190, 200, 201]. The polarization can be rejuvenated for [111]-oriented crystals after annealing the sample above their Curie temperatures, while the fatigue in [011]-oriented crystals are nonrecoverable due to the fact that the fatigue in [011]-oriented crystals is induced by the strong anisotropic symmetry, which gives rise to microcracks under cyclic electric field [202–204]. It is inferred that an engineered domain state "4R" in relaxor-based ferroelectric crystal with spontaneous polarization inclined to the normal of the electrode is associated with negligible or no fatigue at room temperature [190, 201]. Although the mechanism by which engineered domain states mitigate fatigue is not well understood, one possibility is that the inclined polarization state redistribute the space charge accumulation and thereby reduce the fatigue rate at a given temperature and composition [190]. It is also expected that the engineered domain structures with rhombohedral symmetry have a higher percentage of charged domain walls [190]. These charged domain walls could act as sinks to the injected charge in ferroelectric systems. In fact, the mechanisms based on these models affect the fatigue phenomena concurrently. Therefore, it is not favorable to emphasize any individual model to interpret the fatigue.

It is interesting to note that most fatigue phenomena have been studied in the ferroelectric materials without obvious aging effect. However, the study of the fatigue effect in hard ferroelectric perovskite materials (with strong aging effect and pinched hysteresis loop) would shed light on this point. Figure 37 shows the evaluation of the hysteresis loop of a hard 1.0 at.% Fe-doped PZT58/42 ceramics

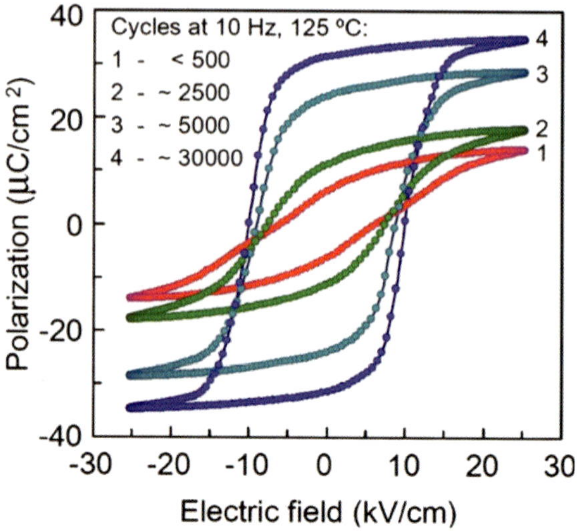

Figure 37. Relaxation of P–E hysteresis loops observed in hard 58/42 PZT ceramics doped with 1.0 at. % Fe with low frequency field cycling of 10 Hz at 125°C. Reprinted with permission from Ref. [160]. Copyright © 2008, AIP Publishing LLC.

with respect to the number of the ac electric field cycle at 125°C [160]. In contrast to previous reports on fatigue phenomena, both saturated and remnant polarizations are increased with respect to increasing the cycle number from 500 to 30000. As we have discussed in aging section, the domain wall motion is pinned by the ordered charged defects in hard ferroelectrics. Thus, not all the domain walls can respond to the external electric field. However, with the disturbance by cyclic field, the ordered distribution of the charged defects gradually smear and their pinning effect to the domain wall disappear simultaneously. This trend is well depicted in Fig. 37, though such a deaging effect is more obvious at high temperature. On the other hand, for acceptor-doped relaxor-PT crystals, such as Mn-modified PIN-PMN-PT, it was reported that the crystals shown improved fatigue behavior when compared to their pure counterparts, due to the enhanced coercive field and the existence of internal bias [202].

For antiferroelectric materials, similar degradation of the polarization after fatigue has been reported, as shown in Fig. 38 [205].

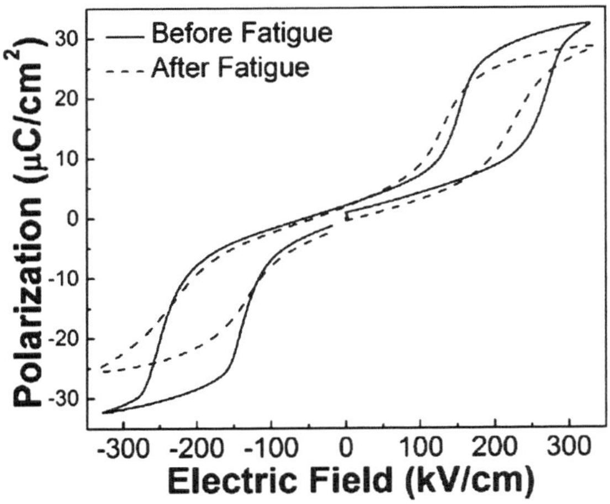

Figure 38. *P–E* hysteresis loops for PLZST thin films before and after fatigue at 10^9 cycles. Reprinted with permission from Ref. [205]. Copyright © 2012, Elsevier.

However, compared to ferroelectrics, antiferroelectrics normally show higher fatigue resistance under bipolar electric cycling [206]. Lou attributes this phenomenon to the lower depolarization field, lower local injected power density, and lower local phase decomposition probability of antiferroelectrics at the phase nuclear sites [206]. Furthermore, microcracks are also reported in antiferroelectrics after fatigue [207]. The microcracks actually pin the domain wall motion, resulting in the lower switchable polarization. By comparing with bipolar and unipolar fatigue processes, Lou and Wang suggested a scenario that polarization fatigue is mainly caused by the switching induced charge injection other than the charge injection during the stable/quasi-stable leakage stage [208].

4.3.3. *Frequency*

It is often observed that the shape of the hysteresis loops is strongly affected by the measuring frequency. Generally, a square hysteresis loop transforms to a slanted one with increasing frequency, where the E_c is found to increase, while the P_r maintains similar value. These phenomena have been observed in ferroelectric materials, including ceramics, single crystals, and thin films [185, 209, 210]. As shown in

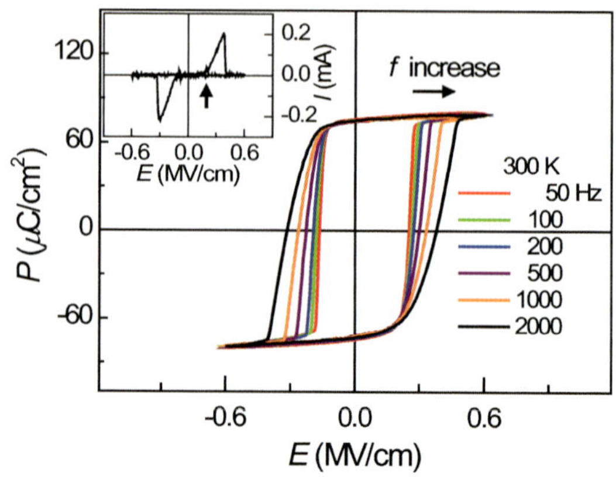

(a)

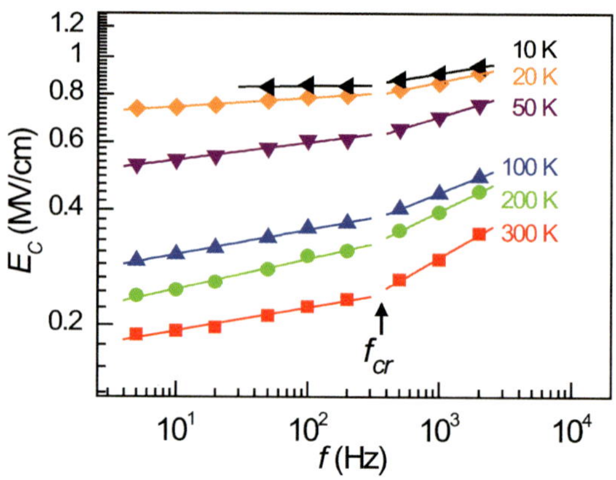

(b)

Figure 39. (a) Freqency-dependent P–E hysteresis loops of the PZT film at 300 K. The inset shows the switching current I–E curve at $f = 1000$ Hz. (b) The E_c values as a function of frequency at various temperature. The f_{cr} indicates the crossover frequency where two scaling regions are separated. The solid lines show the linear fitting results for each frequency region. Reprinted with permission from Ref. [209]. Copyright © 2010, The American Physical Society.

Fig. 39(a), the E_c of $PbZr_{0.2}Ti_{0.8}O_3$ thin film does increase gradually as a function of the frequency, with the enhancement being as much as 80% with the frequency increasing from 50 to 2000 Hz [209].

It was discussed earlier that the domains played an important role in the polarization reversal process for a ferroelectric materials.

Subjected to an external electric field, the volume of the favored domain patterns increases by the nucleation and subsequent growth of the domains [211]. Therefore, the morphologies of the hysteresis loops are strongly affected by the nucleation and domain wall motion, leading to the fact that the P–E loops are frequency-dependent, as nucleation and domain wall motion are time-dependent [209]. There are two models based on the domain dynamics being generally used to interpret the frequency-dependent hysteresis loop. A phenomenological model was developed on considering the domain growth process; the time-dependent fractional volume of the reversed domains was calculated based on the extended Avrami theory, suggesting that the E_c follows a simple power law relationship $E_c \propto f^{\beta}$ [212]. Another model assumed that the limiting step of the frequency-dependent E_c is the nucleation process [209]. According to these models, the E_c is expected to increase with increasing frequency, consistent with the experimental observations. A careful analysis of the characteristic parameters of the PZT thin films shown in Fig. 39(a) suggests that there are two scaling regimes, which are well depicted in the log–log plot of E_c and f, as shown in Fig. 39(b) [209].

In bulk materials, such as soft PZT ceramics [213], [100]-orientated KNN [214] and BT [215] single crystals, only one scaling regime was reported. It should be noted, however, due to the limitation of the voltage amplifier power, the varying frequency range is much narrower for bulk materials than that for thin films. More than one regime is expected if a broader frequency window can be applied to the bulk materials. Due to the fact that the domain reversal process is composed of two stages, that is, nucleation and domain growth, the dynamic process of these two stages will be different as reflected in the log–log plot of E_c and f [209].

4.3.4. *Temperature*

The influence of temperature on hysteresis loops is an inevitable topic from both fundamental and application points of view. Here, we only discuss the temperature effect on ferroelectric materials far below the ferroelectric-to-paraelectric phase transition temperature T_C, above which, the macroscopic net polarization vanishes completely, with the

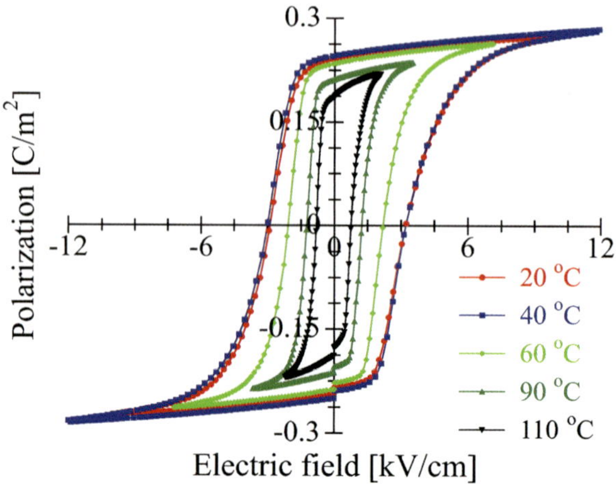

Figure 40. P–E hysteresis loops for PMN-PT crystals measured at different temperature. Reprinted with permission from Ref. [216]. Copyright © 2013, Elsevier.

PE loop being transformed from normal square shape through slim loop (2S loop for first-order ferroelectrics), and finally to a linear nonhysteresis response between the P and E, as discussed in the above section, as shown in Fig. 40 [216].

For ferroelectric materials, temperature increasing is associated with the increase in thermal fluctuation, which will disrupt the long-range order of the polarization and weaken the coupling stabilizing effect between the charged defects and domains [49]. Therefore, both E_c and P_r are generally decreased with increasing the temperature, as shown in Fig. 41 [217].

Generally speaking, the E_c is decreased with respect to increasing temperature. A linear relationship between E_c and T, that is, $(E_{c0} - E_c) \propto T$ (E_{c0} is the coercivity at $T \sim 0\,\mathrm{K}$), was reported for soft PZT ceramics [217], $Pb(Mn_{1/3}Sb_{2/3})$-$Pb(Zr,Ti)O_3$ ceramics [218], $(K,Na)NbO_3$ thin films [219], and Zr-rich PZT ceramics [220]. However, this trend is often disturbed by the PPT. In the vicinity of the ferroelectric phase transition temperature, such a linear relationship is invalid, where an abnormal increase in the E_c with increasing temperature was reported, as shown in Fig. 42(a).

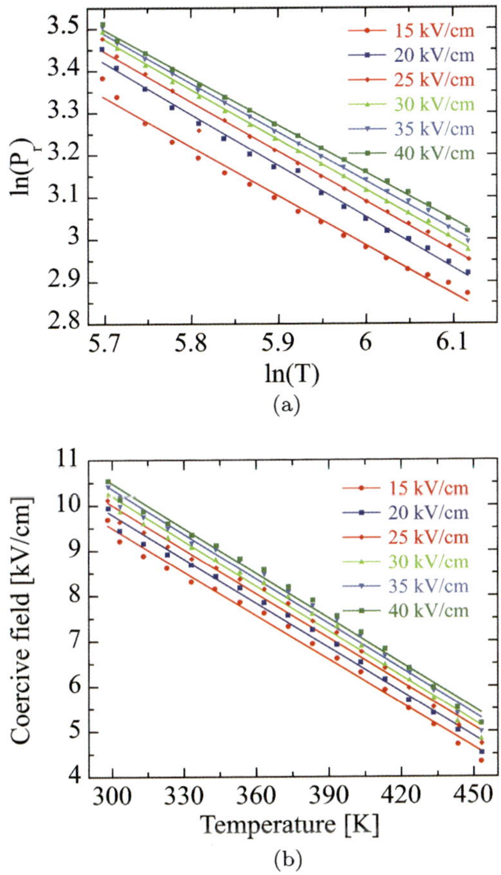

Figure 41. (a) Double logarithmic plot between $\ln(P_r)$ and $\ln(T)$ and (b) linear plot between E_c and T. Reprinted with permission from Ref. [218]. Copyright © 2007, AIP Publishing LLC.

The abnormal E_c observed on the proximity of PPT transition can be explained by the coexistence of different ferroelectric phases. It is interesting to note that for a poled Mn-doped PMN-PZT ceramic, the asymmetric P–E hysteresis loop was found to gradually transform to symmetric shape with increasing temperature [218]. Similar results can also be observed in Mn-modified PIN-PMN-PT crystals, where the internal bias field E_i and E_c follow a similar linear relationship with respect to temperature, as shown in Fig. 42(b) [221]. This is

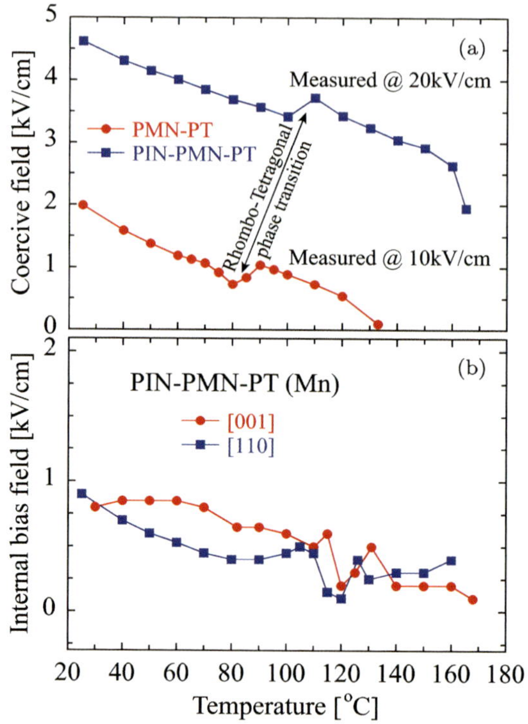

Figure 42. (a) E_c as a function of temperature for PIN-PMN-PT and PMN-PT crystals. (b) Internal bias as a function of temperature. Reprinted with permission from Ref. [221]. Copyright © 2011, Elsevier.

due to the fact that the aligned defect dipoles decoupled gradually by increasing temperature and consequently, an almost symmetric square P–E hysteresis loop was observed at elevated temperature.

The change in P_r as a function of temperature is more subtle compared to E_c. Yimnirum *et al.* reported the effect of temperature on P_r for soft and hard PZT ceramics, suggesting a power law relationship, that is, $P_r \propto T^{\beta}$, with a negative β [217, 222]. In other report, however, the P_r was found to maintain similar values in the temperature range 298–473 K for poled Mn-doped PMN-PZT ceramics [218], which may be due to the existence of PPT in the studied temperature range. In a Zr-rich PZT ceramic, Chen *et al.* found that the P_r–T relationship could only be well described by two power law fittings, and the β changed from -0.62 at low T to -0.31 at high T [220]. The critical temperature separating these two regions is around 313 K, corresponding to a low-temperature ferroelectric rhombohedral $F_{R(LT)}$ to high-temperature ferroelectric rhombohedral $F_{R(HT)}$ phase transition point. It is interesting to note that the $F_{R(LT)}$–$F_{R(HT)}$ phase transition cannot be detected by the dielectric properties as a function of temperature, but very sensitive to the polarization variation, where the β shows distinguished values below and above the transition point. Similar results were also reported for KNN thin film [219], where there are two power law regions for Pr measured from 100 to 340 K, being expressed as $P_r \propto T^{-0.77}$ for $T < 245$ K and $P_r \propto T^{0.88}$ for $T > 245$ K, respectively. Meanwhile, a rhombohedral-to-orthorhombic phase transition was found to occur at 245 K, corresponding to the inversion temperature of the polarization. Although the power law was valid at both regions, the β became positive at high-temperature region, which means the polarization was increased with increasing the temperature.

It is important to point out that there are very limited studies on the hysteresis loop in cryogenic temperature region, which may shed more light on the understanding of the domain behavior as a function of temperature and electric field, especially near the freezing temperature range.

4.3.5. *Stress*

Ferroelectric materials are frequently subjected to external stress for practical applications. According to the loading methods, three kinds of stress effects on the hysteresis loops will be discussed in this section.

(a) Uniaxial stress. Domain switching occurs when the applied electric field exceeds the coercive field, being determined by the crystalline structure and symmetry. If an uniaxial compressive stress is applied to the sample with the direction parallel to the electric field, there are two factors need to be considered, being associated with intrinsic and extrinsic contributions [49, 52]. One is the crystal lattice distortion or spontaneous deformation, which is suppressed by such an external mechanical stress to a certain extent. As a consequence, the intrinsic contribution by the crystal lattices to the polarizations, which can be reflected by P_s and P_r, are reduced. The other one is the domain wall motion (especially the non-180° ferroelastic domain walls), which is clamped or constrained when subjected to compressive stress [72]. In this case, less domains will be involved in the movement responding to the external electric driving field, leading to a decreased contribution to the polarization. Furthermore, even the electric field exceeds the coercive field, the domain wall motion is not easy to occur under the stress, and thus higher field is required to switch the domain walls [72]. Both the intrinsic and extrinsic contributions account for the hysteresis loops being transformed from a square loop to a slanted one. If the stress is large enough, domain wall motions, especially non-180° ferroelastic domains, will be frozen and a nonhysteresis loop is expected. This was confirmed by the investigation on the electromechanical properties of 8/65/35 PLZT ceramics under uniaxial compressive stress, where both P_s and P_r were found to decrease with respect to increasing stress [72]. Later, this observation was also verified in soft PZT [223], Fe^{3+}/Nb^{5+} $BaTiO_3$ [224], $Pb(In_{0.5}Nb_{0.5})O_3$-PT [225], and PZT-885 [226]. A typical evolution of the hysteresis loops with respect to the uniaxial stress is given in Fig. 43(a). However, the effect of the uniaxial stress on the E_c is controversial based on the results reported by different research groups. In most cases, E_c was found to decrease as the stress increases [Fig. 43(c)], with the exception that the E_c was reported to be insensitive to the external compressive stress in soft ceramics [223].

On the contrary, a stable polar phase is preferred with the uniaxial compressive stress applied perpendicular to the electric field direction. Figure 44 shows the evolution of the hysteresis loops for 0.9PMN-0.1PT relaxor ferroelectrics, with increasing perpendicular uniaxial stress from 0 to 130 MPa [226]. It is clear that there is a transition from a slim loop under the stress-free condition to a near square hysteresis loop under a high compressive stress, suggesting that the macroscopic polar phase induced by the electric field becomes more stable with the aid of the perpendicular stress. This will benefit

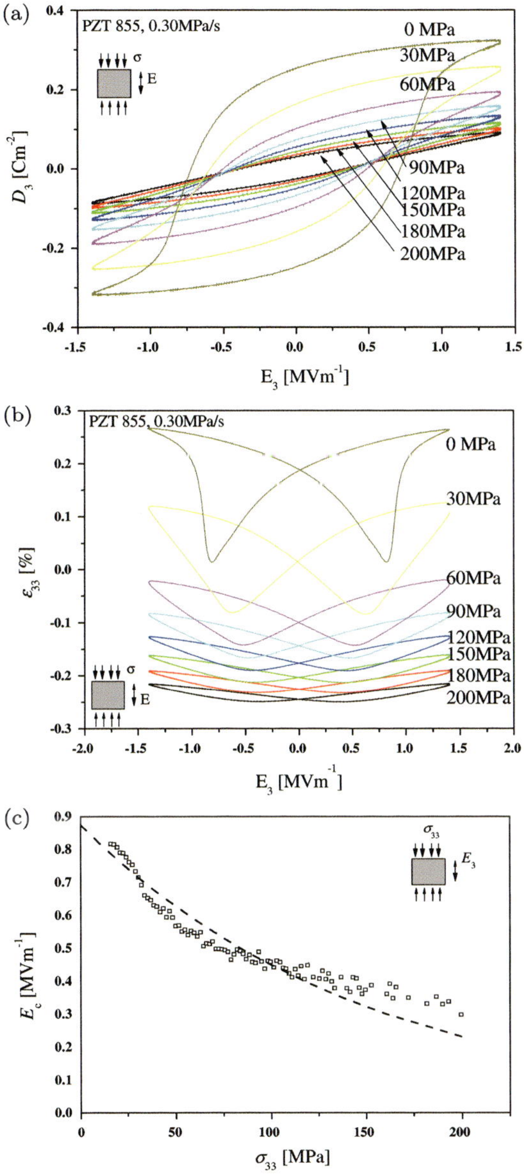

Figure 43. P–E hysteresis loops depend on the applied uniaxial stress for PZT 855 from initially unpoled state under cyclic electric loading of 1.4 MV/m and constant stress ramp of 0.3 MPa/s: (a) P_3 against E_3; (b) ε_{33} against E_3; (c) E_c against σ_{33}. Reprinted with permission from Ref. [226]. Copyright © 2011, Springer-Verlag.

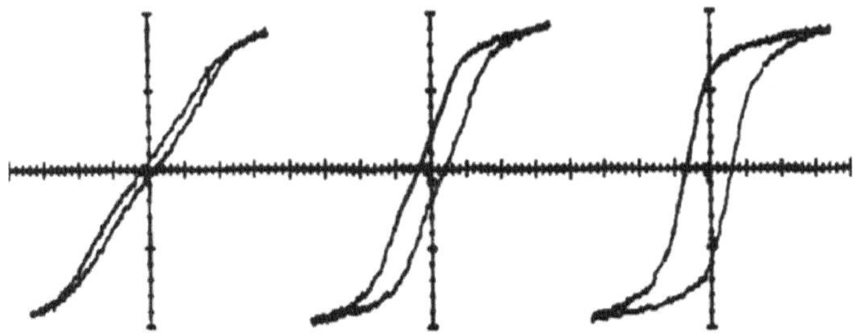

Figure 44. Change in the P–E hysteresis loops with the compressive stress σ_1 from slim loop to square loop characteristic of regular ferroelectric state. The axis scale is 1.5 kV/cm per division for electric field and 0.14 C/m^2 per division for polarization. Reprinted with permission from Ref. [227]. Copyright © 1999, AIP Publishing LLC.

the practical application, especially for shear vibration mode, where the driving electric field is perpendicular to the poling direction. Li *et al.* reported that for application of thickness shear crystals, the main drawback is the low allowable ac electric field, being less than half of its coercive field [89, 228]. The shear piezoelectric response was found to drastically decrease with unexpected interference by other vibration modes once the ac field exceeds the allowable field [89]. With the prestress perpendicular to the poling direction, the polar state was sustained effectively; in addition, it will induce the ferroelectric phase transition if the stress is large enough [229]. As shown in Fig. 45, the hysteresis properties of PIN-PMN-PT crystals were drastically reduced by applying the perpendicular stress to the crystals, due to the fact that the domain wall motions are stabilized by the stress [228]. As a consequence, the allowable drive field is increased by more than 50%.

(b) Radial Stress. It was reported that the domain wall motion of PZT ceramics became facilitated with the aid of a radial mechanical load, leading to the increased P_s and P_r and decreased E_c [230, 231]. However, this mechanical confinement effect seems more obvious in relaxor ferroelectric single crystals. Marsilius *et al.* suggested that the radial compressive stress broaden the hysteresis loops for both

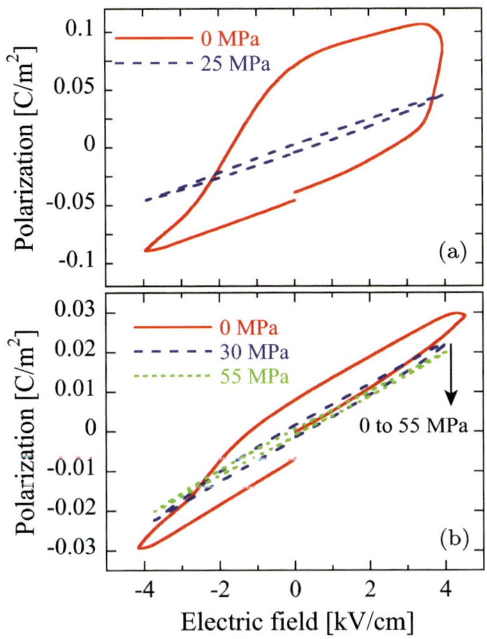

Figure 45. P–E hysteresis loops at various uniaxial stresses for "1T" and (b) "2R" PIN-PMN-PT crystals. "1T" is [001] poled single domain tetragonal crystal, the uniaxial stress was applied along [010] direction, "2R" indicates the [011] poled rhombohedral crystal, the uniaxial stress was along [0–11] direction. Measured at 1 Hz. Reprinted with permission from Ref. [228]. Copyright © 2012, AIP Publishing LLC.

rhombohedral and tetragonal PMN-PT single crystals, as shown in Fig. 46 [232]. A drastic increase of 200% then followed by a saturation plateau of the E_c was reported, with the P_r initially increased and then decreased with increasing radial stress. These observations are interpreted in terms of direct/converse piezoelectric effect and possible phase transition in rhombohedral crystals and by the ferroelastic domain switching in tetragonal crystals [232]. It is noted that due to the ultrahigh piezoelectric response, the coupling between the mechanical and electrical field is much larger than that in polycrystalline ceramics. This may be the reason why the radial stress effect in ferroelectric single crystals is more significant than that in ceramics.

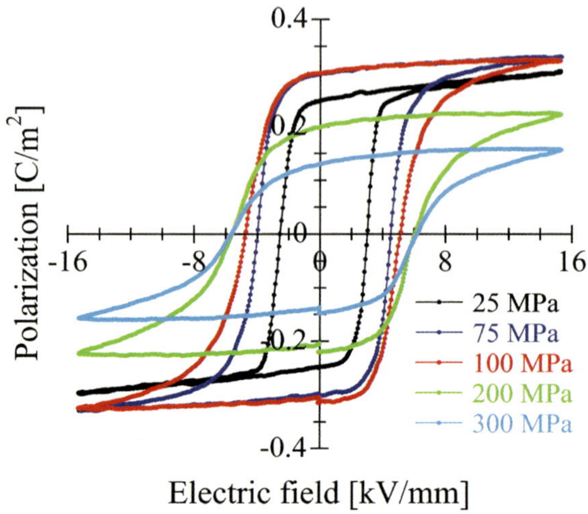

Figure 46. *P–E* hysteresis loops recorded from a pure rhombohedral phase PMN-27.2%PT single crystal under different levels of radial confinements. Reprinted with permission from Ref. [232]. Copyright © 2012, Wiley.

(c) Hydrostatic Stress. The hydrostatic stress effect on the ferroelectric properties for PLZT ceramics with MPB compositions was studied [233]. For the ferroelectric PLZT 2/90/10, both P_s and P_r show a small decrease with increasing the hydrostatic stress up to 200 MPa. However, a drastic decrease in P_r and pronounced broadening of the switching region were observed at 300 MPa, suggesting a transformation form ferroelectric-to-antiferroelectric phase [233]. In contrast, for PLZT4/90/10, typical antiferroelectric double hysteresis loops were obtained up to 200 MPa. Both the forward and backward switching fields increased significantly, followed by a rapid decrease of $P_{\max}$, as P_s cannot be induced with a limited electric field. Importantly, for PLZT3/90/10 sample, a clear ferroelectric-to-antiferroelectric phase transitions occurs when the pressure exceeded 100 MPa. Based on the above, the antiferroelectric phase is more preferred with respect to the hydrostatic stress, due to its smaller volume compared to its corresponding ferroelectric phase [233]. It is interesting to note that for antiferroelectric ceramics, even uniaxial and radial stresses prefer the antiferroelectric phase. Similar observations as in PLZT4/90/10 were also reported by Tan *et al.* [234].

It should be noted that the stress effect on the P–E characteristics of thin film is dependent on the film thickness, which was discussed in the Section 4.1.5

5. Summary and future perspective

In this chapter, the hysteresis loop which is the critical characteristic of ferroic materials has been discussed for ferroelectric-related materials. The ferroelectric loops are divided into four groups according to their morphology features, that is, classical ferroelectric loops, double hysteresis loops, asymmetric hysteresis loops, and slim hysteresis loops. To clarify the comprehension of the loops in ferroelectrics, we discussed the effects of materials, aging, and measuring conditions on hysteresis loops in terms of recent developments of ferroelectrics. By analyzing various examples, the general rules of the evolution of the hysteresis loops were summarized in this study and the hysteresis phenomena in ferroelectric could be understood thoroughly. In the future perspective, it is expected that more information of ferroelectric materials, including macroscopic properties and microscopic structures, can be "read" from the hysteresis loop, by combining the modeling and simulation of materials.

Acknowledgment

The authors gratefully thank Prof. Dragan Damjanovic for his original inspiration at ISAF-PFM-2011 conference to compose this review article and Prof. Thomas R. Shrout for his valuable comments. This work was supported by the National Nature Science Foundation of China (grant nos. 51102193, 51202183, and 51372196), the China Postdoctoral Science Foundation, and the Fundamental Research Funds for the Central Universities.

References

[1] Wadhawan, V. K. (2000). Ferroic materials: A primer, *Resonance*, 7, pp. 15–24.
[2] Wadhawan, V. K. (2000). *Introduction to Ferroic Materials* (Gordon and Breach, Tokyo).

[3] Spaldin, N. A., Cheong, S.-W. and Ramesh, R. (2010). Multiferroics: Past, present, and future, *Phys. Today*, 63, pp. 38–43.
[4] Cheong, S.-W. and Mostovoy, M. (2007). Multiferroics: A magnetic twist for ferroelectricity, *Nature Mater.*, 6, pp. 13–20.
[5] Khomskii, D. (2009). Classifying multiferroics: Mechanisms and effects, *Physics*, 2, p. 20.
[6] Aizu, K. (1969). Possible species of ferroelastic crystals and of simultaneously ferroelectric and ferroelastic crystals, *J. Phys. Soc. Jpn.*, 27, pp. 387–396.
[7] Scott, J. F. (2007). Applications of modern ferroelectrics, *Science*, 315, pp. 954–959.
[8] Spaldin, N. A. and Fiebig, M. (2005). The renaissance of magnetoelectric multiferroics, *Science*, 309, pp. 391–392.
[9] Ma, J., Hu, J., Li, Z. and Nan, C.-W. (2011). Recent progress in multiferroic magnetoelectric composites: From bulk to thin films, *Adv. Mater.*, 23, pp. 1062–1087.
[10] Zhang, S. and Li, F. (2012). High performance ferroelectric relaxor-$PbTiO_3$ single crystals: Status and perspective, *J. Appl. Phys.*, 111, p. 031301.
[11] Schmid, H. (1994). Multi-ferroic magnetoelectrics, *Ferroelectrics*, 162, pp. 317–338.
[12] Wang, K. F., Liu, J. M. and Ren, Z. F. (2009). Multiferroicity: The coupling between magnetic and polarization orders, *Adv. Phys.*, 58, pp. 321–448.
[13] Slater, J. C. (1950). The Lorentz correction in barium titanate, *Phys. Rev.*, 78, pp. 748–761.
[14] Cochran, W. (1959). Crystal stability and the theory of ferroelectricity, *Phys. Rev. Lett.*, 3, pp. 412–414.
[15] Cochran, W. (1960). Crystal stability and the theory of ferroelectricity, *Adv. Phys.*, 9, pp. 387–423.
[16] Cohen, R. E. (1992). Origin of ferroelectricity in perovskite oxides, *Nature*, 358, pp. 136–138.
[17] Valasek, J. (1921). Piezoelectric and allied phenomena in Rochelle salt, *Phys. Rev.*, 17, pp. 475–481.
[18] Valasek, J. (1922). Piezoelectric activity of Rochelle salt under various conditions, *Phys. Rev.*, 19, pp. 478–491.
[19] Valasek, J. (1922). Properties of Rochelle salt related to the piezo-electric effect, *Phys. Rev.*, 20, pp. 639–664.
[20] Kanzig, W. (1987). History of Ferroelectricity, 1938–1955, *Ferroelectrics*, 74, pp. 285–291.
[21] Cross, L. E. and Newnham, R. E. (1987). *Ceramics and Civilization, Vol. III, High-Technology Ceramics-Past, Present, and Future*, "History of ferroelectricity, " (American Ceramic Society, Westerville, OH), pp. 289–305.
[22] Fousek, J. (1991). Ferroelectricity: Remarks on historical aspects and present trends, *Ferroelectrics*, 113, pp. 3–20.
[23] Haertling, G. H. (1999). Ferroelectric ceramics: History and technology, *J. Am. Ceram. Soc.*, 82, pp. 797–818.

[24] Kittel, C. (1951). Theory of antiferroelectric crystals, *Phys. Rev.*, 82, pp. 729–732.
[25] Sawaguchi, E., Maniwa, H. and Hoshino, S. (1951). Antiferroelectric structure of lead zirconate, *Phys. Rev.*, 83, p. 1078.
[26] Shirane, G., Sawaguchi, E. and Takagi, Y. (1951). Dielectric properties of lead zirconate, *Phys. Rev.*, 84, pp. 476–481.
[27] Shirane, G. and Suzuki, K. (1952). Crystal structure of Pb(Zr-Ti)O_3, *J. Phys. Soc. Jpn.*, 7, p. 333.
[28] Shirane, G. and Takeda, A. (1952). Phase transitions in solid solutions of $PbZrO_3$ and $PbTiO_3$ (I) small concentrations of $PbTiO_3$, *J. Phys. Soc. Jpn.*, 7, pp. 5–11.
[29] Shirane, G., Suzuki, K. and Takeda, A. (1952). Phase transitions in solid solutions of $PbZrO_3$ and $PbTiO_3$ (II) X-ray study, *J. Phys. Soc. Jpn.*, 7, pp. 12–18.
[30] Jaffe, B., Roth, R. S. and Marzullo, S. (1954). Piezoelectric properties of lead zirconate-lead titanate solid-solution ceramics, *J. Appl. Phys.*, 25, pp. 809–810.
[31] B. Jaffe, W. R. Cook Jr., and H. Jaffe, *Piezoelectric Ceramics*, Academic, New York, 1971.
[32] Jaffe, H. and Berlincourt, D. A. (1965). Piezoelectric transducer materials, *Proc. IEEE*, 53, pp. 1372–1386.
[33] Matthias, B. T. and Remeika, J. P. (1949). Ferroelectricity in the ilmenite structure, *Phys. Rev.*, 76, pp. 1886–1887.
[34] Shirane, G., Danner, H., Pavlovic, A. and Pepinsky, R. (1954). Phase transitions in ferroelectric $KNbO_3$, *Phys. Rev.*, 93, pp. 672–673.
[35] Cross, L. E. (1958). Electric double hysteresis in $(K_xNa_{1-x})NbO_3$ single crystals, *Nature*, 181, pp. 178–179.
[36] Smolenskii, G. A., Isupov, V. A., Agranovskaya, A. I. and Krainik, N. N. (1961). New ferroelectrics of complex composition. 4, *Sov. Phys. Solid State*, 2, pp. 2651–2654.
[37] Roedel, J., Jo, W., Seifert, K. T. P., Anton, E.-M., Granzow, T. and Damjanovic, D. (2009). Perspective on the development of lead-free piezoceramics, *J. Am. Ceram. Soc.*, 92, pp. 1153–1177.
[38] Goodman, G. (1953). Ferroelectric properties of lead metaniobate, *J. Am. Ceram. Soc.*, 36, pp. 368–372.
[39] Cross, L. E. (1987). Relaxor ferroelectrics, *Ferroelectrics*, 76, pp. 241–267.
[40] Samara, G. A. (2003). The relaxational properties of compositionally disordered ABO_3 perovskites, *J. Phys.: Condens. Matter*, 15, pp. R367–R411.
[41] Ye, Z. G., Relaxor ferroelectric complex perovskites: Structure, properties and phase transitions, Key *Eng. Mater.*, 155–156, pp. 81–122.
[42] Kuwata, J., Uchino, K. and Nomura, S. (1981). Phase-transitions in the $Pb(Zn_{1/3}Nb_{2/3})O_3$-$PbTiO_3$ system, *Ferroelectrics*, 37, pp. 579–582.
[43] Kuwata, J., Uchino, K. and Nomura, S. (1982). Dielectric and piezoelectric properties of $0.91Pb(Zn_{1/3}Nb_{2/3})O_3$-$0.09PbTiO_3$ single-crystals, *Jpn. J. Appl. Phys.*, 21, pp. 1298–1302.

[44] Park, S. E. and Shrout, T. R. (1997). Ultrahigh strain and piezoelectric behavior in relaxor based ferroelectric single crystals, *J. Appl. Phys.*, 82, pp. 1804–1811.
[45] Park, S. E. and Shrout, T. R. (1997). Relaxor based ferroelectric single crystals for electro-mechanical actuators, *Mater. Res. Innov.*, 1, pp. 20–25.
[46] Service, R. F. (1997). Materials science-shape-changing crystals get shiftier, *Science*, 275, p. 1878.
[47] Li, X. and Luo, H. (2010). The growth and properties of relaxor-based ferroelectric single crystals, *J. Am. Ceram. Soc.*, 93, pp. 2915–2928.
[48] von hippel, A. (1950). Ferroelectricity, domain structure, and phase transitions of barium titanate, *Rev. Mod. Phys.*, 22, pp. 221–237.
[49] Damjanovic, D. (1998). Ferroelectric, dielectric and piezoelectric properties of ferroelectric thin films and ceramics, *Rep. Prog. Phys.*, 61, pp. 1267–1324.
[50] Glass, M. L. and Lines, A. M. (1979). *Principles and Applications of Ferroelectrics and Related Materials* (Oxford, New York).
[51] Mayergoyz, I. D. (1991). *Mathematical Models of Hysteresis* (Springer-Verlag, New York).
[52] Damjanovic, D. (2005). *Hysteresis in Piezoelectric and Ferroelectric Materials* (Elsevier, Amsterdam).
[53] Spaldin, N. (2010). *Magnetic Materials: Fundamentals and Applications*, 2nd Ed. (Cambridge, New York).
[54] Scott, J. F. (2008). Ferroelectrics go bananas, *J. Phys.: Condens. Matter*, 20, p. 021001.
[55] Maglione, M. and Subramanian, M. A. (2008). Dielectric and polarization experiments in high loss dielectrics: A word of caution, *Appl. Phys. Lett.*, 93, p. 032902.
[56] Sawyer, C. B. and Tower, C. H. (1930). Rochelle salt as a dielectric, *Phys. Rev.*, 35, pp. 0269–0273.
[57] Xu, Y. (1991). *Ferroelectric Materials and Their Applications* (North-Holland, Amsterdam).
[58] aixACCT. Available at http://www.aixacct.com/index.html.
[59] Radiant Technologies, Available at https://www.ferrodevices.com/1/297/index.asp.
[60] Dawber, M., Rabe, K. M. and Scott, J. F. (2005). Physics of thin-film ferroelectric oxides, *Rev. Mod. Phys.*, 77, pp. 1083–1130.
[61] Cheng, Z., Wang, X., Dou, S., Kimura, H. and Ozawa, K. (2008). Improved ferroelectric properties in multiferroic $BiFeO_3$ thin films through La and Nb codoping, *Phys. Rev.* B, 77, p. 092101.
[62] Rojac, T., Bencan, A., Drazic, G., Kosec, M. and Damjanovic, D. (2012). Piezoelectric nonlinearity and frequency dispersion of the direct piezoelectric response of $BiFeO_3$ ceramics, *J. Appl. Phys.*, 112, p. 064114.
[63] Davis, M. (2006). *Phase Transition, Anisotropy and Domain Engineering: The Piezoelectric Properties of Relaxor-Ferroelectric Single Crystals*, Ph.D. Dissertation, (Swiss Federal Institute of Technology-EPFL, Lausanne, Switzerland).

[64] Introduction of LVDT basics. Available at http://www.macrosensors.com/lvdt_tutorial.html.

[65] MTI Instruments Inc. Available at http://www.mtiinstruments.com/technology/Fotonic.aspx.

[66] Kholkin, A. L., Wutchrich, C., Taylor, D. V. and Setter, N. (1996). Interferometric measurements of electric field-induced displacements in piezoelectric thin films, *Rev. Sci. Instrum.*, 67, pp. 1935–1941.

[67] Li, J. F., Moses, P. and Viehland, D. (1995). Simple, high-resolution interferometer for the measurement of frequency-dependent complex piezoelectric responses in ferroelectric ceramics, *Rev. Sci. Instrum.*, 66, pp. 215–221.

[68] Pan, W. Y. and Cross, L. E. (1989). A sensitive double beam laser interferometer for studying high-frequency piezoelectric and electrostrictive strains, *Rev. Sci. Instrum.*, 60, pp. 2701–2705.

[69] Kalinin, S. V., Shao, R. and Bonnell, D. A. (2005). Local phenomena in oxides by advanced scanning probe microscopy, *J. Am. Ceram. Soc.*, 88, pp. 1077–1098.

[70] Bonnell, D. A., Kalinin, S. V., Kholkin, A. L. and Gruverman, A. (2009). Piezoresponse force microscopy: A window into electromechanical behavior at the nanoscale, *Mrs. Bulletin*, 34, pp. 648–657.

[71] Lynch, C. S. (1995). Strain compensated thin film stress gauges for stress wave measurements in the presence of lateral strain, *Rev. Sci. Instrum.*, 66, pp. 5582–5589.

[72] Lynch, C. S. (1996). The effect of uniaxial stress on the electro-mechanical response of 8/65/35 PLZT, *Acta Mater.*, 44, pp. 4137–4148.

[73] Webber, K. G., Zuo, R. and Lynch, C. S. (2008). Ceramic and single-crystal $(1-x)$PMN-xPT constitutive behavior under combined stress and electric field loading, *Acta Mater.*, 56, pp. 1219–1227.

[74] Devonshire, A. F. (1954). Theory of ferroelectrics, *Adv. Phys.*, 3, pp. 85–130.

[75] Cross, L. (1993). Ferroelectric ceramics: Tailoring properties for specific applications, in Ferroelectric ceramics, edited by Setter, N. (Birkhäuser Verlag, Basel), pp. 1–85.

[76] Zhang, S. J., Xia, R., Randall, C. A., Shrout, T. R., Duan, R. R. and Speyer, R. F. (2005). Dielectric and piezoelectric properties of niobium-modified $BiInO_3$-$PbTiO_3$ perovskite ceramics with high Curie temperatures, *J. Mater. Res.*, 20, pp. 2067–2071.

[77] Berlincourt, D., Jaffe, B. and Krueger, H. H. A. (1964). Stability of phases in modified lead zirconate with variation in pressure, electric field, temperature and composition, *J. Phys. Chem. Solids*, 25, pp. 659–674.

[78] Frederick, J. (2010). Strains and polarization developed during electric field-induced antiferroelectric to ferroelectric phase transformations in lead zirconate-based ceramics. Master Thesis (Iowa State University, Ames, Iowa).

[79] Jin, L., He, Z. and Damjanovic, D. (2009). Nanodomains in Fe^{+3}-doped lead zirconate titanate ceramics at the morphotropic phase boundary do not correlate with high properties, *Appl. Phys. Lett.*, 95, p. 012905.

[80] Merz, W. J. (1953). Double hysteresis loop of $BaTiO_3$ at the Curie point, *Phys. Rev.*, 91, pp. 513–517.
[81] Zhang, N. (2013). Study on the electric field induced strain of antiferroelectric materials, Ph.D. Dissertation (Xi'an Jiaotong University, Xi'an, China).
[82] Ma, C. and Tan, X. (2010). Phase diagram of unpoled lead-free $(1-x)(Bi_{1/2}Na_{1/2})TiO_3$-$x$$BaTiO_3$ ceramics, *Solid State Commun.*, 150, pp. 1497–1500.
[83] Ma, C., Tan, X., Dul'kin, E. and Roth, M. (2010). Domain structure-dielectric property relationship in lead-free $(1-x)(Bi_{1/2}Na_{1/2})TiO_3$-$x$$BaTiO_3$ ceramics, *J. Appl. Phys.*, 108, p. 104105.
[84] Ma, C. and Tan, X. (2011). In situ transmission electron microscopy study on the phase transitions in lead-free $(1-x)(Bi_{1/2}Na_{1/2})TiO_3$-$x$$BaTiO_3$ ceramics, *J. Appl. Phys.*, 94, pp. 4040–4044.
[85] Tan, X., Ma, C., Frederick, J., Beckman, S. and Webber, K. G. (2011). The antiferroelectric ↔ ferroelectric phase transition in lead-containing and lead-free perovskite ceramics, *J. Am. Ceram. Soc.*, 94, pp. 4091–4107.
[86] Hall, D. A. and Stevenson, P. J. (1999). High field dielectric behaviour of ferroelectric ceramics, *Ferroelectrics*, 228, pp. 139–158.
[87] Carl, K. and Härdtl, K. H. (1978). Electrical aftereffects in $Pb(Ti,Zr)O_3$ ceramics, *Ferroelectrics*, 17, pp. 473–486.
[88] Robels, U. and Arlt, G. (1993). Domain-wall clamping in ferroelectrics by orientation of defects, *J. Appl. Phys.*, 73, pp. 3454–3460.
[89] Zhang, S. and Shrout, T. R. (2010). Relaxor-PT single crystals: Observations and developments, *IEEE Trans. Ultrason. Ferroelectr. Freq. Control*, 57, pp. 2138–2146.
[90] Lee, H. J. and Zhang, S. (2012). Design of low-loss 1-3 piezoelectric composites for high-power transducer applications, *IEEE Trans. Ultrason. Ferroelectr. Freq. Control*, 59, pp. 1969–1975.
[91] Lee, H. J., Ural, S. O., Chen, L., Uchino, K. and Zhang, S. (2012). High power characteristics of lead-free piezoelectric ceramics, *J. Am. Ceram. Soc.*, 95, pp. 3383–3386.
[92] Bokov, A. A. and Ye, Z. G. (2006). Recent progress in relaxor ferroelectrics with perovskite structure, *J. Mater. Sci.*, 41, pp. 31–52.
[93] Kleemann, W. (2006). The relaxor enigma — charge disorder and random fields in ferroelectrics, *J. Mater. Sci.*, 41, pp. 129–136.
[94] Li, F., Jin, L., Xu, Z., Wang, D. and Zhang, S. (2013). Electrostrictive effect in $Pb(Mg_{1/3}Nb_{2/3})O_3$-$x$$PbTiO_3$ crystals, *Appl. Phys. Lett.*, 102, p. 152910.
[95] Dmowski, W., Vakhrushev, S. B., Jeong, I. K., Hehlen, M. P., Trouw, F. and Egami, T. (2008). Local lattice dynamics and the origin of the relaxor ferroelectric behavior, *Phys. Rev. Lett.*, 100, p. 137602.
[96] Westphal, V., Kleemann, W. and Glinchuk, M. D. (1992). Diffuse phase-transitions and random-field-induced domain states of the relaxor ferroelectric $PbMg_{1/3}Nb_{2/3}O_3$, *Phys. Rev. Lett.*, 68, pp. 847–850.

[97] Kleemann, W. (1993). Random-field induced antiferromagnetic, ferroelectric and structural domain states, *Int. J. Mod. Phys. B*, 7, pp. 2469–2507.

[98] Park, S. E. and Shrout, T. R. (1997). Characteristics of relaxor-based piezoelectric single crystals for ultrasonic transducers, IEEE Trans. *Ultrason. Ferroelectr. Freq. Control*, 44, pp. 1140–1147.

[99] Chu, B., Zhou, X., Ren, K., Neese, B., Lin, M., Wang, Q., Bauer, F. and Zhang, Q. M. (2006). A dielectric polymer with high electric energy density and fast discharge speed, *Science*, 313, pp. 334–336.

[100] Srivastava, N. and Weng, G. J. (2006). A theory of double hysteresis for ferroelectric crystals, *J. Appl. Phys.*, 99, p. 054103.

[101] Berlincourt, D. (1966). Transducers using forced transitions between ferroelectric and antiferroelectric states, *IEEE Trans. Sonics Ultrason.*, 13, pp. 116–125.

[102] Pan, W., Zhang, Q., Bhalla, A. and Cross, L. E. (1989). Field-forced antiferroelectric-to-ferroelectric switching in modified lead zirconate titanate stannate ceramics, *J. Am. Ceram. Soc.*, 72, pp. 571–578.

[103] Zhang, S. J., Luo, J., Xia, R., Rehrig, P. W., Randall, C. A. and Shrout, T. R. (2006). Field-induced piezoelectric response in $Pb(Mg_{1/3}Nb_{2/3})O_3$-$PbTiO_3$ single crystals, *Solid State Commun.*, 137, pp. 16–20.

[104] Tang, H., Zhang, S., Feng, Y., Li, F. and Shrout, T. R. (2013). Piezoelectric property and strain behavior of $Pb(Yb_{0.5}Nb_{0.5})O_3$-$PbHfO_3$-$PbTiO_3$ polycrystalline ceramics, *J. Am. Ceram. Soc.*, 96, pp. 2857–2863.

[105] Damjanovic, D. (2005). Contributions to the piezoelectric effect in ferroelectric single crystals and ceramics, *J. Am. Ceram. Soc.*, 88, pp. 2663–2676.

[106] Setter, N., Damjanovic, D., Eng, L., Fox, G., Gevorgian, S., Hong, S., Kingon, A., Kohlstedt, H., Park, N. Y., Stephenson, G. B., Stolitchnov, I., Tagantsev, A. K., Taylor, D. V., Yamada, T. and Streiffer, S. (2006). Ferroelectric thin films: Review of materials, properties, and applications, *J. Appl. Phys.*, 100, p. 051606.

[107] Li, F., Xu, Z., Wei, X. and Yao, X. (2010). Temperature- and dc bias field-dependent piezoelectric effect of soft and hard lead zirconate titanate ceramics, *J. Electroceram.*, 24, pp. 294–299.

[108] Hwang, S. C., Lynch, C. S. and Mcmeeking, R. M. (1995). Ferro electric/Ferroelastic interactions and a polarization switching model, *Acta Metall. Mater.*, 43, pp. 2073–2084.

[109] Seelecke, S., Kim, S. J., Ball, B. L. and Smith, R. C. (2005). A rate-dependent two-dimensional free energy model for ferroelectric single crystals, *Continuum Mech. Thermodyn.*, 17, pp. 337–350.

[110] Guyomar, D., Ducharne, B. and Sebald, G. (2007). Dynamical hysteresis model of ferroelectric ceramics under electric field using fractional derivatives, *J. Phys. D: Appl. Phys.*, 40, pp. 6048–6054.

[111] Ball, B. L., Smith, R. C., Kim, S.-J. and Seelecke, S. (2007). A stress-dependent hysteresis model for ferroelectric materials, *J. Intell. Mater. Syst. Struct.*, 18, pp. 69–88.

[112] Hajjaji, A., Guyomar, D., Touhtouh, S., Pruvost, S., Boughaleb, Y., Rguiti, M., Courtois, C., Leriche, A. and Benkhouja, K. (2010). Nonlinearity and

scaling behavior in a soft lead zirconate titanate piezoceramic, *J. Appl. Phys.*, 108, p. 064103.
[113] Scott, J. F., Duiker, H. M., Beale, P. D., Pouligny, B., Dimmler, K., Parris, M., Butler, D. and Eaton, S. (1988). Properties of ceramic KNO_3 thin-film memories, *Phys. B*, 150, pp. 160–167.
[114] Li, S. P., Eastman, J. A., Li, Z., Foster, C. M., Newnham, R. E. and Cross, L. E. (1996). Size effects in nanostructured ferroelectrics, *Phys. Lett. A*, 212, pp. 341–346.
[115] Roelofs, A., Schneller, I., Szot, K. and Waser, R. (2002). Piezoresponse force microscopy of lead-titanate nanograins possibly reaching the limit of ferroelectricity, *Appl. Phys. Lett.*, 81, pp. 5231–5233.
[116] Ishikawa, K., Yoshikawa, K. and Okada, N. (1988). Size effect on the ferroelectric phase-transition in $PbTiO_3$ ultrafine particles, *Phys. Rev. B*, 37, pp. 5852–5855.
[117] Uchino, K., Sadanaga, E. and Hirose, T. (1989). Dependence of the crystal-structure on particle-size in barium-titanate, *J. Am. Ceram. Soc.*, 72, pp. 1555–1558.
[118] Tsunekawa, S., Ito, S., Mori, T., Ishikawa, K., Li, Z. Q. and Kawazoe, Y. (2000). Critical size and anomalous lattice expansion in nanocrystalline $BaTiO_3$ particles, *Phys. Rev. B*, 62, pp. 3065–3070.
[119] Zhao, Z., Buscaglia, V., Viviani, M., Buscaglia, M. T., Mitoseriu, L., Testino, A., Nygren, M., Johnsson, M. and Nanni, P. (2004). Grain-size effects on the ferroelectric behavior of dense nanocrystalline $BaTiO_3$ ceramics, *Phys. Rev. B*, 70, p. 024107.
[120] Arlt, G. (1990). Twinning in ferroelectric and ferroelastic ceramics — Stress relief, *J. Mater. Sci.*, 25, pp. 2655–2666.
[121] Cao, W. W. and Randall, C. A. (1996). Grain size and domain size relations in bulk ceramic ferroelectric materials, *J. Phys. Chem. Solids*, 57, pp. 1499–1505.
[122] Hoffmann, M. J., Hammer, M., Endriss, A. and Lupascu, D. C. (2001). Correlation between microstructure, strain behavior, and acoustic emission of soft PZT ceramics, *Acta Mater.*, 49, pp. 1301–1310.
[123] Jin, B. M., Kim, J. and Kim, S. C. (1997). Effects of grain size on the electrical properties of $PbZr_{0.52}Ti_{0.48}O_3$ ceramics, *Appl. Phys. A*, 65, pp. 53–56.
[124] Kamel, T. A. and De With, G. (2008). Grain size effect on the poling of soft $Pb(Zr,Ti)O_3$ ferroelectric ceramics, *J. Eur. Ceram. Soc.*, 28, pp. 851–861.
[125] Randall, C. A., Kim, N., Kucera, J. P., Cao, W. W. and Shrout, T. R. (1998). Intrinsic and extrinsic size effects in fine-grained morphotropic-phase-boundary lead zirconate titanate ceramics, *J. Am. Ceram. Soc.*, 81, pp. 677–688.
[126] Alguero, M., Ricote, J., Jimenez, R., Ramos, P., Carreaud, J., Dkhil, B., Kiat, J. M., Holc, J. and Kosec, M. (2007). Size effect in morphotropic phase boundary $Pb(Mg_{1/3}Nb_{2/3})O_3$-$PbTiO_3$, *J. Am. Ceram. Soc.*, 91.
[127] Jimenez, R., Amorin, H., Ricote, J., Carreaud, J., Kiat, J. M., Dkhil, B., Holc, J., Kosec, M. and Algueró, M. (2008). Effect of

grain size on the transition between ferroelectric and relaxor states in $0.8Pb(Mg_{1/3}Nb_{2/3})O_3$-$0.2PbTiO_3$ ceramics, *Phys. Rev. B*, 78, p. 094103.

[128] Amorin, H., Jimenez, R., Ricote, J., Hungria, T., Castro, A. and Alguero, M. (2010). Apparent vanishing of ferroelectricity in nanostructured $BiScO_3$-$PbTiO_3$, *J. Phys. D: Appl. Phys.*, 43, p. 285401.

[129] Guo, Y. Y., Yan, Z. B., Zhang, N., Cheng, W. W. and Liu, J. M. (2012). Ferroelectric aging behaviors of $BaTi_{0.995}Mn_{0.005}O_3$ ceramics: Grain size effects, *Appl. Phys. A*, 107, pp. 243–248.

[130] Liu, N., Su, Y. and Weng, G. J. (2013). A phase-field study on the hysteresis behaviors and domain patterns of nanocrystalline ferroelectric polycrystals, *J. Appl. Phys.*, 113, p. 04106.

[131] Wagner, S., Kahraman, D., Kungl, H., Hoffmann, M. J., Schuh, C., Lubitz, K., Murmann-Biesenecker, H. and Schmid, J. A. (2005). Effect of temperature on grain size, phase composition, and electrical properties in the relaxor-ferroelectric-system $Pb(Ni_{1/3}Nb_{2/3})O_3$-$Pb(Zr,Ti)O_3$, *J. Appl. Phys.*, 98, p. 024102.

[132] Ducharme, S., Fridkin, V. M., Bune, A. V., Palto, S. P., Blinov, L. M., Petukhova, N. N. and Yudin, S. G. (2000). Intrinsic ferroelectric coercive field, *Phys. Rev. Lett.*, 84, pp. 175–178.

[133] Merz, W. J. (1954). Domain formation and domain wall motions in ferroelectric batio3 single crystals, *Phys. Rev.*, 95, pp. 690–698.

[134] Jin, L. (2011). Broadband dielectric response in hard and soft PZT: Understanding softening and hardening mechanisms, Ph.D. Dissertation, (Swiss Federal Institute of Technology-EPFL, Lausanne, Switzerland).

[135] Damjanovic, D. and Demartin, M. (1997). Contribution of the irreversible displacement of domain walls to the piezoelectric effect in barium titanate and lead zirconate titanate ceramics, *J. Phys.: Condens. Matter*, 9, pp. 4943–4953.

[136] Yoshii, K., Hiruma, Y., Nagata, H. and Takenaka, T. (2006). Electrical properties and depolarization temperature of $(Bi_{1/2}Na_{1/2})TiO_3$-$(Bi_{1/2}K_{1/2})TiO_3$ lead-free piezoelectric ceramics, *Jpn. J. Appl. Phys.*, 45, pp. 4493–4496.

[137] Zhang, S., Shrout, T. R., Nagata, H., Hiruma, Y. and Takenaka, T. (2007). Piezoelectric properties in $(K_{0.5}Bi_{0.5})TiO_3$-$(Na_{0.5}Bi_{0.5})TiO_3$-$BaTiO_3$ lead-free ceramics, *IEEE Trans. Ultrason. Ferroelectr. Freq. Control*, 54, pp. 910–917.

[138] Huibregtse, E. J. and Young, D. R. (1956). Triple hysteresis loops and the free-energy function in the vicinity of the 5-degrees transition in $BaTiO_3$, *Phys. Rev.*, 103, pp. 1705–1711.

[139] Jin, L., Porokhonskyy, V. and Damjanovic, D. (2010). Domain wall contributions in $Pb(Zr,Ti)O_3$ ceramics at morphotropic phase boundary: A study of dielectric dispersion, *Appl. Phys. Lett.*, 96, p. 242902.

[140] Buixaderas, E., Nuzhnyy, D., Petzelt, J., Jin, L. and Damjanovic, D. (2011). Polar lattice vibrations and phase transition dynamics in $Pb(Zr_{1-x}Ti_x)O_3$, *Phys. Rev. B*, 84, pp. 184302.

[141] Zhang, Q., Zhang, Y., Yang, T., Jiang, S., Wang, J., Chen, S., Li, G. and Yao, X. (2013). Effect of compositional variations on phase transition and electric field-induced strain of (Pb,Ba)(Nb,Zr,Sn,Ti)O_3 ceramics, *Ceram. Int.*, 39, pp. 5403–5406.
[142] Pan, W. Y., Dam, C. Q., Zhang, Q. M. and Cross, L. E. (1989). Large displacement transducers based on electric-field forced phase-transitions in the tetragonal ($Pb_{0.97}La_{0.02}$)(Ti,Zr,Sn)O_3 family of ceramics, *J. Appl. Phys.*, 66, pp. 6014–6023.
[143] Zhang, S., Sherlock, N. P., Meyer, R. J., Jr. and Shrout, T. R. (2009). Crystallographic dependence of loss in domain engineered relaxor-PT single crystals, *Appl. Phys. Lett.*, 94, p. 162906.
[144] Damjanovic, D. D., M.; Budimir, M. (2008). Enhancement of piezoelectric properties in perovskite crystals by thermally, compositionally, electric field and stress-induced instabilities, pp. 304–332.
[145] Davis, M., Damjanovic, D., Hayem, D. and Setter, N. (2005). Domain engineering of the transverse piezoelectric coefficient in perovskite ferroelectrics, *J. Appl. Phys.*, 98, p. 014102.
[146] Suzuki, M., Morishita, A., Kitanaka, Y., Noguchi, Y. and Miyayama, M. (2010). Polarization and piezoelectric properties of high performance bismuth sodium titanate single crystals grown by high-oxygen-pressure flux method, *Jpn. J. Appl. Phys.*, 49, p. 09MD09.
[147] Yilmaz, H., Trolier-Mckinstry, S. and Messing, G. L. (2003). (Reactive) templated grain growth of textured sodium bismuth titanate ($Na_{1/2}Bi_{1/2}TiO_3$-$BaTiO_3$) ceramics-II dielectric and piezoelectric properties, *J. Electroceram.*, 11, pp. 217–226.
[148] Duran, C., Trolier-Mckinstry, S. and Messing, G. L. (2002). Dielectric and piezoelectric properties of textured $Sr_{0.53}Ba_{0.47}Nb_2O_6$ ceramics prepared by templated grain growth, *J. Mater. Res.*, 17, pp. 2399–2409.
[149] Hong, S. H., Trolier-Mckinstry, S. and Messing, G. L. (2000). Dielectric and electromechanical properties of textured niobium-doped bismuth titanate ceramics, *J. Am. Ceram. Soc.*, 83, pp. 113–118.
[150] Li, F., Zhang, S., Xu, Z., Wei, X. and Shrout, T. R. (2011). Critical property in relaxor-$PbTiO_3$ single crystals — shear piezoelectric response, *Adv. Funct. Mater.*, 21, pp. 2118–2128.
[151] Lee, H., Zhang, S., Luo, J., Li, F. and Shrout, T. R. (2010). Thickness-dependent properties of relaxor-$PbTiO_3$ ferroelectrics for ultrasonic transducers, *Adv. Funct. Mater.*, 20, pp. 3154–3162.
[152] Lee, H. J., Zhang, S. and Shrout, T. R. (2010). Scaling effects of relaxor-$PbTiO_3$ crystals and composites for high frequency ultrasound, *J. Appl. Phys.*, 107, p. 124107.
[153] Ma, B., Liu, S., Tong, S., Narayanan, M., Koritala, R. E., Hu, Z. and Balachandran, U. (2013). Residual stress of ($Pb_{0.92}La_{0.08}$)($Zr_{0.52}Ti_{0.48}$)O_3 films grown by a sol-gel process, *Smart Mater. Struct.*, 22, p. 055019.
[154] Shaw, T. M., Trolier-Mckinstry, S. and Mcintyre, P. C. (2000). The properties of ferroelectric films at small dimensions, *Annu. Rev. Mater. Sci.*, 30, pp. 263–298.

[155] Bassiri-Gharb, N., Fujii, I., Hong, E., Trolier-Mckinstry, S., Taylor, D. V. and Damjanovic, D. (2007). Domain wall contributions to the properties of piezoelectric thin films, *J. Electroceram.*, 19, pp. 49–65.
[156] Xu, F., Trolier-Mckinstry, S., Ren, W., Xu, B. M., Xie, Z. L. and Hemker, K. J. (2001). Domain wall motion and its contribution to the dielectric and piezoelectric properties of lead zirconate titanate films, *J. Appl. Phys.*, 89, pp. 1336–1348.
[157] Chaudhuri, A. R., Arredondo, M., Haehnel, A., Morelli, A., Becker, M., Alexe, M. and Vrejoiu, I. (2011). Epitaxial strain stabilization of a ferroelectric phase in $PbZrO_3$ thin films, *Phys. Rev. B*, 84, p. 054112.
[158] Scott, B. A. and Burns, G. (1972). Crystal-growth and observation of ferroelectric phase of $PbZrO_3$, *J. Am. Ceram. Soc.*, 55, pp. 331–333.
[159] Ayyub, P., Chattopadhyay, S., Pinto, R. and Multani, M. S. (1998). Ferroelectric behavior in thin films of antiferroelectric materials, *Phys. Rev.* B, 57, pp. R5559–R5562.
[160] Morozov, M. I. and Damjanovic, D. (2008). Hardening-softening transition in Fe-doped $Pb(Zr,Ti)O_3$ ceramics and evolution of the third harmonic of the polarization response, *J. Appl. Phys.*, 104, p. 034107.
[161] Feng, Z., Cheng, Z., Shi, D. and Dou, S. (2013). Aging effect evolution during ferroelectric-ferroelectric phase transition: A mechanism study, *AIP Adv.*, 3, p. 062105.
[162] Jakes, P., Erdem, E., Eichel, R.-A., Jin, L. and Damjanovic, D. (2011). Position of defects with respect to domain walls in Fe^{3+}-doped Pb $Zr_{0.52}Ti_{0.48}O_3$ piezoelectric ceramics, *Appl. Phys. Lett.*, 98, p. 072907.
[163] Lambeck, P. V. and Jonker, G. H. (1986). The nature of domain stabilization in ferroelectric perovskites, *J. Phys. Chem. Solids*, 47, pp. 453–461.
[164] Zhang, L., Erdem, E., Ren, X. and Eichel, R.-A. (2008). Reorientation of $(Mn_{Ti}^{//}-V_O^{\bullet\bullet})^{\times}$ effect dipoles in acceptor-modified $BaTiO_3$ single crystals: An electron paramagnetic resonance study, *Appl. Phys. Lett.*, 93, p. 202901.
[165] Zhang, L. X. and Ren, X. (2005). *In situ* observation of reversible domain switching in aged Mn-doped $BaTiO_3$ single crystals, *Phys. Rev. B*, 71, p. 174108.
[166] Warren, W. L., Dimos, D., Tuttle, B. A. and Smyth, D. M. (1994). Electronic and ionic trapping at domain-walls in $BaTiO_3$, *J. Am. Ceram. Soc.*, 77, pp. 2753–2757.
[167] Warren, W. L., Dimos, D., Tuttle, B. A., Nasby, R. D. and Pike, G. E. (1994). Electronic domain pinning in $Pb(Zr,Ti)O_3$ thin-films and its role in fatigue, *Appl. Phys. Lett.*, 65, pp. 1018–1020.
[168] Takahashi, M. (1970). Space charge effect in lead zirconate titanate ceramics caused by addition of impurities, *Jpn. J. Appl. Phys.*, 9, pp. 1236–1246.
[169] Lupascu, D. C., Genenko, Y. A. and Balke, N. (2006). Aging in ferroelectrics, *J. Am. Ceram. Soc.*, 89, pp. 224–229.
[170] Jonker, G. H. (1972). Nature of aging in ferroelectric ceramics, *J. Am. Ceram. Soc.*, 55, pp. 57–58.

[171] Zhang, S., Lee, S.-M., Kim, D.-H., Lee, H.-Y. and Shrout, T. R. (2008). Characterization of Mn-modified $Pb(Mg_{1/3}Nb_{2/3})O_3$-$PbZrO_3$-$PbTiO_3$ single crystals for high power broad bandwidth transducers, *Appl. Phys. Lett.*, 93, p. 122908.
[172] Lohkamper, R. H. N. and Arlt, G. (1980). Internal bias in acceptor-doped $BaTiO_3$ ceramics: Numerical evaluation of increase and decrease, *J. Appl. Phys.*, 68, pp. 4220–4224.
[173] Gao, Y., Uchino, K. and Viehland, D. (2006). Time dependence of the mechanical quality factor in "hard" lead zirconate titanate ceramics: Development of an internal dipolar field and high power origin, *Jpn. J. Appl. Phys.*, 45, pp. 9119–9124.
[174] Taylor, D. V. and Damjanovic, D. (1997). Evidence of domain wall contribution to the dielectric permittivity in PZT thin films at sub-switching fields, *J. Appl. Phys.*, 82, pp. 1973–1975.
[175] Eitel, R. E., Shrout, T. R. and Randall, C. A. (2006). Nonlinear contributions to the dielectric permittivity and converse piezoelectric coefficient in piezoelectric ceramics, *J. Appl. Phys.*, 99, p. 124110.
[176] Hall, D. A. (2001). Review nonlinearity in piezoelectric ceramics, *J. Mater. Sci.*, 36, pp. 4575–4601.
[177] Ma, B., Hu, Z., Liu, S., Tong, S., Narayanan, M., Koritala, R. E. and Balachandran, U. (2013). Temperature-dependent dielectric nonlinearity of relaxor ferroelectric $Pb_{0.92}La_{0.08}Zr_{0.52}Ti_{0.48}O_3$ thin films, *Appl. Phys. Lett.*, 102, p. 202901.
[178] Eitel, R. and Randall, C. A. (2007). Octahedral tilt-suppression of ferroelectric domain wall dynamics and the associated piezoelectric activity in $Pb(Zr,Ti)O_3$, *Phys. Rev. B*, 75, p. 094106.
[179] Damjanovic, D. and Demartin, M. (1996). The Rayleigh law in piezoelectric ceramics, *J. Phys. D: Appl. Phys.*, 29, pp. 2057–2060.
[180] Damjanovic, D. (1997). Logarithmic frequency dependence of the piezoelectric effect due to pinning of ferroelectric-ferroelastic domain walls, *Phys. Rev.* B, 55, pp. R649–R652.
[181] Gharb, N. B. and Trolier-Mckinstry, S. (2005). Dielectric nonlinearity of $Pb(Yb_{1/2}Nb_{1/2})O_3$-$PbTiO_3$ thin films with {100} and {111} crystallographic orientation, *J. Appl. Phys.*, 97, p. 064106.
[182] Damjanovic, D. (1997). Stress and frequency dependence of the direct piezoelectric effect in ferroelectric ceramics, *J. Appl. Phys.*, 82, pp. 1788–1797.
[183] Trolier-Mckinstry, S. N. B. G. and Damjanovic, D. (2006). Piezoelectric nonlinearity in ferroelectric thin films, *Appl. Phys. Lett.*, 88, p. 202901.
[184] Pramanick, A., Damjanovic, D., Nino, J. C. and Jones, J. L. (2009). Subcoercive cyclic electrical loading of lead zirconate titanate ceramics I: Nonlinearities and losses in the converse piezoelectric effect, *J. Appl. Phys.*, 92, pp. 2291–2299.
[185] Scott, J. (2000). *Ferroelectric Memories* (Springer, New York).

[186] Tagantsev, A. K., Stolichnov, I., Colla, E. L. and Setter, N. (2001). Polarization fatigue in ferroelectric films: Basic experimental findings, phenomenological scenarios, and microscopic features, *J. Appl. Phys.*, 90, pp. 1387–1402.
[187] Lupascu, D. C. and Rodel, J. (2005). Fatigue in bulk lead zirconate titanate actuator materials, *Adv. Eng. Mater.*, 7, pp. 882–898.
[188] Balke, N., Kungl, H., Granzow, T., Lupascu, D. C., Hoffmann, M. J. and Rodel, J. (2007). Bipolar fatigue caused by field screening in $Pb(Zr,Ti)O_3$ ceramics, *J. Am. Ceram. Soc.*, 90, pp. 3869–3874.
[189] Lou, X. J. (2009). Polarization fatigue in ferroelectric thin films and related materials, *J. Appl. Phys.*, 105, p. 024101.
[190] Ozgul, M., Takemura, K., Trolier-Mckinstry, S. and Randall, C. A. (2001). Polarization fatigue in $Pb(Zn_{1/3}Nb_{2/3})O_3$-$PbTiO_3$ ferroelectric single crystals, *J. Appl. Phys.*, 89, pp. 5100–5106.
[191] Zhang, S., Xia, R., Hao, H., Liu, H. and Shrout, T. R. (2008). Mitigation of thermal and fatigue behavior in $K_{0.5}Na_{0.5}NbO_3$-based lead free piezoceramics, *Appl. Phys. Lett.*, 92, p. 152904.
[192] Stewart, W. C. and Cosentino, L. S. (1970). Some optical and electrical switching characteristics of a lead zirconate titanate ferroelectric ceramic, *Ferroelectrics*, 1, pp. 149–167.
[193] Amanuma, K., Hase, T. and Miyasaka, Y. (1994). Fatigue characteristics of sol–gel derived $Pb(Zr, Ti)O_3$ thin-films, *Jpn. J. Appl. Phys.*, Part 1, 33, pp. 5211–5214.
[194] Scott, J. F. (1996). Three fundamental problems in ferroelectricity, *J. Phys. Chem. Solids*, 57, pp. 1439–1443.
[195] Alshareef, H. N., Bellur, K. R., Kingon, A. I. and Auciello, O. (1995). Influence of platinum interlayers on the electrical-properties of RuO_2/ $Pb(Zr_{0.53}Ti_{0.47})O_3/RuO_2$ capacitor heterostructures, *Appl. Phys. Lett.*, 66, pp. 239–241.
[196] Kushida-Abdelghafer, K., Hiratani, M. and Fujisaki, Y. (1999). Post-annealing effects on antireduction characteristics of $IrO_2/Pb(Zr_xTi_{1-x})O_3$/ Pt ferroelectric capacitors, *J. Appl. Phys.*, 85, pp. 1069–1074.
[197] Yin, J., Zhu, T., Liu, Z. G. and Yu, T. (1999). Enhanced fatigue and retention properties of $Pb(Ta_{0.05}Zr_{0.48}Ti_{0.47})O_3$ films using $La_{0.25}Sr_{0.75}CoO_3$ top and bottom electrodes, *Appl. Phys. Lett.*, 75, pp. 3698–3700.
[198] Cheng, Z., Kannan, C. V., Ozawa, K., Kimura, H. and Wang, X. (2006). Orientation dependent ferroelectric properties in samarium doped bismuth titanate thin films grown by the pulsed-laser-ablation method, *Appl. Phys. Lett.*, 89, p. 032901.
[199] Cheng, Z. X., Wang, X. L., Dou, S. X., Ozawa, K. and Kimura, H. (2007). Ferroelectric properties of $Bi_{3.25}Sm_{0.75}V_{0.02}T_{2.98}O_{12}$ thin film at elevated temperature, *Appl. Phys. Lett.*, 90, p. 222902.
[200] Takemura, K., Ozgul, M., Bornand, V., Trolier-Mckinstry, S. and Randall, C. A. (2000). Fatigue anisotropy in single crystal $Pb(Zn_{1/3}Nb_{2/3})O_3$-$PbTiO_3$, *J. Appl. Phys.*, 88, pp. 7272–7277.

[201] Ozgul, M., Trolier-Mckinstry, S. and Randall, C. A. (2004). Influence of electrical cycling on polarization reversal processes in $Pb(Zn_{1/3}Nb_{2/3})O_3$-$PbTiO_3$ ferroelectric single crystals as a function of orientation, *J. Appl. Phys.*, 95, pp. 4296–4302.

[202] Zhang, S., Luo, J., Li, F., Meyer, R. J., Jr., Hackenberger, W. and Shrout, T. R. (2010). Polarization fatigue in $Pb(In_{0.5}Nb_{0.5})O_3$-$Pb(Mg_{1/3}Nb_{2/3})O_3$-$PbTiO_3$ single crystals, *Acta Mater.*, 58, pp. 3773–3780.

[203] Fang, F., Luo, X. and Yang, W. (2013). Domain structure evolution and polarization degradation of 101-oriented $0.74Pb(Mg_{1/3}Nb_{2/3})O_3$-$0.26PbTiO_3$ single crystal under cyclic electric loadings, *J. Am. Ceram. Soc.*, 96, pp. 228–233.

[204] Fang, F., Yang, W., Zhang, F. C. and Qing, H. (2008). Electric field-induced crack growth and domain-structure evolution for [100]- and [101]-oriented 72%$Pb(Mg_{1/3}Nb_{2/3})O_3$-28%$PbTiO_3$ ferroelectric single crystals, *J. Mater. Res.*, 23, pp. 3387–3395.

[205] Mirshekarloo, M. S., Zhang, L., Yao, K. and Sritharan, T. (2012). Electromechanical properties and fatigue of antiferroelectric $(Pb,La)(Zr,Sn,Ti)O_3$ thin film cantilevers fabricated by micromachining, *Sens. Actuators, A*, 187, pp. 127–131.

[206] Lou, X. J. (2009). Why do antiferroelectrics show higher fatigue resistance than ferroelectrics under bipolar electrical cycling? *Appl. Phys. Lett.*, 94, p. 072901.

[207] Zhou, L. J., Zimmermann, A., Zeng, Y. P. and Aldinger, F. (2004). Fatigue of field-induced strain in antiferroelectric $Pb_{0.97}La_{0.02}(Zr_{0.77}Sn_{0.14}Ti_{0.09})O_3$ ceramics, *J. Am. Ceram. Soc.*, 87, pp. 1591–1593.

[208] Lou, X. J. and Wang, J. (2010). Unipolar and bipolar fatigue in antiferroelectric lead zirconate thin films and evidences for switching-induced charge injection inducing fatigue, *Appl. Phys. Lett.*, 96, p. 102906.

[209] Yang, S. M., Jo, J. Y., Kim, T. H., Yoon, J. G., Song, T. K., Lee, H. N., Marton, Z., Park, S., Jo, Y. and Noh, T. W. (2010). Ac dynamics of ferroelectric domains from an investigation of the frequency dependence of hysteresis loops, *Phys. Rev. B*, 82, p. 174125.

[210] Landauer, R., Young, D. R. and Drougard, M. E. (1956). Polarization reversal in the barium titanate hysteresis loop, *J. Appl. Phys.*, 27, pp. 752–758.

[211] Gysel, R., Stolichnov, I., Tagantsev, A. K., Setter, N. and Mokry, P. (2008). Restricted domain growth and polarization reversal kinetics in ferroelectric polymer thin films, *J. Appl. Phys.*, 103, p. 084120.

[212] Avrami, M. (1941). Granulation, phase change, and microstructure — kinetics of phase change. III, *J. Chem. Phys.*, 9, pp. 177–184.

[213] Yimnirun, R., Laosiritaworn, Y., Wongsaenmai, S. and Ananta, S. (2006). Scaling behavior of dynamic hysteresis in soft lead zirconate titanate bulk ceramics, *Appl. Phys. Lett.*, 89, p. 162901.

[214] Gupta, S. and Priya, S. (2011). Ferroelectric properties and dynamic scaling of ⟨100⟩ oriented $(K_{0.5}Na_{0.5})NbO_3$ single crystals, *Appl. Phys. Lett.*, 98, p. 242906.

[215] Wongdamnern, N., Ngamjarurojana, A., Laosiritaworn, Y., Ananta, S. and Yimnirun, R. (2009). Dynamic ferroelectric hysteresis scaling of $BaTiO_3$ single crystals, *J. Appl. Phys.*, 105, p. 044109.

[216] Wan, Y., Li, Z., Xu, Z., Fan, S. and Yao, X. (2013). Phase transition characteristics of the relaxor-based 0.24PIN-0.51PMN-0.25PT single crystals, *J. Alloy. Compd.*, 558, p. 244–247.

[217] R. Yimnirun, R. W., S. Wongsaenmai, A. Ngamjarurojana, S. Ananta, and Y. Laosiritaworn (2007). Temperature scaling of dynamic hysteresis in soft lead zirconate titanate bulk ceramic, *Appl. Phys. Lett.*, 90, p. 112906.

[218] Du, G., Liang, R., Wang, L., Li, K., Zhang, W., Wang, G. and Dong, X. (2013). Linear temperature scaling of ferroelectric hysteresis in Mn-doped $Pb(Mn_{1/3}Sb_{2/3})O_3$-$Pb(Zr,Ti)O_3$ ceramic with internal bias field, *Appl. Phys. Lett.*, 102, p. 142903.

[219] Li, T., Wang, G., Du, G., Li, K., Chen, Y., Zhou, Z., Cao, F. and Dong, X. (2013). Temperature scaling behavior of dynamic hysteresis for $(K,Na)NbO_3$ lead-free ferroelectric films, *J. Appl. Phys.*, 113, p. 214103.

[220] Chen, X. D. X. and Zhang, H. (2009). Temperature scaling of dynamic hysteresis in Zr-rich $PbZr_{1-x}Tix O_3$ ceramics, *J. Appl. Phys.*, 105, p. 096104.

[221] Zhang, S., Li, F., Sherlock, N. P., Luo, J., Lee, H. J., Xia, R., Meyer, R. J., Jr., Hackenberger, W. and Shrout, T. R. (2011). Recent developments on high Curie temperature PIN-PMN-PT ferroelectric crystals, *J. Cryst. Growth*, 318, pp. 846–850.

[222] Yimnirun, R. A. N., Wongmaneerung, R., Wongsaenmai, S. and Ananta, S. A. Y. L. (2007). Temperature Scaling of Ferroelectric Hysteresis in Hard Lead Zirconate Titanate Bulk Ceramic, *Appl. Phys. A*, 89, pp. 737–741.

[223] Yimnirun, R., Laosiritaworn, Y. and Wongsaenmai, S. (2006). Effect of uniaxial compressive pre-stress on ferroelectric properties of soft PZT ceramics, *J. Phys. D: Appl. Phys.*, 39, pp. 759–764.

[224] Unruan, M., Sareein, T., Tangsritrakul, J., Prasertpalichatr, S., Ngamjarurojana, A., Ananta, S. and Yimnirun, R. (2008). Changes in dielectric and ferroelectric properties of Fe^{3+}/Nb^{5+} hybrid-doped barium titanate ceramics under compressive stress, *J. Appl. Phys.*, 104, p. 124102.

[225] Unruan, M., Wongsaenmai, S., Ngamjarurojana, A., Laosiritaworn, Y., Ananta, S., Guo, R., Bhalla, A. and Yimnirun, R. (2010). Changes in ferroelectric properties of lead indium niobate-lead titanate ceramics under compressive stress applied perpendicular to an electric field, *Phys. Lett. A*, 374, pp. 1147–1153.

[226] Liu, Q. D. (2011). Investigation into the creeping polarization and strain in PZT-855 under combined mechanical and electrical loadings, *Acta Mech.*, 220, pp. 1–14.

[227] Zhao, J., Glazounov, A. E. and Zhang, Q. M. (1999). Change in electromechanical properties of 0.9PMN: 0.1PT relaxor ferroelectric induced by uniaxial compressive stress directed perpendicular to the electric field, *Appl. Phys. Lett.*, 74, pp. 436–438.

[228] Li, F., Zhang, S., Xu, Z., Lin, D., Gao, J., Li, Z. and Wang, L. (2012). An efficient way to enhance output strain for shear mode $Pb(In_{1/2}Nb_{1/2})O_3$-$Pb(Mg_{1/3}Nb_{2/3})O_3$-$PbTiO_3$ crystals: Applying uniaxial stress perpendicular to polar direction, *Appl. Phys. Lett.*, 100, p. 192901.

[229] Zhang, S., Taylor, S., Li, F., Luo, J. and Meyer, R. J., Jr. (2013). Piezoelectric property of relaxor-$PbTiO_3$ crystals under uniaxial transverse stress, *Appl. Phys. Lett.*, 102, p. 172902.

[230] Granzow, T., Kounga, A. B., Aulbach, E. and Roedel, J. (2006). Electromechanical poling of piezoelectrics, *Appl. Phys. Lett.*, 88, p. 252907.

[231] Granzow, T., Leist, T., Kounga, A. B., Aulbach, E. and Rödel, J. (2007). Ferroelectric properties of lead zirconate titanate under radial load, *Appl. Phys. Lett.*, 91, p. 142904.

[232] Marsilius, M., Frederick, J., Hu, W., Tan, X., Granzow, T. and Han, P. (2012). Mechanical Confinement: An Effective Way of Tuning Properties of Piezoelectric Crystals, *Adv. Funct. Mater.*, 22, pp. 797–802.

[233] Pelaiz-Barranco, A. and Hall, D. A. (2009). Influence of composition and pressure on the electric field-induced antiferroelectric to ferroelectric phase transformation in lanthanum modified lead Zirconate Titanate ceramics, *IEEE Trans. Ultrason. Ferroelectr. Freq. Control*, 56, pp. 1785–1791.

[234] Tan, X., Frederick, J., Ma, C., Aulbach, E., Marsilius, M., Hong, W., Granzow, T., Jo, W. and Roedel, J. (2010). Electric-field-induced antiferroelectric to ferroelectric phase transition in mechanically confined $Pb_{0.99}Nb_{0.02}[(Zr_{0.57}Sn_{0.43})_{0.94}Ti_{0.06}]_{0.98}O_3$, *Phys. Rev.* B, 81, p. 014103.

Chapter 3

Recent Progress in Relaxor Ferroelectrics with Perovskite Structure*

A. A. Bokov and Z.-G. Ye[†]

Department of Chemistry, Simon Fraser University, 8888 University Drive, Burnaby, BC, V5A 1S6, Canada

[†] *zye@sfu.ca*

Relaxor ferroelectrics were discovered almost 50 years ago among the complex oxides with perovskite structure. In recent years this field of research has experienced a revival of interest. In this Chapter we review the progress achieved. We consider the crystal structure including quenched compositional disorder and polar nanoregions (PNR), the phase transitions including compositional order-disorder transition, transition to nonergodic (probably spherical cluster glass) state and to ferroelectric phase. We discuss the lattice dynamics and the peculiar (especially dielectric) relaxation in relaxors. Modern theoretical models for the mechanisms of PNR formation and freezing into nonergodic glassy state are also presented.

1. Introduction

Relaxor ferroelectrics or relaxors are a class of disordered crystals possessing peculiar structure and properties. At high temperature they exist in a non-polar paraelectric (PE) phase, which is similar in many respects to the PE phase of normal ferroelectrics. Upon cooling they transform into the ergodic relaxor (ER) state in

*This article was firstly published as "Bokov, A. A. and Ye, Z.-G. (2006). Recent progress in relaxor ferroelectrics with perovskite structure, *J. Mater. Sci.*, 41, pp. 31–52." Reprinted with the permission of Springer Science + Business Media, Inc.

which polar regions of nanometer scale with randomly distributed directions of dipole moments appear. This transformation which occurs at the so-called Burns temperature (T_{B}) cannot be considered a structural phase transition because it is not accompanied by any change of crystal structure on the macroscopic or mesoscopic scale. Nevertheless, the polar nanoregions (PNRs) affect the behavior of the crystal dramatically, giving rise to unique physical properties. For this reason the state of crystal at $T < T_{\mathrm{B}}$ is often considered as the new phase different from the PE one.

At temperatures close to T_{B} the PNRs are mobile and their behavior is ergodic. On cooling, their dynamics slows down enormously and at a low enough temperature, T_{f} (typically hundreds degrees below T_{B}), the PNRs in the canonical relaxors become frozen into a nonergodic state, while the average symmetry of the crystal still remains cubic. Similar kind of nonergodicity is characteristic of a dipole glass (or spin glass) phase. The existence in relaxors of an equilibrium phase transition into a low-temperature glassy phase is one of the most interesting hypotheses which has been intensively discussed. Freezing of the dipole dynamics is associated with a large and wide peak in the temperature dependence of the dielectric constant (ε) with characteristic dispersion observed at all frequencies practically available for dielectric measurements (Fig. 1). This peak is of the same order of magnitude as the peaks at the Curie point in the ordinary ferroelectric (FE) perovskites, but in contrast to ordinary ferroelectrics it is highly diffuse and its temperature $T_{\mathrm{m}}(> T_{\mathrm{f}})$ shifts with frequency due to the dielectric dispersion. Because of the diffuseness of the dielectric anomaly and the anomalies in the temperature dependences of some other properties, relaxors are often called (especially in early literature) the "ferroelectrics with diffuse phase transition," even though no transition into FE phase really occurs.

The nonergodic relaxor (NR) state existing below T_{f} can be irreversibly transformed into a FE state by a strong enough external electric field. This is an important characteristic of relaxors which distinguishes them from typical dipole glasses. Upon heating the FE phase transforms to the ER one at the temperature T_{C} which is

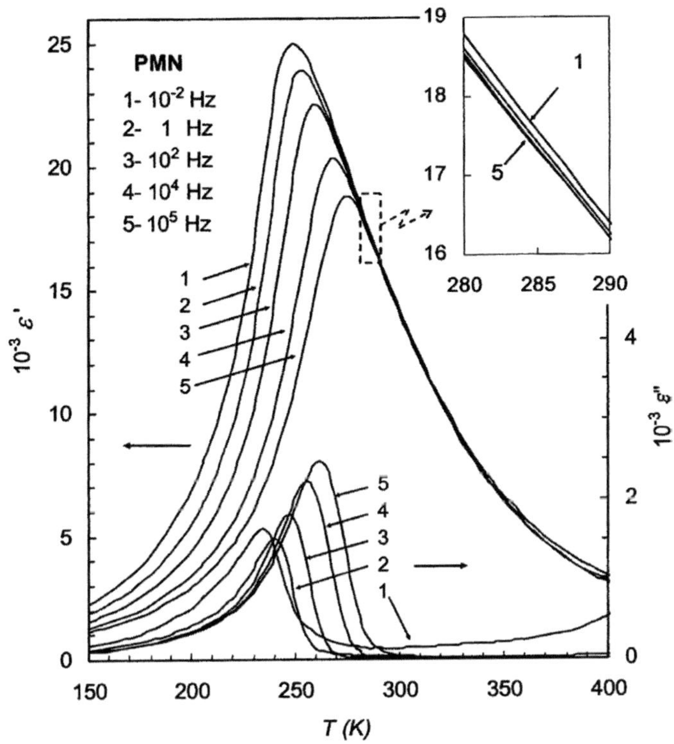

Figure 1. Temperature dependences of the real and imaginary parts of the relative dielectric permittivity measured at different frequencies in a crystal of the prototypical relaxor $Pb(Mg_{1/3}Nb_{2/3})O_3$. Enlarged view in the insert shows the universal relaxor dispersion (URD) (after Bokov and Ye, unpublished).

very close to T_f. In many relaxors the spontaneous (i.e. without the application of an electric field) phase transition from the ER into a low-temperature FE phase still occurs at T_C and thus the NR state does not exist.

Compositional disorder, i.e. the disorder in the arrangement of different ions on the crystallographically equivalent sites, is the common feature of relaxors. The relaxor behavior was first observed in the perovskites with disorder of non-isovalent ions, including the stoichiometric complex perovskite compounds, e.g., $Pb(Mg_{1/3}Nb_{2/3})O_3$ (PMN) [1] or $Pb(Sc_{1/2}Ta_{1/2})O_3$ (PST) [2] (in which Mg^{2+}, Sc^{3+}, Ta^{5+} and Nb^{5+} ions are fully or partially disordered in the B-sublattice of the perovskite ABO_3 structure) and nonstoichiometric solid solutions, e.g., $Pb_{1-x}La_x(Zr_{1-y}Ti_y)_{1-x/4}O_3$ (PLZT) [3, 4] where the substitution of La^{3+} for Pb^{2+} ions necessarily leads to vacancies on the A-sites. Recently an increasing amount of data reported has shown that many homovalent solid solutions, e.g. $Ba(Ti_{1-x}Zr_x)O_3$ (BTZ) [5, 6] and $Ba(Ti_{1-x}Sn_x)O_3$ [7] can also exhibit relaxor behavior. Other examples of relaxor ferroelectrics are complex perovskites $Pb(Zn_{1/3}Nb_{2/3})O_3$ (PZN) $Pb(Mg_{1/3}Ta_{2/3})O_3$ (PMT), $Pb(Sc_{1/2}Nb_{1/2})O_3$ (PSN), $Pb(In_{1/2}Nb_{1/2})O_3$ (PIN), $Pb(Fe_{1/2}Nb_{1/2})O_3$ (PFN), $Pb(Fe_{2/3}W_{1/3})O_3$ (PFW) and the solid solutions: $(1-x)Pb(Mg_{1/3}Nb_{2/3})O_{3-x}PbTiO_3$ (PMN-PT) and $(1-x)Pb(Zn_{1/3}Nb_{2/3})O_3-xPbTiO_3$ (PZN-PT).

Although relaxor ferroelectrics were first reported nearly half a century ago, this field of research has experienced a revival of interest in recent years. In this chapter, we try to provide an overview of the current understanding of the various issues of relaxors. Emphasis is put on the latest developments. For a review of the earlier studies, readers can refer to Refs. [8–11].

2. Compositional order–disorder phase transitions and quenched disorder in complex perovskites

As mentioned above, the disordered distribution of different ions on the equivalent lattice sites (i.e., compositional disorder, also called chemical, ionic or substitutional disorder) is the essential structural characteristic of relaxors. The ground state of the complex perovskites should be compositionally ordered, e.g., in the $A(B'_{1/2}B''_{1/2})O_3$ compounds each type of the cations, B' or B'', should be located in its own sublattice, creating a superstructure with complete translational symmetry. This is because the electrostatic and elastic energies of the structure are minimized in the ordered state due to the difference in both the charge and the size of B' and B'' ions. Thermal motion is capable of destroying the order at a certain nonzero temperature (T_t). This occurs in the form of structural phase transition, the order parameter (compositional long-range order, s) of which can be measured by the X-ray or other diffraction methods. Such kind of phase transitions had been known long ago (e.g., in many metallic alloys) and was also discovered comparatively recently at $T_t \sim 1500\,\mathrm{K}$ in PST, PSN [12] and several other complex perovskites. Ordering implies the site exchange between B′ and B″ cations via diffusion. It is a relaxation process with a nearly infinite characteristic time at low temperatures, but at 1500 K it can be quite fast. As a result, in some perovskites (e.g., in PST, PSN, PIN), by annealing at temperatures around T_t and subsequent quenching, one can obtain the metastable states with different s at low temperatures. In some other materials (e.g., in PFN and PMN), the compositional disorder cannot be changed by any heat treatment because the relaxation time of ordering is too long. However, in all known relaxors, at T_B and below, the compositional order is frozen (quenched), i.e., cannot vary during practically reasonable time.

In the real complex perovskite crystals and ceramics, the quenched compositional disorder is often inhomogeneous, e.g., small regions of the ordered state are embedded in a disordered matrix. These regions can be regarded as a result of incomplete compositional order–disorder phase transformation or as quenched phase fluctuations. In the prototypical relaxor PMN this kind of inhomogeneous structure always exists and cannot be changed by any heat treatment.

In $A(B'_{1/2}B''_{1/2})O_3$ perovskites the ordering of B-site ions converts the disordered PE $Pm\bar{3}m$ structure into the ordered $Fm\bar{3}m$ structure in which B′ ions alternate with B″ ions along the ⟨100⟩ directions (1:1 ordering). In the ordered phase of many non-ferroelectric $A(B^{2+}_{1/3}B^{5+}_{2/3})O_3$ perovskites, B^{2+} ions alternate with two B^{5+} ions along the ⟨100⟩ directions (1:2 ordering). The type of ordering in lead-containing relaxor perovskites, $Pb(B^{2+}_{1/3}B^{5+}_{2/3})O_3$, has been the subject of debates. In the early works only inhomogeneous ordering (ordered regions within disordered surroundings) was found in the samples studied. High-resolution electron microscopy of PMN revealed nano-size (∼2–5 nm) regions in which the ordering of 1:1 type ($Fm\bar{3}m$) was observed (see e.g., Refs. [13, 14]). These chemical nanoregions (CNR) give rise to weak superlattice reflections (the so-called F-spots). The results of anomalous X-ray scattering measurements [15] showed that the CNRs in PMN exhibit an isotropic shape and a temperature-independent (as expected for the quenched order) size in the temperature interval of 15–800 K. Alternating Mg^{2+} and Nb^{5+} ions, i.e., the same type of ordering as in the ordered $Pb(B'_{1/2}B''_{1/2})O_3$ perovskites, were initially supposed to exist in these regions. This structural model was called "space charge model" because it implies the existence of the negatively charged compositionally ordered non-stoichiometric nanoregions, and the positively charged disordered non-stoichiometric matrix. Later, by means of appropriate high temperature treatments, Davies and Akbas [16] were able to increase the size of CNRs and obtained highly 1:1 ordered samples without the disordered matrix in the PMT and modified PMN ceramics. The existence of such ordering in overall stoichiometric samples is obviously inconsistent with the space charge model. The results of X-ray energy dispersive spectroscopy with a nanometer probing size revealed that the Mg/Nb ratio is the same in the CNRs as in the disordered regions of PMN [17], which also disagrees with the space charge model. A charge-balanced "random-site" model has been suggested in which one of the B-sublattices is occupied exclusively by B^{5+} ions while the other one contains a random distribution of B^{2+} and B^{5+} ions in a 2:1 ratio so that the local stoichiometry is preserved [16]. The inhomogeneous compositional disorder, characteristic of the $Pb(B^{2+}_{1/3}B^{5+}_{2/3})O_3$ perovskites, are shown schematically in Fig. 2.

The degree of compositional disorder can greatly influence the FE properties. For example, the disordered PIN crystals are relaxor ferroelectrics, but in the ordered state, they are antiferroelectrics with a sharp phase transition [18, 19], confirming the general rule that the relaxor behavior can only be observed in disordered crystals. The possibility for real perovskite samples to have different states of compositional disorder, depending on crystal growth or ceramic sintering conditions, should be taken into account in research work. More detailed discussion on the compositional ordering and its impact on FE properties can be found in Refs. [20–22], and the literature therein.

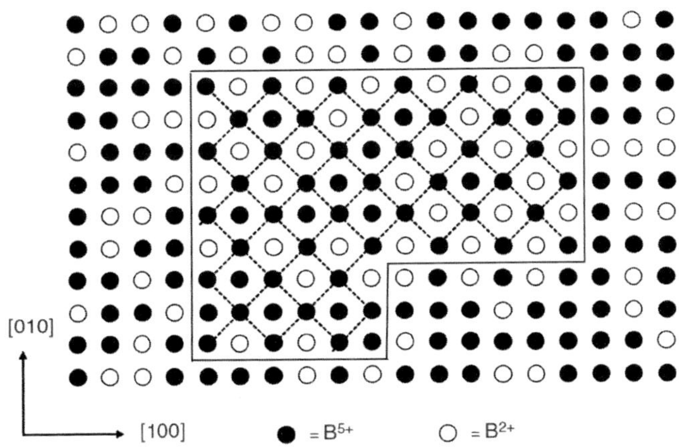

Figure 2. Schematic representation of the ordered chemical nanoregion, CNR (the area delimited by the solid line) within the disordered matrix in $Pb(B^{2+}_{1/3}B^{5+}_{2/3})O_3$ perovskites according to the random-site model. One of the two sublattices inside CNR (shown by dashed lines) is formed by B^{5+} ions only. Pb and O ions are not shown.

3. Relaxors in the ergodic state

3.1. *Paraelectric structure*

The PE phase of all perovskite ferroelectrics has the cubic $m\bar{3}m$ average symmetry, but locally the ion configuration can be distorted, i.e., the ions are not located in the special crystallographic sites of the ideal perovskite structure. For example, in the classical ferroelectrics $BaTiO_3$, the random displacements of Ti cations along the $\langle 111 \rangle$ directions caused by the multiple-well structure of potential surface were found [23, 24]. Such kind of displacements is due to the hybridization between electronic states of cations and the $2p$ states of oxygen (and should not exist in the case of purely ionic bonds). This effect is an important factor in the FE instability [25] and is also expected to occur in perovskite relaxors. Moreover, owing to the different sizes of the compositionally disordered cations and the random electric fields created because of the different charges of these cations in relaxors, all ions are expected to be displaced from special positions. These shifts should exist in the PE phase and also at lower temperatures.

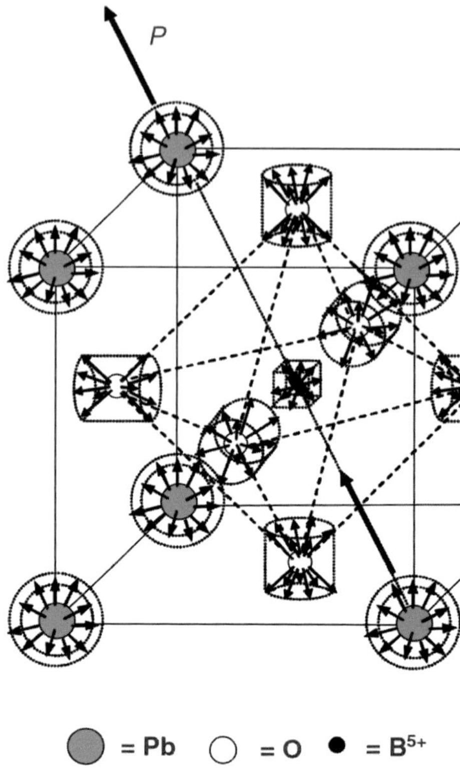

Figure 3. Typical uncorrelated ion displacements (shown by small arrows) in the unit cell of the lead-containing complex perovskite relaxor. Thick arrows show the direction of the local spontaneous polarization P caused by the correlated displacements of ions inside PNRs.

Permanent uncorrelated displacements of ions from the high-symmetry positions of the (fully or partially compositionally disordered) cubic perovskite-type structure were indeed found in relaxors at temperatures much higher, as well as lower, than T_B. They are shown schematically in Fig. 3. The displacements of Pb^{2+} were detected by X-ray and neutron diffraction in PMN [26–28], PZN, PSN, PST, PIN, PFN, PZN-PT, and PMN-PT with small x (see Refs. [29–31], and references therein). To describe the Pb distribution, a spherical layer model has been proposed [28] according to which the shifts of ions are random both in length and direction so that they are distributed isotropically within the spherical layer centered on the special Pb site.[1] The typical radius of the sphere is ~0.3 Å. It decreases slightly with increasing temperature. The off-symmetry displacements of Pb ions in PMN were found to vanish at $T > 925\,\mathrm{K}$ [26] (for other relaxors no data up to so high temperatures are available). The spherical layer model for Pb displacements in PMN was confirmed by the NMR investigations [32] and by the pulsed neutron atomic pair-distribution function (PDF) analysis [33]. Note that the significant random off-centering of Pb ions in perovskites is not the result of compositional disorder. It was also found in the PE phase of the ordinary perovskite $PbZrO_3$ [34]. On the other hand, in the PMN-PT solid solution with $x = 0.4$ which is still compositionally disordered, the Pb displacements from the special perovskite positions were not observed at $T > T_C$ [31].

According to neutron diffraction data [27, 29, 35] the shifts of oxygen ions in the planes parallel to the corresponding faces of the perovskite cubic cell are isotropic (in PMN the shifts are close to 0.2 Å). The oxygen ions are also shifted (by about 0.06 Å in PMN) in the perpendicular direction so that the distribution of shifts forms two rings parallel to the face of the cube. The displacements of B-site ions (Nb^{5+}, Mg^{2+}, Zn^{2+} etc.) from the ideal positions were not noticed in diffraction experiments [29, 35] (some authors found small seemingly isotropic displacements of about 0.1 Å in PMN [27]). Nevertheless the investigations of the extended X-ray absorption fine structure (EXAFS) and the pre-edge regions of absorption spectra revealed the off-center random displacements of Nb in the direction close to $\langle 110 \rangle$ in PMN, PZN, PSN, and PIN [36]. These displacements are not sensitive to the change of temperature (in the range of 290–570 K), nor to the degree of compositional disorder (in PSN and PIN). The pulsed neutron PDF studies confirm [33] that the Nb displacements (in PMN at room temperature) are comparatively small (much smaller than in $KNbO_3$).

[1]For PZN the shifts of Pb from the ideal positions along the eight quivalent $\langle 111 \rangle$ directions were reported, instead of spherical layer distribution [29].

In the canonical relaxors such as PMN, the average crystal symmetry seems to remain cubic with decreasing temperature (see, however, the discussion in Sections 7.1.1 and 7.2), but the local structure changes. In addition to the uncorrelated local distortions described above, the clusters of FE order (i.e., PNRs) appear at $T < T_B$ ($T_B \approx 620\,K$ in PMN). Due to their extremely small (nanometric) size, these clusters cannot be detected from the profiles of the X-ray and neutron Bragg diffraction peaks. Other experiments are needed to validate their existence.

3.2. *Experimental evidence for PNRs*

The first experimental (although indirect) evidence for the PNRs came from the temperature dependences of the optic index of refraction (n) which appear to be linear at $T > T_B$, as shown in Fig. 4 [37]. At lower temperatures a deviation from linearity was observed which was attributed to the variation of n induced (via quadratic electrooptic effect) by local spontaneous polarization inside

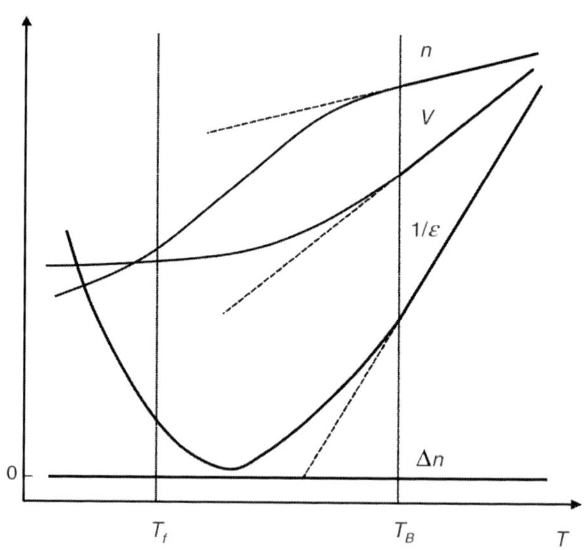

Figure 4. Schematical typical temperature dependences of the refractive index, n, unit cell volume, V, reciprocal dielectric permittivity, $1/\varepsilon$, and birefringence, Δn, in the canonical relaxor.

the PNRs. The existence of PNRs was later confirmed by elastic diffuse neutron and X-ray scattering around the reciprocal lattice points [38–41]. In the PMN crystals, significant diffuse scattering appears at $T < T_B$ with the intensity increasing with decreasing temperature. This effect resembles the scattering caused by FE critical fluctuations, but an important difference (found in synchrotron X-ray experiments [41]) is that the shape of wavevector dependence of scattering intensity at large distances from the reciprocal lattice point deviates from the Lorentzian. This means [41] that the PNRs are more compact than the usual FE critical fluctuations and have better defined borders. The correlation length (ξ) of the atomic displacements contributing to the diffuse scattering, which is a measure of the size of PNR, can be derived from the experiment: it is inversely proportional to the width of the diffuse (Lorentzian) peak. According to the recent high-resolution neutron elastic diffuse scattering study of PMN [42], the size of the emerging PNRs is very small (ξ is around 1.5 nm) and practically temperature independent at high temperatures (Fig. 5). The perovskite unit cell parameter

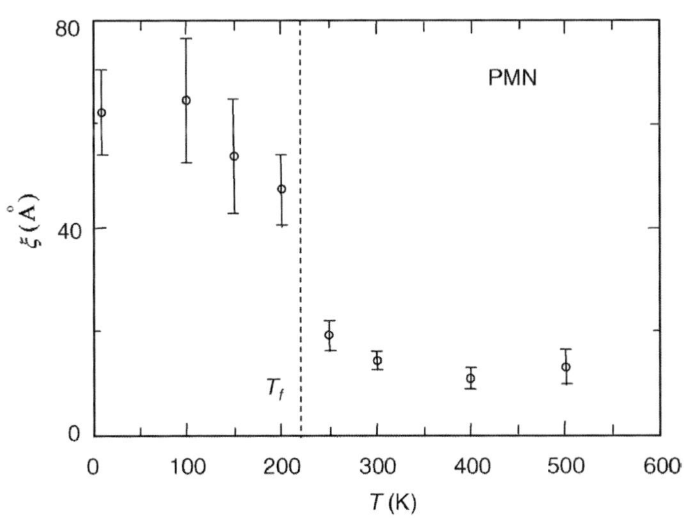

Figure 5. Average size of PNRs in the $Pb(Mg_{1/3}Nb_{2/3})O_3$ crystal (determined from diffuse neutron scattering) as a function of temperature. Vertical dashed line corresponds to T_f (after Xu *et al.* [42]).

being ~0.4 nm, each PNR is composed of only a few unit cells. Below about 300 K, ξ begins to grow on cooling, reaching ~7 nm at 10 K. The most significant growth is found around T_f. Qualitatively the same behavior was observed in the bulk of PZN crystals (the structure of PZN surface layers is different, see Section 7.2) but the size of PNRs is larger: they grow from ~7 nm at high temperatures to ~18 nm at 300 K [43, 44]. From the analysis of the relation between ξ and the integrated intensity of scattering, it was concluded [42] that the number of PNRs also increases on cooling, but in contrast to the temperature evolution of ξ, the increase begins right from T_B and at $T \approx T_f$ a sharp decrease of this number occurs (presumably due to the merging of smaller PNRs into larger ones). Below T_f the number of PNRs remains practically the same at any temperature.

Emergence of PNRs below T_B was also observed in the PMN crystal by means of transmission electron microscopy (TEM) [14], but their size was an order of magnitude larger than that determined from the neutron diffuse scattering, probably because of the influence of electron beam irradiation.

The directions of ionic displacements responsible for the spontaneous dipole moment of PNRs were investigated in several works. By means of dynamic structural analysis of diffuse neutron scattering in PMN crystals it was found that the B-site cations (Nb and Mg) and the O anions are displaced with respect to the Pb cations in the opposite directions along the body diagonal (i.e. the [111] direction) of the perovskite unit cell, forming a rhombohedral polar structure [45]. The rhombohedral $R3m$ symmetry was also derived from the analysis of ion-pair displacement correlations obtained by an X-ray diffuse scattering technique [46], but according to this study, O displacements deviate from the body diagonal and remain parallel to the $\langle 110 \rangle$ direction. The shape of PNR was found to be ellipsoidal [46]. The same shape was revealed by TEM [14].

Besides the structural features, many properties of relaxors can be adequately explained on the basis of the idea of PNRs. For example, in contrast to ordinary ferroelectrics, where a sharp anomaly of specific heat is known to appear at phase transition, in relaxors such anomaly is smeared over a wide temperature range and thus is hardly distinguishable from the background of the lattice contribution. The excess specific heat (total minus lattice contribution) has been determined in PMN and PMT crystals using precise adiabatic and thermal relaxation techniques [47]. It appears as a diffuse symmetric maximum located within the same temperature interval where PNRs nucleate and grow (between 150 K and 500 K in PMN). Therefore the anomaly is likely to be caused by the formation of PNRs and/or by dipolar interactions among them.

Brillouin spectra of PMN-PT at $T \gg T_C$ revealed significant relaxation mode (central peak) which was attributed to the thermally activated fast (10–100 GHz) relaxation of PNRs [48]. The intensity and the width of the peak increase with decreasing temperature, indicating an increase of the number of PNRs and a slowing-down of their dynamics, respectively. The hypersonic damping was also observed. It increases upon cooling, and is attributed to the scattering of acoustic mode by PNRs [48].

PNRs can be thought as unusually large dipoles whose direction and/or magnitude are dependent on an external electric field. Therefore the related properties are expected to be unusual. Indeed, at those temperatures where PNRs exist, relaxors are characterized by giant electrostriction [49–51], remarkable electrooptic effect [50] and extremely large dielectric constant (see Fig. 1 and Section 6).

Even though no unambiguous structural confirmations for the phase transition at T_B are known, the anomalies of properties at this temperature were reported. The frequency-independent maximum of the dielectric loss tangent was found at this temperature in PMN [52]. In the course of thermal cycling of PMN and PMN-PT crystals unannealed after growth, a narrow maximum of the acoustic emission activity is observed (and decreases with the increase of number of cycles) in the vicinity of T_B [53]. Not only the temperature dependence of the index of refraction deviates from linearity at $T < T_B$ (as discussed above in this section), but the temperature dependences of the reciprocal dielectric constant, lattice parameter [51] (see Fig. 4 and (consequently) thermal strain [10, 53] also do the same).

Little is known about the relation between the CNRs and the PNRs in relaxors, although such relation can *a priori* be expected.

Based on the TEM data, it was concluded that the PNRs in PMN may contain CNRs inside and in this case the regions in which PNRs and CNRs overlap remain non-polar [14]. In the framework of the theoretical models discussed in Sections 5 and 7.3, the CNRs can be considered as one of the factors influencing the formation and behavior of PNRs, but not necessarily the determining factor.

4. Lattice dynamics in relaxors

Phase transitions in displacive ferroelectrics (including perovskites) are known to be caused by softening and condensation of transverse optic (TO) phonon mode at the Brillouin zone center. Since the frequency of this mode (ω_0) is connected to the static lattice dielectric constant (through the Lyddane–Sachs–Teller relation, $1/\varepsilon \propto \omega_0^2$), the divergence of ε in ferroelectrics at T_{CW} according to the Curie–Weiss law ($1/\varepsilon \propto T - T_{\mathrm{CW}}$) implies that the mode condenses (its frequency tends to zero) at T_{CW}, too. In relaxors, the Curie–Weiss law also holds in the PE phase, i.e. at $T > T_{\mathrm{B}}$ (see Fig. 4 and Section 6), but the corresponding softening of phonon modes had not been detected until recently. During the last few years neutron inelastic scattering technique was applied to investigate the lattice dynamics in relaxor crystals. In the PE phase of PMN far above T_{B}, the dispersion of the transverse acoustic (TA) and low-energy TO phonons were found to be very similar to that existing in the PE phase of classical displacive ferroelectrics $PbTiO_3$ [39, 54] (see the curves for 1100 K in Fig. 6). On cooling down to $T \approx T_{\mathrm{B}}$, the optic branch softens in the same manner as in displacive ferroelectrics, i.e., the frequency of the mode at wave vector $\vec{q} = 0$ (zone center mode) follows the Cochran law [54, 55]:

$$\omega_0^2 = A(T - T_0), \tag{1}$$

with $A > 0$, as shown in Fig. 7. At $T < T_{\mathrm{f}}$, the well-defined TO modes are also observed [55] (see the curve for 150 K in Fig. 6). Once again, the temperature evolution is consistent with the typical behavior of a FE soft mode below the Curie temperature (Fig. 7), i.e. Eq. (1) holds with $A < 0$ (note that T_0 is close to T_{B} in this case). The same dependence is observed at low temperatures in

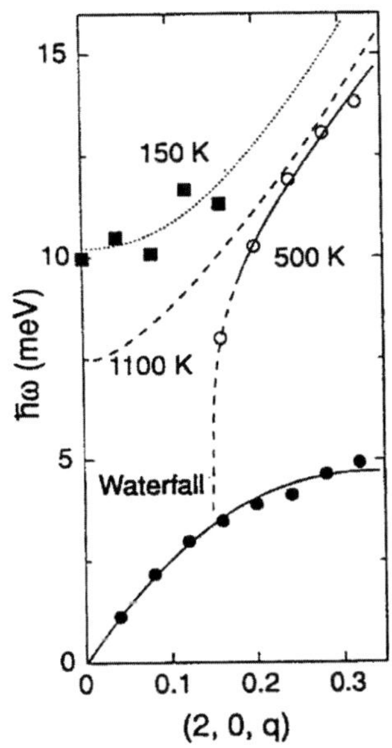

Figure 6. Phonon dispersion curves in PMN crystal for the TA branch at 1100 K (solid circles) and the lowest-energy TO branch at three different temperatures: $T = 1100\,\mathrm{K} > T_\mathrm{B}$ (dashed line), $T = 500\,\mathrm{K} < T_\mathrm{B}$ (open circles), and $T = 150\,\mathrm{K} < T_\mathrm{f}$ (squares). (after Wakimoto *et al.* [55]).

PZN crystal [44]. But in the temperature range between T_f and T_B, the lattice dynamics is different. The propagating TO modes are observed here only for the wave vectors larger than q_wf. For $q < q_\mathrm{wf}$, the modes are overdamped. The TO phonon branch drops sharply into the TA branch at q_wf, resembling a waterfall (as shown in Fig. 6 for 500 K), and for this reason, the phenomenon is called "waterfall." It has been observed not only in PMN but also in other relaxor materials for which neutron inelastic experiments were performed, i.e., in PZN [56] and PZNT [57]. In the same temperature range where the TO mode is overdamped (between T_f and T_B), the damping of TA phonons is also enhanced (e.g., neutron inelastic experiments in PMN revealed a large maximum of the TA-phonon linewidth at ~400 K [55]).

It was initially proposed that the large damping of TO modes is due to the presence of PNRs which prevent the propagation of

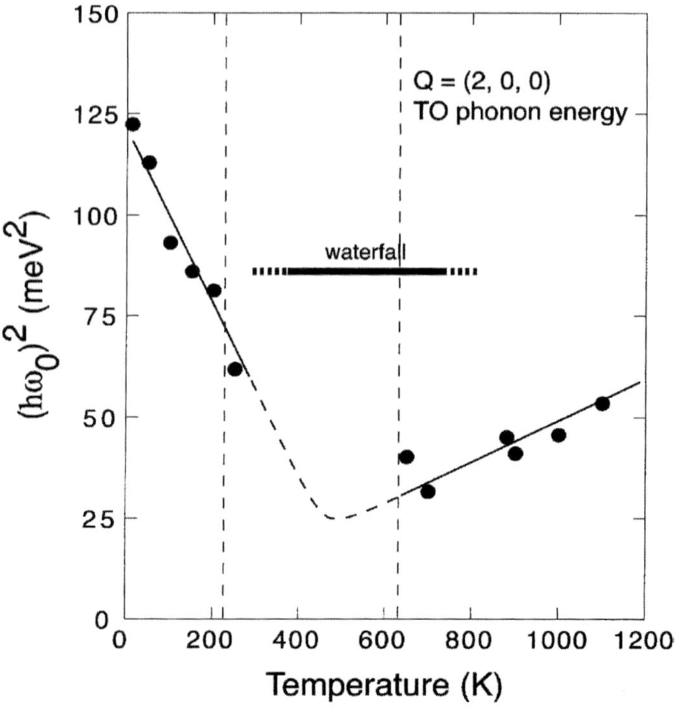

Figure 7. Temperature dependence of TO-phonon energy squared measured at (200) in PMN crystal. Vertical dashed lines correspond to T_{f} and T_{B}. The temperature range in which the waterfall feature appears is indicated by the thick horizontal line. The other dashed and solid lines are guides to the eye. (after Wakimoto *et al.* [55])

phonons with wavelength larger than the size of PNR, and thus q_{wf} is the measure of the average size of the PNRs [54, 57]. Later it was shown [58] that q_{wf} depends on the choice of the Brillouin zone and the relation of q_{wf} to the size of PNRs is improbable. The waterfall effect was then explained by the interactions of acoustic and optic branches. It was also noticed that damping of the soft mode near phase transition is not the unique feature of relaxors; similar effect can be found in some ordinary ferroelectrics [58].

An important question arises as to whether the low-energy TO mode found in relaxors is really the FE soft mode, i.e. whether the frozen mode displacements are responsible for the spontaneous dipole moment of PNRs and for the spontaneous polarization of the low-temperature FE phase. Initially the answer to the question was negative. Nabereznov *et al.* [39] compared the correlated atomic displacements in PNRs (found by elastic diffuse scattering experiments) and the TO-mode atomic displacements and concluded that they are incompatible. Besides, Vakhrushev and Shapiro [59] noticed that the value of T_0 derived from Eq. (1) does not coincide with T_{CW} found in dielectric measurements and therefore this mode could not be the FE soft mode. They also identified an additional "quasioptic" branch lying significantly lower in energy than the TO branch and having the temperature variation consistent with the Curie–Weiss behavior of the dielectric constant. The quasioptic mode was proposed to be the true soft mode. This opinion was later disputed. The apparent conflict between diffuse and soft mode scattering experiments was reconciled with the help of the "phase-shifted condensed soft mode" model proposed by Hirota *et al.* [40]. According to this model the total displacement of atoms inside a PNR consists of two components. The first component is created by the TO soft-mode condensation and gives rise to the spontaneous polarization of PNR. The second one results from a uniform displacement of all atoms leading to the shift of the PNR along their polar direction relative to the surrounding non-polar matrix. Wakimoto *et al.* [60] described the lattice dynamics by a coupling between TA and TO modes without the need for considering any additional quasioptic modes. The concept of soft coupled optic mode was introduced. Being condensed, this mode has the optic component responsible for the dipole moment of PNR and the acoustic component giving rise to the uniform displacement of PNR as a whole. The microscopic origin of the uniform component is not yet clear.

Infrared (IR) spectroscopy of PMN between 20 K and 300 K [61] and of PLZT between 10 K and 530 K [62] revealed three main zone-center TO modes as typical of cubic perovskites. High-energy TO modes do not show any pronounced temperature dependence both in PMN and PLZT. The lowest-energy mode (which was also resolved by time-domain THz spectrometry) was found to soften following Eq. (1) with $A < 0$ and $T_0 \approx T_{\mathrm{B}}$, i.e., in agreement with the above-discussed neutron scattering data. Nevertheless, in contrast to the neutron data, the mode remains underdamped in the whole temperature range. This discrepancy is related to the fact that phonons with different wave vectors are probed in IR ($q \approx 10^{-5}$ Å^{-1}) and neutron ($q \geq 10^{-2}$ Å^{-1}) experiments.

Raman scattering, which is known to be an effective tool for studying the optic soft modes in the crystals with normal phase transitions, was also widely applied to relaxors (see Refs. [63–65], for a review), but no soft mode has been found here. Light scattering spectra in relaxors appear to be quite complex and their interpretation is not straightforward. In particular, it is not clear what kind of disorder gives rise to the observed first-order lines that are forbidden by the $Pm\bar{3}m$ (O_h^I) average symmetry of relaxors.

Recently, vibration spectra in PMN were determined theoretically from the first principles [66]. Although the computations were performed for the case of compositionally ordered structure (the real structure is disordered with the inclusions of ordered CNRs, see Section 2), it appeared to be possible to assign the calculated phonons to the main peculiarities of Raman and IR spectra.

5. Origin and evolution of PNRs: Models and theories

Although the very existence of PNRs in relaxors seems to be doubtless, the cause and mechanisms of their formation are not conclusively understood. At temperatures higher than T_B the structure and properties of relaxors closely resemble those of normal displacive ferroelectrics. When a relaxor becomes compositionally ordered after high-temperature annealing (without changing the chemical composition), a sharp ferro- or antiferroelectric (AFE) phase transition is observed (see Section 2). These facts seemingly suggest that the relaxor crystal tends to be ferro- or antiferroelectric at low temperatures, but the quenched compositional disorder somehow prevents the normal transition into the phase with macroscopic FE or AFE order from happening. Instead, the PNRs appear. There exist different approaches to explain the formation of PNRs. All of them can be schematically subdivided into two categories. The

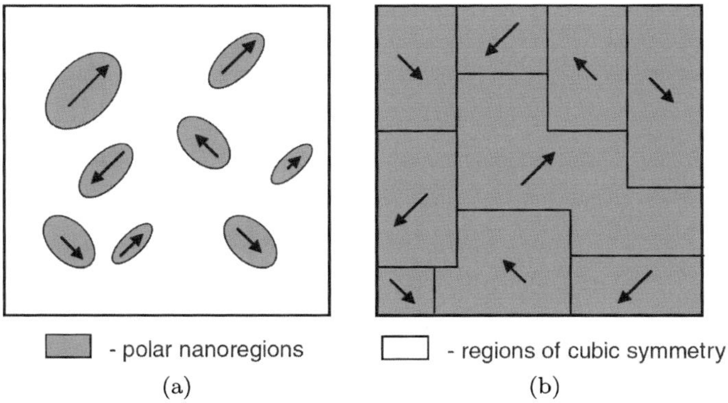

Figure 8. Schematic representation of PNRs in relaxors according to the different models.

models of the first category [8, 67–70] consider the PNRs as a result of local "phase transitions" or phase fluctuations so that the crystal consists of nanosize polar islands embedded into a cubic matrix in which the symmetry remains unchanged (as shown in Fig. 8(a)). The models of the second category assume the transition to occur in all regions of the crystal and the crystal consists of low-symmetry nanodomains separated by the domain walls but not by the regions of cubic symmetry [71, 72] (the example is shown in Fig. 8(b)). Note that these two situations can hardly be distinguished experimentally by structural examinations [73] because the *local* symmetry of cubic matrix is not expected to be cubic and the thickness of domain walls (i.e., the regions where polarization is not well-defined) is comparable with the size of nanodomains.

The second category is represented by the random-field model proposed by Westphal, Kleemann, and Glinchuk (WKG model) [71, 74], who applied the results of a theoretical work by Imry and Ma [75] to the relaxors. It was shown in Ref. [75] that in the systems with a continuous symmetry of order parameter, a second-order phase transition should be destroyed by quenched random local fields conjugate to the order parameter. Below the Curie temperature the system becomes broken into small-size domains (analogy of PNRs) instead of forming a long-range ordered state. It should be emphasized that this model does not consider the trivial case of the local spontaneous polarization which is directed parallel to the quenched field when the field is strong enough. Instead the situation is determined by the interplay of the surface energy of domain walls and the bulk energy of domains in the presence of *arbitrary weak* random fields [75].

For displacive transitions, continuous symmetry means that the spontaneous deformation is incommensurate with the PE lattice. However, this is not the case for the perovskite ferroelectrics in which the spontaneous deformation and the polarization (order parameter) are aligned along definite crystallographic directions (e.g., the $\langle 111 \rangle$ directions for the rhombohedral phase). Nevertheless, when the number of allowed directions is large (e.g., eight for the rhombohedral phase), the symmetry of order parameter can be considered quasi-continuous and the approach appears to be applicable. The disordered distribution of the heterovalent ions inherent to the compositionally disordered structure (e.g., Nb^{5+} and Mg^{2+} ions in PMN) provides the source for quenched random electric fields.

Ishchuk [72] analyzed the thermodynamic potential in the framework of Landau phenomenological theory for the systems in which the energies of the FE and AFE phases are close to each other. It was shown that the state with coexisting FE and AFE domains may have lower thermodynamic potential than the homogeneous (FE or AFE) state. This effect is due to the interactions (electrostatic and elastic) between the FE and AFE domains. It was suggested that relaxors are just the crystals in which this effect occurs. In other words, the nonpolar regions, coexisting with PNRs (FE domains), are the domains of AFE structure.

The best-known model of the first category was developed in the early works by Isupov and Smolenskii [8, 22]. Due to the compositional disorder, the concentrations of different ions (e.g., Mg^{2+} and Nb^{5+} in PMN) are subject to quenched spatial fluctuations. As the FE Curie temperature (T_C) depends on the concentration, spatial fluctuations of local T_C are expected. It was suggested that upon cooling, local FE phase transitions occur first in those regions where

T_C is higher, whereas the other parts of the crystal remain in the PE phase. Therefore, PNRs are simply the regions with elevated Curie temperature.

Several other models use the microscopic approach and consider the structural evolution and formation of PNRs in terms of interatomic interactions. The FE lattice distortion in the ordinary perovskites is known to be determined by a delicate balance between the electrostatic (dipole–dipole) interactions and the short-range repulsions. Hybridization between the oxygen $2p$ states and electronic states of cations (covalent bonding) is able to change this balance, influencing thereby the phase transition temperature [25]. In the compositionally ordered (translationally symmetric) crystals, exactly the same forces affect all the atoms of a certain type because they have the same coordination neighborhood. In the case of compositional disorder, the ions of different types may be found in the neighboring unit cells on the same crystallographic positions (e.g., in the B-sublattice of PMN, both Mg and Nb ions are the nearest neighbors of Nb ions). The interatomic interactions which would cause ferro- or antiferroelectric order in the compositionally ordered state become random in this case, and as a result, the long-range polar order is disturbed. The models described below emphasize the importance of different interactions: the interactions under random local electric fields only (including dipole–dipole interactions) [67], and the dipole–dipole interactions together with random short-range repulsions [9, 68] or random covalent bonding [33].

In the random field theory developed for relaxors by Glinchuk and Farhi (GF model) [67] (see also Ref. [76] in which the related papers are reviewed), the transition is regarded as an order–disorder one, i.e., at high temperature the crystal is represented by a system of reorientable dipoles (dipoles caused by the shifts of ferroactive ions from their ideal perovskite positions, see Section 3.1). These random-site dipoles are embedded in highly polarizable "host lattice" (the high polarizability is due to the transverse optic soft mode existing in relaxors, see Section 4). The dipole–dipole interactions are indirect (they occur via the host lattice) and random. Nevertheless, according to the theory, they should lead to uniformly directed local fields and thus to FE ordering at low temperature (in contrast to *direct* dipole–dipole interactions which can lead to a dipole glass state). Thus to explain the absence of macroscopic FE order in relaxors, additional sources of random local electric fields are considered. These additional fields can be static (coming from quenched compositional disorder, lattice vacancies, impurities and other imperfections) or dynamic (associated with shifts of non-ferroactive ions from the special positions). In contrast to the fields considered in the WKG model, these fields should be rather large (larger than critical value) to destroy the long-range FE order. The FE order parameter, phase transition temperature T_C, linear and nonlinear dielectric susceptibilities are calculated within the framework of statistical theory using the distribution function for local fields. It is found that depending on the model parameters (concentration of dipoles, other field sources and the host lattice correlation length), the low-temperature phase can be FE, dipole glass or mixed ferroglass. In the temperature interval between T_C and T_B, the short-range clusters may appear, in which the reorientable dipoles are ferroelectrically correlated (i.e., PNRs). In the ferroglass state these clusters coexist with the macroscopic regions in which the dipoles are coherently ordered.

Note that the GF model for relaxors is the extension of the analogous theory for incipient ferroelectrics with off-center impurities (e.g., $KTaO_3$: Li, Nb, or Na). In the later case the off-center impurities are the interacting dipoles. Due to their small concentration the crystal can be considered as a system of identical dipoles with random long-range interactions. In the case of complex perovskite relaxors, the dipole concentration cannot be considered small. The random interactions of different (short-range) nature are also involved and thus the dipoles are not identical. It was first recognized in the model proposed by one of the co-authors of the present review [9, 68]. In this model the PNRs are the result of local condensation of the soft phonon mode (which exists in relaxors as discussed in Section 4). The consideration is based on the model of coupled anharmonic oscillators which is often applied to ordinary ferroelectrics. The effective Hamiltonian is given by the sum of Hamiltonians of the individual unit cells:

$$H = \sum_l \left[0.5 \prod_l^2 + A_l \xi_l^2 + B_l \xi_l^4 - \sum_{l'} \nu_{ll'} \xi_l \langle \xi_{l'} \rangle \right],$$

where $\prod_l$ and ξ_l are the generalized momentum and coordinate of the soft mode displacements, A_l and B_l are parameters of one-particle potential, which are determined by the interactions (mainly short-range repulsive) between ions of the lth unit cell, and v'_{ll} are parameters characterizing the interactions (long-range dipole–dipole) between the different cells. In the translationally invariant crystal, all the parameters, A_l, B_l, and v'_{ll}, would be the same. In the case of compositional disorder they are different. The distribution function for these parameters is introduced in the model. This distribution gives rise to the spatial distribution of local "Curie temperature" T_C. PNRs appear in the regions with enhanced local T_C. The model parameters are linked to the parameters of real structure (in particular, the size of ions). Based on the crystal composition, this model is able to predict quantitatively the degree of "diffusion" of the transition, i.e., the extent of temperature interval in which the PNRs develop before the crystal transform into the low-temperature nonergodic phase. In particular, the degree of diffusion increases with increasing difference in the radii of ions in the ferroactive sublattice (A or B perovskite sublattice) or with increasing compositional disorder in this sublattice.[2] On the other hand, the diffusion is much less sensitive to the disorder in the non-ferroactive sublattice. The influence of the degree of compositional disorder on T_C is also explained. Based on the arguments similar to those used in the original model [68] it was recently suggested [77] that, because of the randomness of microscopic forces responsible for the onset of spontaneous polarization, each PNR can consist of unit cells polarized in different directions. This model of "soft nanoregions" also implies that, due to thermally activated reorientations of some unit cells inside PNR, not only the direction (as believed before),

[2]In lead-containing complex perovskites the Nb and Ta cations are supposed to be ferroactive.

but also the magnitude of the spontaneous dipole moment of individual PNR can strongly change with time (due to fluctuations or under the external field), while the size of PNR remains the same.

The Hamiltonian considered in the model by Egami [33] consists of two terms,

$$H = H_1 + H_2. \tag{2}$$

The first term is written in a standard form

$$H_1 = -\sum_{ij} J_{ij} J_{ij} \vec{S}_i \cdot \vec{S}_j, \tag{3}$$

where $\vec{S}_i$ is the local polarization caused by the displacement of i-th Pb ion from its special position (as discussed in Section 3.1), J_{ij} describes the random interaction between local polarizations mediated by oxygen and B-site ions. It is explained that in PMN the Pb ions cannot form the covalent bonds with those O ions which are bonded to Nb. On the other hand, Mg ions create purely ionic bonds and do not prevent the Pb–O bonding. Consequently the direction toward Mg is an "easy" direction for Pb displacement. This directional dependence of the energy of Pb displacements resembles the crystalline anisotropy in magnetic systems. It is random in compositionally disordered crystal and can be described by model Hamiltonian H_2. This model was established to account for the relaxor properties in ER as well as in NR phases, but the appearance of PNRs was not derived.

Timonin [69] suggested that the ergodic phase in relaxors is an antilog of Griffiths phase theoretically predicted long ago (but not yet experimentally found) for dilute ferromagnetics. Ferroelectric clusters of various sizes (i.e. PNRs) appear in this model at $T < T_C$ (where T_C is the Curie temperature for non-dilute crystal) and specific non-exponential relaxation is predicted.

Specific temperature evolution of PNRs can be explained in terms of the phenomenological kinetic theory of phase transitions in compositionally disordered crystals [70]. The emergence of PNR, i.e., the region of polar crystal symmetry within the cubic surrounding, should be accompanied by the creation of electric and elastic fields around PNR, which increase the total energy of the system. Due to the similar effects in the compositionally ordered crystals undergoing a first-order phase transition, the regions of the new phase (nuclei) are not stable. They tend to grow if their size is larger than the critical one or disappear otherwise. As follows from the theory [70], in disordered crystals the nuclei of the new phase can be stable and the equilibrium size of newly formed nuclei can be arbitrary small. The PNRs in relaxors which are really small (contain several unit cells) and stable can be regarded as such kind of nuclei. The theory predicts that PNRs begin to appear in the PE phase at T_{B} as a result of local "phase transitions" (e.g. condensation of phonon soft mode). Upon cooling, the number of PNRs increases but the equilibrium size of each PNR remains unchanged within a certain temperature interval just below the temperature at which it appears. Upon further cooling, the PNR grows slowly with decreasing temperature while remaining in a stable equilibrium, and finally at $T = T_{\text{C}}$, becomes metastable so that the size of PNR increases steeply due to phase instability. In other words, the behavior predicted by this model is the same as experimentally observed in PMN (see Fig. 5). But this theory is unable to describe quantitatively the real behavior at $T < T_{\text{C}}$, because it does not take into account the interactions between different PNRs, which are obviously significant at low temperatures. It was further explained [70, 78] that depending on the model parameters (in particular, the mean T_{C} and the width of the distribution of local transition temperatures), a sharp phase transition can occur, resulting in large FE domains at $T < T_{\text{C}}$ (in the case of a small width and a comparatively high T_{C}) or the transition is diffuse and the low-temperature polar regions are of nanometer size. The dipole–dipole interactions between them can lead to the formation of a glass-type phase at a certain temperature T_{f}. The intermediate situations are also possible with moderately diffuse transition and mesoscopic polar regions (domains). Note that these different types of behavior have indeed been observed experimentally in different perovskite materials (see Section 7.2).

6. Dielectric response in relaxors

Small-signal dielectric response has been intensively studied in a large number of relaxor materials, but most investigations were restricted within the frequency range of 10–10^9 Hz or narrower. In the past few years, modern measurement facilities with enlarged frequency range have been applied to relaxors. It has been found that significant dielectric dispersion exists in the whole spectrum starting from the frequency of lattice vibrations down to the lowest practically measurable frequency of $f \sim 10^{-5}$ Hz. The present section will focus on these works.

The field-induced polarization in relaxors can be divided into several qualitatively different parts so that the total relative permittivity in the temperature range of permittivity maximum can be written as

$$\varepsilon = 1 + \chi_e + \chi_{Ph} + \chi_R + \chi_U + \chi_{LF}, \tag{4}$$

where χ_e, χ_{Ph}, χ_R, χ_U, and χ_{LF} are the susceptibilities (complex numbers) describing the electronic, phonon, "conventional relaxor" (CR), "universal relaxor" (UR) and "low-frequency" contributions, respectively. All the contributions are frequency dependent (in Eq. (4) the susceptibilities are ranked in the order of increasing typical characteristic time). As in any materials, electronic contribution persists in relaxors at all temperatures and at frequencies up to 10^{15}–10^{17} Hz, but at lower frequencies, the value of $\chi_e' = (n^2-1) \sim 10$ is small as compared with other susceptibilities.

The phonon (lattice) susceptibility (caused by the mutual displacements of cation and anion sublattices) is active up to the frequencies of 10^{12}–10^{14} Hz. To separate χ_{Ph} from other contributions, measurements at these frequencies are necessary. In relaxors χ_{Ph} has been determined from IR reflectivity spectra at temperatures lower than T_B only. In PMN crystals, χ_{Ph} increases from ~40 at 20 K to ~100 at 300 K [61]. In other words, it constitutes less than 1% of the total low-frequency permittivity measured at T_m (see Fig. 1). This is an important difference of relaxors from ordinary displacive ferroelectrics in which phonon polarization totally accounts for the permittivity peak at the phase transition.

The susceptibilities χ_R and χ_U related to the relaxation-type polarizations are the main contributions giving rise to the peculiar relaxor peaks in the temperature dependences of permittivity (shown in Figs. 1 and 9). The real part of χ_R is constant at low enough frequencies and decreases to zero when the frequency reaches the (temperature-dependent) characteristic value. This decrease is accompanied by the peak in the frequency dependence of the imaginary part.[3] Both real and imaginary parts of the UR susceptibility continuously (without any loss peak) decrease in the whole frequency range practically available for measurements according to the power law

$$\chi'_U = \tan(n\pi/2)\chi''_U \propto f^{n-1}, \tag{5}$$

where n is close to but smaller than unity. Note that the same empiric classification (i.e., the monotonic frequency variation versus the variation with loss maximum) applies not only to relaxors, but also to the relaxation processes found in many other solids [79]. Nevertheless, the values of χ_R and χ_U in relaxors are extraordinary large as compared to other dielectrics. The CR dispersion (CRD) is observed at the low-temperature slope of permittivity peak giving rise to the frequency shift of T_m, while the UR dispersion (URD) exists at temperatures lower, as well as higher, than T_{m} (as shown in Figs. 1 and 9).

The last term χ_{LF} in Eq. (4) combines all possible relaxation contributions not related to the relaxor ferroelectricity, which may include the polarization of hopping charge carriers [79, 80], Maxwell–Wagner-type polarization, etc. Typically, these contributions become significant in good-quality samples of relaxor perovskites at comparatively high temperatures and/or low frequencies. In the PMN crystal presented in Fig. 1 the contribution of χ_{LF} at lowest frequency $f = 10^{-2}$ leads to the noticeable increase of ε'' at temperatures above ~300 K.

In the radio- and audio-frequency ranges, where the dielectric properties of relaxors are most often studied, the value of χ'_R is much

[3]The well-known Debye relaxation is an example of such kind of the behaviour.

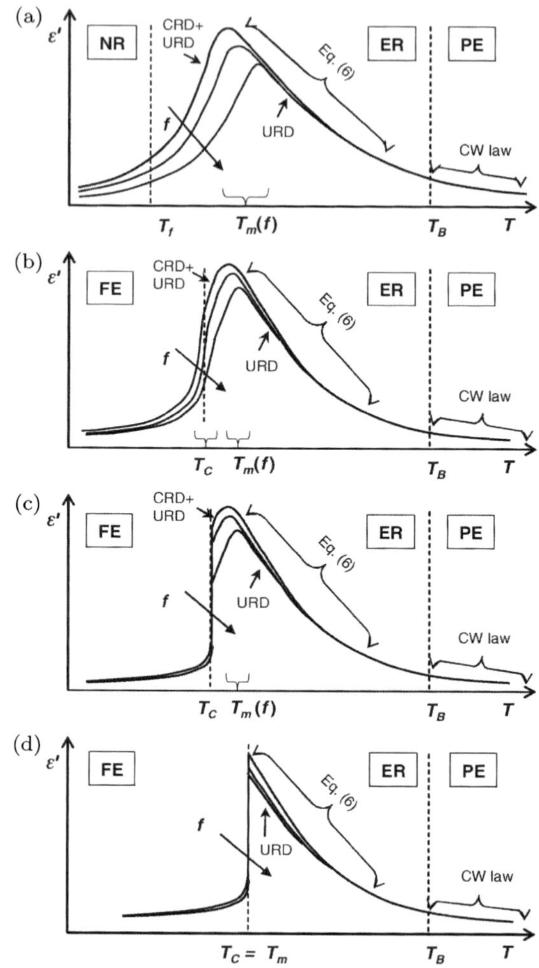

Figure 9. Different possibilities for the temperature evolution of structure and dielectric properties in compositionally disordered perovskites: (a) canonical relaxor; (b) crystal with a diffuse relaxor-to-ferroelectric phase transition at $T_{\rm C} < T_{\rm m}$; (c) crystal with a sharp relaxor-to-ferroelectric phase transition at $T_{\rm C} < T_{\rm m}$; (d) crystal with a sharp relaxor-to-ferroelectric phase transition at $T_{\rm C} = T_{\rm m}$. The temperature dependences of the dielectric constant at different frequencies $f$are schematically shown. The temperature intervals in which the Lorenz-type Equation 6 and the Curie–Weiss law hold, the regions of conventional relaxor dispersion (CRD) and universal relaxor dispersion (UR) and the types of structure (paraelectric (PE), nonergodic relaxor (NR), ergodic relaxor (ER), and ferroelectric (FE)) are identified. Note the similar behavior at high temperatures in all cases.

larger than χ'_U. That is probably the reason why the UR contribution has been discovered only recently with the help of the frequency response analyser that is able to work at ultra-low frequencies (as one can see from Eq. (5), χ'_U increases with decreasing f). The χ_U component was separated from the χ_R one by means of the analysis of dielectric spectra at $T > T_{\mathrm{C}}$ (or $T > T_{\mathrm{f}}$) first in PMN-PT [80–82] and then in PMN [77], PSN [83], and BTZ [84]. Since CR is the dominant contribution giving rise to the diffuse ε'(T) peak, $\chi'_R \approx \varepsilon'$ in the vicinity of T_m (at least for the frequencies that are not very low or very high). Therefore, most of the dielectric investigations of relaxors dealt in fact with the CR contribution, even though it was not identified by the authors.

As shown recently for many relaxors [85, 86], the high temperature slope of the diffuse $\varepsilon'(T) \approx \chi'_R(T)$ peak can be scaled with the empirical Lorenz-type relation,

$$\frac{\varepsilon_A}{\varepsilon'} - 1 = \frac{(T - T_A)^2}{2\delta^2}, \tag{6}$$

where $T_A(< T_m)$ and $\varepsilon_A(> \varepsilon_m)$ are the fitting parameters defining the temperature and magnitude of the Lorenz peak (6), and δ is as a measure of the degree of diffuseness of the peak. This formula gives a more adequate description of the experimental data than the previously used relation, $\varepsilon_m/\varepsilon'' - 1 \propto (T - T_m)^\gamma$ (where $1 < \gamma < 2$, and ε_m is the value of ε' at T_{m}). Equation (6) holds from temperature T_1, which is typically several degrees higher than T_{m}, to temperature T_2, which is a few dozens of degrees lower than T_{B} (see Fig. 9). The diffuse peak of ε'' $(T) \approx \chi''_R(T)$ can be scaled with more complex relations (see Refs. [19, 87] for details).

At $T > T_{\mathrm{B}}$, the dielectric constant is described by the Curie–Weiss law,

$$\varepsilon' = C/(T - T_{\mathrm{CW}}),$$

where the Curie constant, C, has the same order of magnitude ($\sim 10^5$ K) as in ordinary displacive ferroelectrics and T_{CW} is typically higher than the low-frequency value of T_{m} (as shown in Fig. 4) but at high frequencies (e.g. in PMN at $f > 20$ GHz) T_{m} can become larger [52].

The CR contribution can consist of several components in itself involving different polarization mechanisms. Each of the mechanisms gives rise to the corresponding dispersion and can be seen in the $\varepsilon''(f)$ curve as an individual maximum (or an anomaly if neighboring maxima overlap each other). For example, in PMN crystals three components (dispersion regions) have been found at $T < T_B$ [52, 61, 88], which were resolved simultaneously between 210 K ($\approx T_f$) and 290 K. The first component appears at the (temperature-independent) frequency of ~1 THz and gives rise to a comparatively small input to the static dielectric constant (about 130). The two other dispersion regions become broadened on cooling and their mean relaxation time increases (i.e., the frequency of the corresponding $\varepsilon''(f)$ peak decreases), so that at $T < T_f$ the low-frequency component shifts out of the measurement frequency range and the higher-frequency component develops into a constant (frequency-independent in the range of 10^2–10^{11} Hz) loss. This effect of constant loss is a noticeable property of the low-temperature nonergodic phase in relaxors. It can also be observed in other relaxors, e.g., in the PLZT ceramics [89] and the compositionally disordered PIN crystals [19]. The magnitude of constant loss decreases exponentially on cooling, but still remains measurable at liquid-helium temperature. At extremely low frequencies ε'' in NR phase is no longer constant and slightly increases with decreasing frequency (see Fig. 1).

To describe the CR dielectric spectra, the same empirical expressions as used for other dielectrics were applied. The Kohlrausch–Williams–Watts [80, 90], the Havriliak–Negami [91], and the simpler Cole–Cole [19, 89] formulae have been employed by different authors to fit the experimental $\varepsilon(f)$ data. The alternative way to analyze the dispersion is to find the appropriate function for the distribution of relaxation times. For example, Rychetsky *et al.* [89] fitted the relaxation in PLZT to a uniform distribution that broadens upon cooling.

A remarkable feature which was observed first in PMN [92] and then in many other relaxor ferroelectrics is the Vogel–Fulcher (VF) law connecting the temperature and the frequency of the $\varepsilon'(T)$ peak:

$$f = (2\pi\tau_0)^{-1}\exp[-E_a/(T_m - T_{VF})], \tag{7}$$

where f is the measurement frequency, τ_0, E_a, and T_{VF} are the fitting parameters. The same relation but with slightly different parameters has also been reported for the peak temperature in the $\varepsilon''(T)$ dependences (T_{mi}). Investigations of PIN crystals showed that the parameters of Eq. (7) can be different in different frequency intervals [19]. The VF law was known in structural and spin glasses. When revealed in relaxors, it became one of the main reasons to postulate the existence of a dipole glass phase at $T < T_{\mathrm{VF}}$. Equation (7) might (but not necessarily, see below) signify the similar VF relation for the characteristic relaxation time τ of the corresponding relaxation process:

$$\tau = \tau_\infty \exp[E_b/(T - T_f)], \tag{8}$$

where τ_∞, E_b are the parameters and T_{f} is the freezing temperature (i.e., the temperature below which the relaxation time becomes infinite). This divergence of τ indicates that the thermally activated reorientations of dipoles responsible for polarization slow down with decreasing temperature and become impossible (consequently dipoles cannot respond to the electric field) at $T = T_{\mathrm{f}}$, but not at $T = 0$ as prescribed by the Arrhenius law for the dynamics of independent dipoles. In dipole glasses the interactions among the dipoles are the cause for such kind of freezing. These interactions (bonds) are frustrated (i.e., can be either FE or AFE but cannot be satisfied simultaneously) and thus favor the configurations with random directions of dipoles, in contrast to the ferroelectrics and antiferroelectrics in which the dipole directions are parallel and antiparallel, respectively.

The relations between Eqs. (7) and (8) in relaxors were studied by several authors. In the case of Debye relaxation (which can be expected in the system of identical non-interacting dipoles) this relation should be simple: τ follows the Arrhenius law (which is the same as the VF law (8), but with $T_f = 0$) and this automatically means that law (7) also holds for T_{mi} with the same parameters, i.e., $\tau_0 = \tau_\infty$, $E_a = E_b$, and $T_{\mathrm{VF}} = 0$. The relaxation in relaxors is much more complex and can be characterized by a wide spectrum for the distribution of the relaxation times. Simple relations between the parameters of Eqs. (7) and (8) are not evident. Furthermore, different relaxation times from the spectrum may have different freezing temperatures. It was also shown theoretically that the situations are possible in which Eq. (7) holds with $T_{VF} \neq 0$, but no freezing at a non-zero temperature really takes place [93], i.e. the VF relationships for T_m and T_{mi} do not necessarily imply glass-type dipole dynamics. The problem seems to be solved in some relaxors by means of a special analysis of the frequency-temperature dependences of the real part of permittivity. It was shown that the longest relaxation time in the spectra of PMN [94, 95], PST [94], and PLZT [96] diverges according to the relationship (8) with $T_f = T_{VF}$ ($\approx$220 K in PMN), while the bulk of the distribution of relaxation times remains finite even below T_f [95, 96]. The divergence of the longest relaxation time means that, at least empirically, the behavior of relaxors in small-signal electric field is similar to the behavior of dipole glasses. However, a microscopic interpretation of this fact is not so clear. In contrast to ordinary dipole (or spin) glasses in which the susceptibility can by unambiguously attributed to the reorientation of certain permanent dipoles (spins), the structure of relaxors is more complex and the polarization mechanisms responsible for the large and diffuse $\varepsilon(T)$ peak have not been definitely identified.

Most of the existing explanations relate the dielectric relaxation in relaxors to the PNRs. The PNRs are very small and can be considered as individual thermally activated dipoles giving rise to the orientational polarization. Thus, the dominant contribution to the measured $\varepsilon(T)$ relaxor peak (i.e., the CR contribution according to the classification described above) may be attributed to the thermally activated reorientation of dipole moments of PNRs (local spontaneous polarization vectors). Many authors proceeded upon this assumption when analyzing the dielectric data (see e.g. Refs. [97–99]). The dipole moments of PNRs are considered in many models as interacting (directly or via surrounding matrix) entities constituting a glassy system (see Sections 5 and 7.3). The reorientations may be affected by the random anisotropy and (in contrast to magnetic spin glasses) by an environment of random electric and elastic fields.

The second possible mechanism associated with PNRs is the sideway motion of their boundaries without the change of the orientation. In the course of such motion, the volume (and thereby the dipole moment) of the polar region changes, giving rise to the characteristic polarization response. This looks like breathing of PNRs and therefore, the corresponding model developed by Glazounov and Tagantsev is called "breathing" model [100]. The model considers the vibrating PNR boundaries in terms of the theory of randomly pinned interface, which was developed earlier for magnetic materials. In the case of relaxors the internal random local fields induced by charge disorder act as the pinning centers. Another approach was used by Rychetsky *et al.* who proposed a thermodynamic model for the polarization reversal near the PNR boundary, which is equivalent to the displacement of the boundary [89]. In particular, this model describes well the constant loss effect at low temperatures. From the analysis of the behavior of PMN crystals in large dc and ac electric fields, it was suggested [101] that the dielectric response in the ergodic phase (i.e., in the vicinity of the $\varepsilon(T)$ maximum) is controlled by the vibration of PNR boundaries, rather than by the thermally activated reorientations of PNRs.

Note that in PMN and some other relaxors two main components determine the CR dielectric response in the ergodic phase (see above in this section). The low-frequency component may result from the reversal of the spontaneous dipole moments of PNRs and the high-frequency one may originate from the PNR boundary motion [88].

The value of the universal susceptibility (i.e., the susceptibility whose dispersion is described by Eq. (5) at all frequencies) in relaxors is several orders of magnitude larger than in non-relaxor materials with the same n [80]. Thus it is reasonable to suggest that the UR polarization mechanism is also connected with PNRs which are inherent only in relaxors. Within the scope of the soft nanoregions model (see Section 5), the UR response has been attributed to the thermally activated reorientations of dipole moments of individual unit cells inside PNRs [77].

The Curie–Weiss law in relaxors can be treated in two different ways, depending on the polarization mechanism which is supposed to be valid in the temperature range of the law (i.e., at $T > T_B$). The first way (see e.g. Refs. [52, 59, 60, 88, 102]) implies that, as in the case of normal perovskite (displacive) ferroelectrics, the field-induced polarization is due to the phonon contribution. The second approach suggests that the polarization mechanism in the temperature range of the Curie–Weiss law is qualitatively the same as at T_m (by analogy to order–disorder ferroelectrics and spin glasses) and involves the relaxation of individual dipoles (see e.g. Ref. [103] in which the Sherrington–Kirkpatrick model was used to analyze the susceptibility in PMN). To determine which way is adequate, the experimental investigation of high-frequency (IR) dispersion at $T > T_B$ is needed.

7. Relaxors at low temperatures: A glassy state or a ferroelectric phase

In the previous sections we have mainly considered the relaxors at comparatively high temperatures, i.e., in the PE and ER phases. We have also discussed some basic aspects of the low-temperature behavior of canonical relaxors, i.e., those in which the structure remains macroscopically cubic at all temperatures and the FE phase can be achieved only by poling (e.g. by applying an external electric field). In the canonical relaxors (e.g. PMN, PMT, PLZT with large x), a nonergodic (glassy) state appears at low temperatures. In many other materials (e.g. PSN, PST, PLZT with small x, and PMN-PT with large x) that exhibit relaxor properties and related structural features (e.g., PNRs) at high temperatures, a spontaneous (i.e., without poling) structural phase transition into the FE phase occurs. These two different paths of temperature evolution are shown schematically in Fig. 9. In this section, we describe the low-temperature behavior of relaxors in more detail.

7.1. *Glassy nonergodic relaxor phase*

7.1.1. *Structure*

As mentioned in Section 4, the soft mode in the prototypical relaxor PMN recovers below T_f so that the temperature dependence of the mode frequency shows the behavior characteristic of a normal ferroelectric phase (i.e., follows Eq. (1) with $A < 0$). A sharp peak of hypersonic dumping was observed at T_f [104]. However, no other evidence of the structural phase transition at T_f has been detected. The average cubic symmetry of PMN at low temperatures was confirmed in many structural studies by the absence of any splitting of X-ray and neutron Bragg reflections (which means that the shape of unit cell is cubic) as well as by the analysis of the intensities of the reflections (which are sensitive to the positions of atoms in the cell). For instance, in Refs. [27, 73], the unit cell was determined to be cubic by X-ray and neutron powder diffraction experiments performed down to 5 K, but due to the limited number of reflections analyzed, the positions of atoms and the thermal parameters could not be refined simultaneously. In Refs. [105, 106], the analysis of a large number of reflections obtained from X-ray diffraction of PMN single crystals confirmed the $Pm\bar{3}m$ space group in the range of 100–300 K. The cubic structure is also confirmed by the absence of birefringence [107–109][4]

Even though the structural phase transition in PMN is not definitely observed, some important structural changes not affecting the average symmetry are still found. With decreasing temperature, the average size of PNRs increases significantly around T_f (Fig. 5). The synchrotron X-ray scattering revealed the emergence of very weak and wide 1/2(hk0) superlattice reflections (α spots) in the vicinity of T_f [110]. These reflections were attributed to the antiferroelectric nanoregions (AFNR) formed by the correlated antiparallel (static or dynamic) displacements of Pb ions along the $\langle 110 \rangle$ directions with a magnitude of ~0.2 Å. Significant enhancement of the intensity of α spots below T_f is believed to arise from an increase in the total number of the AFNRs, whose average size of ~30 Å (determined from the width of reflections) remains constant down to the lowest measured temperature of 10 K [110]. AFNRs appear

[4]While most researchers agree that the average structure of PMN is cubic, the rhombohedral structure was also reported [109]. The possible reason for this discrepancy will be discussed in Section 7.2.

to be different from PNRs and CNRs, and unrelated to either of them [110].

7.1.2. *Broken ergodicity in relaxors*

Relaxors show nonergodic behavior resembling the behavior of spin (or dipole) glasses. In the high temperature (ergodic) phase of glasses, the spins (or dipoles, which can be considered as pseudospins) are weakly correlated and free to rotate, so that after any excitation (e.g. after application and removal of an external field) the system quickly comes back to the state with the lowest free energy, i.e., the state with zero total magnetization. It is always the same state regardless of the initial conditions (i.e., the strength and direction of the field in our example). At lower temperatures, due to the correlations between spins, the free energy surface has very many minima of almost the same depth separated by energy barriers of different heights (each minimum corresponds to a specific configuration of spins). In the glass phase, some of these barriers are so high that the time needed to overcome them is larger than any practically reasonable observation time. Therefore, during this time the system cannot reach all the configuration states, and consequently, the usual thermodynamic averaging and the time averaging give different results, i.e., the system is in a nonergodic state. On its way to a new state of minimum free energy required by the changed external conditions, the system should pass many barriers of different heights. This leads to a process with a wide distribution of relaxation times. The maximum relaxation time from this distribution may be so large (infinite for an infinite crystal) that the system cannot effectively reach the equilibrium. As a result, the state and the physical properties of the material depend on the history (i.e. the external field applied, the temperature variations, the observation time, etc.). In particular, substantial ageing effects should be observed, i.e., the change of properties with time spent by the sample at certain fixed external thermodynamic parameters (temperature, field, etc.).

All the main (mutually related) characteristics of nonergodic behavior typical of spin glasses, i.e., anomalously wide relaxation spectrum, ageing, dependence of the thermodynamic state on the thermal and field history of a sample, are observed in relaxors at temperatures around and below T_f. The slowing down of dipole dynamics was already discussed in terms of small-signal dielectric response in Section 6. Slow relaxation manifests itself also in other properties related to the local and/or macroscopic polarization. In particular, the relaxation of optical linear birefringence induced in PMN by a weak ($E < E_{cr}$) external electric field was studied [107] (E_{cr} is the critical field needed to induce the transition to the FE phase). The Kohlrausch–Williams–Watts-type and the Curie–von Schweidler-type relaxations were found in the temperature intervals of $180 < T < 210\,K$ and $210 < T < 230\,K$, respectively. The results were successfully described in terms of Chamberlin's approach to dynamic heterogeneity [111], implying a broad relaxation spectrum. Application of a strong ($E > E_{cr}$) d.c. field to the PMN crystal at $T < T_f$ results in a nearly logarithmic decay of dielectric permittivity [112] and a slow evolution of X-ray Bragg peaks reflecting the change of crystal symmetry [113].[5] The effects of ageing of susceptibility in the NR phase of PMN and in the typical spin glass phase were found to be very similar (and much stronger than in typical dipole glasses) [114]. The example of the dependence of properties on the thermal and electrical history of sample is shown in Fig. 10. The other examples are the splitting in the temperature dependences of the field-cooled and zero-field-cooled quasistatic dielectric constants in PMN and PLZT [95, 96] and the $P(E)$ hysteresis loops (see Section 7.1.3).

The ergodicity is clearly broken in relaxors at low temperatures, but this does not necessarily mean that relaxors are really dipole glass systems. Many other systems may also be nonergodic [115]. In particular, an ordinary FE phase is also nonergodic, but its potential landscape contains only a few minima (which are symmetric and correspond to the different directions of spontaneous polarization). As a result, the properties are easily distinguishable from those of nonergodic spin glass (or relaxor) phase. Wide relaxation spectrum

[5] After a long (several hours) waiting time the entire crystal suddenly transforms to the FE phase via a first-order transition.

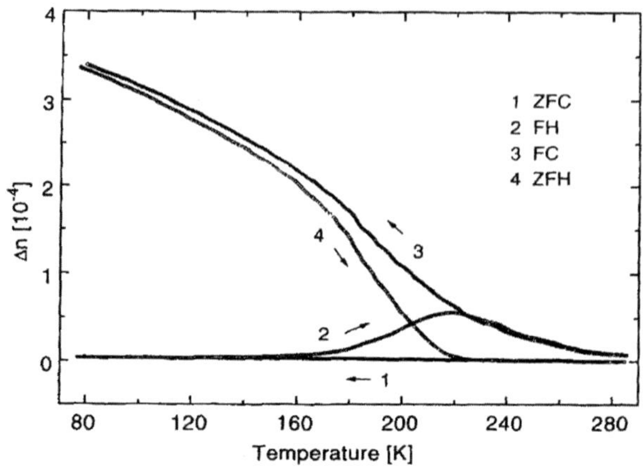

Figure 10. Linear birefringence measured subsequently as a function of temperature on zero-field cooling (ZFC), field heating with E $= 1.2\,\mathrm{kV/cm} < E_{cr}$ (FH), field cooling (FC), and zero-field heating (ZFH), illustrating the nonergodic behavior of PMN crystal. (after Kleemann *et al.* [107]).

and ageing phenomena are absent in the ideal FE crystal. But in the compositionally disordered perovskite crystal the situation is very different and different explanations for the nonergodic behavior are possible. For instance, the above-mentioned Kohlrausch–Williams–Watts-type relaxation of birefringence was explained by domain wall displacements, rather than by the reorientations of dipoles [107]. Furthermore, some peculiarities of the relaxor behavior have never been observed in spin and dipole glasses. In particular, the Barkhausen jumps during poling process (detected optically in PMN) are not compatible with the glassy reorientation of dipoles, which takes place on a microscopic length scale and hence should be continuous and monotonic [71]. Field-induced FE phase and FE hysteresis loops have not been observed in typical dipole glasses. Thus, the nature of the nonergodic phase in relaxors remains the subject of intensive discussion. In particular, the WKG model suggests that the low-temperature phase of canonical relaxors is a ferroelectric state, but broken into nanodomains by quenched random fields. We will discuss the origin of nonergodic phase in more detail in Section 7.3.

Note also that in terms of compositional disorder, relaxors are frozen in a metastable state, as discussed in Section 2. The degree of compositional disorder can depend on thermal prehistory. This is also an effect of nonergodicity.[6] However, at temperatures around T_B and below, the compositional disorder remains unchanged on the experimental time scales (i.e. frozen), and at the same time, the motion of dipoles (at $T > T_f$) is fast. Thus, when considering the subsystem of dipoles at $T > T_f$, one can believe that the crystal reaches the equilibrium[7] and the phase is effectively ergodic. On the other hand, if the sample has been annealed during experiment at high temperatures (~700 K or higher) the possible effects of nonergodicity related to the compositional disorder should be taken into account.

7.1.3. *Electric-field-induced ferroelectric phase in relaxors*

An important feature of the NR state is that it can be irreversibly transformed to the phase with the FE dipole order when poling by an electric field larger than the critical strength (in PMN the minimal E_{cr} is about 1.7 kV/cm at $T_C \simeq 210$ [108]). This feature points to the common nature of relaxor and normal ferroelectrics. The FE hysteresis loops, which are known to be the determinative characteristic of FE phase, are observed in relaxors with the values of remnant polarization and coercive field typical of normal ferroelectrics. Pyro- and piezoelectric effects are also observed after poling. X-ray diffraction [113, 117] and optical [108] investigations of poled PMN crystals showed that the field-induced phase has the rhombohedral 3m symmetry, i.e. the same symmetry as in several normal perovskite ferroelectrics. On the other hand, locally the structure is inhomogeneous, i.e. different from normal

[6]In contrast to spin (dipole) or FE state where the relaxation time is expected to become infinite in infinite crystal, the rate of compositional ordering does not depend on a crystal size. The corresponding relaxation time at nonzero temperature can be very large, but not infinite.

[7]According to Feynman [116] R.P. Feynman, Statistical mechanics, Reading, Benjamin, 1972, a system is in equilibrium if "all the fast things have happened and all the slow things have not".

ferroelectric structure. The traces of cubic phase were observed at low temperature by X-ray diffraction experiments in poled PMN crystal [113]. The NMR investigations of PMN crystal poled by a field almost two times as large as E_{cr}, revealed that only about 50% of Pb ions are displaced parallel to the [111] poling direction in a FE manner, while the other 50% exhibit spherical layer-type displacements characteristic of PE phase [32]. The size and number of AFNRs found in PMN in the unpoled state (see Section 7.1.1) remain unchanged in the FE phase [110].

Upon heating, the FE phase transforms to the cubic (ER) phase at a well-defined temperature, TC ($\approx$210 K in PMN). This first-order phase transition is accompanied by a step-like drop of spontaneous polarization (as determined from pyroelectric current), sudden vanishing of birefringence, and sharp peak of dielectric constant. A more detailed description of the field-induced transition and FE phase in relaxors is given in Ref. [11].

7.2. *Spontaneous relaxor-to-ferroelectric phase transition*

The transition from the ER to the FE phase typically takes place at temperature T_{C}, which is several degrees or several dozens of degrees lower than T_{m}, as schematically shown in Figs. 9(b) and (c). Usually the transition is observed in those relaxors where the $\varepsilon'(T)$ peak is not very diffuse (i.e., the diffuseness parameter δ is relatively small). X-ray and neutron diffraction experiments unambiguously indicate the change of symmetry at T_{C} from the high-temperature cubic to a low-temperature tetragonal or rhombohedral (in most cases) one [118, 119]. The symmetry breaking is also confirmed by the appearance of Brillouin scattering peaks which are forbidden in cubic phase [48], the appearance of optical birefringence [120], and the formation of FE domains which are clearly observed by optical polarizing microscopy [121, 122], electron microscopy [123], and scanning force microscopy [124–126]. The FE phase in relaxors exhibits typical FE properties, namely the large dielectric constant, the FE hysteresis loops [118, 121, 127], the pyro- [128] and piezoelectric (see below) effects, etc. At temperatures slightly above T_C, double hysteresis loops can be observed [118], as is typical of normal ferroelectrics.

The spontaneous relaxor-to-ferroelectric phase transition can be accompanied by significant anomalies in the temperature dependences of structural parameters [118, 119, 128], dielectric [4, 82, 118], optical [4, 120, 128, 129], thermal [118, 128], and other properties. The transition can be very sharp: e.g., in PFN crystals, the related jump of $\varepsilon(T)$ occurs in a temperature interval smaller than 0.1 K [130]. In many other cases, it is smeared for different reasons. The change of $\varepsilon(T)$ in ceramics is usually not as sharp as in single crystals, probably because of the inhomogeneity related to the existence of grains and boundaries [131]. Another type of inhomogeneity that can smear the transition in the relaxor-based solid solution crystals (e.g., PMN-PT, PZN-PT) is the macroscopic variation of x across the sample [120, 132]. Besides, some "intrinsic" causes for the smearing of phase transition also exist so that a clear boundary between the canonical relaxors (in which the anomalies of structure and properties are diffuse or absent) and the relaxors with a sharp FE transition cannot be defined. The intermediate behavior can appear in different ways. X-ray diffraction studies of PZN crystals revealed the coexistence of the mesoscopic domains of FE phase and the regions of cubic (relaxor) phase in a temperature range of about 70 K around the mean T_C [78, 133]. The concentration of the cubic phase gradually decreases on cooling, i.e. the transition is highly diffuse. The size of FE domains in PZN (40–200 nm) [78, 133] and in disordered PST (25–75 nm) [123] is smaller than in a normal FE phase, but larger than the size of typical PNRs. The average domain size (at room temperature) in the $(1-x)$PMN$-x$PT solid solutions was found, by scanning force microscopy, to gradually increase from $\sim$40 nm in the rhombohedral phase with $x = 0.1$, to $\sim 2\,\mu$m (which is comparable to the domain size in ordinary ferroelectrics) in the tetragonal phase with $x = 0.4$ [126].

Because of the similarity between the FE phase in ordinary ferroelectrics and the FE phase in relaxor ferroelectrics, the transition at T_C was initially called "spontaneous relaxor-to-normal ferroelectric transition," but later investigations showed that the low-temperature phase in relaxors is not exactly a "normal" ferroelectric phase, even in those relaxors where the FE transition is relatively sharp and FE domains are large. In particular, in the PMN-PT crystals with $x = 0.35$, the central peak in the Brillouin spectra, which is related to the relaxation of PNRs, was observed not only at $T > T_C$ (see Section 3.2), but also at $T < T_C$, indicating that the PNRs persist in the FE phase [48]. Furthermore, macroscopic (1–2 μm) areas of average cubic symmetry were found alongside with the areas of FE phase [48]. In PMN-PT crystals with $x = 0.20$, the piezoresponse force microscopy revealed a continuous distribution of the sizes of polar regions starting from ~5 nm (resolution limit). The complex structure of the micron-size FE domains with the PNRs of the opposite polarity embedded in them was observed [125]. A mixture of the rhombohedral domains and the domains of a different low-symmetry (presumably monoclinic) phase was observed by synchrotron X-ray diffraction in the FE phase of PZN crystal [78]. Unlike the plane walls in ordinary ferroelectrics, the domain walls in relaxor ferroelectrics are usually diffuse and irregular [122, 124]. The IR spectroscopy of PZN-PT and PMN-PT crystals (with $x = 0.08$ and $x = 0.29$, respectively) did not reveal any phonon softening that was expected for normal FE phase transitions at T_C and below [134]. Accordingly, the phonon contribution to the dielectric constant at these temperatures is small (~100), i.e., much less than the low-frequency value that reaches $\sim 5 \times 10^4$ at $T = T_C$ and $\sim 5 \times 10^3$ at $T \ll Tc$. Therefore, similarly to the case of canonical relaxors (and in contrast to ordinary displacive ferroelectrics), the dielectric response is determined by the relaxation polarization at all temperatures around and below the dielectric peak. It was concluded that the transition into the FE phase consists in a stepwise increase in the size of PNRs which transform into FE domains [134].

Another phenomenon, which is unusual for ordinary ferroelectrics, is the specific macroscopic phase inhomogeneities discovered recently in good-quality crystals of some relaxor ferroelectrics. Diffraction experiments performed in PZN with X-ray of different energies (and thus different penetration lengths) revealed that the outer layer (an estimated thickness of ~10–50 μm) undergoes a structural phase transition into the FE phase while the lattice inside the crystal maintains the cubic unit cell [135, 136].[8] Another interesting point is that at all temperatures the lattice parameter of the outer layers is slightly (∼0.2%) smaller than that of the bulk (inside) [136]. The same feature, i.e., a FE "skin" (observed by low-energy X-ray diffraction) and a cubic phase in the bulk (observed by high-energy X-ray or neutron diffraction), was also found in PZN-PT [137, 138] and PMN-PT [139] crystals with small x. It was suggested that this cubic phase (named X-phase) is similar to the average cubic phase in pure PMN [140], i.e., it is a NR phase. As a typical NR phase, the X-phase can be irreversibly transformed into the FE phase by poling [137].

Using spatially resolved neutron diffraction technique, it was found that even in PMN crystals, where the rhombohedral phase has not been detected, the near-surface layer (of ∼100 μm thick) has the lattice constant noticeably smaller than the bulk structure [141]. It was supposed [140] that a very thin rhombohedral skin possibly exists in PMN also, with a thickness much smaller than the penetration length of X-rays, so that the skin could not be detected in the usual diffraction experiments. Note in this connection that, as reported in Ref. [109], the Rietveld refinement of neutron diffraction data collected on PMN powder revealed a rhombohedral macroscopic symmetry, namely $R3m$ at 300 K (i.e., above T_f) and $R3c$ at 10 K. Second harmonic generation signal was also detected, indicating a non-centric symmetry. In these experiments, very fine powder (of 4–5 μm particle size) synthesized by a special route was used, so the whole material can be considered a near-surface region. This is the possible reason why the rhombohedral phase was found instead of the cubic phase usually observed by other authors in crystals and large-grain powders of PMN.

The behavior unusual for relaxor as well as for ordinary ferroelectrics was recently found in PMN-PT crystals with large x (∼0.5) [132]. The dielectric properties typical of the ER phase are

[8]A tetragonal unit cell was initially reported [135], but more elaborate investigations [136] later showed that it is indeed cubic.

observed at $T > T_m$, namely the deviation from the Curie–Weiss law at $T < T_B$ (where $T_B \gg T_m$) according to Eq. (6) and the behavior permittivity, indicating the UR dispersion. Nevertheless, the CR dispersion is absent, and consequently, T_m coincides with the temperature of the FE phase transition, T_C. This is shown in Fig. 9(d). Both the deviation from the Curie–Weiss law and the UR dispersion are believed to result from the existence of PNRs, so the high-temperature phase is the ER state. PMN-PT with high x seems to be the first examples of relaxor without the characteristic CR dispersion. To confirm this opinion, more direct identification of PNRs (e.g., by neutron scattering) is desirable.

While an external electric field transforms a NR state to a FE one, the hydrostatic pressure is able to induce the reverse transformation. In the crystals exhibiting the relaxor-to-FE phase transition, the FE phase does not appear if the sample is cooled under a high enough pressure and the behavior typical of the canonical relaxors is observed. The pressure-induced crossover from a ferroelectric to a relaxor state was discussed in detail in the recent reviews by Samara [142, 143].

Excellent piezoelectric properties were found near the morphotropic phase boundary in PMN-PT, PZN-PT and some other solid solutions of complex perovskite relaxors with $PbTiO_3$. The transition from FE to ER phase is observed in these crystals at temperatures much higher than room temperature. Thus, they are considered and investigated as promising materials for practical applications. This field of research, currently very active, has been reviewed in a number of recent papers [144–148].

7.3. *Theoretical description of nonergodic phase in relaxors*

Early works on relaxors (e.g., the composition fluctuations model by Smolenskii and Isupov [8, 97] and the super-paraelectric model by Cross [10]) considered the PNRs to be relatively independent noninteracting entities. It was later understood that the specific nonergodic behavior of relaxors at low temperatures cannot be explained without taking into account the interactions among PNRs and/or quenched random local fields existing in the compositionally disordered structure. The interactions among PNRs may lead to anomalous slowing-down of their dynamics (nonergodicity effects) or, when becoming frustrated, even to the formation of the glass state in which the dipole moments of individual PNRs are randomly fixed in different directions. Note that these interactions are of dipole–dipole nature and can be considered as dynamic local fields. Additionally PNRs can be influenced (or probably even fixed) by quenched local random fields stemming from the compositional disorder or other types of lattice defects.

In Section 5 we have already discussed the modern theories explaining the formation of PNRs. Some of these theories can also explain the transition from the ergodic to nonergodic relaxor state. In particular, in the GF model, the PNRs naturally appear in the temperature interval between the PE and the low-temperature dipole glass or mixed ferroglass phase. In the WKG model, the formation of PNRs as well as the transition to the NR state is ascribed to the quenched random fields exclusively.

However, the mechanisms leading to the formation of PNRs at high temperatures are not necessarily responsible for their freezing and for the development of the low-temperature nonergodic state. The formation and freezing of PNRs are possibly two distinct phenomena requiring different approaches. The "semi-microscopic" models [98, 149, 150] of glass state in relaxors describe only the latter phenomenon, while PNRs are believed to be already-existing objects and the mechanisms of their formation are not examined.

In the spherical random-bond-random-field (SRBRF) model proposed by Pirc and Blinc [149, 150], the Hamiltonian is formally written with Eqs. (2) and (3), but the meanings of the parameters are different from those discussed in Section 5. Pseudospins $\vec{S}_i$ proportional to the dipole moments of PNRs are introduced so that the relation

$$\Sigma_i(\vec{S}_i)^2 = {}^3N \tag{9}$$

is satisfied (N is the number of pseudospins in the crystal). It is assumed that each component of $\vec{S}_i$ can fluctuate continuously and take any value,[9] i.e.,

$$-\infty < S_{i\mu} < +\infty. \tag{10}$$

J_{ij} in Eq. (3) are the random interactions (bonds) between PNRs which are assumed to be infinitely ranged. The second term in the Hamiltonian in Eq. (2) describes the interaction of pseudospins with quenched random electric fields $\vec{h}_i$

$$H_2 = -\sum_i \vec{h}_i \cdot \vec{S}_i.$$

Both random bonds J_{ij} and random fields $\vec{h}_i$ obey the (uncorrelated) Gaussian probability distributions with an rms variance of $J/\sqrt{N}$ and Δ, respectively. The mean value of the distributions equals J_0/N (for random bonds) and zero (for random fields).

In the absence of random fields ($\Delta = 0$), if $J < J_0$, the theory predicts the transition from the PE phase (in the model this phase is equivalent to the ER phase) into an inhomogeneous FE phase with a nonzero spontaneous polarization; if $J > J_0$, the system transforms, at a well-defined temperature $T = J$, from the PE to a spherical glass phase without long range order, and the glass order parameter (which is equivalent in this model to the well-known Edwards–Anderson order parameter, q_{EA}) decreases linearly from 1 at $T = 0$ to zero at $T = J$. The presence of random fields ($\Delta \neq 0$) destroys the phase transition so that q_{EA} remains nonzero at $T = J$, and approaches zero when the temperature further increases. Figure 11a shows the temperature dependence of q_{EA} determined experimentally from the NMR data of PMN (q_{EA} is shown to be proportional to the second moment, M_2, of the frequency distribution corresponding to the narrow ^{93}Nb NMR line) [150]. The solid line represents the fit with the parameters $J/k = 20\,\mathrm{K}$ and $\Delta/J^2 = 0.002$, confirming the applicability of the model. The local polarization distribution function $W(\vec{p})$ (where $\vec{p} = \langle\vec{S}\rangle$) predicted by the model and determined experimentally from the NMR lineshape also appears

[9]Models in which the order parameter satisfies conditions (9), (10) are called "spherical" models. Due to these conditions the model is exactly solvable by the replica method.

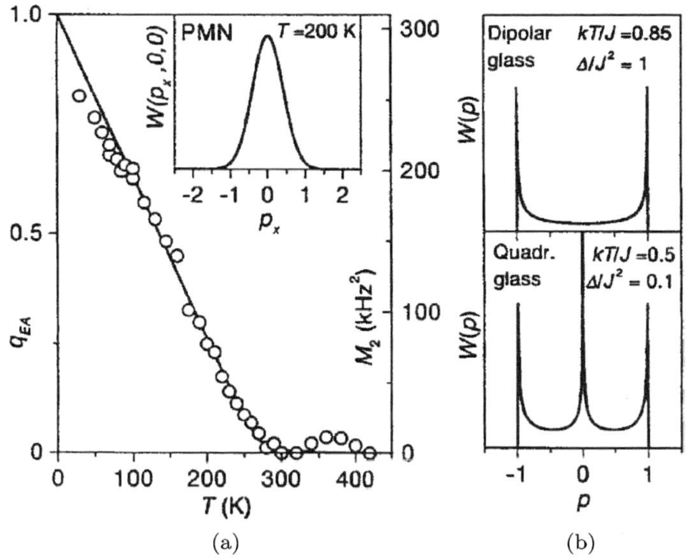

Figure 11. (a) Temperature dependence of the Edwards–Anderson glass order parameter q_{EA} in PMN. The solid line is the fit to the "spherical random bond random field" (SRBRF) model. The inset shows the local polarization distribution function $W(p)$ along the p_x axis according to the SRBRF model. (b) Examples of the $W(\vec{p})$ functions for dipolar and quadrupolar glasses are shown for comparison (after Blinc *et al.* [150])

to be the same as shown in the inset of Fig. 11a [150]. The $W(\vec{p})$ shape observed in dipolar and quadrupolar glasses look very different, as shown in Fig. 11b. These results suggest that the NR phase in PMN cannot be described as a dipolar or quadrupolar glass. It is a new type of glass which can be called "spherical cluster glass" [151]. The SRBRF model is also able to explain the dielectric non-linearity in PMN. The dynamic version of SRBRF model describing the dispersion of linear and nonlinear dielectric susceptibility has been developed [152]. In the coupled SRBRF-phonon model [99], the coupling of PNRs with soft TO phonons leads to the modification of interactions among PNRs. The effect of pressure on the relative stability of different phases in relaxors are explained.

Vugmeister and Rabitz [98, 153] considered in their model the hopping of PNRs in multi-well potentials. The PNRs exist in a highly polarizable PE host lattice with a displacive dielectric response. The theory takes into account the broad distribution of the potential barriers controlling PNR dynamics and the effect of interactions between PNRs mediated by highly polarizable host. These two aspects are described in terms of the local field distribution function. In this model, the dipole glass freezing is believed to be accompanied by the critical FE slowing-down. It is shown that the true glass state in which all dipoles (PNRs) are frozen is not achieved in relaxors: the degree of the local freezing is rather small even at low temperatures.[10] The role of the critical slowing-down is shown to be significant in the dynamics of the system due to the closeness of FE instability. In other words, relaxors can be considered incipient ferroelectrics. This explains their very large dielectric constant. In the framework of this model, the shape of the frequency-dependent permittivity as a function of temperature in typical relaxors is explained qualitatively. The glasslike freezing of the dynamics of PNRs is characterized by the non-equilibrium spin-glass order parameter, the temperature behavior of which is consistent with the NMR experiments (shown in Fig. 11). The kinetics of the electric field induced transition from the NR to FE phase was also successfully reproduced [154] (while the glass models experience difficulties in explaining this transition).

The behavior of PNRs can be influenced by the electronic subsystem. In particular, the thermo-localization of charge carriers on the defects in the temperature range of phase transition can change the relaxation dynamics. The direction of spontaneous polarization of PNRs can be pinned by localized charge carriers, preventing the alignment of PNRs in the external electric field. The related phenomena are studied in Refs. [155, 156]

As mentioned above, the models so far discussed in this section consider PNRs (pseudospins) to be already existing entities. In order to describe the process of their formation and development (which begins from $T_B \gg T_f$), other models are needed. Recently, it has

[10]This is in agreement with the experimental finding that in relaxors only the longest relaxation time diverges at T_f, while the bulk of the relaxation spectrum remains active at low temperatures (see Section 6).

been proposed that quenched random fields give rise to the formation of PNRs in the PE phase, as prescribed by the WKG model, and then, upon further cooling, the crystal undergoes a transition into the spherical cluster glass state due to random interactions between PNRs [151]. Alternatively, some other models can be used to describe the formation of SRBRF pseudospins, in particular, the soft nanoregions model [77] (which justifies the fulfilment of condition (10)) together with the kinetic model [70] (as discussed in Section 5).

Let us now discuss the mechanisms of the spontaneous relaxor-to-ferroelectric phase transition. There are two ways to explain the formation of FE phase at T_C from the system of disordered PNRs in relaxors. The first one suggests that the dipole–dipole interactions between PNRs (or individual ions) lead to their FE-type arrangement (as, e.g., in the SRBRF model discussed above). The second mechanism arises from the kinetic model of phase transitions in compositionally disordered crystals [70] (see Section 5) and suggests the thermally activated growth of PNRs at T_C. It is not easy to discriminate these two mechanisms from each other. In fact, it is also possible that both mechanisms contribute to the process of the formation of FE phase.

8. Conclusions

In this chapter, we have analyzed the peculiar behavior of relaxor ferroelectrics that occurs in compositionally disordered perovskites. The quenched compositional disorder in these compounds gives rise to another type of disorder, i.e. the glassy nonergodic state that can be observed at low temperatures, instead of a FE or AFE ordering that exists in many simple perovskites. The research in this field has undergone such a tremendous growth that it was not possible to review all the important works in this short paper. Some subjects were discussed only briefly just to give the examples characterizing the peculiarities of the behavior. Some other important topics have been left out, in particular, the materials technology of crystals, ceramics and thin films and applications of relaxor ferroelectrics.

Despite the remarkable progress achieved in the recent years, fundamental physics of the relaxors remains a fascinating puzzle. Some key questions, such as what the origin of relaxor behavior is, still have no definite answers. Several theoretical models have been proposed; some of them contradict each other. Further experiments need to be performed in order to prove or reject these models, while new and more satisfactory theories are yet to be worked out. With their complex structures and intriguing properties, relaxors represent truly a frontier of research in ferroelectrics and related materials, offering great opportunities both for fundamental research and for technological applications.

Acknowledgments

This work was supported by the Natural Science and Engineering Research Council of Canada (NSERC) and the U.S. Office of Naval Research (Grant# N00014-99-1-0738).

References

[1] Smolenskii, G. A., Isupv, V. A. and Agranovskaya, A. I. (1961). The ferroelectric properties of strontium-bismuth titanate, *Sov. Phys., Solid State*, 2, pp. 2584–2594.
[2] Chu, F., Setter, N. and Tagantsev, A. K. (1993). The spontaneous relaxor-ferroelectric transition of $Pb(Sc_{0.5}Ta_{0.5})O_3$, *J. Appl. Phys.*, 74, pp. 5129–5134.
[3] Dai, X. H., Xu, Z. and Viehland, D. (1994). The spontaneous relaxor to normal ferroelectric transformation in La-modified lead-zirconate-titanate, *Phil. Mag. B*, 70, pp. 33–48.
[4] Krumins, A., Shiosaki, T. and Koizumi, S. (1994). Spontaneous transition between relaxor and ferroelectric states is lanthanum-modified lead-zirconate-titanate-(6-7)/65/35, *Jpn. J. Appl. Phys.*, 33, pp. 4940–4945.
[5] Sciau, P., Calvarin, G. and Ravez, J. (1999). X-ray diffraction study of $BaTi_{0.65}Zr_{0.35}O_3$ and $Ba_{0.92}Ca_{0.08}Ti_{0.75}Zr_{0.25}O_3$ compositions: Influence of electric field, *Solid State Commun.*, 113, pp. 77–82.
[6] Simon, A., Ravez, J. and Maglione, M. (2004). The crossover from a ferroelectric to a relaxor state in lead-free solid solutions, *J. Phys.: Condens. Matter*, 16, pp. 963–970.
[7] Yasuda, N., Ohwa, H. and Asano, S. (1996). Dielectric properties and phase transitions of $Ba(Ti_{1-x}Sn_x)O_3$ solid solution, *Jpn. J. Appl. Phys.*, 35, pp. 5099–5103.

[8] Smolenskii, G. A. (1970). Physical phenomena in ferroelectrics with diffused phase transition, *J. Phys. Soc. Jpn.*, 28, pp. 26–37.

[9] Bokov, A. A. (1992). Recent advances in diffuse ferroelectric phase-transitions, *Ferroelectrics*, 131, pp. 49–55.

[10] Cross, L. E. (1994). Relaxor ferroelectrics: An overview, *Ferroelectrics*, 151, pp. 305–320.

[11] Ye, Z. G. (1998). Relaxor ferroelectric complex perovskites: Structure, properties and phase transitions, *Key Eng. Mater.* 155–156, pp. 81–122.

[12] Stenger, C. G. F. and Burggraaf, A. J. (1980). Order-disorder reaction in the ferroelectric perovskites $Pb(Sc_{1/2}Nb_{1/2})O_3$ and $Pb(Sc_{1/2}Ta_{1/2})O_3$ .2. Relation between ordering and properties, *Phys. Stat. Sol.* (a), 61, pp. 653–664.

[13] Boulesteix, C., Varnier, F., Llebaria, A. and Husson, E. (1994). Numerical determination of the local ordering of $PbMg_{1/3}Nb_{2/3}O_3$ (PMN) from high-resolution electron-microscopy images, *J. Solid State Chem.*, 108, pp. 141–147.

[14] Yoshida, M., Mori, S., Yamamoto, N., Uesu, Y. and Kiat, J. M. (1998). TEM observation of polar domains in relaxor ferroelectric $Pb(Mg_{1/3}Nb_{2/3})O_3$, *Ferroelectrics*, 217, pp. 327–333.

[15] Tkachuk, A. and Chen, H. (2001). Anomalous x-ray scattering study of chemical and polar nanodomains in $Pb(Mg_{1/3}Nb_{2/3})O_3$ single crystal, *Ferroelectrics*, 253, pp. 557–565.

[16] Davies, P. K. and Akbas, M. A. (2000). Chemical order in PMN-related relaxors: structure, stability, modification, and impact on properties, *J. Phys. Chem. Solids*, 61, pp. 159–166.

[17] Jin, H. Z., Zhu, J., Miao, S., Zhang, X. W. and Cheng, Z. Y. (2001). Ordered domains and polar clusters in lead magnesium niobate $Pb(Mg_{1/3}Nb_{2/3})O_3$, *J. Appl. Phys.*, 89, pp. 5048–5052.

[18] Bokov, A. A., Raevskii, I. P. and Smotrakov, V. G. (1984). Composition, ferroelectric and antiferroelectric ordering in Pb_2InNbO_6 crystals, *Sov. Phys. Solid State*, 26, pp. 2824–2828.

[19] Bokov, A. A., Leshchenko, M. A., Malitskaya, M. A. and Raevski, I. P. (1999). Dielectric spectra and Vogel-Fulcher scaling in $Pb(In_{0.5}Nb_{0.5})O_3$ relaxer ferroelectric, *J. Phys.: Condens. Matter*, 11, pp. 4899–4911.

[20] Bokov, A. A. (1996). Compositional order-disorder in mixed ferroelectrics, *Ferroelectrics*, 183, pp. 65–73.

[21] Davies, P. K. (1999). Cation ordering in complex oxides, *Curr. Opin. Sol. Stat. Mater. Sci.*, 4, pp. 467–471.

[22] Isupov, V. A. (2003). Ferroelectric and antiferroelectric perovskites $PbB'_{0.5}B''_{0.5}O_3$, *Ferroelectrics*, 289, pp. 131–195.

[23] Comes, R., Lambert, M. and Guinier, A. (1970). Linear disorder in crystals (silicon, quartz and ferroelectric perovskites), *Acta Crystallogr.: B*, 26, p. 244.

[24] Stern, E. A. (2004). Character of order-disorder and displacive components in barium titanate, *Phys. Rev. Lett.*, 93, p. 037601.

[25] Cohen, R. E. (1992). Origin of ferroelectricity in perovskite oxides, *Nature*, 358, pp. 136–138.
[26] Bonneau, P., Garnier, P., Husson, E. and Morell, A. (1989). Structural study of pmn ceramics by X-ray-diffraction between 297 and 1023 K, *Mater. Res. Bull.*, 24, pp. 201–206.
[27] Bonneau, P., Garnier, P., Calvarin, G., Husson, E., Gavarri, J. R., Hewat, A. W. and Morell, A. (1991). X-ray and neutron-diffraction studies of the diffuse phase-transition in $PbMg_{1/3}Nb_{2/3}O_3$ ceramics, *J. Solid State Chem.*, 91, pp. 350–361.
[28] Vakhrushev, S., Zhukov, S., Fetisov, G. and Chernyshov, V. (1994). The high-temperature structure of lead magnoniobate, *J. Phys. Condens. Matter*, 6, pp. 4021–4027.
[29] Fujishiro, K., Iwase, T., Uesu, Y., Yamada, Y., Dkhil, B., Kiat, J. M., Mori, S. and Yamamoto, N. (2000). Optical and structural studies of long-range order development in relaxor ferroelectrics, *J. Phys. Soc. Jpn.*, 69, pp. 2331–2338.
[30] Dkhil, B., Kiat, J. M., Calvarin, G., Baldinozzi, G., Vakhrushev, S. B. and Suard, E. (2001). Local and long range polar order in the relaxor-ferroelectric compounds $PbMg_{1/3}Nb_{2/3}O_3$ and $PbMg_{0.3}Nb_{0.6}Ti_{0.1}O_3$, *Phys. Rev. B*, 65, p. 024104.
[31] Zhukov, S. G., Chernyshov, V. V. and Vakhrushev, S. B. (1999). Structural peculiarities of $(PbMg_{1/3}Nb_{2/3}O_3)_{(1-x)}$-$(PbTiO_3)_{(x)}$ solid solutions, *Ferroelectrics*, 235, pp. 143–149.
[32] Blinc, R., Laguta, V. and Zalar, B. (2003). Field cooled and zero field cooled Pb-207 NMR and the local structure of relaxor $PbMg_{1/3}Nb_{2/3}O_3$, *Phys. Rev. Lett.*, 91, p. 247601.
[33] Egami, T. (2002). Atomistic mechanism of relaxor ferroelectricity, *Ferroelectrics*, 267, pp. 101–111.
[34] Aoyagi, S., Kuroiwa, Y., Sawada, A., Tanaka, H., Harada, I., Nishibori, E., Takata, M. and Sakata, M. (2002). Direct observation of covalency between O and disordered Pb in cubic $PbZrO_3$, *J. Phys. Soc. Jpn.*, 71, pp. 2353–2356.
[35] Uesu, Y., Yamada, Y., Fujishiro, K., Tazawa, H., Enokido, S., Kiat, J. M. and Dkhil, B. (1998). Structural and optical studies of development of the long-range order in ferroelectric relaxor $Pb(Zn_{1/3}Nb_{2/3})O_3$/9%$PbTiO_3$, *Ferroelectrics*, 217, pp. 319–325.
[36] Shuvaeva, V. A., Pirog, I., Azuma, Y., Yagi, K., Sakaue, K., Terauchi, H., Raevskii, I. P., Zhuchkov, K. and Antipin, M. Y. (2003). The local structure of mixed-ion perovskites, *J. Phys.: Condens. Matter*, 15, pp. 2413–2421.
[37] Burns, G. and Dacol, F. H. (1983). Glassy polarization behavior in ferroelectric compounds $Pb(Mg_{1/3}Nb_{2/3})O_3$ and $Pb(Zn_{1/3}Nb_{2/3})O_3$, *Sol. Stat. Commun.*, 48, pp. 853–856.
[38] Vakhrushev, S. B., Kvyatkovsky, B. E., Naberezhnov, A. A., Okuneva, N. M. and Toperverg, B. P. (1989). Glassy phenomena in disordered perovskite-like crystals, *Ferroelectrics*, 90, pp. 173–176.

[39] Naberezhnov, A., Vakhrushev, S., Dorner, B., Strauch, D. and Moudden, H. (1999). Inelastic neutron scattering study of the relaxor ferroelectric $PbMg_{1/3}Nb_{2/3}O_3$ at high temperatures, *Eur. Phys. J.*, 11, pp. 13–20.
[40] Hirota, K., Ye, Z. G., Wakimoto, S., Gehring, P. M. and Shirane, G. (2002). Neutron diffuse scattering from polar nanoregions in the relaxor $Pb(Mg_{1/3}Nb_{2/3})O_3$, *Phys. Rev. B*, 65, pp. 104105.
[41] Vakhrushev, S., Nabereznov, A., Sinha, S. K., Feng, Y. P. and Egami, T. (1996). Synchrotron X-ray scattering study of lead magnoniobate relaxor ferroelectric crystals, *J. Phys. Chem. Solids*, 57, pp. 1517–1523.
[42] Xu, G. Y., Shirane, G., Copley, J. R. D. and Gehring, P. M. (2004). Neutron elastic diffuse scattering study of $Pb(Mg_{1/3}Nb_{2/3})O_3$, *Phys. Rev. B*, 69, pp. 064112.
[43] La-Orauttapong, D., Toulouse, J., Robertson, J. L. and Ye, Z. G. (2001). Diffuse neutron scattering study of a disordered complex perovskite $Pb(Zn_{1/3}Nb_{2/3})O_3$ crystal, *Phys. Rev. B*, 64, pp. 212101.
[44] Stock, C., Birgeneau, R. J., Wakimoto, S., Gardner, J. S., Chen, W., Ye, Z. G. and Shirane, G. (2004). Universal static and dynamic properties of the structural transition in $Pb(Zn_{1/3}Nb_{2/3})O_3$, *Phys. Rev. B*, 69, p. 094104.
[45] Vakhrushev, S. B., Naberezhnov, A. A., Okuneva, N. M. and Savenko, B. N. (1995). Determination of polarization vectors in lead magnoniobate, *Phys. Solid State*, 37, pp. 1993–1997.
[46] Takesue, N., Fujii, Y. and You, H. (2001). X-ray diffuse scattering study on ionic-pair displacement correlations in relaxor lead magnesium niobate, *Phys. Rev. B*, 64, pp. 184112.
[47] Moriya, Y., Kawaji, H., Tojo, T. and Atake, T. (2003). Specific-heat anomaly caused by ferroelectric nanoregions in $Pb(Mg_{1/3}Nb_{2/3})O_3$ and $Pb(Mg_{1/3}Ta_{2/3})O_3$ relaxors, *Phys. Rev. Lett.*, 90, pp. 205901.
[48] Jiang, F. M. and Kojima, S. (2000). Microheterogeneity and relaxation in $0.65Pb(Mg_{1/3}Nb_{2/3})O_3$-$0.35PbTiO_3$ relaxor single crystals, *Appl. Phys. Lett.*, 77, pp. 1271–1273.
[49] Lemanov, V. V., Yushin, N. K., Smirnova, E. P., Sotnikov, A. V., Tarakanov, E. A. and Maksimov, A. Y. (1992). Giant electrostriction of ferroelectrics with diffuse phase-transition — physics and applications, *Ferroelectrics*, 134, pp. 139–144.
[50] Uchino, K. (1994). Relaxor ferroelectric devices, *Ferroelectrics*, 151, pp. 321–330.
[51] Zhao, J., Glazounov, A. E., Zhang, Q. M. and Toby, B. (1998). Neutron diffraction study of electrostrictive coefficients of prototype cubic phase of relaxer ferroelectric $PhMg_{1/3}Nb_{2/3}O_3$, *Appl. Phys. Lett.*, 72, pp. 1048–1050.
[52] Poplavko, Y. M., Bovtun, V. P., Krainik, N. N. and Smolenskii, G. A. (1985). Microwave dielectric dispersion of lead magnoniobate, *Sov. Phys.: Sol. Stat.*, 27, pp. 1903–1904.
[53] Dul'kin, E., Raevskii, I. P. and Emel'yanov, S. M. (2003). Acoustic emission and thermal expansion of $Pb(Mg_{1/3}Nb_{2/3})O_3$ and $Pb(Mg_{1/3}Nb_{2/3})O_3$-$PbTiO_3$ crystals, *Phys. Sol. Stat.*, 45, pp. 158–162.

[54] Gehring, P. M., Wakimoto, S., Ye, Z. G. and Shirane, G. (2001). Soft mode dynamics above and below the burns temperature in the relaxor $Pb(Mg_{1/3}Nb_{2/3})O_3$, *Phys. Rev. Lett.*, 87, pp. 277601.

[55] Wakimoto, S., Stock, C., Birgeneau, R. J., Ye, Z. G., Chen, W., Buyers, W. J. L., Gehring, P. M. and Shirane, G. (2002). Ferroelectric ordering in the relaxor $Pb(Mg_{1/3}Nb_{2/3})O_3$ as evidenced by low-temperature phonon anomalies, *Phys. Rev. B*, 65, pp. 172105.

[56] Gehring, P. M., Park, S. E. and Shirane, G. (2001). Dynamical effects of the nanometer-sized polarized domains in $Pb(Zn_{1/3}Nb_{2/3})O_3$, *Phys. Rev. B*, 63, pp. 224109.

[57] Gehring, P. M., Park, S. E. and Shirane, G. (2000). Soft phonon anomalies in the relaxer ferroelectric $Pb(Zn_{1/3}Nb_{2/3})_{0.92}Ti_{0.08}O_3$, *Phys. Rev. Lett.*, 84, pp. 5216–5219.

[58] Hlinka, J., Kamba, S., Petzelt, J., Kulda, J., Randall, C. A. and Zhang, S. J. (2003). Origin of the "Waterfall" effect in phonon dispersion of relaxor perovskites, *Phys. Rev. Lett.*, 91, pp. 107602.

[59] Vakhrushev, S. B. and Shapiro, S. M. (2002). Direct evidence of soft mode behavior near the Burns temperature in the $PbMg_{1/3}Nb_{2/3}O_3$ relaxor ferroelectric, *Phys. Rev. B*, 66, pp. 214101.

[60] Wakimoto, S., Stock, C., Ye, Z. G., Chen, W., Gehring, P. M. and Shirane, G. (2002). Mode coupling and polar nanoregions in the relaxor ferroelectric $Pb(Mg_{1/3}Nb_{2/3})O_3$, *Phys. Rev. B*, 66, pp. 224102.

[61] Bovtun, V., Kamba, S., Pashkin, A., Savinov, M., Samoukhina, P., Petzelt, J., Bykov, I. P. and Glinchuk, M. D. (2004). Central-peak components and polar soft mode in relaxor $PbMg_{1/3}Nb_{2/3}O_3$ crystals, *Ferroelectrics*, 298, pp. 23–30.

[62] Kamba, S., Bovtun, V., Petzelt, J., Rychetsky, I., Mizaras, R., Brilingas, A., Banys, J., Grigas, J. and Kosec, M. (2000). Dielectric dispersion of the relaxer PLZT ceramics in the frequency range 20 Hz–100 THz, *J. Phys.: Condens. Matter*, 12, pp. 497–519.

[63] Siny, I. G., Lushnikov, S. G. and Katiyar, R. S. (1999). Light scattering spectroscopy of relaxer ferroelectrics, *Ferroelectrics*, 231, pp. 703–708.

[64] Siny, I. G., Lushnikov, S. G., Katiyar, R. S. and Schmidt, V. H. (1999). $PbMg_{1/3}Nb_{2/3}O_3$ as a model object for light scattering experiments, *Ferroelectrics*, 226, pp. 191–215.

[65] Svitelskiy, O., Toulouse, J., Yong, G. and Ye, Z. G. (2003). Polarized Raman study of the phonon dynamics in $Pb(Mg_{1/3}Nb_{2/3})O_3$ crystal, *Phys. Rev. B*, 68, pp. 104107.

[66] Prosandeev, S. A., Cockayne, E., Burton, B. P., Kamba, S., Petzelt, J., Yuzyuk, Y., Katiyar, R. S. and Vakhrushev, S. B. (2004). Lattice dynamics in $PbMg_{1/3}Nb_{2/3}O_3$, *Phys. Rev. B*, 70, pp. 134110.

[67] Glinchuk, M. D. and Farhi, R. (1996). A random field theory based model for ferroelectric relaxors, *J. Phys.: Condens. Matter*, 8, pp. 6985–6996.

[68] Bokov, A. A. (1997). Influence of disorder in crystal structure on ferroelectric phase transitions, *JETP*, 84, pp. 994–1002.

[69] Timonin, P. N. (1997). Griffiths' phase in dilute ferroelectrics, *Ferroelectrics*, 199, pp. 69–81.

[70] Bokov, A. A. (1997). Ferroelectric domains and polar clusters in disordered structures, *Ferroelectrics*, 190, pp. 197–202.

[71] Westphal, V., Kleemann, W. and Glinchuk, M. D. (1992). Diffuse phase-transitions and random-field-induced domain states of the relaxor ferroelectric $PbMg_{1/3}Nb_{2/3}O_3$, *Phys. Rev. Lett.*, 68, pp. 847–850.

[72] Ishchuk, V. M. (2001). Was it necessary to introduce the notion "relaxor ferroelectrics"? The problem of phase transitions in $(Pb,Li_{1/2}La_{1/2})(Zr,Ti)O_3$, $(Pb,La)(Zr,Ti)O_3$, $Pb(Mg_{1/3}Nb_{2/3})O_3$, $Pb(In_{1/2}Nb_{1/2})O_3$ and related materials. 1. Model conceptions, *Ferroelectrics*, 255, pp. 73–109.

[73] DE Mathan, N., Husson, E., Calvarn, G., Gavarri, J. R., Hewat, A. W. and Morell, A. (1991). A structural model for the relaxor $PbMg_{1/3}Nb_{2/3}O_3$ at 5 K, *J. Phys.: Condens. Matter*, 3, pp. 8159–8171.

[74] Kleemann, W. (1993). Random-field induced antiferromagnetic, ferroelectric and structural domain states, *Int. J. Mod Phys B*, 7, pp. 2469–2507.

[75] Imry, Y. and Ma, S. (1975). Random-field instability of ordered state of continuous symmetry, *Phys. Rev. Lett.*, 35, pp. 1399–1401.

[76] Glinchuk, M. D. (2004). Relaxor ferroelectrics: from cross superparaelectric model to random field theory, *British Ceram. Trans.*, 103, pp. 76–82.

[77] Bokov, A. A. and Ye, Z. G. (2002). Universal relaxor polarization in $Pb(Mg_{1/3}Nb_{2/3})O_3$ and related materials, *Phys. Rev. B*, 66, pp. 064103.

[78] Bing, Y. H., Bokov, A. A., Ye, Z. G., Noheda, B. and Shirane, G. (2005). Structural phase transition and dielectric relaxation in $Pb(Zn_{1/3}Nb_{2/3})O_3$ single crystals, *J. Phys.: Condens. Matter*, 17, pp. 2493–2507.

[79] Jonscher, A. K., (1996). *Universal Relaxation Law:A Sequel to Dielectric Relaxation in Solids*, (Chelsea Dielectrics Press, London).

[80] Bokov, A. A. and Ye, Z. G. (2002). Low-frequency dielectric spectroscopy of the relaxor ferroelectric $Pb(Mg_{1/3}Nb_{2/3})O_3$-$PbTiO_3$, *Phys. Rev. B*, 65, pp. 144112.

[81] Bokov, A. A. and Ye, Z. G. (2000). Freezing of dipole dynamics in relaxor ferroelectric $Pb(Mg_{1/3}Nb_{2/3})O_3$-$PbTiO_3$ as evidenced by dielectric spectroscopy, *J. Phys.: Condens. Matter*, 12, pp. L541–L548.

[82] Bokov, A. A. and Ye, Z. G. (2000). Dielectric dispersion and critical behavior in relaxor ferroelectric $Pb(Mg_{1/3}Nb_{2/3})O_3$-$PbTiO_3$, *Appl. Phys. Lett.*, 77, pp. 1888–1890.

[83] Bing, Y. H., Bokov, A. A. and Ye, Z. G. (2011). Diffuse and sharp ferroelectric phase transitions in relaxors, *Curr. Appl. Phys.*, 11, pp. S14–S21.

[84] Bokov, A. A., Maglione, M., Simon, A. and Ye. Z.-G. (2006). Dielectric behavior of $Ba(Ti_{1-x}Zr_x)O_3$ solid solution, *Ferroelectrics*, 337, pp. 1341–1350.

[85] Bokov, A. A. and Ye, Z. G. (2000). Phenomenological description of dielectric permittivity peak in relaxor ferroelectrics, *Sol. Stat. Commun.*, 116, pp. 105–108.

[86] Bokov, A. A., Bing, Y. H., Chen, W., Ye, Z. G., Bogatina, S. A., Raevski, I. P., Raevskaya, S. I. and Sahkar, E. V. (2003). Empirical scaling of the dielectric permittivity peak in relaxor ferroelectrics, *Phys. Rev. B*, 68, pp. 052102.

[87] Viehland, D., Jang, S., Cross, L. E. and Wuttig, M. (1991). The dielectric-relaxation of lead magnesium niobate relaxor ferroelectrics, *Phil. Mag.: B*, 64, pp. 335–344.

[88] Bovtun, V., Petzelt, J., Porokhonskyy, V., Kamba, S. and Yakimenko, Y. (2001). Structure of the dielectric spectrum of relaxor ferroelectrics, *J. Eur. Ceram. Soc.*, 21, pp. 1307–1311.

[89] Rychetsky, I., Kamba, S., Porokhonskyy, V., Pashkin, A., Savinov, M., Bovtun, V., Petzelt, J., Kosec, M. and Dressel, M. (2003). Frequency-independent dielectric losses (1/f noise) in PLZT relaxors at low temperatures, *J. Phys.: Condens. Matter*, 15, pp. 6017–6030.

[90] Tsurumi, T., Soejima, K., Kamiya, T. and Daimon, M. (1994). Mechanism of diffuse phase-transition in relaxor ferroelectrics, *Jpn. J. Appl. Phys.*, 33, pp. 1959–1964.

[91] Christen, H. M., Sommer, R., Yushin, N. K. and Vanderklink, J. J. (1994). Small-signal dielectric-relaxation in the disordered perovskite $Pb(Mg_{1/3}Nb_{2/3})O_3$, *J. Phys.: Condens. Matter*, 6, pp. 2631–2642.

[92] Viehland, D., Jang, S. J., Cross, L. E. and Wuttig, M. (1990). Freezing of the polarization fluctuations in lead magnesium niobate relaxors, *J. Appl. Phys.*, 68, pp. 2916–2921.

[93] Tagantsev, A. K. (1994). Vogel-fulcher relationship for the dielectric permittivity of relaxor ferroelectrics, *Phys. Rev. Lett.*, 72, pp. 1100–1103.

[94] Glazounov, A. E. and Tagantsev, A. K. (1998). Direct evidence for Vogel–Fulcher freezing in relaxor ferroelectrics, *Appl. Phys. Lett.*, 73, pp. 856–858.

[95] Levstik, A., Kutnjak, Z., Filipic, C. and Pirc, R. (1998). Glassy freezing in relaxor ferroelectric lead magnesium niobate, *Phys. Rev. B*, 57, pp. 11204–11211.

[96] Kutnjak, Z., Filipic, C., Pirc, R., Levstik, A., Farhi, R. and El Marssi, M. (1999). Slow dynamics and ergodicity breaking in a lanthanum-modified lead zirconate titanate relaxor system, *Phys. Rev. B*, 59, pp. 294–301.

[97] Kirillov, V. V. and Isupov, V. A. (1973). Relaxation polarization of $PbMg_{1/3}Nb_{2/3}O_3$ (PMN) — A ferroelectric with a diffused phase-transition, *Ferroelectrics*, 5, pp. 3–9.

[98] Vugmeister, B. E. and Rabitz, H. (1998). Dynamics of interacting clusters and dielectric response in relaxer ferroelectrics, *Phys. Rev. B*, 57, pp. 7581–7585.

[99] Blinc, R., Bobnar, V. and Pirc, R. (2001). Coupled spherical pseudospin-phonon model and the pressure-temperature phase diagram of relaxor ferroelectrics, *Phys. Rev. B*, 64, pp. 132103.

[100] Glazounov, A. E. and Tagantsev, A. K. (1999). A "breathing" model for the polarization response of relaxer ferroelectrics, *Ferroelectrics*, 221, pp. 57–66.

[101] Tagantsev, A. K. and Glazounov, A. E. (1998). Mechanism of polarization response in the ergodic phase of a relaxor ferroelectric, *Phys. Rev. B*, 57, pp. 18–21.

[102] Ye, Z. G. and Bokov, A. A. (2004). Dielectric and structural properties of relaxor ferroelectrics, *Ferroelectrics*, 302, pp. 473–477.

[103] Viehland, D., Jang, S. J., Cross, L. E. and Wuttig, M. (1992). Deviation from Curie–Weiss behavior in relaxor ferroelectrics, *Phys. Rev. B*, 46, pp. 8003–8006.

[104] Tu, C. S., Schmidt, V. H. and Siny, I. G. (1995). Hypersound anomalies and elastic-constants in single-crystal $PbMg_{1/3}Nb_{2/3}O_3$ by brillouin-scattering, *J. Appl. Phys.*, 78, pp. 5665–5668.

[105] Verbaere, A., Piffard, Y., Ye, Z. G. and Husson, E. (1992). Lead magnoniobate crystal-structure determination, *Mater. Res. Bull.*, 27, pp. 1227–1234.

[106] Lebedinskaya, A. R. and Kupriyanov, M. F. (2002). A study of PMN crystal structure below the phase transition temperature, *Phase Trans.*, 75, pp. 289–299.

[107] Kleemann, W. and Lindner, R. (1997). Dynamic behavior of polar nanodomains in $PbMg_{1/3}Nb_{2/3}O_3$, *Ferroelectrics*, 199, pp. 1–10.

[108] Ye, Z.-G. and Schmid, H. (1993). Optical, dielectric and polarization studies of the electric field-induced phase transition in $Pb(Mg_{1/3}Nb_{2/3})O_3$ [PMN], *Ferroelectrics*, 145, pp. 83–108.

[109] Thomas, N. W., Ivanov, S. A., Ananta, S., Tellgren, R. and Rundlof, H. (1999). New evidence for rhombohedral symmetry in the relaxer ferroelectric $Pb(Mg_{1/3}Nb_{2/3})O_3$, *J. Eur. Ceram. Soc.*, 19, pp. 2667–2675.

[110] Tkachuk, A. and Chen, H. (2003). *Fundamental Physics of Ferroelectrics*, AIP Conference Proceedings No. 677, pp. 55–64.

[111] Chamberlin, R. V. (1996). Universalities in the primary response of condensed matter, *Europhys. Lett.*, 33, pp. 545–550.

[112] Colla, E. V., Koroleva, E. Y., Okuneva, N. M. and Vakhrushev, S. B. (1995). Long-time relaxation of the dielectric response in lead magnoniobate, *Phys. Rev. Lett.*, 74, pp. 1681–1684.

[113] Vakhrushev, S. B., Kiat, J. M. and Dkhil, B. (1997). X-ray study of the kinetics of field induced transition from the glass-like to the ferroelectric phase in lead magnoniobate, *Sol. Stat. Commun.*, 103, pp. 477–482.

[114] Colla, E. V., Chao, L. K., Weissman, M. B. and Viehland, D. D. (2000). Aging in a relaxor ferroelectric: Scaling and memory effects, *Phys. Rev. Lett.*, 85, pp. 3033–3036.

[115] Palmer, R. G. (1982). Broken ergodicity, *Adv. Phys.*, 31, pp. 669–735.

[116] Feynman, R. P., (1972). *Statistical Mechanics* (Reading, Benjamin).

[117] Calvarin, G., Husson, E. and Ye, Z. G. (1995). X-ray study of the electric field induced phase transition in single crystal $Pb(Mg_{1/3}Nb_{2/3})O_3$, *Ferroelectrics*, 165, pp. 349–358.

[118] Chu, F., Reaney, I. M. and Setter, N. (1995). Spontaneous (zero-field) relaxor-to-ferroelectric-phase transition in disordered $Pb(Sc_{1/2}Nb_{1/2})O_3$, *J. Appl. Phys.*, 77, pp. 1671–1676.

[119] Ye, Z. G., Bing, Y., Gao, J., Bokov, A. A., Stephens, P., Noheda, B. and Shirane, G. (2003). Development of ferroelectric order in relaxor (1-x)$Pb(Mg_{1/3}Nb_{2/3})O_3$-$x$$PbTiO_3$ ($0 \leq x \leq 0.15$), *Phys. Rev. B*, 67, pp. 104104.
[120] Ye, Z. G. and Dong, M. (2000). Morphotropic domain structures and phase transitions in relaxor-based piezo-/ferroelectric (1-x)$Pb(Mg_{1/3}Nb_{2/3})O_3$-$x$$PbTiO_3$ single crystals, *J. Appl. Phys.*, 87, pp. 2312–2319.
[121] Tu, C. S., Tsai, C. L., Schmidt, V. H., Luo, H. S. and Yin, Z. W. (2001). Dielectric, hypersonic, and domain anomalies of $(PbMg_{1/3}Nb_{2/3}O_3)_{1-x}$ $(PbTiO_3)_x$ single crystals, *J. Appl. Phys.*, 89, pp. 7908–7916.
[122] Bokov, A. A. and Ye, Z. G. (2004). Domain structure in the monoclinic Pm phase of $Pb(Mg_{1/3}Nb_{2/3})O_3$-$PbTiO_3$ single crystals, *J. Appl. Phys.*, 95, pp. 6347–6359.
[123] Chu, F., Reaney, I. M. and Setter, N. (1995). Role of defects in the ferroelectric relaxer lead scandium tantalate, *J. Am. Ceram. Soc.*, 78, pp. 1947–1952.
[124] Abplanalp, M., Barosova, D., Bridenbaugh, P., Erhart, J., Fousek, J., Gunter, P., Nosek, J. and Sulc, M. (2002). Scanning force microscopy of domain structures in $Pb(Zn_{1/3}Nb_{2/3})O_3$-8% $PbTiO_3$ and $Pb(Mg_{1/3}Nb_{2/3})O_3$-29% $PbTiO_3$, *J. Appl. Phys.*, 91, pp. 3797–3805.
[125] Shvartsman, V. V. and Kholkin, A. L. (2004). Domain structure of $0.8Pb(Mg_{1/3}Nb_{2/3})O_3$-$0.2PbTiO_3$ studied by piezoresponse force microscopy, *Phys. Rev. B*, 69, pp. 014102.
[126] Bai, F. M., Li, J. F. and Viehland, D. (2004). Domain hierarchy in annealed (001)-oriented $Pb(Mg_{1/3}Nb_{2/3})O_3$-x%$PbTiO_3$ single crystals, *Appl. Phys. Lett.*, 85, pp. 2313–2315.
[127] Bokov, A. A. and Ye, Z. G. (2002). Ferroelectric properties of monoclinic $Pb(Mg_{1/3}Nb_{2/3})O_3$-$PbTiO_3$ crystals, *Phys. Rev. B*, 66, p. 094112.
[128] Perrin, C., Menguy, N., Bidault, O., Zahra, C. Y., Zahra, A. M., Caranoni, C., Hilczer, B. and Stepanov, A. (2001). Influence of B-site chemical ordering on the dielectric response of the $Pb(Sc_{1/2}Nb_{1/2})O_3$ relaxor, *J. Phys.: Condens. Matter*, 13, pp. 10231–10245.
[129] Kamzina, L. S. and Krainik, N. N. (2000). Optical study of the spontaneous ferroelectric phase transition in lead scandoniobate single crystals, *Phys. Solid State*, 42, pp. 1712–1715.
[130] Bokov, A. A. and Emelyanov, S. M. (1991). Electrical-properties of $Pb(Fe_{0.5}Nb_{0.5})O_3$ crystals, *Phys. Stat. Sol.: B*, 164, pp. K109–K112.
[131] Bao, P., Yan, F., Li, W., Dai, Y. R., Shen, H. M., Zhu, J. S., Wang, Y. N., Chan, H. L. W. and Choy, C. L. (2002). Mechanical properties related to the relaxor-ferroelectric phase transition of titanium-doped lead magnesium niobate, *Appl. Phys. Lett.*, 81, pp. 2059–2061.
[132] Bokov, A. A., Luo, H. and Ye, Z. G. (2005). Polar nanodomains and relaxor behaviour in (1-x)$Pb(Mg_{1/3}Nb_{2/3})O_3$-$x$$PbTiO_3$ crystals with x = 0.3-0.5, *Mater. Sci. Eng. B*, 120, pp. 206–209.

[133] Lebon, A., Dammak, H., Calvarin, G. and Ahmedou, I. O. (2002). The cubic-to-rhombohedral phase transition of $Pb(Zn_{1/3}Nb_{2/3})O_3$: A high-resolution x-ray diffraction study on single crystals, *J Phys.: Condens. Matter*, 14, pp. 7035–7043.

[134] Kamba, S., Buixaderas, E., Petzelt, J., Fousek, J., Nosek, J. and Bridenbaugh, P. (2003). Infrared and Raman spectroscopy of $[Pb(Zn_{1/3}Nb_{2/3})O_3]_{0.92}$-$[PbTiO_3]_{0.08}$ and $[Pb(Mg_{1/3}Nb_{2/3})O_3]_{0.71}$-$[PbTiO_3]_{0.29}$ single crystals, *J. Appl. Phys.*, 93, pp. 933–939.

[135] Xu, G. Y., Zhong, Z., Bing, Y., Ye, Z. G., Stock, C. and Shirane, G. (2003). Ground state of the relaxor ferroelectric $Pb(Zn_{1/3}Nb_{2/3})O_3$, *Phys. Rev. B*, 67, pp. 104102.

[136] Xu, G. Y., Zhong, Z., Bing, Y., Ye, Z. G., Stock, C. and Shirane, G. (2004). Anomalous phase in the relaxor ferroelectric $Pb(Zn_{1/3}Nb_{2/3})O_3$, *Phys. Rev. B*, 70, pp. 064107.

[137] Ohwada, K., Hirota, K., Rehrig, P. W., Fujii, Y. and Shirane, G. (2003). Neutron diffraction study of field-cooling effects on the relaxor ferroelectric $Pb(Zn_{1/3}Nb_{2/3})_{0.92}Ti_{0.08}O_3$, *Phys. Rev. B*, 67, pp. 094111.

[138] Xu, G. Y., Hiraka, H., Shirane, G. and Ohwada, K. (2004). Dual structures in (1-x)$Pb(Zn_{1/3}Nb_{2/3})O_3$-$x$$PbTiO_3$ ferroelectric relaxors, *Appl. Phys. Lett.*, 84, pp. 3975–3977.

[139] Gehring, P. M., Chen, W., Ye, Z. G. and Shirane, G. (2004). The non-rhombohedral low-temperature structure of PMN-10% PT, *J. Phys.: Condens. Matter*, 16, pp. 7113–7121.

[140] Xu, G. Y., Viehland, D., Li, J. F., Gehring, P. M. and Shirane, G. (2003). Evidence of decoupled lattice distortion and ferroelectric polarization in the relaxor system PMN-xPT, *Phys. Rev. B*, 68, pp. 212410.

[141] Conlon, K. H., Luo, H., Viehland, D., Li, J. F., Whan, T., Fox, J. H., Stock, C. and Shirane, G. (2004). Direct observation of the near-surface layer in $Pb(Mg_{1/3}Nb_{2/3})O_3$ using neutron diffraction, *Phys. Rev. B*, 70, pp.172204.

[142] Samara, G. A. (2002). Pressure induced crossover and mechanism for the ferroelectric-to-relaxor (glass-like) transition in compositionally-disordered soft mode systems, *Ferroelectrics*, 274, pp. 183–202.

[143] Samara, G. A. (2003). The relaxational properties of compositionally disordered $ABO_{(3)}$ perovskites, *J. Phys.: Condens. Matter*, 15, pp. R367–R411.

[144] Yamashita, Y., Hosono, Y., Harada, K. and Yasuda, N. (2002). Present and future of piezoelectric single crystals and the importance of B-site cations for high piezoelectric response, *IEEE Transactions on Ultrasonics Ferroelectrics and Frequency Control*, 49, pp. 184–192.

[145] P. W. Rehrig, W. S. Hackenberger, Park, S.-E. and Shrout, T. R. (2002). *Piezoelectric Materials in Devices* (*EPFL*, Lausanne), p. 433.

[146] Ye, Z. G. (2002). Crystal chemistry and domain structure of relaxor piezoerystals, *Curr. Opin. Solid State Mat. Sci.*, 6, pp. 35–44.

[147] Noheda, B. (2002). Structure and high-piezoelectricity in lead oxide solid solutions, *Curr. Opin. Solid State Mat. Sci.*, 6, pp. 27–34.
[148] Bokov, A. A. and Ye, Z. G. (2003). Morphotropic phase boundary and related properties in relaxor-based piezoelectric perovskite solid solutions, *Ceram. Transact.*, 136, pp. 37–53.
[149] Pirc, R. and Blinc, R. (1999). Spherical random-bond-random-field model of relaxer ferroelectrics, *Phys. Rev. B*, 60, pp. 13470–13478.
[150] Blinc, R., Dolinsek, J., Gregorovic, A., Zalar, B., Filipic, C., Kutnjak, Z., Levstik, A. and Pirc, R. (1999). Local polarization distribution and Edwards-Anderson order parameter of relaxor ferroelectrics, *Phys. Rev. Lett.*, 83, pp. 424–427.
[151] Kleemann, W., Dec, J., Blinc, R., Zalar, B. and Pankrath, R. (2002). Random fields at transitions from relaxor to glassy and ferroelectric states, *Ferroelectrics*, 267, pp. 157–164.
[152] Blinc, R. and Bobnar, V. (2001). Dynamics of relaxor ferroelectrics, *Phys. Rev. B*, 63, pp. 054203.
[153] Vugmeister, B. E. and Rabitz, H. (2000). Coexistence of the critical slowing down and glassy freezing in relaxor ferroelectrics, *Phys. Rev. B*, 61, pp. 14448–14453.
[154] Vugmeister, B. E. and Rabitz, H. (2002). Kinetics of electric-field-induced ferroelectric phase transitions in relaxor ferroelectrics, *Phys. Rev. B*, 65, pp. 024111.
[155] Mamin, R. F. (2001). To the theory of phase transitions in relaxors, *Phys. Solid States*, 43, pp. 1314–1319.
[156] Mamin, R. F. and Blinc, R. (2003). Time delay in the low-temperature phase of relaxors, *Phys. Sol. Stat.*, 45, pp. 942–945.

Chapter 4

Relaxor Ferroelectrics: Cluster Glass Ground State via Random Fields and Random Bonds*

Wolfgang Kleemann†

†*wolfgang.kleemann@uni-due.de*

Quenched random fields (RFs) are well-known to be a basic driving force of the *relaxor* behavior in disordered ferroelectrics containing either random charges as in $PbMg_{1/3}Nb_{2/3}O_3$ or random cation vacancies as in $Sr_xBa_{1-x}Nb_2O_6$. They give rise to the formation of polar nanoregions in the paraelectric regime, which freeze into a nanodomain state at low temperatures thus precluding the ferroelectric phase transition. On the other hand, isovalent relaxors like bond-disordered $BaTi_{1-x}Zr_xO_3$ prevalently experience random bonds (RBs), which are responsible for glassy features such as polydispersivity and non-ergodic aging and rejuvenation processes. Inspection shows, however, that all of the above characteristics apply generally to relaxor systems *albeit* with different weight. Subsequent formation of ferroic nanoregions via RFs followed by mesoscopic cluster glassy freezing via RBs upon cooling are universal signatures.

Keywords: Cluster glass; random bonds; random fields; relaxor ferroelectrics.

1. Introduction

Relaxor ferroelectrics (*relaxors* for short) have intensely been debated as a specific class of disordered ferroelectrics since the discovery of

*This article was firstly published as "Kleemann, W. (2014). Relaxor ferroelectrics: Cluster glass ground state via random fields and random bonds, *Phys. Status Solidi B* 251, pp. 1993–2002". Reprinted with the permission of WILEY-VCH Verlag GmbH & Co. KGaA, Weinheim.

the random compound $PbMg_{1/3}Nb_{2/3}O_3$ (PMN) by Smolenskii and Isupov [1] in 1954. They exhibit a rounded ("smeared") permittivity response with strongly frequency dependent peak temperatures $T_m = T_m(f)$ as shown for the prototypical relaxors PMN [2] (Fig. 1) and SBN75 ($Sr_{0.75}Ba_{0.25}Nb_2O_6$) [3] (Fig. 2). Both, smearing and frequency dispersion clearly evidence that regular phase transitions are absent. Their "*diffuse*" phase transitions [4] even remind of dipolar glassy ground states in analogy to those of *spin glasses* as

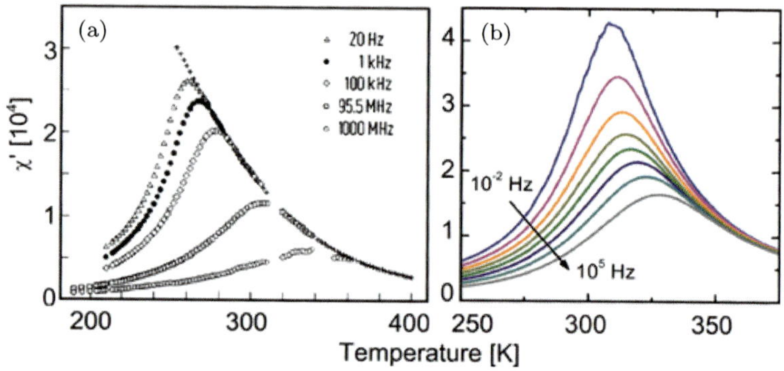

Figure 1. Electric susceptibility, $\chi'(T)$, of (a) PMN for frequencies $20 < f < 10^9$ Hz [2] and (b) SBN75 for frequencies $10^{-2} < f < 10^5$ Hz [3]. Reprinted with permission from Refs. [2, 3].

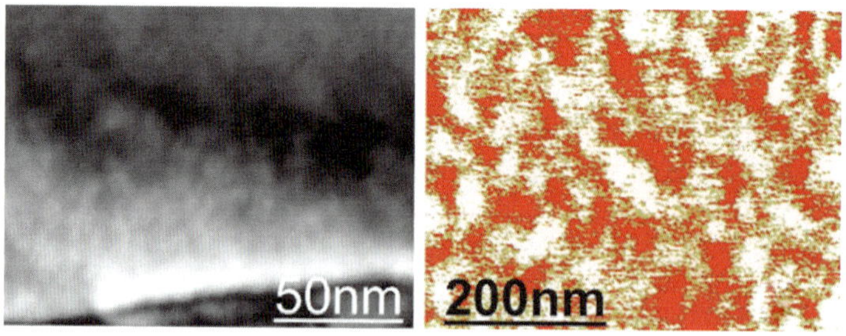

(a) (b)

Figure 2. Spatial distribution of polar domains in (a) PMN [9] and (b) SBN75 [10] recorded by TEM and PFM at $T = 70$ K and 293 K, respectively. In (a) also chemically ordered regions with cubic symmetry are contained. In (b), the red/dark and white areas refer to $\pm P_z$, respectively, while yellow/gray designates non-polar interface regions and unresolved fine-grained PNRs, respectively. Reproduced with permission from Refs. [9, 10].

observed in disordered magnetic materials with competing ferroic and antiferroic interactions [5].

The glass conjecture has dominated the early work on relaxors [6–8]. However, it was simultaneously noticed that their specific properties are also associated with the presence of *polar nanoregions* (PNRs) [6] as shown in Figs. 2(a) and (b) for PMN [9] and SBN75 [10], respectively. Their granular structure resembles the well-known domain states forming in magnetic random field (RF) systems [11] as realized by dilute antiferromagnets in an external uniform magnetic field [12]. An illustrating example in Fig. 3 shows a model calculation on a three-dimensional Ising antiferromagnet with 50% diamagnetic dilution and volume $V = 99 \times 99 \times 98$ elementary cells after cooling to low temperatures in a constant magnetic field $\boldsymbol{H}$ parallel to the z-axis. Owing to the randomly missing spins, $\boldsymbol{H}$ acts onto the antiferromagnetic order parameter like a random staggered field [13]. The differently registered domains (blue vs. yellow) are forming according to local excess field fluctuations along the $+z$ and $-z$ directions, respectively, by optimizing both Zeeman and exchange free energies in the bulk and domain walls, respectively.

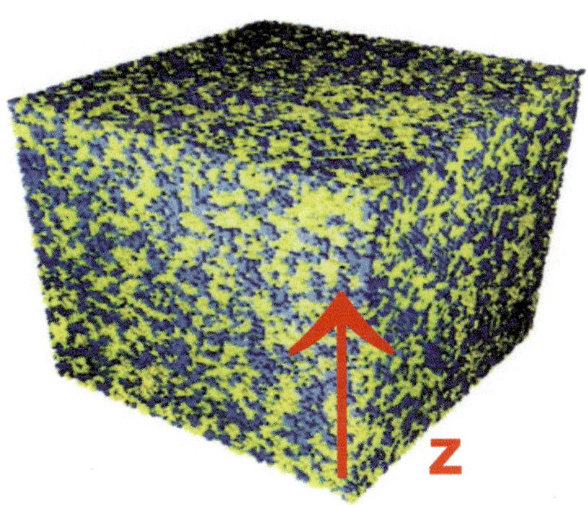

Figure 3. Simulated random field-induced conversely registered (blue vs. yellow) domains in a 50% diamagnetically dilute uniaxial antiferromagnet cooled to below T_N under an external magnetic field $H \| z$. Reprinted with permission from Ref. [13].

Being inspired by the action of random local fields on magnetic moments in RF magnets, we conjectured about 20 years ago that charge-disordered cubic relaxors like PMN might be subject to random fields in a similar way [14]. Much evidence in favor of the RF model became readily available, although the observed glassy features clearly called for careful inspection. The present review extends our previous one [15] by proposing a model adopting both random field and random bond interactions in accordance with the observed phenomenology of a mesoscopic cluster glass ground state.

2. Random field-induced domain states

Fundamental conditions of the RF model theory [11] are obviously met in prototypical *relaxors* such as PMN [1, 2], where inherent charge disorder due to the random distribution of Mg^{2+} and Nb^{5+} ions at B sites of this ABO_3-type perovskite crystal (Fig. 4(a)) offers largely uncorrelated and quenched electric fields at the sites of the *ferroelectric-active* ions, Pb^{2+} and Nb^{5+}. Within a pseudospin

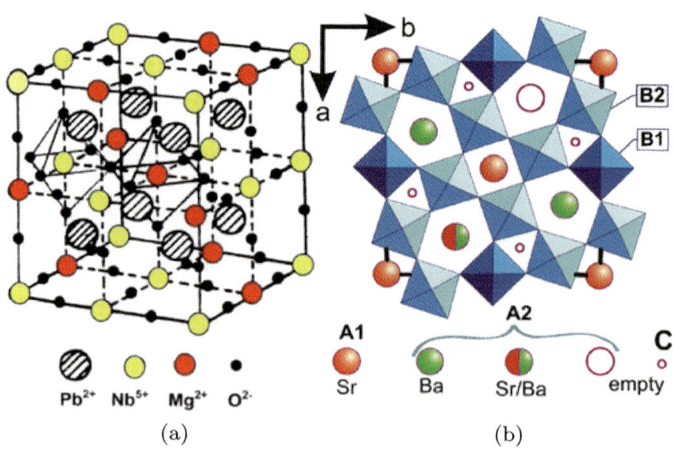

Figure 4. Ionic distribution in unit cells of (a) $PbMg_{1/3}Nb_{2/3}O_3$ (PMN): Pb^{2+} = hatched circles, O^{2-} = small solid circles, Nb^{5+} = large solid circles, Mg^{2+} = open circles; (b) $Sr_xBa_{1-x}Nb_2O_6$ (SBN): Sr^{2+} = red/light gray spheres, Ba^{2+} = green/dark gray spheres, vacancies = open spheres, empty channels = small open circles. Reprinted with permission from Ref. [15].

treatment known from order–disorder ferroelectrics [16], the typical quasi-glassy slowing-down of the polar relaxation on approaching the critical temperature [7] is readily explained by RF theory.

In the case of uniaxial RF systems such as $Sr_xBa_{1-x}Nb_2O_6$ (SBN100x), the RF Ising model (RFIM) [11] is applicable because of the uniaxial polarity in the ferroelectric phase and the random absence of one cation (Ba^{2+} or Sr^{2+}) from five A2-sites (Fig. 4(b)). It predicts a similar tremendous slowing down of the order parameter dynamics as in *spin glasses* [5]. Here, the critical *order parameter fluctuations* (referring to temperature-dependent correlated regions with a preferential uniform direction of the order parameter) become pinned to *spatial fluctuations* of the RFs (i.e., temperature independent regions with an excess of the corresponding field direction) as the correlation length grows in the vicinity of T_C [14]. This is experimentally shown at the nanoscale by transmission electron microscopy (TEM) on a single crystal of PMN-PT in Fig. 2(a) [9] and by piezoforce microscopy (PFM) on a single crystal of SBN75 with cluster-like up- and down-polarized PNRs in Fig. 2(b) [10].

Neutron scattering [17] has lent particular support in favor of the existence of PNRs below the so-called Burns temperature, T_d [18]. A close relationship was established between their size and the polar correlation length, r_c, which increases with decreasing T. Diffuse X-ray and Raman scattering [19] enabled the discovery of the "*intermediate*" temperature, T_L, indicating the thermal stabilization of the PNRs on cooling. This "local phase transition" being due to RF pinning takes place between T_m and T_d. Both, T_d and T_L, have very precisely been measured by acoustic emission signals due to the strain anomalies occurring at these temperatures [20]. As a rule, the static local phase transitions are observed at lower temperature, $T_L < T_d$, e.g., T_L (alias $T*$) = 500 K and T_d = 630 K for the cubic relaxor PMN [21], and T_L = 466 K and T_d = 623 K for the uniaxial relaxor SBN75 [22].

While T_d marks the nucleation temperature of the PNRs, T_L is widely agreed to characterize their mesoscopic phase transition, where static stability is gained in one of the possible local field directions thanks to the support of RF fluctuations. Following the fundamental ideas of RF interactions [12, 14], Qian and Bursill [23] were the first to evidence the thermal evolution and stabilization of PNRs on cooling by Monte Carlo simulations on a two-dimensional (2D) pseudospin 4-state Potts model. Irregular domain patterns are observed to become gradually space-filling and coarsening upon cooling (Figs. 5(a) and (b)). Exchange-like (instead of long-range dipolar) interactions and localized RFs were considered similarly as in simulations of magnetic RF systems [24]. This situation appears immediately acceptable in order–disorder systems such as the uniaxial SBN relaxors, being hailed as the first materialization of the RFIM [25]. In the case of cubic relaxors like PMN, the applicability of a pseudospin approach is justified as well within the concept of soft-mode ferroelectricity with deep double well potentials [23].

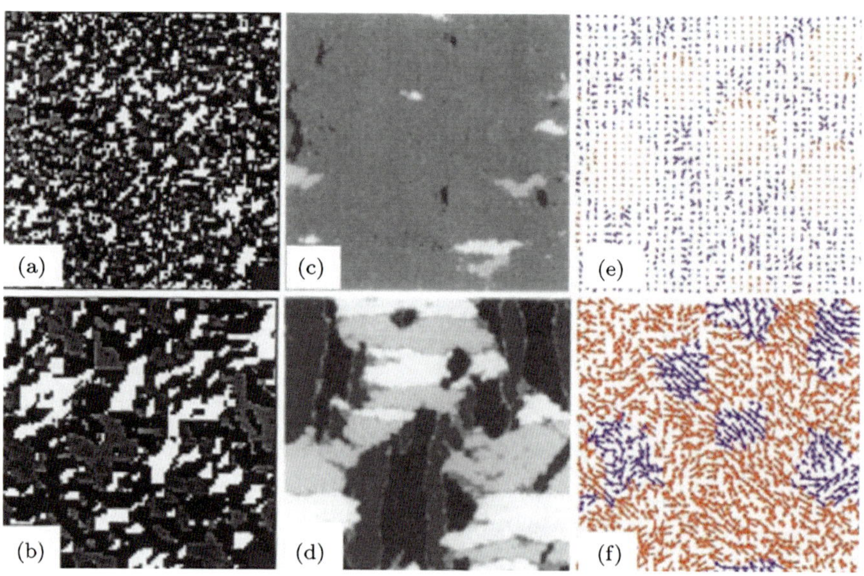

Figure 5. Simulated domain distribution with polarizations $\pm P_x$ and $\pm P_y$ as indicated by different gray tones in a 2D 4-state random field Potts model at temperatures $k_B T/J = 1$ (a) and 0.3 (b) [23], from 3D solutions of a time-dependent stochastic field equation in the paraelectric regime at high (c) and low temperatures (d), respectively [26], from first-principles molecular dynamics solutions (e,f) of an effective Hamiltonian including quenched RFs in a 403 simulation box [31]. Reproduced with permission from Ref. [15].

A further step toward experimental reality was the stochastic field kinetic equation approach of Semenovskaya and Khachaturyan [26], which accounts for the dipole–dipole interaction between local polar order parameter components and their interaction with the quenched RFs of static defects under the equilibrium condition of stress-free strain. As a result, a non-percolating PNR structure emerges as shown in Fig. 5(c), which coarsens and gradually replaces the paraelectric background on cooling (Fig. 5(d)). The quasi-stability of this mixed state was stressed [26] to be a unique feature of both the dipolar interaction and the local RF interaction. As compared to this first simplified approach including merely dipolar and RF interaction, additionally included elastic energy of PNRs with the paraelectric background [27] yields increased irregularity of the domain walls and superficial jaggedness of the PNRs and microdomains, respectively (Fig. 5(d)).

Owing to their disordered nature, the directions of the RFs fluctuate in space and statistically create local majorities of distinct field directions, e.g., up and down in the case of an uniaxial (*Ising*) system [23]. These fluctuations favor local phase transitions even above the — in most cases fictitious — transition temperature in the absence of RFs. As an example, we note that $T_d = 466$ K of the relaxor SBN75 [22] clearly exceeds $T_c = 450$ K of the ferroelectric SBN40 [28].

Alternatively, Burton *et al.* [29] modeled the PNR structure of PMN-like relaxors using molecular dynamics (MD) simulations. They chose first-principles effective Hamiltonian model for $PbSc_{1/2}Nb_{1/2}O_3$ (PSN) and a 40^3 unit-cell simulation box [30]. Random local field interactions were added to an effective Hamiltonian model for a normal FE. In order to mimic the microscopic experience of real samples so-called *chemically ordered domains* (COR) with regular Sc^{3+}–Nb^{5+} ordering on next-neighboring perovskite A sites (red/light gray regions) were built-in intentionally as shown by the local field vectors in a (110) plane in Fig. 5(e). As a result, Fig. 5(f) shows blue/dark gray regions, where the weak, but homogeneous local field density at the CORs clearly gives rise to preferential local ordering of the ferroelectric sample.

Contrastingly, *chemically disordered regions* (CDR) with stronger disordered local fields in Fig. 5(e) (blue/dark gray regions) are observed to cause larger disorder in the matrix of Fig. 5(f) (red/light gray regions), which suppresses the phase transition. The abrupt difference of the field distribution, hence, favors phase transitions at the COR locations, but suppresses them at the percolating CDRs. By using a more realistic model of local fields it was later [31] shown that PNRs form again at the COR length scale without long-range polar order (ferroelectricity) below some temperature, which was associated with the Burns temperature, T_d. Regrettably the localization of RFs in $(PMN)_{1-x}(PbTiO_3)_x$ by using highly-resolved piezoelectric force microscopy (PFM) failed [32]. This is to explain by the poor resolution of the PFM tip, which does not reach the atomic scale. Only recently, by using TEM Fu *et al.* [9] have clarified that the appearing nanodomain fine structure in PMN (Fig. 2) is a merger of PNRs and CORs, which stresses the complexity of structural heterogeneity at multiple length scales in this system.

Recently, a novel effort to model a typical 3D relaxor [33] containing quenched random electric fields [11] has revived the successful [26, 27] time-dependent Ginzburg–Landau method in conjunction within a phase-field approach. Here, the destruction of regular domain patterns has clearly been demonstrated at increasing disorder, viz. strength of the RFs.

Topologically very similar $\pm P_z$ domain patterns are formed in as-cooled single crystals of SBN probed at their polar surfaces with scanning piezoforce microscopy (PFM) (Fig. 6 [10]). Different $[Sr^{2+}]$:$[Ba^{2+}]$ ratios from 40:60 to 75:25 are chosen in order to tune the effective RFs. They are primarily due to the unfilled tungsten bronze structure (Fig. 4(b)) with randomly distributed negative effective charges at empty A2 sites at a 1:4 ratio, whose entropic intensity increases as $[Sr^{2+}]$ increases [34]. In addition, extra random dipolar fields are due to the off-centrality of the small Sr^{2+} ions. They become increasingly important at increasing admixture [10]. Consequently, the PFM images taken at room temperature reveal decreasing average domain sizes at increasing Sr^{2+} content. This

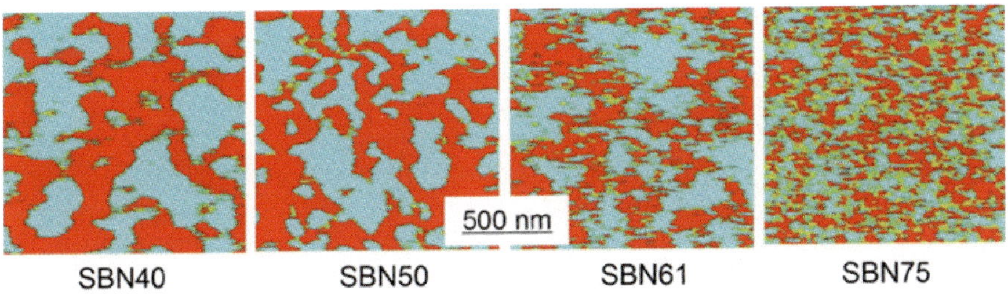

Figure 6. Spatial distributions of the (001) surface polarization $\pm P_z$ and vanishing ("red/dark", "green/white", and "yellow/gray", respectively) of SBN 40, 50, 61, and 75, respectively, recorded by PFM at $T = 295$ K. Adapted from Ref. [10].

evidences an increasing average RF, $h_{rf} = \left\langle \sqrt{h_{loc}^2} \right\rangle$, since the average free energy gain, $\Delta F \propto h_{RF} L^{3/2}$ [10], maximizes at decreasing length scale of the domains, L.

This is illustrated by the PFM image of the PNR distribution in a (001) surface of the relaxor SBN75 taken at $T = 293$ K (Fig. 2(b) [15]), hence, close to, but above its transition temperature $T_c = 289$ K [10]. A power law known from the domain distribution in the RFIM [35] describes the cross-sections S of *up* (red) and *down* (white) domains,

$$N(S) = N_0 S^{-\delta} \exp\left(\frac{-S}{S_0}\right), \tag{1}$$

involving an exponent $\delta = 1.7(2)$ and exponential cutoff at $S_0 = 0.009(2)$ μm^2 [34]. The yellow "interfaces" are due to either broadened domain walls or unresolved very small PNRs [10]. Note that the interfaces of the PNRs are very rough thanks to vanishing interfacial strain at the interfaces of tetragonal *up* and *down* domains in SBN, which can thus fully exploit the local energy gain due to the RFs [35].

3. Relaxation toward a glassy ground state

The ground state of RF perturbed systems has been heavily debated in the magnetic case. Proof had been given for ferromagnetic standard systems that only the 3D RF Ising model (RFIM) system should attain long-range FM order, but not the 2D RFIM or the *xy*- and Heisenberg systems in any dimension [11]. The reason lies in the delicate balance of the volume versus interface free energies, which are different in different spatial and spin dimensions. However, the as-cooled 3D RFIM system will not immediately reach thermodynamic long-range ordered equilibrium, but get stuck in metastable domain states upon cooling at finite rates.

The slow equilibration of the RF induced domain state is shown for SBN61:1.1 mol% Ce^{3+} by the coarsening of the nanodomain pattern due to room temperature aging in Figs. 7(a) and (b) [36]. In particular, the neutral or fine-grained interfacial regions between up- and down-domains are considerably shrinking in agreement with RFIM predictions and simulations (Fig. 3). While these observations reflect the extremely slow relaxation processes under RF interaction, they do not necessarily mirror the predicted [7, 8] phase transition into a cluster glass state. This requires an experiment, which evidences aging not just as a global coarsening event (Fig. 7), but as a fingerprint of the chaotic nature of the dipole glass ground state involved. This scenario is well-known from "superspin glass" properties of interacting magnetic nanoparticle systems [37], which are known to undergo a magnetic dipolar glass state at a suitable

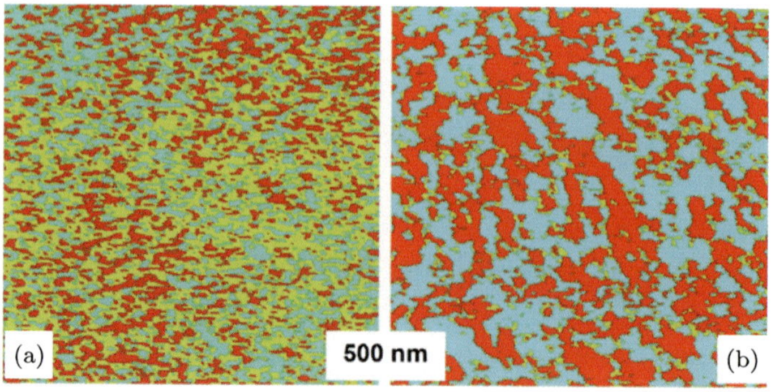

Figure 7. Piezoresponse images of a SBN61:Ce single crystal acquired at room temperature after cooling down from the paraelectric state: (a) after 1 h, (b) after 72 h (color code: see Fig. 6). Reproduced with permission from Ref. [35].

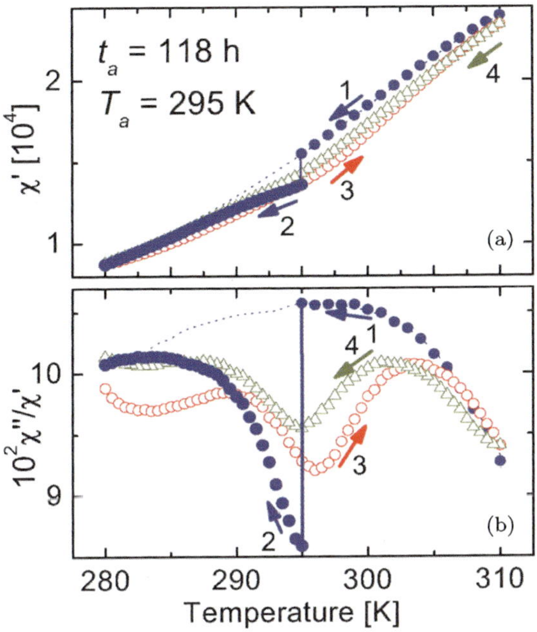

Figure 8. Aging of χ' (a) and χ''/χ' (b) versus T of SBN75 measured after ZFC from $T = 450$ K at rates $\mathrm{d}T/\mathrm{d}t = \pm 0.2$ K min^{-1} and frequency $f = 100$ Hz on first cooling to 280 K with intermittent halt at 295 K (curves 1 and 2, respectively), continuous reheating to 310 K (curve 3), and subsequent continuous recooling to 280 K. The dotted line refers to continuous cooling from 450 K with the same rate. Reproduced with permission from Ref. [38].

particle concentration — dense enough in order to obtain a sizable glass transition temperature T_g.

Pertinent non-ergodic effects are also observed in relaxor ferroelectrics in their *domain state*: This is shown in Figs. 8(a) and (b) for the uniaxial relaxor system SBN75 [38] by a classic stop-and-wait procedure. Aging of the susceptibility components χ'(a) and χ''/χ'(b) of SBN75 is recorded versus T after zero-field cooling (ZFC) from $T = 450$ K at rates $\mathrm{d}T/\mathrm{d}t = \pm 0.2$ K min^{-1} and frequency $f = 100$ Hz, on first cooling to halt for 118 h at the wait temperature $T_w = 295$ K (curves 1 and 2, respectively), continuous reheating to 310 K (curve 3), and subsequent continuous cooling back to 280 K. The dotted line refers to continuous cooling from 450 K under identical cooling rate.

Distinct drops of both χ' and χ''/χ' are observed at T_w. They remain clearly visible albeit slightly smeared on subsequent heating (curve 3) and recooling (curves 4). Obviously a distinct memory effect due to glassy domain growth (in the sense of "*locally optimized disorder*") is encountered, which confirms that the experiment has been carried out in the regime $T_w < T_g$. Stretched exponential temporal relaxation of χ' during T_w as measured after ZFC at $T = 295$ K with an exponent $\beta = 0.367(9)$ s (not shown) confirms the polydispersive glassy relaxation processes.

Another crucial condition of glassiness of the PNR systems is the spatial randomness of their local polarization, i.e., its statistical distribution lacking any long-range correlation. An appropriate tool for checking this condition is the autocorrelation analysis of the polarization provided by scanning piezoforce microscopy. It allows monitoring the distribution of polarization within a given sample plane. This is most reliably performed in Ising-type materials, such as SBN, where merely the distribution of *up*- and *down* polarization components have to be checked [10, 28, 34]. As an example we show the PFM image of SBN80 recorded at $T = 295$ K in Fig. 9(a), which delivers the PNR size distribution of the 3D RF-Ising model, Eq. (1), with a cutoff-area $S_0 = 9(2)\,\text{nm}^2$ and exponent $\delta = 1.7(2)$ (Fig. 9(c), solid line). The autocorrelation image (Fig. 9(b)) is obtained from the transformation [10]

$$C(r_1, r_2) = \sum_{xy} D(x, y) D(x + r_1, y + r_2 r_2), \tag{2}$$

where $D(x, y)$ denotes the PFM signal. The decay of the central peak at $r_1 = r_2 = 0$ (Fig. 9(d)) measures the average size of PNRs with correlated polarization according to

$$\langle C(r) \rangle = \sigma^2 \exp\left(-\left[\frac{r}{\langle \xi \rangle}\right]^{2h}\right), \tag{3}$$

where $r = \sqrt{r_1^2 + r_2^2}$, $\langle \xi \rangle = 41(2)$ nm the best fitted average correlation length, and $h = 1.05(3)$ (solid line). It is seen that correlations do not exceed ≈ 100 nm thus confirming the conjectured randomness of the PNR system.

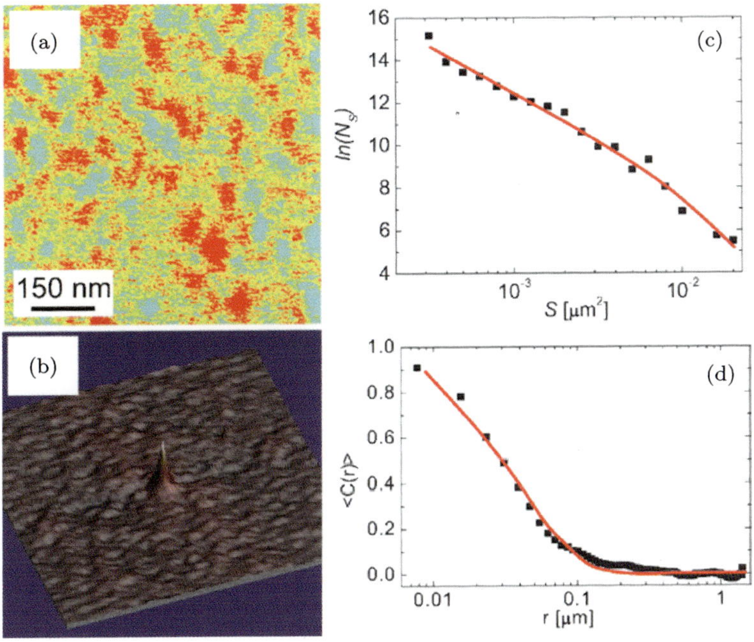

Figure 9. (a) PFM image of SBN80 (001) (color code: see Fig. 2(b)). (b) Autocorrelation image at 295 K. (c) PNR size distribution function, $N_S(S)$, best-fitted to Eq. (1). (d) Distance dependence of the autocorrelation function, $\langle C(r)\rangle$, averaged over all in-plane directions at 295 K and best-fitted to Eq. (3). (a) and (c) reproduced with permission from Ref. [34]. (b) and (d) reproduced with permission from Ref. [39].

4. Glassy ground state of isovalent relaxors

Novel ideas about the origin of the relaxor properties are required in the case of isovalent relaxors lacking net charge disorder such as $BaTi_{1-x}Sn_xO_3$ (BTS) [40] and $BaTi_{1-x}Zr_xO_3$ (BTZ) [41]. Here, the substitution of Ti^{4+} by non-off-central Sn^{4+} and Zr^{4+}, respectively, causes random breaking of Ti–O bonds, the spatial fluctuations of which give rise to fluctuations of the polar correlations.

Upon increasing the amount of broken bonds a redistribution of charges due to local modifications of the polarizability is supposed to give rise to (weak) *random dipolar fields*. Their quenched spatial fluctuations act as pinning centers of the thermally fluctuating polarization [10, 42]. Further, the distortions arising around, e.g., the

Zr^{4+} ions [43] will cause redistribution of the charges and the local formation of charged centers. Obviously, these RFs are much weaker than those stemming from heterovalent cation substitution as in PMN (Fig. 2(a)) or from structural cation vacancies as in SBN (Fig. 2(b)), respectively. As a consequence, the crossover to relaxor behavior occurs only at relatively high doping levels, $x \geq 0.30$ and 0.35 for BTS and BTZ as indicated via PFM by the formation of precursor clusters [40, 41].

Another important source of RFs is *random nanostress* due to ionic volume constraints. They are supposed to provoke the formation of tetragonal domains following local majority rules similarly as in the case of electric RFs [14, 42]. Laulhé *et al.* [43, 44] came to this conclusion in the case of BTZ on the basis of EXAFS and neutron scattering studies, as well as on supercell *ab initio* calculations.

It was observed that different numbers and conformations of nearest neighboring ZrO_6 octahedra cause different displacements of a given Ti^{4+} ion, while different Zr–O bond lengths and chemical environments give rise to different distortions of the TiO_6 octahedra. Random nanostress fields are thus able to originate ferroelastic domain states as conjectured in early discussions of RF-induced domain state phenomena [45]. Random nanostress interaction may, indeed, be a crucial ingredient to form domain states in displacive ferroelectrics such as $BaTiO_3$. They are related to those found in relaxors under the caveat that random stress fields can only control structural disorder, while polar (dis)ordering will be unaffected.

Recently, Akbardazeh *et al.* [46] argued from first-principles calculations that "*antiferroelectric-like interactions between Ti-rich nanopolar clusters gain importance for the relaxor behavior*" at higher Zr^{4+} contents. This statement complies closely with our previous assumption of a secondary dipolar glassy freezing process after thermal stabilization of polar nanoregions [41]. We conjectured this in particular for BTZ with $x = 0.35$ (BTZ35 for short) from its spectrally flat dielectric loss spectra, $\chi_1''(\mathrm{f})$, for $\mathrm{T} \leq 130$ K [47] as confirmed for frequencies $31 \leq f \leq 10^6$ Hz in Fig. 10(a).

We have discussed this conjecture, again, on the basis of non-linear dielectric susceptibility data revealing novel details of the

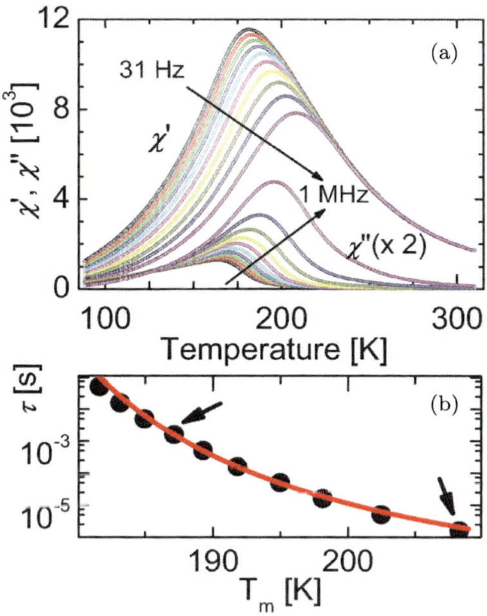

Figure 10. (a) Temperature dependence of χ' and χ'' of BTZ35 measured within $31 \leq f \leq 10^6$ Hz. (b) Relaxation time τ versus $T_{\rm m}(\chi')$ from (a). Reprinted with permission from Ref. [47].

crossover from the ferroelectric to the relaxor via a mixed state in BTZ [47]. The linear and non-linear dielectric susceptibilities χ_1, χ_2, and χ_3 are measured on ceramic samples of BTZ25 and BTZ35 at frequencies $37 \leq f \leq 10^3$ Hz and temperatures $90 \leq T \leq 350$ K as shown in Fig. 11. As x increases the compositional disorder increases, hence, the rounded χ_1 curves (a), are gradually increasing their FWHM by up to ~250% and flattening by up to ~10%. At all concentrations the peak heights (i) increase by ~8% upon decreasing the frequency, and (ii) decrease by typically 1–2% between cooling and heating runs, while (iii) reversibility is regained after heating to $T \approx (1 + \varepsilon)\ T_{\rm m}$ with $\varepsilon \sim 0.03$ at maximum. We attribute these features to the growth and stabilization of PNRs upon cooling as evidenced by calculations [46]. However, only the BTZ35 curves reveal a marked low-T shift of $T_{\rm m}$ by ~8% as f decreases, or by ~15% when extending frequencies up to $f = 1$ MHz (Fig. 10(a)).

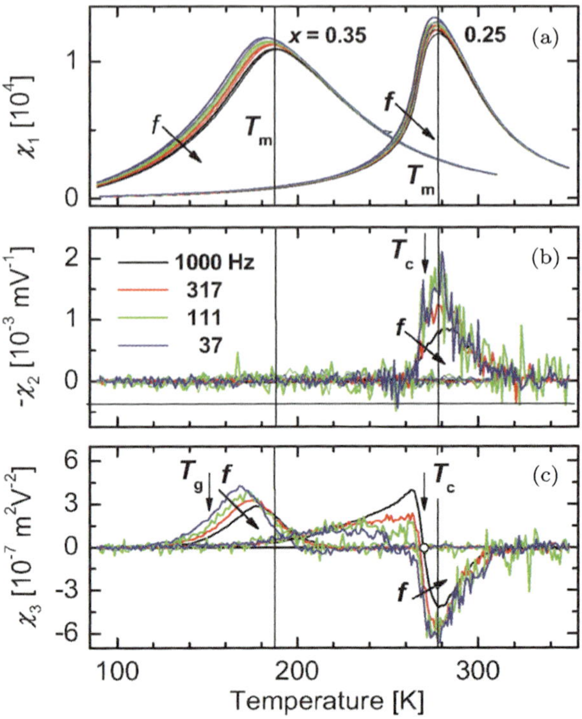

Figure 11. Temperature dependences of χ_1' (a), χ_2' (b), and χ_3' (c) of BTZ25 and BTZ35, respectively, for frequencies $37 \leq f \leq 10^3$ Hz. Reprinted with permission from Ref. [47].

This unambiguous feature of relaxor behavior [1, 6, 14] is expected to obey a Vogel–Fulcher relationship [48].

$$\tau \equiv (2\pi f)^{-1} = \tau_0 \exp\left(\frac{E_{\mathrm{B}}}{T_{\mathrm{m}} - T_{\mathrm{VF}}}\right), \tag{4}$$

where τ_0 is a microscopic attempt time, E_{B} an average energy barrier, and T_{VF} the V_{F} freezing temperature, which must not, however, necessarily imply "freezing" of the system [48]. Fitting the T_{m} data from Fig. 10(a) within 1 kHz $\leq f \leq$ 1 MHz (Fig. 10(b); marked by arrows) to Eq. (4) yields reasonable parameters: $\tau_0 = 1.0 \times 10^{-10}$ s, $E_{\mathrm{B}} = 346$ K and $T_{\mathrm{VF}} = 162.8$ K (Fig. 10(b); solid line). However, when extending the frequency span by three more data points down to $f = 31$ Hz, thus entering a lower T_{m} regime, the reliability of the fit decreases by four orders of magnitude yielding substantially shifted parameters with poorer precision. Obviously, this characterizes the relaxor-to-glass crossover, which occurs at sufficiently low temperatures (see below). This is readily confirmed by the collapse of all χ_1'' data for $T \leq 130$ K in Fig. 10(a) in agreement with the flat loss spectra [41].

More insight into the relaxation processes of the different samples is gained from the second harmonic plotted as $-\chi_2(T)$ in Fig. 11(b), which is an unequivocal signature of macroscopic "spontaneous" polarization P [49]. In case of a second-order transition, it can be expressed as

$$\chi_2 = -3\varepsilon_0^2 BP\chi_1^3, \tag{5}$$

where $B > 0$ refers to the P^4 expansion term of the Ginzburg–Landau–Devonshire free energy [50]. Figure 11(b) indicates large contributions to $\chi_2(T)$ for BTZ25, while it virtually vanishes for BTZ35, obviously due to thermal instability of remnant polarization in too small TiO_6 clusters.

The phase transition temperature is marked by the third harmonic contributions, $\chi_3(T)$, in Fig. 11(c). For BTZ25, the negative high temperature tails transform into positive signals below T_{C}, which clearly signifies a second-order phase transitions, where χ_3 is expected to jump from $-3\varepsilon_0^3 B\chi_1^4$ to $8\varepsilon_0^2 B\chi_1^4$ within mean field theory [50]. The large χ_2 signals observed above T_{C} in the paraelectric regime of BTZ25 (Fig. 11(c)) hint at occurrence of frozen polarization prior to its percolation into a coherent ferroelectric phase. Pretransitional freezing of the polarization supported by local fluctuations of RFs is encountered in PNRs, before they coalesce into the ferroelectric phase.

No homogeneous PNR freezing at all is encountered in BTZ35, where χ_2 virtually disappears, but a substantial positive χ_3 peak emerges (Fig. 11(c)). Obviously, the locally frozen PNRs in BTZ35 do not undergo "superferroelectric" alignment, but arrange cluster glass-like as conjectured for SBN75 [38]. Theoretically [51, 52], a negative divergence of the third order susceptibility is expected at the glass transition temperature in the static limit, $T_g = \lim_{f \to 0} T_f$. However, negative χ_3 values are found only at the high-T edge and at low frequencies, while positive peaks dominate at low T.

This is ascribed to the fact that the nanopolar moments do not flip under the ac field, but just change both shape and center of gravity in the sense of a breathing mode. The same phenomenon and explanation were also encountered in conventional relaxor ferroelectrics within the (*compressible*) *spherical random-bond random-field* model [51, 52]. Obviously locally frozen PNRs in BTZ35 do not undergo "superferroelectric" alignment, but prefer to organize themselves cluster glass-like as we conjectured recently for the extreme relaxor SBN75 [34]. This event is indicated at the lowest frequency, $f = 37$ Hz, by a sharp rise of χ_3 (Fig. 12(a)). Its peak temperature, $T_m(\chi_3) \approx 169$ K, lies well below $T_m(\chi_1) \approx 182$ K, hence, confirming glassy cluster freezing.

It should be noted that the peak rounding of $\chi_3(T)$ is due to presence of random fields. This has to be respected when fitting

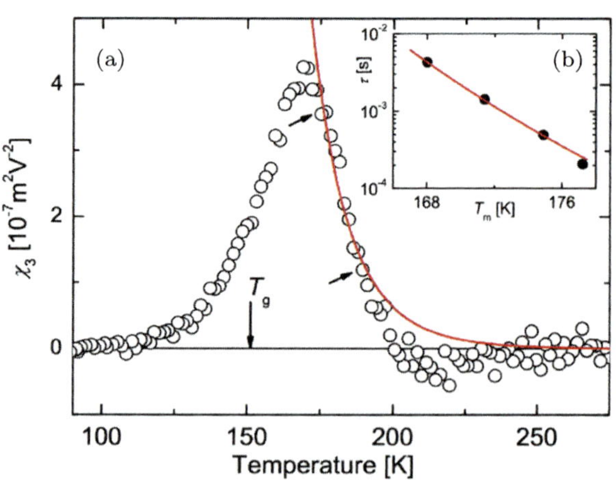

Figure 12. (a) χ_3 versus T measured on BTZ35 on cooling at $f = 37$ Hz, and fitted to Eq. (6) (solid line) for the data range delimited by arrows and yielding $T_g = 151$ K (arrow). (b) Selected relaxation times τ versus T_m of $\chi_3(T)$ for BTZ35 (Fig. 10(a)) best fitted to Eq. (7) (solid line). Reprinted with permission from Ref. [47].

pertinent data (marked by arrows in Fig. 12(a)) by

$$|\chi_3| \propto (T - T_{\mathrm{g}})^{-\gamma}. \tag{6}$$

Best-fitting yields reasonable values of both the glass temperature, $T_{\mathrm{g}} = 151(3)$ K, and the exponent $\gamma = 1.25(0.3)$, which compares well with values from orientational ($\gamma = 0.95 - 1.7$) [51] and magnetic spin and cluster glasses ($\gamma = 0.95 - 1.7$) [5].

Another way to extrapolate the value of T_{g} is offered by the peak shifts of the $\chi_3(T, f)$ data in Fig. 11(c) by assuming glassy critical dynamics [53],

$$\tau \equiv (2\pi f)^{-1} = \tau_0 \exp\left(\frac{T_{\mathrm{m}}}{T_{\mathrm{g}}} - 1\right)^{-zv}. \tag{7}$$

The best-fitted attempt time $\tau_0 = 2.3(0.5) \times 10 - 9$ s agrees with relaxation frequencies typical of nanoparticular systems, while $z\nu =$ 7.5(2.5) complies with literature [51, 52]. The glass temperature $T_{\mathrm{g}} =$ 146(7) K complies within errors with that obtained from Eq. (6), but is definitely much lower than $T_{\mathrm{VF}} \approx 162$ K as extrapolated via Eq. (4) from the $\chi_1(T, f)$ data (Fig. 10(b)). This strongly supports the relaxor-to-cluster glass conjecture of BTZ35.

Hence, the applicability of Eq. (4) is definitely limited to $T'_{\mathrm{m}}(\chi_1) \approx$ 185 K (Fig. 10(b)). On further cooling and passing through the (analytically obscure) crossover regime, cluster glass freezing finally dominates and determines the final glass scenario below ≈ 175 K [47]. Since the different data sets exploited by Eqs. (4) and (7) are centered at higher and lower temperatures, respectively, automatically the most suitable temperature ranges are selected, where the different approaches asymptotically apply.

In conclusion, we have evidenced the Zr^{4+} concentration to control the crossover from ferroelectric to relaxor and cluster glassy behavior in isovalent BTZ ceramics with $x = 0.25$ and 0.35 via dielectric harmonics response. While a ferroelectric transition clearly appears for the $x = 0.25$ compound, a relaxor state is manifest for $x = 0.35$, which finally converts into a cluster glassy ground state via thermally blocked PNRs as conjectured previously [41].

A similar conclusion was recently reached by Novak *et al.* [54] in the case of the prototypical cubic relaxor PMN [1, 2, 9, 14]. Indeed the zero-field behavior of PMN is compatible with our scenario, viz. coarse graining of random dipolar moments via RF stabilization toward PNRs and cluster glass formation via frustrated dipolar interaction of "*blocked superparaelectric*" moments as in SBN and BTZ35. Such behavior was previously conjectured by Tachibana *et al.* [55] from heat capacity and thermal conductivity data.

5. Hierarchical random-field random-bond model

As we have seen in Section 2, charge-disordered relaxor ferroelectrics such as PMN and SBN see primarily RF disorder, which clearly favors the formation of PNRs. By contrast, as stressed in Section 4, the isovalent relaxor BTZ35 primarily seems to undergo RB disorder, but nevertheless also develop PNRs at $x_{\mathrm{Zr}} > 0.3$ as clearly deduced from Raman and linear dielectric spectroscopies [56, 57] and from non-linear dielectric spectroscopy (Fig. 11(c)). Obviously, also in this case the residual RF interaction is powerful enough to create similar polar nanostructures as in PMN (Fig. 2(a)).

Hence, we believe that the essential relaxor properties are initiated by RF-induced generation of metastable PNRs at the Burns temperature, i.e., far above room temperature. Statistical fluctuations of arbitrarily weak quenched RFs are able to stabilize the PNRs above the Curie temperature. This mesoscopic event determines the further development on cooling. This concerns in particular the frustrated dipolar interaction between PNRs and their freezing into a cluster glass as reported on PMN [9, 54], SBN [3, 34, 38], and BTZ35 [47]. Here, we adopt the idea of relaxors being "ferroelectrics with multiple inhomogeneities" [9] rather than that of breeding ferroic PNRs by virtue of Anderson localization in relaxors such as BTZ [46, 58] and PMN [59] treated within pseudospin glass models. Note that unlike PNR freezing via statistical RF fluctuations the Anderson mechanism is supposed to localize also *pseudospin-glassy* extended states [58] in contrast to *polar* ones as being observed and crucial for relaxors.

The simultaneous relevance of RFs and RBs seems to call for the spherical random bond-random field (SRBRF) model of Pirc and Blinc [60] in which the dipole moments of individual spins of variant size interact via a spin-glass-type random exchange coupling, and are subject to quenched random local electric fields. However, this model yields negative χ_3 in contrast to experimental experience (see Fig. 11(c)). An improved SRBRF model with "displaceable superspins" can remedy this deficiency [61]. Alternatively we suggested [62, 63] a positive P_0^2 term due to PNRs to be decisive in the expression of χ_3. This is a clear signature of the mesoscopic regime, which paves the way to cluster glassiness as observed in all of our relaxor prototypes, PMN, SBN, and BTZ.

Recent theoretical attempts to solve the relaxor problem on the basis of small supercell [46] or microscopic Hamiltonians [58, 59] have come to the conclusion that RB-based spin glass solutions may deliver basically correct descriptions. In particular, they seem to have shown that the influence of quenched RFs is virtually negligible. However, we feel that this result is just reflecting the neglect of truly mesoscopic scales in these treatments, a well-known artifact in numerical investigations of the thermodynamic limit for ground states in models with quenched disorder [64]. In our opinion, more adequate answers (Fig. 5) have been given by more realistic attempts to model relaxors [26, 31], which consider mesoscopicity from the beginning. Another argument in favor of the primary importance of RFs is their linear coupling to the order parameter, which on principle should outrival the quadratic one of the RBs [65].

6. Conclusions

Early ideas of a glassy ground state of relaxor ferroelectrics [6–8] at the atomic level have finally enjoyed a revival at the mesoscale. The underlying argumentation at different length scales resides on two subsequent steps: (i) the formation of mesoscopic dipolar clusters (PNRs) and (ii) their frustrated dipolar interaction, which finally enable the cluster glass transition. We have collected fingerprints of such behavior in uniaxial and cubic systems such as SBN75 (Fig. 8 [38]) and BTZ35 (Fig. 11 [47]), respectively. Comparable values of the static glass transition temperature T_g were extrapolated in BTZ35 from both dynamic scaling and the divergence of $\chi_3(T, f)$.

The micro-to-mesoscopic crossover into a cluster glass ground state with non-ergodic properties has independently also been proposed for the archetypical relaxor system PMN as corroborated by caloric anomalies [54].

We are convinced that in all of these cases quenched RFs play a basic role in destroying the conventional ferroelectric phase transition, nucleating mesoscopically ordered, but uncorrelated PNRs, and thus enable cluster glassy ground states. Similar, if not identical, hierarchical rules are very probably imposed to all relaxor systems. It will be challenging to develop a proper theory involving both the microscopic RF interaction and the mesoscopic dipolar coarse graining with frustrated interaction and subsequent cluster glass transition.

Acknowledgments

Thanks are due to Jan Dec, Seweryn Miga, and Vladimir V. Shvartsman for their invaluable cooperation during many years of common research on relaxor physics. Illuminating discussions with David Sherrington are acknowledged.

References

[1] Smolenskii, G. A., Agranovskaya, A. I. and Krainik, N. N. (1953). Segnetoelektricheskie svoistva tverdykh rastvorov $BaTiO_3$-$PbZrO_3$, *Dokl. Akad. Nauk SSSR*, 97, pp. 55–58.

[2] Levstik, A., Kutnjak, Z., Filipic, C. and Pirc, R. (1998). Glassy freezing in relaxor ferroelectric lead magnesium niobate, *Phys. Rev. B*, 57, pp. 11204–11211.

[3] Dec, J., Kleemann, W. and Lukasiewicz, T. (2006). Critical slowing down in the single crystalline relaxor SBN75, *Phase Transit.*, 79, pp. 505–511.

[4] Smolenskii, G. A., Isupov, V. A., Agranovskaya, A. I. and Popov, S. N. (1961). Ferroelectrics with diffuse phase transitions, *Phys. Solid State*, 2, pp. 2584–2594.

[5] Binder, K. and Young, A. P. (1986). Spin glasses: Experimental facts, theoretical concepts, and open questions, *Rev. Mod. Phys.*, 58, pp. 801–976.

[6] Cross, L. E. (1987). Relaxor ferroelectrics, *Ferroelectrics*, 76, pp. 241–267.

[7] Vakhrushev, S. B., Kvyatkovsky, B. E., Naberezhnov, A. A., Okuneva, N. M. and Toperverg, B. P. (1989). Glassy phenomena in disordered perovskite-like crystals, *Ferroelectrics*, 90, pp. 173–176.

[8] Viehland, D., Wuttig, M. and Cross, L. E. (1991). The glassy behavior of relaxor ferroelectrics, *Ferroelectrics*, 120, pp. 71–77.
[9] Fu, D., Taniguchi, H., Itoh, M., Koshihara, S.-Y., Yamamoto, N. and Mori, S. (2009). Relaxor $Pb(Mg_{1/3}Nb_{2/3})O_3$: A ferroelectric with multiple inhomogeneities, *Phys. Rev. Lett.*, 103, pp. 207601.
[10] Shvartsman, V. V., Kleemann, W., Lukasiewicz, T. and Dec, J. (2008). Nanopolar structure in $Sr_xBa_{1-x}Nb_2O_6$ single crystals tuned by Sr/Ba ratio and investigated by piezoelectric force microscopy, *Phys. Rev. B*, 77, pp. 054105.
[11] Imry, Y. and Ma, S. (1975). Random-field instability of ordered state of continuous symmetry, *Phys. Rev. Lett.*, 35, pp. 1399–1401.
[12] Belanger, D. P., Rezende, S. M., King, A. R. and Jaccarino, V. (1985). Hysteresis, metastability, and time-dependence in $d = 2$ and $d = 3$ random-field Ising systems, *J. Appl. Phys.*, 57, pp. 3294–3296.
[13] Nowak, U. (1993). Relaxation of the fractal domain state of diluted antiferromagnets, *Fractals*, 1, pp. 992–997.
[14] Westphal, V., Kleemann, W. and Glinchuk, M. D. (1992). Diffuse phase-transitions and random-field-induced domain states of the relaxor ferroelectric $PbMg_{1/3}Nb_{2/3}O_3$, *Phys. Rev. Lett.*, 68, pp. 847–850.
[15] Kleemann, W. (2012). Random fields in relaxor ferroelectrics — A jubilee review, *J. Adv. Dielectr.*, 2, pp. 1241001.
[16] Lines, M. E. and Glass, A. M. (1977). *Principles and Applications of Ferroelectrics and Related Materials* (Clarendon Press, Oxford), pp. 28–56.
[17] Jeong, I. K., Darling, T. W., Lee, J. K., Proffen, T., Heffner, R. H., Park, J. S., Hong, K. S., Dmowski, W. and Egami, T. (2005). Direct observation of the formation of polar nanoregions in $Pb(Mg_{1/3}Nb_{2/3})O_3$ using neutron pair distribution function analysis, *Phys. Rev. Lett.*, 94, pp. 147602.
[18] Burns, G. and Dacol, F. H. (1983). Glassy polarization behavior in ferroelectric compounds $Pb(Mg_{1/3}Nb_{2/3})O_3$ and $Pb(Zn_{1/3}Nb_{2/3})O_3$, *Solid State Commun.*, 48, pp. 853–856.
[19] Dkhil, B., Kiat, J. M., Calvarin, G., Baldinozzi, G., Vakhrushev, S. B. and Suard, E. (2002). Local and long range polar order in the relaxor-ferroelectric compounds $PbMg_{1/3}Nb_{2/3}O_3$ and $PbMg_{0.3}Nb_{0.6}Ti_{0.1}O_3$, *Phys. Rev. B*, 65, pp. 024104.
[20] Dul'kin, E., Raevskii, I. P. and Emel'yanov, S. M. (2003). Acoustic emission and thermal expansion of $Pb(Mg_{1/3}Nb_{2/3})O_3$ and $Pb(Mg_{1/3}Nb_{2/3})O_3$-$PbTiO_3$ crystals, *Phys. Solid State*, 45, pp. 158–162.
[21] Dkhil, B., Gemeiner, P., Al-Barakaty, A., Bellaiche, L., Dul'kin, E., Mojaev, E. and Roth, M. (2009). Intermediate temperature scale T in lead-based relaxor systems, *Phys. Rev. B*, 80, pp. 064103.
[22] Dul'kin, E., Kojima, S. and Roth, M. (2011). Dielectric maximum temperature non-monotonic behavior in uniaxial $Sr_{0.75}Ba_{0.25}Nb_2O_6$ relaxor seen via acoustic emission, *J. Appl. Phys.*, 110, pp. 044106.
[23] Qian, H. and Bursill, L. A. (1996). Random-field Potts model for the polar domains of lead magnesium niobate and lead scandium tantalate, *Int. J. Mod. Phys. B*, 10, pp. 2027–2047.

[24] Esser, J., Nowak, U. and Usadel, K. D. (1997). Exact ground-state properties of disordered Ising systems, *Phys. Rev. B*, 55, pp. 5866–5872.

[25] Kleemann, W., Dec, J., Lehnen, P., Blinc, R., Zalar, B. and Pankrath, R. (2002). Uniaxial relaxor ferroelectrics: The ferroic random-field Ising model materialized at last, *Europhys. Lett.*, 57, pp. 14–19.

[26] Semenovskaya, S. and Khachaturyan, A. G. (1998). Ferroelectric transition in a random field: Possible relation to relaxor ferroelectrics, *Ferroelectrics*, 206, pp. 157–180.

[27] Semenovskaya, S. and Khachaturyan, A. G. (1998). Development of ferroelectric mixed states in a random field of static defects, *J. Appl. Phys.*, 83, pp. 5125–5136.

[28] Shvartsman, V. V. and Kleemann, W. (2006). Evolution of nanodomains in the uniaxial relaxor $Sr_{0.61}Ba_{0.39}Nb_2O_6$:Ce, *IEEE Trans. Ultrason. Ferroelectr. Freq. Contr.*, 53, pp. 2275–2279.

[29] Burton, B. P., Cockayne, E. and Waghmare, U. V. (2005). Correlations between nanoscale chemical and polar order in relaxor ferroelectrics and the lengthscale for polar nanoregions, *Phys. Rev. B*, 72, pp. 064113.

[30] Waghmare, U. V., Cockayne, E. J. and Burton, B. P. (2003). Ferroelectric phase transitions in nano-scale chemically ordered $PbSc_{0.5}Nb_{0.5}O_3$ using a first-principles model hamiltonian, *Ferroelectrics*, 291, pp. 187–196.

[31] Tinte, S., Burton, B. P., Cockayne, E. and Waghmare, U. V. (2006). Origin of the relaxor state in $Pb(B_xB'_{1-x})O_3$ perovskites, *Phys. Rev. Lett.*, 97, pp. 137601.

[32] Rodriguez, B. J., Jesse, S., Bokov, A. A., Ye, Z. G. and Kalinin, S. V. (2009). Mapping bias-induced phase stability and random fields in relaxor ferroelectrics, *Appl. Phys. Lett.*, 95. pp. 092904.

[33] Kholkin, A., Morozovska, A., Kiselev, D., Bdikin, I., Rodriguez, B., Wu, P., Bokov, A., Ye, Z.-G., Dkhil, B., Chen, L.-Q., Kosec, M. and Kalinin, S. V. (2011). Surface domain structures and mesoscopic phase transition in relaxor ferroelectrics, *Adv. Funct. Mater.*, 21, pp. 1977–1987.

[34] Dec, J., Kleemann, W., Shvartsman, V. V., Lupascu, D. C. and Lukasiewicz, T. (2012). From mesoscopic to global polar order in the uniaxial relaxor ferroelectric $Sr_{0.8}Ba_{0.2}Nb_2O_6$, *Appl. Phys. Lett.*, 100, pp. 052903.

[35] Nowak, U., Esser, J. and Usadel, K. D. (1996). Dynamics of domains in diluted antiferromagnets, *Phys. A*, 232, pp. 40–50.

[36] V. V. Shvartsman, W. Kleemann, D. Kiselev, I. K. Bdikin and Kholkin, A. L. (2010). Polar structures in relaxors by piezoresponse force microscopy, in *Scanning Probe Microscopy of Functional Materials: Nanoscale Imaging and Spectroscopy*, edited by Kalinin, S. V. and Gruverman, A. (Springer Verlag Press, Berlin), pp. 345–384.

[37] Bedanta, S. and Kleemann, W. (2009). Supermagnetism, *J. Phys. D, Appl. Phys.*, 42, pp. 013001.

[38] Dec, J., Kleemann, W., Miga, S., Shvartsman, V. V., Lukasiewicz, T. and Swirkowicz, M. (2007). Aging, rejuvenation, and memory effects in the domain state of $Sr_{0.75}Ba_{0.25}Nb_2O_6$, *Phase Trans.*, 80, pp. 131–140.

[39] Shvartsman, V. V. (unpublished).

[40] Shvartsman, V. V., Dec, J., Xu, Z. K., Banys, J., Keburis, P. and Kleemann, W. (2008). Crossover from ferroelectric to relaxor behavior in $BaTi_{1-x}Sn_xO_3$ solid solutions, *Phase Trans.*, 81, pp. 1013–1021.

[41] Shvartsman, V. V., Zhai, J. and Kleemann, W. (2009). The Dielectric Relaxation in Solid Solutions $BaTi_{1-x}Zr_xO_3$, *Ferroelectrics*, 379, pp. 301–309.

[42] Kleemann, W. (2006). The relaxor enigma — charge disorder and random fields in ferroelectrics, *J. Mater. Sci.*, 41, pp. 129–136.

[43] Laulhé, C., Hippert, F., Kreisel, J., Maglione, M., Simon, A., Hazemann, J. L. and Nassif, V. (2006). EXAFS study of lead-free relaxor ferroelectric $BaTi_{1-x}Zr_xO_3$ at the Zr K edge, *Phys. Rev. B*, 74, pp. 014106.

[44] Laulhé, C., Hippert, F., Kreisel, J., Pasturel, A., Simon, A., Hazemann, J. L., Bellissent, R. and Cuello, G. J. (2011). Random local strain effects in the relaxor ferroelectric $BaTi_{1-x}Zr_xO_3$: Experimental and theoretical investigation, *Phase Transit.*, 84, pp. 438–452.

[45] Kleemann, W. (1993). Random-field induced antiferromagnetic, ferroelectric and structural domain states, *Int. J. Mod. Phys. B*, 7, pp. 2469–2507.

[46] Akbarzadeh, A. R., Prosandeev, S., Walter, E. J., Al-Barakaty, A. and Bellaiche, L. (2012). Finite-temperature properties of $Ba(Zr,Ti)O_3$ relaxors from first principles, *Phys. Rev. Lett.*, 108, pp. 257601.

[47] Kleemann, W., Miga, S., Dec, J. and Zhai, J. (2013). Crossover from ferroelectric to relaxor and cluster glass in $BaTi_{1-x}Zr_xO_3$ (x = 0.25 – 0.35) studied by non-linear permittivity, *Appl. Phys. Lett.*, 102, pp. 232907.

[48] Tagantsev, A. K. (1994). Vogel-Fulcher relationship for the dielectric permittivity of relaxor ferroelectrics, *Phys. Rev. Lett.*, 72, pp. 1100–1103.

[49] Ikeda, S., Kominami, H., Koyama, K. and Wada, Y. (1987). Nonlinear dielectric-constant and ferroelectric-to-paraelectric phase-transition in copolymers of vinylidene fluoride and trifluoroethylene, *J. Appl. Phys.*, 62, pp. 3339–3342.

[50] Miga, S., Dec, J. and Kleemann, W. (2011). Non-linear dielectric response of ferroelectrics, relaxors and dipolar glasses, in *Ferroelectrics: Characterization and Modeling*, edited by Lallart, M. (IntechOpen, London), pp. 181–202.

[51] Binder, K. and Reger, J. D. (1992). Theory of orientational glasses models, concepts, simulations, *Adv. Phys.*, 41, pp. 547–627.

[52] Pirc, R., Tadic, B. and Blinc, R. (1994). Nonlinear susceptibility of orientational glasses, *Phys. B*, 193, pp. 109–115.

[53] Parisi, G. and Rizzo, T. (2013). Critical dynamics in glassy systems, *Phys. Rev. E*, 87, pp. 012101.

[54] Novak, N., Pirc, R., Wencka, M. and Kutnjak, Z. (2012). High-resolution calorimetric study of $Pb(Mg_{1/3}Nb_{2/3})O_3$ single crystal, *Phys. Rev. Lett.*, 109, pp. 037601.

[55] Tachibana, M. and Takayama-Muromachi, E. (2009). Thermal conductivity and heat capacity of the relaxor ferroelectric $(1-x)PbMg_{1/3}Nb_{2/3}O_{3-x}PbTiO_3$, *Phys. Rev. B*, 79, pp. 100104.

[56] Farhi, R., El Marssi, M., Simon, A. and Ravez, J. (1999). A Raman and dielectric study of ferroelectric $Ba(Ti_{1-x}Zr_x)O_3$ ceramics, *Eur. Phys. J. B*, 9, pp. 599–604.
[57] Maiti, T., Guo, R. and Bhalla, A. S. (2008). Structure-property phase diagram of $BaZr_xTi_{1-x}O_3$ system, *J. Am. Ceram. Soc.*, 91, pp. 1769–1780.
[58] Sherrington, D. (2013). BZT: A soft pseudospin glass, *Phys. Rev. Lett.*, 111, pp. 227601.
[59] Sherrington, D. (2014). $Pb(Mg_{1/3}Nb_{2/3})O_3$: A minimal induced-moment soft pseudospin glass perspective, *Phys. Rev. B*, 89, pp. 064105.
[60] Pirc, R. and Blinc, R. (1999). Spherical random-bond-random-field model of relaxer ferroelectrics, *Phys. Rev. B*, 60, pp. 13470–13478.
[61] Pirc, R., Blinc, R. and Kutnjak, Z. (2002). Nonlinear dielectric response of relaxor ferroelectrics, *Phys. Rev. B*, 65, pp. 214101.
[62] Kleemann, W., Miga, S. and Dec, J. (2007). Non-linear dielectric response of relaxor ferroelectrics within ginzburg-landau-devonshire approximation, *Ferroelectrics*, 354, pp. 13–18.
[63] Dec, J., Miga, S., Kleemann, W. and Dkhil, B. (2008). Nonlinear dielectric properties of PMN relaxor crystals within landau-ginzburg-devonshire approximation, *Ferroelectrics*, 363, pp. 141–149.
[64] Middleton, A. A. (1999). Numerical investigation of the thermodynamic limit for ground states in models with quenched disorder, *Phys. Rev. Lett.*, 83, pp. 1672–1675.
[65] Guzman-Verri, G. G., Littlewood, P. B. and Varma, C. M. (2013). Paraelectric and ferroelectric states in a model for relaxor ferroelectrics, *Phys. Rev. B*, 88, pp. 134106.

Chapter 5

Multiferroic Magnetoelectric Composites: Historical Perspective, Status, and Future Directions*

Ce-Wen Nan[†,||], M. I. Bichurin[‡], Shuxiang Dong[§,**], D. Viehland[§] and G. Srinivasan[¶]

[†]*Department of Materials Science and Engineering, State Key Laboratory of New Ceramics and Fine Processing, Tsinghua University, Beijing 100084, People's Republic of China*
[‡]*Institute of Electronic and Informative Systems, Novgorod State University, B. S.-Peterburgskaya st. 41, 173003 Veliky Novgorod, Russia*
[§]*Department of Materials Science and Engineering, Virginia Technology, Blacksburg, Virginia 24061, USA*
[¶]*Physics Department, Oakland University, Rochester, Michigan 48309, USA*
[||]*cwnan@mail.tsinghua.edu.cn*
[**]*sdong@mse.vt.edu*

Multiferroic magnetoelectric materials, which simultaneously exhibit ferroelectricity and ferromagnetism, have recently stimulated a sharply increasing number of research activities for their scientific interest and significant technological promise in the novel multifunctional devices. Natural multiferroic single-phase compounds are rare, and their magnetoelectric responses are either relatively weak or occurs at temperatures too low for practical applications. In contrast, multiferroic composites, which incorporate both ferroelectric and ferri-/ferromagnetic phases, typically yield giant magnetoelectric coupling response above room

*This article was firstly published as "Nan, C.-W., Bichurin, M. I., Dong, S., Viehland, D. and Srinivasan, G. (2008). Multiferroic magnetoelectric composites: Historical perspective, status, and future directions, *J. Appl. Phys.*, 103, pp. 031101". Reprinted with the permission of American Institute of Physics.

temperature, which makes them ready for technological applications. This review of mostly recent activities begins with a brief summary of the historical perspective of the multiferroic magnetoelectric composites since its appearance in 1972. In such composites the magnetoelectric effect is generated as a product property of a magnetostrictive and a piezoelectric substance. An electric polarization is induced by a weak ac magnetic field oscillating in the presence of a dc bias field, and/or a magnetization polarization appears upon applying an electric field. So far, three kinds of bulk magnetoelectric composites have been investigated in experimental and theoretical, i.e., composites of (a) ferrite and piezoelectric ceramics (e.g., lead zirconate titanate), (b) magnetic metals/alloys (e.g., Terfenol-D and Metglas) and piezoelectric ceramics, and (c) Terfenol-D and piezoelectric ceramics and polymer. The elastic coupling interaction between the magnetostrictive phase and piezoelectric phase leads to giant magnetoelectric response of these magnetoelectric composites. For example, a Metglas/lead zirconate titanate fiber laminate has been found to exhibit the highest magnetoelectric coefficient, and in the vicinity of resonance, its magnetoelectric voltage coefficient as high as 10^2 V/cm Oe orders has been achieved, which exceeds the magnetoelectric response of single-phase compounds by many orders of magnitude. Of interest, motivated by on-chip integration in microelectronic devices, nanostructured composites of ferroelectric and magnetic oxides have recently been deposited in a film-on substrate geometry. The coupling interaction between nanosized ferroelectric and magnetic oxides is also responsible for the magnetoelectric effect in the nanostructures as was the case in those bulk composites. The availability of high-quality nanostructured composites makes it easier to tailor their properties through epitaxial strain, atomic-level engineering of chemistry, and interfacial coupling. In this review, we discuss these bulk and nanostructured magnetoelectric composites both in experimental and theoretical. From application viewpoint, microwave devices, sensors, transducers, and heterogeneous read/write devices are among the suggested technical implementations of the magnetoelectric composites. The review concludes with an outlook on the exciting future possibilities and scientific challenges in the field of multiferroic magnetoelectric composites.

1. Introduction

Multiferroic materials [1–3] with coexistence of at least two ferroic orders (ferroelectric, ferromagnetic, or ferroelastic) have drawn increasing interest due to their potential for applications as multifunctional devices. In multiferroic materials, the coupling interaction between the different order parameters could produce new effects, such as magnetoelectric (ME) effect [4–7]. The magnetoelectric response is the appearance of an electric polarization $\boldsymbol{P}$ upon applying a magnetic field $\boldsymbol{H}$ (i.e., the direct ME effect, designated as ME_H effect: $P = \alpha H$) and/or the appearance of a magnetization $\boldsymbol{M}$ upon applying an electric field $\boldsymbol{E}$ (i.e., the converse ME effect, or ME_E: $M = \alpha E$). Magnetoelectricity has been observed as an intrinsic effect in some natural material systems at low temperature, which have been under intensive study recently [8–15], motivated by potential applications in information storage, spintronic, and multiple-state memories. Although over ten different compound families have been widely investigated as multiferroic ME materials, a high inherent coupling between multiferroic order parameters (especially above room temperature) has not yet been found in a single-phase compound, which hinders their applications. Research progress in single-phase multiferroic ME materials have been summarized and reviewed in a series of conference proceedings on ME interaction phenomena in crystals [4–7] and especially in recent review articles [2, 14, 15]. Accordingly, this review will not discuss single-phase multiferroic materials.

Alternatively and with greater design flexibility, multiferroic ME composites [16] made by combining piezoelectric and magnetic substances together have drawn significant interest in recent years due to their multifunctionality, in which the coupling interaction between piezoelectric and magnetic substances could produce a large ME response [17] (e.g., several orders of magnitude higher than that in those single-phase ME materials so far available) at room temperature. These ME composites provide opportunities for potential applications as multifunctional devices such as magnetic-electric transducers, actuators, and sensors.

The ME effect in composite materials is known as a product tensor property [16, 18], which results from the cross interaction between different orderings of the two phases in the composite. Neither the piezoelectric nor magnetic phase has the ME effect, but composites of these two phases have remarkable ME effect. Thus the ME effect is a result of the product of the magnetostrictive effect (magnetic/mechanical effect) in the magnetic phase and the piezoelectric effect (mechanical/electrical effect) in the piezoelectric one, namely [17],

$$\text{ME}_H \text{ effect} = \frac{\text{magnetic}}{\text{mechanical}} \times \frac{\text{mechanical}}{\text{electric}}, \tag{1a}$$

$$\text{ME}_H \text{ effect} = \frac{\text{electric}}{\text{mechanical}} \times \frac{\text{mechanical}}{\text{magnetic}}. \tag{1b}$$

This is a coupled electrical and magnetic phenomenon via elastic interaction. That is, for the ME_H effect, when a magnetic field is applied to a composite, the magnetic phase changes its shape magnetostrictively. The strain is then passed along to the piezoelectric phase, resulting in an electric polarization. Thus, the ME effect in composites is extrinsic, depending on the composite microstructure and coupling interaction across magnetic-piezoelectric interfaces.

Due to technologically viable ME response observed in multiferroic ME composites above room temperature, various ME composites in different systems have been investigated in recent years, including (1) bulk ceramic ME composites of piezoelectric ceramics and ferrites (see, e.g., Refs. [19–46]); (2) two-phase ME composites of magnetic alloys and piezoelectric materials (see, e.g., Refs. [47–68]); (3) three-phase ME composites (see, e.g., Refs. [69–78]); and (4) thin films (nanostructured composites) of ferroelectric and magnetic oxides (see, e.g., Refs. [79–94]). Using the concept of phase connectivity introduced by Newnham *et al.* [18], we can describe the structures of a two-phase composite using the notations 0–3, 2–2, 1–3, etc., in which each number denotes the connectivity of the respective phase. For example, a 0-3–type particulate composite means one-phase particles (denoted by 0) embedded in the matrix of another phase (denoted by 3). So far, 0-3–, 3-3–, 2-2–, and 1-3–type structured multiferroic ME composites of ferroelectric and magnetic phases have been developed. Some prototype ME devices based on the ME composites have been proposed due their large ME effect at room temperature. In this chapter, we will give a comprehensive review on the major research topics and developments of multiferroic ME composites from bulk materials to thin films.

This review is divided into eight sections including the Introduction. Section 2 provides a literature review of prior work since van Suchtelen [16] originally proposed the concept of the product properties in 1972. Sections 3–6 are devoted to the current status of fundamental knowledge and experiments of these four kinds of ME composites mentioned above. Applications as novel ME devices are discussed in Section 7. Section 8 highlights some of the key issues that need to be addressed in the future.

2. Historical perspective

van Suchtelen [16] proposed the concept of a product property in two-phase composite materials in 1972, arising from an elastic coupling between two phases of different properties. The ME effect in a composite material having one magnetostrictive and one piezoelectric phase is one such product tensor property, as described by Eqs. (1a) and (1b) above. In this case, the coupling is mechanical. If an input magnetic field is applied to this composite material, then

$$\frac{\partial S}{\partial H} = e^m, \tag{2a}$$

for the magnetic phase, and

$$\frac{\partial p}{\partial s} = e, \tag{2b}$$

for the piezoelectric phase; where S is the strain and e^m and e are, respectively, piezomagnetic and piezoelectric coefficients. As a result, this two-phase composite material can be characterized by

$$\frac{\partial p}{\partial H} = \alpha = k_c e^m e, \tag{2c}$$

where k_c is a coupling factor ($0 \leq |k_c| \leq 1$) between the two phases [95], and α is the ME coefficient of the composite. Thus, a new property (i.e., nonzero ME coefficient α) appears in the composite consisting of magnetic and piezoelectric phases, since neither constituent phase is magnetoelectric. This new ME product response is due to elastic coupling between the two constituent phases. High piezomagnetic and piezoelectric coefficients and strong coupling (large k_c) favor a large ME coefficient.

Soon after the product ME property in the composite combining magnetostrictive and piezoelectric phases was proposed, scientists in Philips Laboratory experimentally found that a large ME effect could be produced in such composites [19–22]. They prepared ceramic composites of $BaTiO_3$–$CoFe_2O_4$ by unidirectional solidification of eutectic compositions in the quinary Fe–Co–Ti–Ba–O system, and obtained ME voltage coefficients as high as $\partial E/\partial H = 0.13\,\mathrm{V/cm}$ Oe at room temperature. The magnetoelectric effect obtained in this way can reach about 100 times larger than that in single-phase multiferroic one. But unidirectional solidification is complex and requires critical control over composition and processing. At that time, ME composites did not attract attention, and the field of ME composite research went dormant for almost 20 years across the world. Then in the early 1990s, Newnham's group [23] and Russia scientists [24–26] prepared particulate ceramic composites of ferrites and $BaTiO_3$ or $Pb(ZrTi)O_3$ (PZT) by a conventional sintering processing. The sintered ceramic composites were much more easier and cost effective to fabricate in comparison to eutectic composites; and in additional provided the opportunity to combine phases with widely different crystal structures. However, these sintered ceramic composites had lower ME coefficients than the prior eutectic composites in Philips. Although experimental studies of these ME composites in the 1990s did not represent a great step forward, the experiments did inspire significant theoretical work on ME ceramic composites. A few models, such as Green's function approach [17, 96] and various micromechanical approaches (see, e.g., Refs. [97–113]), were developed to understand coupling between the two ferroic phases and to predict the resultant ME tensor responses in bulk ceramic composites. An upsurge in the multiferroic ME composite research appeared in the early 2000s. In the past few years, various ceramic composites with different connectivity schemes including 0–3–type particulate composites and 2–2–type laminate composites have been reported. The maximum ME coefficient in the 2–2–type laminate ceramic composites is several hundreds mV/cm Oe at room temperature [31].

A milestone in the development of ME bulk composites was the appearance of ones containing the giant magnetostrictive rare-earth-iron alloy $Tb_{1-x}Dy_xFe_2$(Terfenol-D) in 2001 [47–49, 114]. In 2001, by developing Green's function technique [114, 115], it was predicted that both particulate composites with Terfenol-D embedded in a piezoelectric polymer matrix such as poly(vinylidene fluoride-trifluoroethylene) copolymer [P(VDF-TrFE)] or a piezoelectric ceramic matrix such as PZT, and laminate composites of Terfenol-D/P(VDF-TrFE) or Terfenol-D/PZT, could exhibit a giant ME (GME) effect. Subsequently, the laminate Terfenol-D/PZT (Refs. [47–49]) and Terfenol-D/PVDF composites [116] have been experimentally found to exhibit such a GME effect. In particular, since 2003, Dong and co-workers [49–66] have reported various laminate composites of Terfenol-D and piezoelectric ceramics, and developed various prototype ME devices based on these bulk composites. The GME response (with a ME coefficient of >1V/cm Oe) of such Terfenol-D based bulk composites made them particularly attractive for technological applications as ME devices.

Recently, in order to overcome brittleness and high eddy current loss of the Terfenol-D disks used in the two-phase Terfenol-D/piezoceramic composites, three-phase ME bulk composites of Terfenol-D/piezoceramics/polymer have been developed [69, 70]. Various three-phase composites with different connectivity schemes (including quasi-0-3, 1-3, and 2-2 types) have been reported [69–78]. The maximum ME coefficient in these three-phase composites is 0.1 V/cm Oe orders at room temperature, and over 1 V/cm Oe at resonance.

Most recently, nanostructured composite thin films of magnetic and ferroelectric oxides have been developed. In 2004, Zheng *et al.* [79] reported a pioneering experiment on nanostructured films of the $BaTiO_3/CoFe_2O_4$ system with 1-3 or 2-2 connectivity schemes. In the last two years, a series of experimental [79–93] and theoretical work [117–121] on such multiferroic nanostructured films of ferroelectric and magnetic oxides has been reported. Such multiferroic nanostructures have become the topic of the day in the multiferroic composites field [94], and they promise potential applications of ME composites in microelectronic devices.

3. Bulk ceramic composites

3.1. *Theories of the ME composites*

3.1.1. *General description*

As for piezoelectric composites [18], the ME composites could have various connectivity schemes, but the common connectivity schemes are 0–3–type particulate composites of piezoelectric and magnetic oxide grains, 2–2–type laminate ceramic composites consisting of piezoelectric and magnetic oxide layers, and 1–3–type fiber composites with fibers of one phase embedded in the matrix of another phase, as shown in Fig. 1. $BaTiO_3$, PZT, $Pb(MgNb)O_3$–$PbTiO_3$, etc., are usually chosen as the piezoelectric ceramic phase, and ferrites usually as the magnetic phase.

The constitutive equation for describing coupling mechanical–electric–magnetic response in the ME composites to linear approximation can be written by direct notation for tensors as

$$\begin{aligned} \boldsymbol{\sigma} &= \boldsymbol{cS} - \boldsymbol{e}^T\boldsymbol{E} - \boldsymbol{q}^T\boldsymbol{H}, \\ \boldsymbol{D} &= \boldsymbol{eS} + \boldsymbol{\varepsilon E} + \boldsymbol{\alpha H}, \\ \boldsymbol{B}&\boldsymbol{qS} + \boldsymbol{a}^T\boldsymbol{E} + \boldsymbol{\mu H}, \end{aligned} \tag{3}$$

where σ, $\boldsymbol{S}$, $\boldsymbol{D}$, $\boldsymbol{E}$, $\boldsymbol{B}$, and $\boldsymbol{H}$ are the stress, strain, electric displacement, electric field, magnetic induction, and magnetic field, respectively; $\boldsymbol{c}$, $\boldsymbol{\varepsilon}$, and $\boldsymbol{\mu}$ are, respectively, the stiffness, dielectric constant, and permeability; $\boldsymbol{e}$ and $\boldsymbol{q}$ are the piezoelectric and piezomagnetic coefficients, respectively; $\boldsymbol{\alpha}$ is the ME coefficient. The superscript T means the transpose of the tensor. The tensors $\boldsymbol{c}$, $\boldsymbol{e}$, $\boldsymbol{q}$, $\boldsymbol{\varepsilon}$, $\boldsymbol{\mu}$, and $\boldsymbol{\alpha}$ are (6×6), (3×6), (3×6), (3×3), (3×3), and (3×3) matrices, respectively, by means of the compressive representation. For the piezoelectric phase (e.g., $BaTiO_3$ and PZT) in the composites, $\boldsymbol{q} = 0$ and $\boldsymbol{\alpha} = 0$; and for the magnetic phase (e.g., Co ferrites and Ni ferrites) in the composites, $\boldsymbol{e} = 0$ and $\boldsymbol{\alpha} = 0$. But for their composites, the effective ME coefficient $\boldsymbol{\alpha}^* \neq 0$, which depends on details of the composite microstructures, i.e., component phase properties, volume fraction, grain shape, phase connectivity, etc.

The first analysis for the ME effect in such bulk ceramic composites was performed by Harshe *et al.* [23]. They assumed a relatively simple cube model, in which the 0–3 particulate composites were considered as consisting of small cubes, and then solved the fields in one cube for which the boundary value problem involved is tractable.

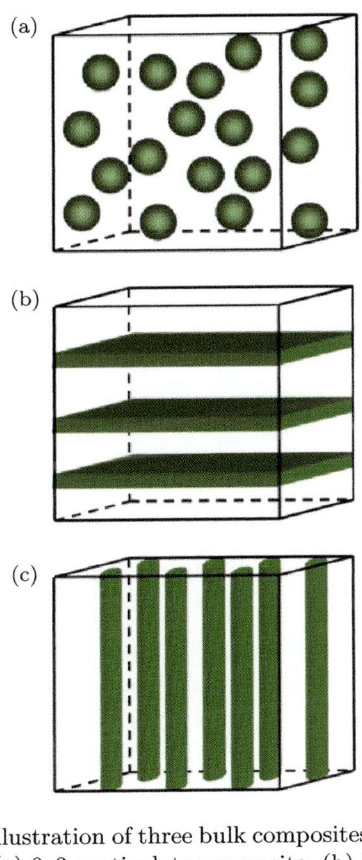

Figure 1. Schematic illustration of three bulk composites with the three common connectivity schemes: (a) 0–3 particulate composite, (b) 2–2 laminate composite, and (c) 1–3 fiber/rod composite.

This simple cube model is an elementary single-grain model. While conceptually straightforward and providing physical insight, this simple model cannot be generalized to other phase topologies. More rigorous treatments of the ME behavior of the composites were performed by using a physically based, Green's function technique (multiple-scattering approach) [17, 96] and micromechanics models [97–113].

3.1.2. *Modeling of particulate ceramic composites*

Green's function technique [17, 96] was developed for solving the constitutive Eq. (3) above for the ME composites, and derived the effective properties of the composites, defined as usual in terms of averaged fields. All effective properties of the composites can be got by this approach. Among the effective ME coefficient tensor α^* of the composites was obtained as

$$a^* = \langle(e^* - \boldsymbol{e})\boldsymbol{T}^{13}\rangle\langle\boldsymbol{T}^{33}\rangle^{-1} = \langle(\boldsymbol{q}^* - \boldsymbol{q})\boldsymbol{T}^{12}\rangle\langle\boldsymbol{T}^{22}\rangle^{-1}, \tag{4}$$

where angular brackets denote the microstructural average; $\boldsymbol{e}^*$ and $\boldsymbol{q}^*$ are, respectively, the effective piezoelectric and piezomagnetic coefficients of the composites. $\boldsymbol{T}^{ij}$ are so-called t-matrix tensors [17]. The theory developed is formally straightforward and is universal.

As an example, firstly consider a 1-3–type composite (Fig. 1) with piezoelectric (or magnetic) rods aligned in a magnetic (or piezoelectric) matrix. Let the piezoelectric phase be poled along x_3 axis of the composite. Thus the composite has a ∞ mm symmetry with ∞ denoting the x_3 axis. The magnetic field is also along the symmetric x_3 axis. In the limit case that the aspect ratio p of the rods approaches infinite, Eq. (4) gives the following simple expression for the ME coefficient along the symmetric x_3 axis:

$$\alpha_{33}^* = -f\frac{e_{33}^* q_{31}}{{}^m k + m^0} = -(1-f)\frac{q_{31}^* e_{31}}{{}^p k + m^0}, \tag{5}$$

where f is the volume fraction of the magnetic phase; $k = (c_{11} + c_{12})/2$ and $m = (c_{11} - c_{12})/2$ are, respectively, transverse in-plane bulk modulus and transverse shear modulus (the superscripts m and p denoting the magnetic and piezoelectric phases; the superscript o denoting a homogeneous reference medium); q_{31} and e_{31} are, respectively, piezomagnetic and piezoelectric coefficients. Different approximations pertain and can be easily obtained from the general solution [Eq. (5)], depending on the choice made for m^0 of the homogeneous reference medium. An intuitive and common choice is to take the host matrix phase as the reference medium. In essence it is a non-self-consistent approximation (NSCA) which is generally valid for matrix-based composites such as a 0-3-type particulate microstructure. Naturally, for the choice $m^0 = m^*$, i.e., the constituent phases are embedded into an effective medium with yet unknown m^*, a self-consistent effective medium approximation (SCA) is captured [96].

For a 1-3-type composite with piezoelectric rods aligned in the magnetic matrix (denoting as 1-3 p/m), the NSCA of Eq. (5) gives

$$\alpha_{33}^* = -\frac{f(1-f)q_{31}e_{31}}{{}^{m}k + {}^{m}m + f({}^{p}k - {}^{m}k)}. \tag{6a}$$

Inversely, for a 1-3-type composite with magnetic rods aligned in the piezoelectric matrix (denoting as 1-3p/m), the NSCA of Eq. (5) becomes

$$\alpha_{33}^* = -\frac{f(1-f)q_{31}e_{31}}{{}^{m}k + {}^{p}m + f({}^{p}k - {}^{m}k)}. \tag{6b}$$

For these two kinds 1-3 microstructures, the SCA of Eq. (5) gives

$$\alpha_{33}^* = -\frac{f(1-f)q_{31}e_{31}}{{}^{m}k + m^* + f({}^{p}k - {}^{m}k)}, \tag{7}$$

Similarly, the constitutive equation [Eq. (3)] for the ME composites can also be solved by using micromechanics methods. Among them, Li [101–105] and Huang [106–108] gave more details about these micromechanics simulations. The micromechanics models are also formally straightforward and universal. It has been already known that micromechanics methods give almost the same approximations as the NSCA from Green's function technique. Figure 2 shows such a comparison between two approaches [122].

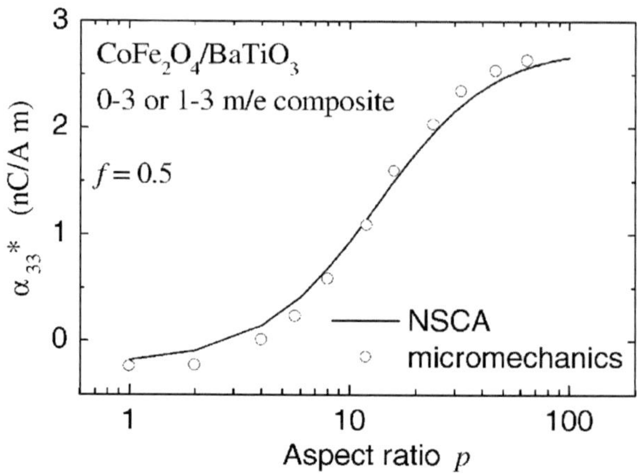

Figure 2. Comparison of calculated α_{33}^{*} for 0–3 or 1–3 m/p $CoFe_2O_4/BaTiO_3$ ceramic composites ($BaTiO_3$ being as the matrix phase) by the micromechanical approximation and NSCA [Ref. 122].

For the extreme case of a 1–3 fiber composites with infinite aspect ratio, Benveniste [97] and Chen [99] proposed a set of relationships between the effective properties including the effective ME coefficients by generalizing Hill's method [123] for the purely elastic case of such a fiber composite. They paid attention to the internal consistency of the solutions for the constitutive coefficient tensors. All these physically based technique and micromechanics models were further expanded by including pyroelectric and pyromagnetic effects [96, 97].

Although it is hard to get such simple expressions for $\boldsymbol{\alpha}^*$ from the general solution [Eq. (4)] for other connectivity schemes, the general expression [Eq. (4)] is easily programmed for numerical calculations of $\boldsymbol{\alpha}^*$ of the composites. Figure 3 further shows a numerical example for the ME response of the 0-3 or 1-3 m/p $CoFe_2O_4/BaTiO_3$ ceramic composites. Here the ME voltage coefficient α_{E33}, describing the MEH output voltage (on open circuit condition) developed cross the composites along the x_3-axis, is used, i.e.,

$$\alpha_{E33} = \alpha_{E33}^{*}/\varepsilon_{E33}^{*}(=E_3/H_3), \tag{8}$$

which is the figure of merit used to assess the performance of a ME material for a magnetic device. A few points can be drawn from

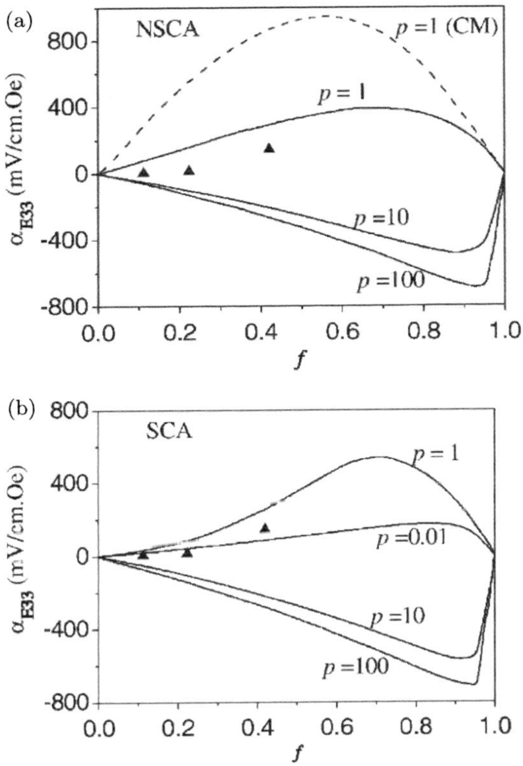

Figure 3. Calculated α_{E33} for 0–3 or $1-3\,m/p$ $CoFe_2O_4/BaTiO_3$ ceramic composites ($BaTiO_3$ being as the matrix phase). The results from the simple cube model (CM) and three experimental data are also shown for comparison [Ref. 17].

Fig. 3, i.e., (1) the simple cube model [23] overestimates the ME effect of the composites [Fig. 3(a)]; (2) NSCA and SCA predict similar results for this system [Figs. 3(a) and 3(b)]; (3) for 0-3 particulate ceramic composites, the ME voltage coefficient α_{E33} reaches a maximum in the middle concentration region around $f \sim 0.6$, but for 1-3 composites the maximum α_{E33} appear at $f \sim 0.9$. Thus remarkable ME effect could be achieved in the composites with high concentration of particulate magnetic phase well dispersed in the piezoelectric phase (see Fig. 4).

It is obvious that the grain shape and phase connectivity have a significant effect on the ME response of the composites. Let us

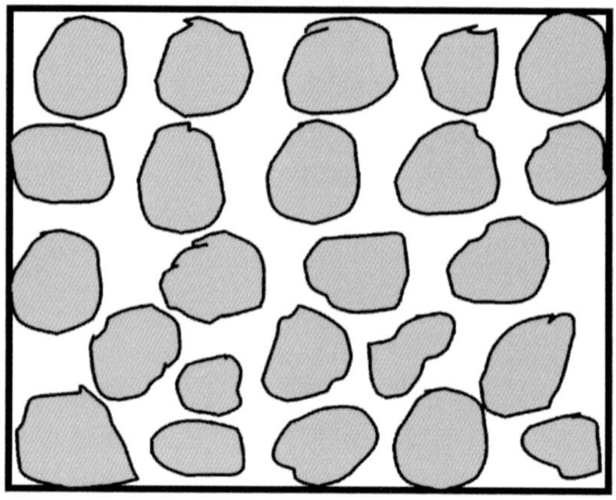

Figure 4. Schematic illustration of a 0–3 particulate ceramic composite with high concentration of particulate magnetic phase well dispersed in the piezoelectric phase.

still consider the case as above that the piezoelectric phase is poled along x_3 axis and the magnetic field is also along the symmetric x_3 axis. Figure 5 shows such an example. From 1-3 to 0-3, α_{E33} of this composite system changes its sign around $p = 3$. Due to strong anisotropy, the 1-3 structure has maximum α_{E33} along the symmetric x_3 axis, but the present case (i.e., both the magnetic field and electrical poling being along the symmetric x_3 axis) is not optimal one for the 2-2 structure. For example, if the magnetic field and/or electrical poling are perpendicular to the symmetric axis, then we can get more larger ME effect in the 2-2 structure (i.e., α_{E31} or $\alpha_{E11} > \alpha_{E33}$), as shown by Getman [98] who also gave the same results for the extreme 1-3 ($p \to \infty$) and 2-2 ($p \to 0$) composites in 1994. Getman gave explicit expressions for the ME coefficients in the two extreme cases, and especially for the 2-2 ($p \to 0$) composite with the magnetic and electrical polings perpendicular to the symmetric axis, the calculated α_{E11} can reach over 1 V/cm Oe, which is higher than that calculated for the 0-3 particulate ceramic composites.

Although the calculations above focus on m/p ferrites/ piezoelectric ceramic composites with the piezoelectric phase as the

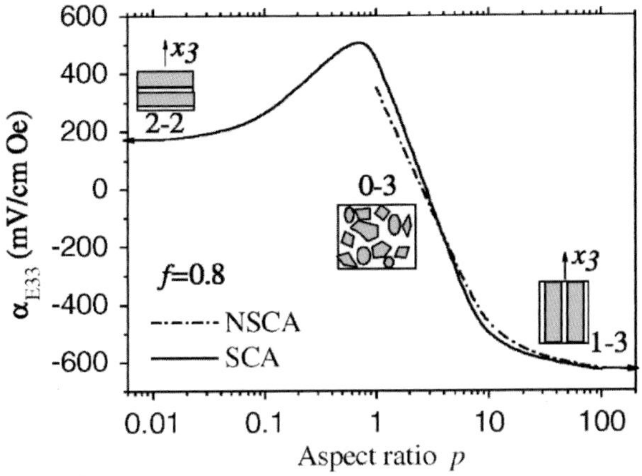

Figure 5. Effect of the grain shape and phase connectivity on the ME response of m/p $CoFe_2O_4$ (denoted by gray areas; volume fraction $f = 0.8$)/$BaTiO_3$ (denoted by white areas) ceramic composites [Ref. 17].

matrix, not high enough resistivity of the ferrite phase would make it hard to obtain high ME response in the 1-3 composites and 0-3 particulate composites with high concentration of ferrite grains (i.e., large f) as expected, due to their large leakage. In comparison, the 2-2 laminate ceramic composites have no such limitation along x_3 because the ferrite layers are separated by the piezoelectric layers. On the other hand, it is easy to prepare the 2-2 laminate ceramic composites in experimental, while such 1-3 ceramic composites are very hard to be prepared. Thus, next we discuss more about modeling of the 2-2 laminate ceramic composites.

3.1.3. *Modeling of laminate ceramic composites*

Static ME effects. Consider a simple bilayered composite of piezoelectric and magnetic ceramic phases with free boundary (Fig. 6). The polarization direction coincides with the axis x_3. For such a simple bilayer structure, the boundary value problem involved is tractable [98, 124]. Thus the constitutive equation [Eq. (3)] can be directly solved by an averaging method to estimate effective material parameters [124–126]. For example, when a magnetic field is applied along x_3(i.e., out-of-plane mode), then the longitudinal ME

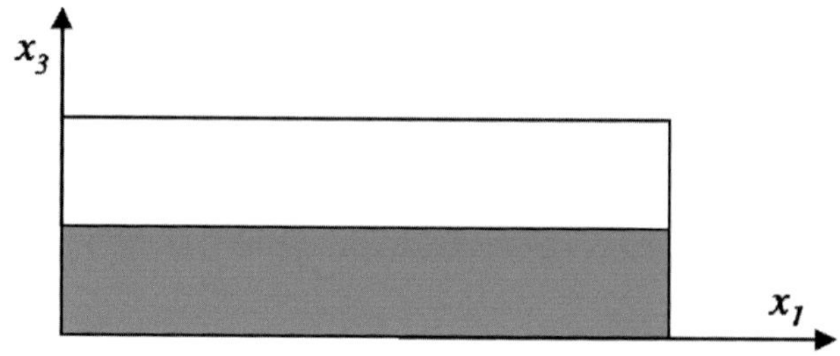

Figure 6. A simple bilayer composite of piezoelectric and magnetostrictive layers.

coefficient is given as [124]

$$\alpha_{E33} = \frac{E_3}{H_3} = \frac{2f(1-f)d_{31}q_{31}\mu_0\overline{s}}{(2fd_{31}^2 - \overline{s})[\overline{\mu}\ \overline{s} + 2q_{31}^2(1-f)^2]} \tag{9}$$

with $\overline{s} = p(f({}^p s_{11} + {}^p s_{12}) + (1-f)({}^m s_{11} + {}^m s_{12})$ and $\overline{\mu} = f\mu_0 + (1-f){}^m\mu_{33}$, where s_{ij} are compliances; d_{33} is the d-type piezoelectric coefficient; q_{ij} are the piezomagnetic coefficients; μ_0 and ${}^m\mu_{33}$ are, respectively, the permeabilities of the free space and magnetic phase. When the magnetic field is applied along x_1 (i.e., in-plane mode), then the transverse ME coefficient is given as [124]

$$\alpha_{E31} = \frac{E_3}{H_3} = \frac{-(1-f)(q_{11}+q_{21})d_{31}}{{}^p\varepsilon_{33}\overline{s} - 2fd_{31}^2}, \tag{10}$$

From Eqs. (9) and (10), one can easily calculate the effective ME coefficients of the 2-2 laminate composites.

Figure 7 shows variation of the ME coefficients with the volume fraction f of the magnetic phase in the 2-2 laminate $CoFe_2O_4$–$BaTiO_3$ ceramic composites. These calculated results are similar to those shown above. For the out-of-plane field mode, a maximum ME coupling appears around $f = 0.4$. For the transverse field mode, this maximum shifts to a bit higher f, and the transverse ME coefficient is larger than the longitudinal ME coefficient, which indicates a stronger transverse coupling than the longitudinal case in such 2-2 laminate composites.

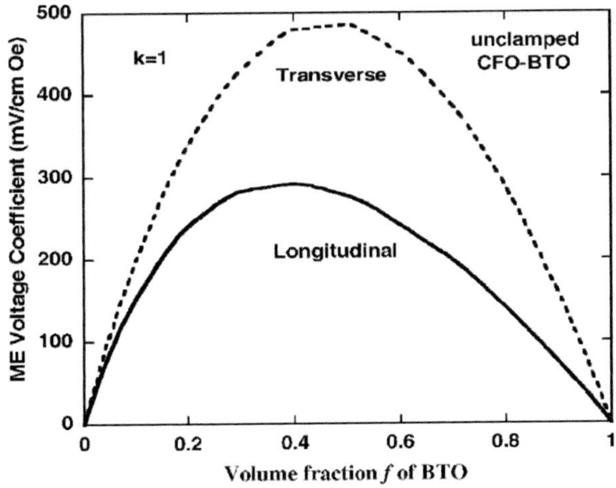

Figure 7. Transverse and longitudinal ME voltage coefficients for a perfectly bonded 2–2 structure of $CoFe_2O_4/BaTiO_3$ [Ref. 125].

In all above simulations, a perfect interface between the magnetic and piezoelectric phases was assumed. This is an ideal case. For such laminate composites, Bichurin *et al.* [125] developed a generalized effective medium method for calculating the composite ME effect by introducing an interface coupling parameter k for characterizing actual bonding conditions at the interface, i.e., $k = ({}^pS_i - {}^pS_{i0})/({}^mS_i - {}^mS_{i0})(i = 1, 2)$, where ${}^pS_{i0}$ is a strain tensor component without frictions between layers [21]. It depends on interface quality and is a measure of differential deformation between piezoelectric and magnetic layers. $k = 1$ for an ideal interface and $k = 0$ for the case without frictions. In this case, the ME coefficients of the laminate composites can be obtained as [125]

$$\alpha_{E31} = \frac{E_3}{H_3} = \frac{-k(1-f)(q_{11}+q_{21})d_{31}}{{}^p\varepsilon_{33}\,\underline{s} - 2kfd_{31}^2}, \tag{11}$$

$$\alpha_{E33} = \frac{E_3}{H_3} = \frac{2k(1-f)d_{31}q_{31}\mu_0\,\underline{s}}{(2fd_{31}^2 - {}^p\varepsilon_{33}\,\underline{s})[\bar{\mu}\underline{s} + 2kq_{31}^2(1-f)^2]}, \tag{12}$$

with $\underline{s} = f({}^ps_{11} + {}^ps_{12}) + k(1-f)({}^ms_{11} + {}^ms_{12})$. Obviously, as $k = 1$, these two expressions reduce to Eqs. (9) and (10); as $k = 0$,

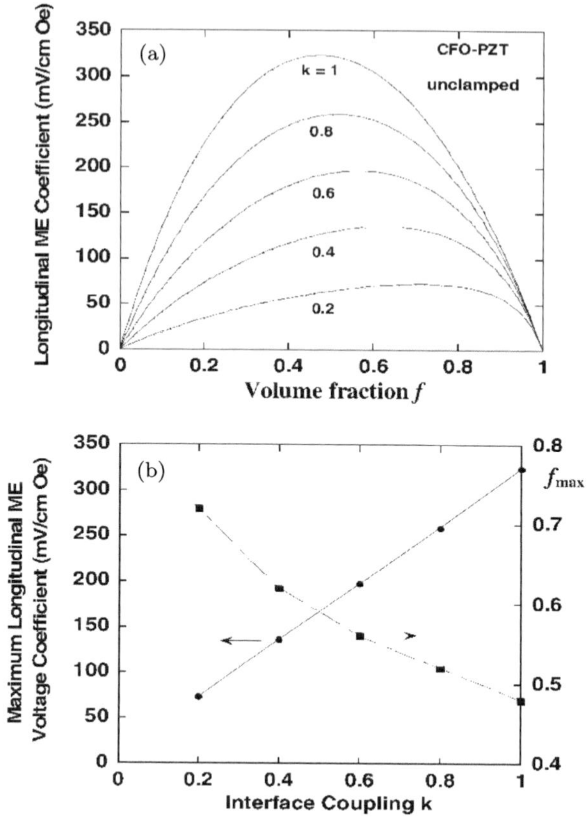

Figure 8. (a) Estimated dependence of longitudinal ME voltage coefficient on interface coupling k and volume fraction f for $CoFe_2O_4$-PZT bilayer. (b) Variation with k of maximum α_{E33} and the corresponding f_{max} [Ref. 125].

the ME effect disappears due to no mechanical coupling between completely unbonded two layers. Any other k values correspond to actual interfaces, and the ME coefficients are between those values for the two extreme cases of $k = 0$ and 1. Figure 8 shows an example calculated from the laminate 2-2 composite of PZT and $CoFe_2O_4$ with different k values. Equations (11) and (12) can well fit experiments by choosing k values or facilitate characterization of the interface bonding status by determining k from the measured ME coefficients.

In these longitudinal and transverse modes, the laminate composite is poled along x_3 and measured in the same x_3 direction. If the composite could be poled with an in-plane electric field (i.e., along x_1) and measured in the same x_1 direction, an in-plane magnetic field would induce a large in-plane longitudinal ME coefficient [125] due to the absence of demagnetization fields and enhanced piezoelectric and piezomagnetic coefficients in this case, as shown in Fig. 9. The maximum ME effect appears at a high f value. This in-plane longitudinal mode in the 2-2 laminate composites is somewhat like the longitudinal mode in the 1-3 fiber composites.

Furthermore, the modeling above for free boundary condition can be generalized to consider the clamped composites [125]. For example, consider the composite be clamped in the x_3 direction. Then the in-plane stresses of the composite $T_1 = T_2 = 0$ (i.e., still free in plane) but the out-of-plane stress $T_3 \neq 0$ ($T_3 \to \infty$ for the rigidly clamped case). A significant change in the ME coupling can be expected when the bilayer is subjected to a uniform out-of-plane stress, as shown in Fig. 9. This x_3-clamping, in general, leads to a large increase in α_{E33}. But for the transverse mode, this clamping leads to a substantial reduction in α_{E31} in comparison to the unclamped case. The clamping induced changes in the in-plane longitudinal ME coupling are quite weak compared to the longitudinal or transverse cases.

Enhanced ME effects at resonance. Bichurin*et al.* first developed a theory for the ME effect at electromechanical, ferromagnetic, and magnetoacoustic resonances in the composites [112, 126–131]. As the ME effect in the composites is due to mechanical coupling between the piezoelectric and magnetic phases, the ME effect would be greatly enhanced when the piezoelectric or magnetic phase undergoes resonance [127–131], i.e., electromechanical resonance (EMR) for the piezoelectric phase and ferromagnetic resonance (FMR) for the magnetic phase. Mechanical oscillations of a medium are induced either by alternating magnetic or electric fields, and the wavelength is tens of meters and much larger than the composite sizes. Thus it is possible to neglect space changing of the electric and magnetic

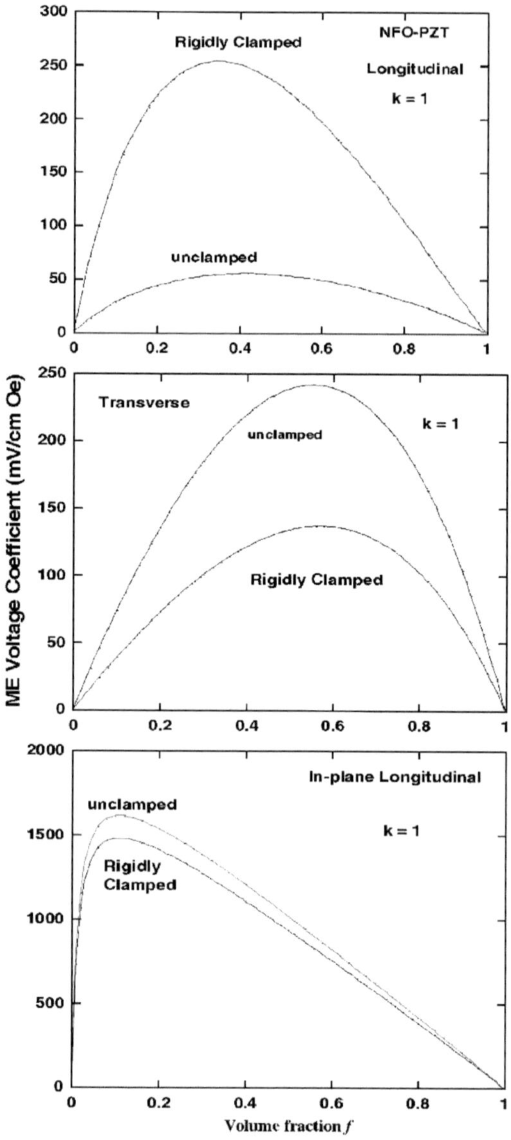

Figure 9. Comparison between f-dependence of the longitudinal, transverse, and in-plane longitudinal ME voltage coefficients for unclamped and rigidly clamped nickel ferrite ($NiFe_2O_4$)-PZT bilayer with $k = 1$ [Ref. 125].

fields within the sample volume. In order to describe frequency-dependent ME effect, the equations of elastodynamics are needed in addition to the constitutive equations [Eq. (3)]. The equations of elastic dynamics are

$$\rho \frac{\partial^2 u_i}{\partial t^2} = \frac{\partial T_{ij}}{\partial x_j}, \tag{13}$$

where u_i is ith projection of the displacement vector, T_{ij} still denote stress tensors which connect with the strain, electric, and magnetic fields by Eq. (3). The joint solutions of Eqs. (3) and (13) under boundary conditions yield the frequency-dependent ME coefficients.

Solution of Eq. (13) depends on the shape of the sample and orientations of the electric and magnetic fields. Solutions for the composites with different shapes can be ready got. For example, consider a rectangular bilayer shown in Fig. 6 with thickness t, width w, and length L. $L \gg t$ and w. The polarization is still along x_3. Under their boundary conditions, by expressing stress components through strain components and substituting in Eq. (13), we can get the displacement u_x, as [129]

$$u_x(x) = \frac{1}{k}\left\{\left[\frac{\cos(kL) - 1}{\sin(kL)}\right]\cos(kx) + \sin(kx)\right\}(d_{31}E_3 + q_{31}H_3), \tag{14}$$

where $k = \omega\,(\rho s_{11})^{1/2}$; ω is the angular frequency; ρ, s_{11}, d_{31}, and q_{31} are, respectively, the effective density, compliance, and piezoelectric and piezomagnetic coefficients of the composite. Replacement of q_{31} with q_{11} gives the expression for the case of the transverse orientation of fields. Further by using the open circuit condition, i.e., $\int_w dy \int_L D_3 dx = 0$, one can work out the induce voltage E_3 in the composite. Using the definition of ME coefficient as $E_3 = \alpha_{E,L}H_3 (=\alpha_{E33}H_3)$ at longitudinal and $E_3 = \alpha_{E,T}H_1 (=\alpha_{E31}H_1)$ at transverse orientation, we obtain the dynamic ME coefficients as [129]

$$\alpha_{E,L} = \frac{d_{31}q_{31}N - \alpha_{33}s_{11}}{\varepsilon_{33}s_{11} - d_{31}^2 N}, \tag{15a}$$

$$\alpha_{E,T} = \frac{d_{31}q_{11}N - \alpha_{31}s_{11}}{\varepsilon_{33}s_{11} - d_{31}^2 N}, \tag{15b}$$

with $N = 1 - (2/kL)\tan(kL/2)$. Here all the property parameters denote the effective ones for the composite. As seen from Eqs. (15a) and (15b) at a so-called frequency of an antiresonance where $\varepsilon_{33}s_{11} - Nd_{31}^2 = 0$, the ME coefficients sharply grow. This antiresonance frequency is dependent on the effective parameters of the composite and its geometrical sizes. For example, for a bilayer of spinel ferrites and PZT with L of about centimeter, this resonance frequency is around 300 kHz.

For illustration, Fig. 10 shows an example for the laminate PZT-$CoFe_2O_4$ composite rectangular plate with a length of 7.3 mm and

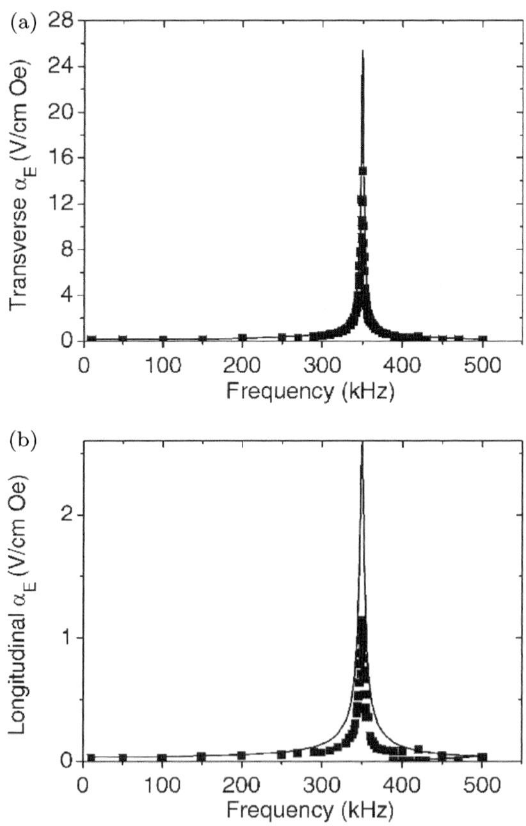

Figure 10. EMR frequency dependence of the transverse and longitudinal ME voltage coefficients of $CoFe_2O_4$-PZT bilayer [Ref. 129].

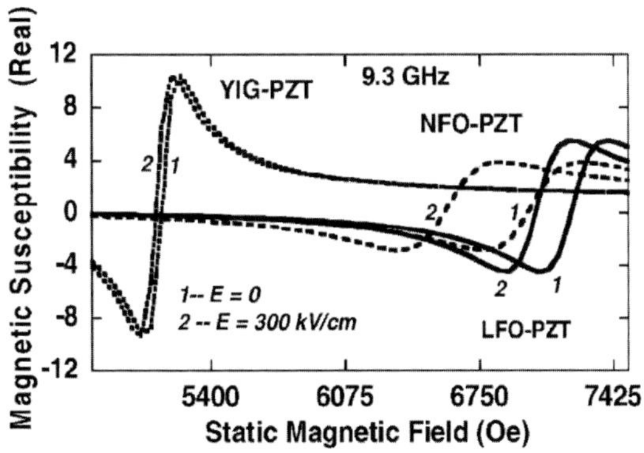

Figure 11. Theoretical static magnetic field H dependence of the magnetic susceptibility for the bilayers of lithium ferrite (LFO)-PZT, nickel ferrite (NFO)-PZT, and yttrium iron garnet (YIG)-PZT. The susceptibilities at 9.3 GHz are for H and E perpendicular to sample plane and for electric fields of (1) $E = 0$ and (2) $E = 300\,\text{kV/cm}$. Notice the downshift in the resonance field when E is increased from 0 to 300 kV/cm [Ref. 127].

width of 2.15 mm. At the frequency of about 300 kHz, the resonance-induced enhancement in the ME effects is observed. The maximal value of the ME coefficient is observed at the transverse orientation of fields, and the resonant value of the ME coefficient is almost ten times higher than its longitudinal orientation value.

The second resonance phenomenon is the FMR. At the FMR, spin-lattice coupling and spin waves that couple energy to phonons through relaxation processes are also expected to enhance the ME interactions. An electric field produces a mechanical deformation in the piezoelectric phase, resulting in a shift in the resonance field for the magnetic phase. Figure 11 shows the estimated resonance profiles with and without electric fields for bilayers of nickel ferrite/PZT, lithium ferrite/PZT, and yttrium iron garnet (YIG)/PZT. The calculations predict a large resonance field shift for the $NiFe_2O_4$ (NFO)/PZT bilayer, and a small shift for the YIG/PZT bilayer [127].

The third resonance phenomenon of importance is the ME interactions at the coincidence of the FMR and EMR, i.e., at the magnetoacoustic resonance (MAR) [112]. Figure 12 provides theoretical estimates on the variation of the ME voltage coefficient

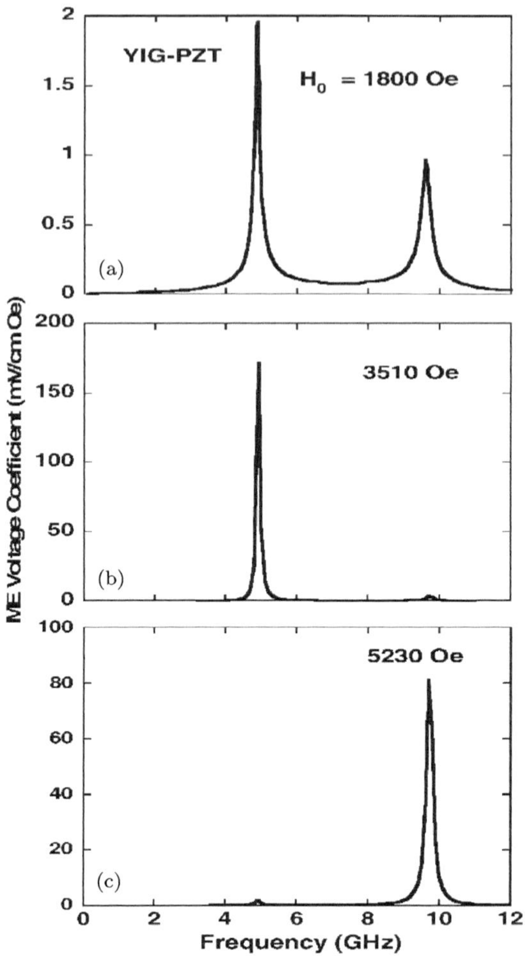

Figure 12. Theoretical estimates on the variation of ME voltage coefficient αE with the frequency of α_E magnetic field for a bilayer of NFO-PZT. (a) The peaks in α_E occur at the fundamental and second harmonic in EMR for thickness modes for the bilayer; (b) the ME coupling at the coincidence of EMR and FMR, and MAR; (c) similar results as in (b), but for MAR at the second harmonic of acoustic modes [Ref. 112].

α_E with frequency of the ac magnetic field applied to a bilayer of NFO/PZT. The thickness of NFO and PZT layers are 100 and 200 nm, respectively. In Fig. 12(a), the bias field H_0 is smaller than the field H_r for FMR in NFO. The peaks in α_E occur at the fundamental and second harmonic of the EMR for the thickness modes of the bilayer. In Fig. 12(b), H_0 is selected so that FMR in NFO coincides with the fundamental EMR mode, resulting in the enhancement in αE at the MAR. In Fig. 12(c), the second harmonic of acoustic modes in NFO-PZT coincides with FMR. Thus one observes a dramatic increase in α_E at MAR.

The modeling provides significant implications, e.g., (1) coincidence of the FMR and EMR allows energy transfer between phonons, spin waves, and electric and magnetic fields, and, for example, this energy transfer is found to be very efficient in the ferrite-PZT bilayer; (2) ultrahigh ME coefficients are expected at MAR. This effect is important for miniature/nanosensors and transducers at microwave frequencies.

3.2. *Experiments of bulk ceramic composites*

In the past decades, various ceramic composites consisting of piezoelectric and magnetic oxide ceramics have been investigated in experimental, mainly including 0–3 particulate and 2–2 laminate ceramic composites (Fig. 1) in the various systems (see Table 1) of $BaTiO_3$ based ferroelectric ceramics and ferrites (or doped ferrites), PZT and ferrites (or doped ferrites), $(SrBa)Nb_2O_5$ (Refs. [43, 132, 133]) and ferrites, and so on. Among them, the particulate ceramic composites are most easily prepared via conventional sintering technique. The powders of piezoelectric ceramics and magnetic oxides (e.g., ferrites)

Table 1. Various constituent phases used for ME ceramic composites

Piezoelectric Phase	Magnetic Phase
$BaTiO_3$ (BTO)	Ni ferrites (e.g., NFO)
PZT	Co ferrites (e.g., CFO)
$Pb(Mg,Nb)O_3$ (PMN)	Li ferrites (LFO)
$PbTiO_3$ (PTO)	Cu ferrite, Mn ferrite
$(Sr,Ba)Nb_2O_5$	Ytttrium iron garnet (YIG)
	$(La,M)MnO_3$ (M=Ca, Sr)

are mixed firstly together, and then the mixed powders are pressed into green bulks followed by sintering at high temperature.

3.2.1. *Particulate ceramic composites*

The original work on *in situ* formation of the ME ceramic composites was done at Philips Laboratories [19–22], which were prepared by unidirectional solidification of a eutectic composition of the quinary system Fe–Co–Ti–Ba–O. Unidirectional solidification helps in the decomposition of the eutectic liquid into alternate layers of a piezoelectric perovskite phase and a piezomagnetic spinel phase. Their results showed that excess of TiO_2 (1.5 wt %) gives a high ME voltage coefficient $\alpha_E = 50\,\text{mV/cm Oe}$. However, other compositions showed a lower αE of about 1–4mV/cm Oe. In a subsequent work, a high ME coefficient of 130 mV/cm Oe was obtained in a eutectic composition of $BaTiO_3$–$CoFe_2O_4$ by unidirectional solidification. Unidirectional solidification requires critical control over the composition, especially when one of the components is a gas (oxygen). Directional solidifications, such as a Bridgman technique or floating zone method using single ellipsoid furnace, are complex involving tight control over the composition, cooling rate, and temperature.

In comparison, sintering processing is much easier and cheaper for fabrication of ME composite ceramics of piezoelectric ceramic phase and ferrites. The sintered composites exhibit several advantages such as freedoms in the selection of constituent phases, their starting particle sizes, processing parameters. Furthermore, sintering does not require the presence of eutectic or eutectoid transformations and also provides the opportunity to combine phases with widely different crystal structures.

A few groups, for example, from USA [23], Russia [25, 26], and India [27], started the preparation of these particulate composite ceramics of BTO or PZT and ferrites via conventional sintering processing in early 1990s. However, the ME coefficients of these particulate composite ceramics were small and around 1mV/cm Oe orders of magnitude [23, 27, 29, 30, 134–142]. Until 2000s, by carefully controlling the sintering processing and composition, higher ME coefficients of about 10–100 mV/cm Oe have been obtained. For example,

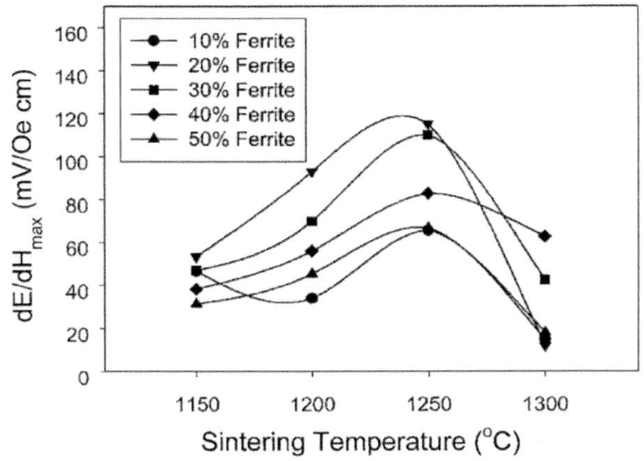

Figure 13. The maximum ME voltage coefficient of NFO-PZT ceramic composites with various compositions as a function of sintering temperature [Ref. 28].

Ryu *et al.* [28] investigated the effect of the sintering temperature on the sintering behaviors, microstructures, and ME properties of particulate composites of PZT and Ni-ferrite doped. It was found that the sintering temperature has a significant effect on the ME coefficient. Figure 13 shows the maximum ME voltage coefficient of various compositions as a function of sintering temperature. A high ME voltage coefficient of 115 mV/cm Oe at 1 kHz was reported for the 0–3 particulate NFO/PZT composites. This high ME coefficient was attributed to a homogeneous and well-dispersed microstructure, and large grain size of the matrix PZT phase.

The low-frequency ME coefficients of the sintered 0–3 composite ceramics vary from about 10 to 100 mV/cm Oe in the literature [25, 26, 28, 39, 40, 132, 133, 143, 144], depending on compositions, powders used, and processing. Their ME coefficients can reach up to V/cm Oe orders of magnitude at resonance. So far, these measured values of the ME coefficients are lower than theoretically predicted coefficients. In spite of the promising high ME coefficients in such simple 0–3 particulate composite ceramics, it is difficult to achieve the predicted ones in experimental, which is due mainly to interdiffusion and/or chemical reactions between the piezoelectric and ferrite phases during high-temperature sintering and to thermal expansion mismatch between two phases. The application of high sintering temperature in conventional sintering processes yields the formation of unwanted phases by chemical reaction (for example, undesired phases such as $BaFe_{12}O_{19}$, $BaCo_6Ti_6O_{19}$, or hexagonal $BaTiO_3$ could appear in the case of $BaTiO_3/CoFe_2O_4$ composites) and the interphase diffusion of the constitutional atoms, though difficult to verify, lowers the local eutectic point around the boundary region thus to facilitates the formation of high concentration of defects and liquid phases. These would deteriorate the piezoelectricity and/or magnetostriction of constituent phases and the strain transfer between two phases. Large thermal expansion mismatch between the piezoelectric and ferrite phases harms the densification and leads to the formation of microcracks.

On the other hand, the theories predict that high ME response appears in the 0–3 particulate composite ceramics with high concentration (e.g., $f > 0.5$) of ferrites (see Fig. 3). However, the percolation of high concentration of the randomly dispersed ferrite phase with low resistivity makes the electric poling of the composites become difficult and the ME properties are reduced because of leakage problem. Therefore, good dispersion of the low-resistance ferrite particles in the piezoelectric matrix is required in order to suppress percolation. Besides, the porosity would influence the ME effect of bulk composite ceramics. For example, the results [143] show a 60%–70% decrease in the low-frequency ME effect and 96% decrease in the ME effect at EMR, as the porosity increases from 5% to 40%.

Recently, in order to achieve sufficient bulk density while avoid possible reactions between the constituent phases to occur, hot pressing [145] and spark plasma sintering [146] (SPS) techniques have been employed to replace the conventional sintering. Hot-pressed and SPS samples exhibited a large improvement in the ME voltage coefficient, as compared to the conventionally sintered samples. Especially, SPS is an efficient sintering method that allows rapid consolidation (e.g., 5 min) at comparatively low temperatures [147], although the underlying mechanisms for an enhanced mass transport within a limited period of time still remain to be understood. By the dynamic features this SPS method was considered to be feasible for the fabrication of the ME ceramic composites with high density (e.g., as high as 99% of theoretical density [146]) and purity. The short time and low temperature required for densification during SPS process diminish the possibility for unwanted reaction to occur. For example, dense composite NFO-PZT ceramics with relative density of 99% have recently been prepared by SPS consolidation of mechanically mixed powder mixtures [146]. The composite ceramics via SPS exhibit fine microstructure (see, e.g., Fig. 14), and the SPS condition has an essential influence on the ME properties. In comparison to the conventional sintering, SPS leads to an obvious improvement in the

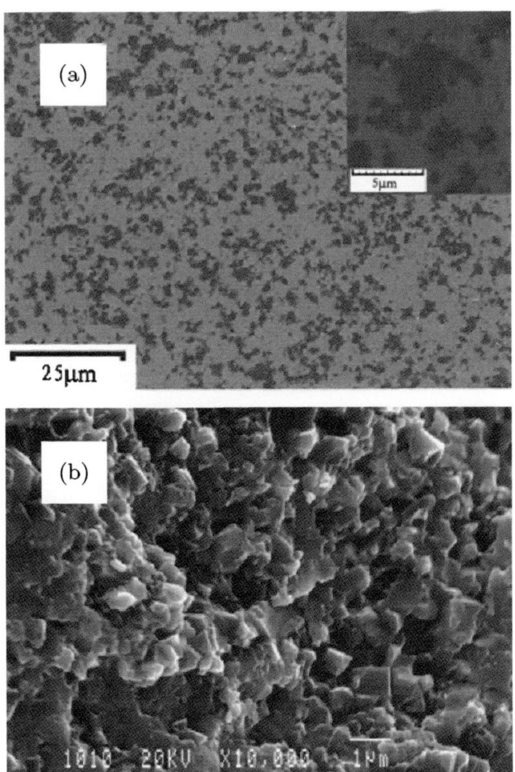

Figure 14. Scanning electron microscopy (SEM) images of (a) polished surface and (b) fractured surface of the NFO-PZT composite ceramics via SPS [Ref. 146].

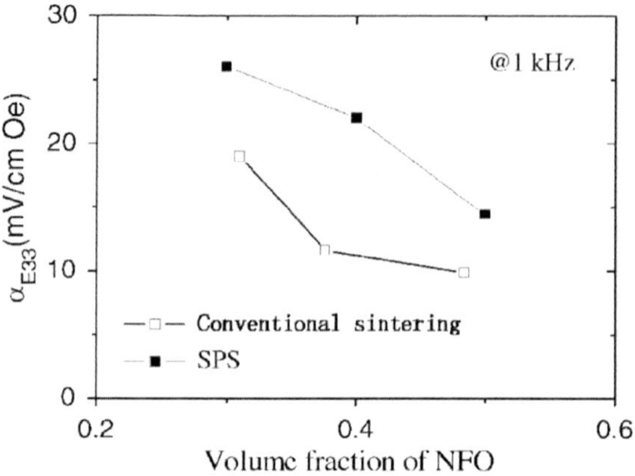

Figure 15. Comparison of the ME coefficients of the 0-3–type NFO-PZT ceramic composites via SPS and conventional sintering.

ME voltage coefficient, as shown in Fig. 15, though the grain size in the ceramics via SPS is much smaller than that via conventional sintering.

In the composite ceramics synthesized by directly mechanical mixing of the two-phase powders, the randomly mixed ferrite particles have low percolation threshold. In order to achieve good dispersion (see Fig. 4) of a high concentration of ferrite particles in the composite ceramics, wet-chemical processing (e.g., sol–gel method) has recently been employed to *in situ* synthesize the homogeneously mixed composite powders of the piezoelectric and ferrite oxides [144, 148, 149]. For example, Ren *et al.* [144] added cobalt ferrite nanoparticles obtained by coprecipitation into the precursor gel of BTO, allowing the *in situ* formation of ferrite-core-BTO-shell and thereby restricting the contact of the ferrite particles during sintering. The *in situ* synthesized samples showed a piezoelectric d_{33} constant approximately six times larger and a ME voltage coefficient approximately three times larger than the corresponding mechanically milled samples. The core–shell structured composites with ferrite cores and piezoelectric shell exhibit enhanced ME effect [144, 148–150].

3.2.2. *Laminate ceramic composites*

The leakage problem due to high concentration of the ferrite phase with low resistivity in the particulate composite ceramics can be eliminated in the laminate 2–2 composite ceramics. Laminate composite ceramics are generally fabricated by cofiring ferrite and piezoelectric ceramic layers at high temperature. ME behavior in laminate composite ceramics has been reported for various material couples (see Table 1) including PZT or $Pb(Mg,Nb)O_3$–$PbTiO_3$ (PMN-PT) layers laminated with Ni-ferrite (e.g., NFO), or Co-ferrite (e.g., CFO), or $(La,Ca)MnO_3$ (LCMO) ones. A laminate composite ceramic exhibits higher ME response than its corresponding particulate composite ceramic.

Srinivasan *et al.* [31, 32, 35, 36] prepared such laminate composite ceramics by tape-casting technique. Both bilayers and multilayers of ferrite piezoelectrics were synthesized from thick films prepared by tape casting. The process involves (i) preparation of submicron size powder of NFO and PZT, (ii) thick film tapes by doctor-blade techniques, and (iii) lamination and sintering of bilayers and multilayers. Ferrite powder obtained by standard ceramic techniques and commercial PZT were used. For tape casting, powders of ferrite or PZT was mixed with a solvent (ethyl alcohol) and a dispersant (Blown Menhaden Fish Oil) and ball milled for 24 h, followed by a second ball milling with a plasticizer (butyl benzyl phthalate) and a binder (polyvinyl butyral) for 24 h. The slurries thus obtained were cast into tapes on silicon coated Mylar sheets using a tape caster consisting of a pair of stationary micrometer controlled blades and a movable casting bed. It was possible to obtain $10 \times 20\,cm^2$ tapes with the thickness in the range of 10–200 μm. The tapes were arranged to obtain the desired structure, laminated under high pressure (3000–5000 psi) and high temperature (400 K), and heated at 1000 K for binder evaporation. The final sintering was carried out at 1400–1500 K.

A high ME voltage coefficients of up to 0.4 V/cm Oe was observed on a laminate NFO/PZT multilayer stack, where NFO was considered to exhibit nearly ideal interface coupling ($k = 1$)

with PZT. Therefore, an important parameter in the selection of the composite constituent phases is interface coupling [see Eqs. (11) and (12)] between the piezoelectric and ferrite phases, which can vary significantly with the dopant concentration. But the composites with cobalt ferrite or lanthanum manganites and PZT exhibit lower ME effects that are attributable to poor interface coupling ($k \leq 0.1$). The interface coupling parameter is dependent on surface inhomogeneities and chemical reaction between constituents during sintering as well as misfit strains.

Another remarkable difference between the laminate and particulate composite ceramics is that the laminate ceramics exhibit much larger anisotropy than the particulate ones. Srinivasan *et al.* [31, 32, 35, 36] analyzed the longitudinal and transverse ME responses for PZT in combination with a few magnetic oxides. The ME coefficient α_E was measured for three conditions as discussed earlier: (i) transverse, (ii) longitudinal, and (iii) in-plane longitudinal. Figure 16 shows representative data on H dependence of α_E for nickel zinc ferrite (NZFO) and PZT. They found that the transverse ME effect yields an up to one order of magnitude higher maximum value than the longitudinal ME effect. The dependence of the transverse and the longitudinal ME voltage coefficients on the magnetic bias

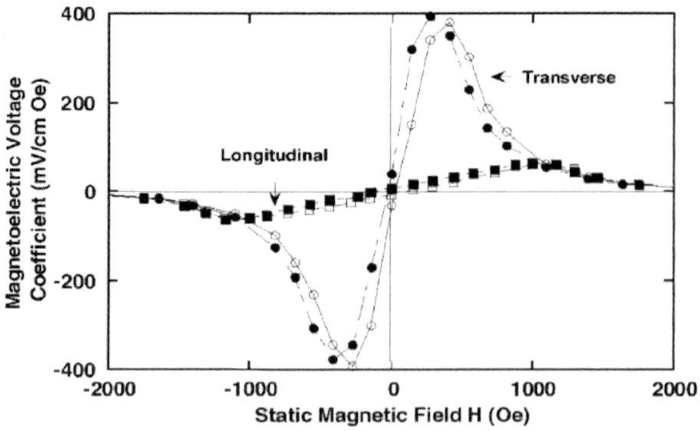

Figure 16. Transverse and longitudinal ME voltage coefficients vs H at 100 Hz for a multilayer sample with 21 layers of Zn-doped NFO and 20 layers of PZT and a layer thickness of 20 μm [Ref. 31].

field is qualitatively different with much higher bias fields for peak longitudinal than for peak transverse ME voltage coefficients. The transverse coefficient is much higher than the longitudinal values due to the absence of demagnetization effects. Further, the strong piezomagnetic coefficient for in-plane magnetic fields is the cause of giant transverse ME coefficients. Saturation of magnetostriction at high field leads to significant decline in the ME voltage coefficient.

The most common 2–2 laminate ME composite is a simple bilayer structure of a ferrite and piezoelectric ceramic layer [151], or sandwich structure [44], with a ferrite layer (or piezoelectric ceramic layer) sandwiched between two piezoelectric (or ferrite) ceramic layers, which is easily obtained by a conventional solid state reaction method. The bilayer and sandwich composite ceramics presented similar dependence of the ME voltage coefficients on the magnetic bias field to the multilayer composite ceramics, but a few times smaller ME voltage coefficients than the latter. The ME voltage coefficients of the sandwich ceramics are strong dependent on their relative thicknesses of the two phases.

As in the particulate composite ceramics, there also easily occurs interdiffusion of elements or chemical reactions [44, 152] between piezoelectric ceramics layer and ferrite layer during the sintering process at high temperature (Fig. 17), which leads to the change in the properties of the piezoelectric and magnetic phases. Different sintering behaviors of two phases lead to porous in the ceramics (Fig. 17). All these deteriorate the ME response of the composite ceramics.

In order to suppress interdiffusion and chemical reaction of the constituents and improve the quality of the laminate composites moderately, the hot-pressing technique [145] has also been employed. Recently a direct comparison of samples made by conventional sintering and by hot pressing was made [145]. For PZT/NFO bulk composites hot pressing enhanced the ME response by more than one order of magnitude, while for PZT/NZFO multilayers a 100% increase was achieved.

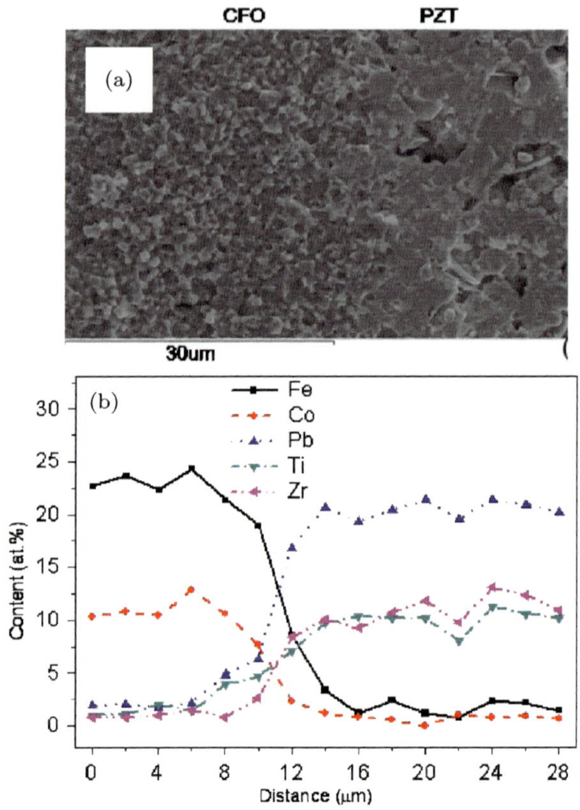

Figure 17. Element distribution across the boundary between CFO and PZT in the sandwich PZT/CFO/PZT ceramics [Ref. 44].

Compared with the particulate composite ceramics, the laminate composite ceramics exhibit high ME coefficients due to elimination of leakage problem. However, high-temperature cofiring processing of the piezoelectric and ferrite ceramic layers is still a big challenge due to different shrinkages, thermal expansion mismatch, and atom interdiffusion and/or chemical reactions, between two ceramic layers during the sintering process at high temperature. Recently, a simple bilayer ceramic composite has been prepared by simply bonding PZT and ferrite ceramic layers together with epoxy [153], not by cofiring at high temperature. In this case, the ME effect is strongly dependent on the interfacial binder.

4. Two-phase composites of alloys and piezoelectric materials

4.1. *Theories*

4.1.1. *Physically-based modeling*

Rare-earth-iron alloys (e.g., $SmFe_2$, $TbFe_2$, or Terfenol-D) are the best known and most widely used giant magnetostrictive alloys exhibiting much higher magnetostriction (over 10^3 ppm) than the magnetic oxide (e.g., ferrite) ceramics. Thus the composites of these alloys and piezoelectric materials should have much large ME response. By generalizing Green's function technique to treat the composites containing these alloys (e.g., Terfenol-D), Nan *et al.* [114, 115, 154] calculated the ME response of such composites and predicted their GME effect.

For Terfenol-D based composites, more generally, by considering the coupling interaction between magnetostriction (*a nonlinear* magnetomechanical effect) and piezoelectricity, the coupling response can be described by the following modified constitutive equation as:

$$\begin{aligned} \boldsymbol{\sigma} &= \mathbf{cS} - \mathbf{e}^{\mathrm{T}}\mathbf{E} - \mathbf{cS}^{\mathrm{ms}} \\ \mathbf{D} &= \mathbf{eS} + \boldsymbol{\varepsilon}\mathbf{E} + \boldsymbol{\alpha}\mathbf{H} \\ \mathbf{B} &= \boldsymbol{\mu}(\mathbf{S}, \mathbf{E}, \mathbf{H})\mathbf{H}, \end{aligned} \tag{16}$$

where the permeability $\boldsymbol{\mu}$ strongly depends on $\mathbf{S}$ and electric and magnetic fields; and $\mathbf{S}^{\mathrm{ms}}$ is the magnetostrictively induced strain related with the magnetic field dependent magnetostriction constants, λ_{100} and λ_{111}, of the magnetostrictive alloy. The effective ME coefficient of the composites can be solved by using Green's function technique as

$$\boldsymbol{\alpha}^*\langle H\rangle = f\boldsymbol{e}^*\langle[\mathbf{I} - \mathbf{G}^u(\mathbf{c} - \mathbf{c}_0)]^{-1}\mathbf{G}^u\mathbf{cS}^{\mathrm{ms}}\rangle_{\mathrm{orient}}, \tag{17}$$

where f is still the volume fraction of the magnetic phase; $\mathbf{c}$ and $\mathbf{c}_0$ are, respectively, the stiffness tensors of the magnetic phase and the homogeneous reference medium; $\mathbf{I}$ is the unit tensor and $\mathbf{G}^u$ is the modified displacement Green's function; $\langle\rangle_{\mathrm{orient}}$ denotes averaging over all possible orientations of the magnetic phase in the composites.

Under the open circuit measurement condition ($\langle \mathbf{D} \rangle = 0$) and a completely mechanically clamped boundary condition ($\langle \mathbf{S} \rangle = 0$), the primary ME output voltage can be obtained from equations above as

$$\bar{\mathbf{E}} = -(\boldsymbol{\kappa}^*)^{-1}\boldsymbol{\alpha}^*\langle \mathbf{H} \rangle = \alpha_E \langle \mathbf{H} \rangle$$
$$= -f\mathbf{h}^*\langle [\mathbf{I} - \mathbf{G}^u(\mathbf{c} - \mathbf{c}_0)]^{-1}\mathbf{G}^u\mathbf{c}\mathbf{S}^{\mathrm{ms}} \rangle_{\mathrm{orient}}, \quad (18a)$$

where $\mathbf{h}^*$ ($=\boldsymbol{\kappa}^{*-1}\mathbf{e}^*$) is the effective piezoelectric stress coefficient tensor. To the other extreme, there is a mechanical free boundary condition, i.e., $\langle \boldsymbol{\sigma} \rangle = 0$, and in this case, the ME output voltage is

$$\overline{\mathbf{E}}^{\mathrm{free}} = \overline{\mathbf{E}} - (\boldsymbol{\kappa}^*)^{-1}\boldsymbol{e}^*\overline{\mathbf{S}}^{\mathrm{ms}}. \quad (18b)$$

The first term in the right side of Eq. (18b) corresponds to the primary ME response, and the second term is the secondary ME response measuring the additional ME response produced by the magnetostrictive strain of the composite through the piezoelectric effect of the composite. These equations are the desired formula for the extrinsic linear ME_H coefficient resulting from the coupling interaction between the *nonlinear magnetostrictive and linear piezoelectric effects.* The equations also show that a strong ME_H response of the composite could be achieved with the large magnetostrictively induced strain, the high piezoelectricity, and the perfect coupling between the phases (i.e., transferring elastic strains without appreciable losses).

For simple 0-3 particulate composites, the low resistitive Terfenol-D grains should be well dispersed in the piezoelectric matrix to keep the composite insulating, since the conductive Terfenol-D grain percolation path can make it difficult to polarize the composites and cause the charges developed in the piezoelectric phase to leak through this conductive path. Thus the Terfenol-D grains must be not in contact with each other, while the piezoelectric matrix is self-connected thus forming a 0-3 connectivity of phases, and the volume fraction of the alloy grains in the piezoelectric matrix is limited by the percolation. For illustrative purpose, Fig. 18 shows the numerical results in the high magnetic field saturation in the whole volume fraction range for the 0-3 particulate Terfenol-D/P(VDF-TrFE) composite with Terfenol-D grains randomly oriented in the

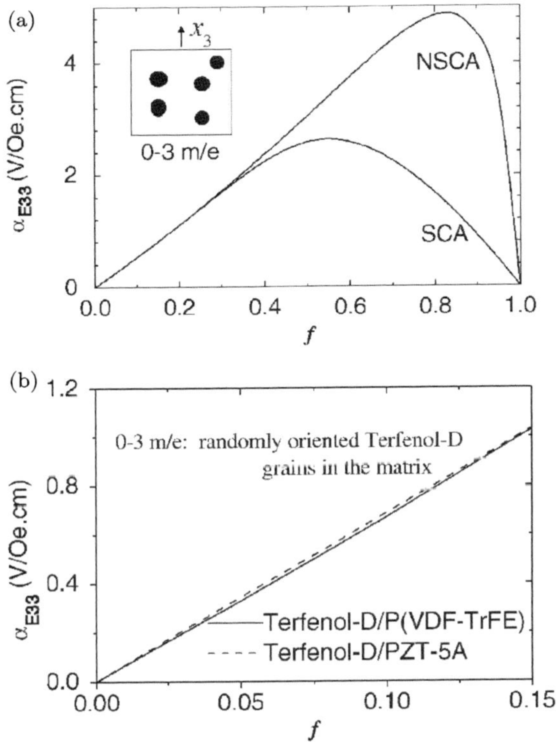

Figure 18. (a) Calculated α_{E33} for 0–3 m/e Terfenol-D/P(VDF-TrFE) particulate composites by NSCA and SCA. (b) Comparison of calculated α_{E33} for the flexible polymer-matrix composite and brittle PZT ceramic matrix composite filled with randomly oriented Terfenol-D particles [Ref. 114].

polymer matrix. The NSCA and SCA predict different maximum α_{E33} at different volume fractions (e.g., at $f \sim 0.85$ and 0.55 for NSCA and SCA, respectively) and different results at high concentration, but they give very close results in the volume fraction range of $f < 0.4$, which means that the theoretical predictions in this volume fraction range of interest are reasonable. The ME coefficients of the particulate composites increase with increasing f in the volume fraction range of $f < 0.4$ of interest. The comparison between the flexible Terfenol-D/P(VDF-TrFE) composite and brittle Terfenol-D/PZT composite [Fig. 18(b)] shows that both systems exhibit similar α_{E33} due to their comparable magnetostriction and piezoelectric stress coefficient h_{33}.

Similarly to the ferrites/piezoelectric ceramic composites discussed in Section 3, although the 0-3–type piezoelectric-matrix composites with Terfenol-D grains embedded are simple, the 2-2 laminate composites are more realizable, since high concentration of conductive Terfenol-D can be easily separated by the piezoelectric layers in the 2-2 composites. Figure 19 shows numerical results for the two 2-2 sandwich composites with perfectly interfacial bonding between PZT and Terfenol-D disks. For the sandwich PZT/Terfenol-D/PZT composite, the ME effect nonmonotonically depends on f with a maximum at $f \sim 0.85$, while the ME effect of the sandwich Terfenol-D/PZT/Tefenol-D composite nearly linearly increases with f. Of particularly interesting to note is that a GME effect is produced in these two laminate composites with thick Terfenol-D layers but thin PZT layers.

In the calculations above, a perfectly bonded interface between magnetic alloys and piezoelectric phases is assumed, which ensures the perfect coupling between these two phases (i.e., transferring elastic strains without appreciable losses) to obtain exceptional control of the predictability of the ME effect in such ME composites.

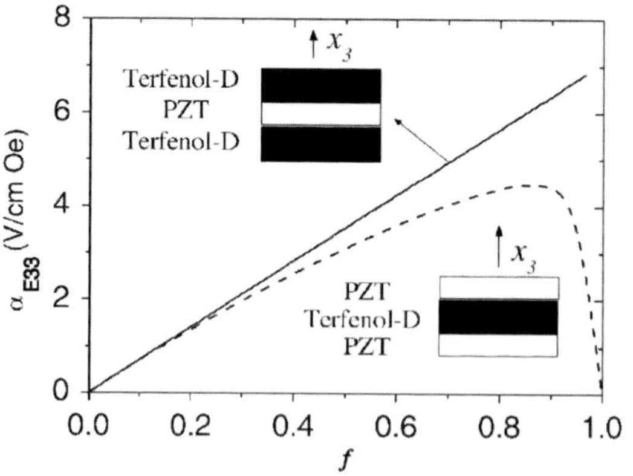

Figure 19. α_{E33} at high-field saturation of these two sandwiched composites as a function of the volume fraction (or relative thickness ratio) of the Terfenol-D layers [Refs. 114 and 155].

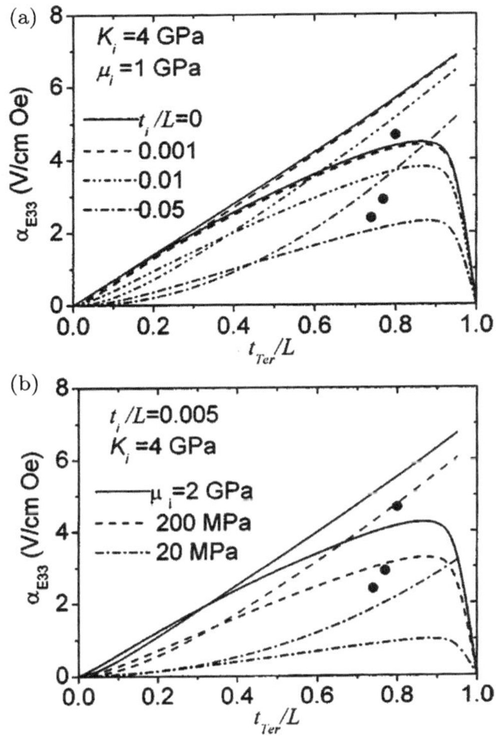

Figure 20. Effect of (a) relative thickness t_i/L(L being the thickness of the composite) and (b) shear modulus μ_i of the interfacial binder layers on α_{E33} at high-field saturation of these sandwich composites. Recent experimental data available for a Terfenol-D/PZT/Terfenol-D composite are also shown for comparison [Ref. 155].

Any imperfect interfaces will more or less decrease the displacement transfer capability, thereby leading to a decrease in the ME response of the composites. In order to treat the interfacial bonding effect, we change the shear modulus μ_i and the relative thickness t_i of the interfacial binder [155]. Figure 20 shows the influence of interface properties on α_{E33} of these sandwich composites. The comparison in Fig. 20 shows that the predictions are in reasonable agreement with experimental data available, though the interfacial bonding status in the experiment is not clear. The increase in the thickness ti of the interfacial binder films between the PZT and Terfenol-D leads to a decrease in the α_{E33} values [Fig. 20(a)]. Figure 20(a) shows that a very thin binder film (e.g., $t_i/L \sim 0.001$) with bonding capability is good enough for producing the GME response. The elastic modulus of the interfacial binder films has a significant effect on the α_{E33} values [Fig. 20(b)]. With decreasing the shear modulus μ_i of the thin binder films (i.e., using a very flexible binder), the interfacial bonding between the PZT and Terfenol-D becomes weak due to the formation of a sliding interface. The weak interfacial contact would lead to appreciable losses of transferring elastic strain/stress from Terfenol-D to PZT, and thus the decrease in the shear modulus μ_i of the thin interfacial binder films results in a large decrease in the ME response of the composites.

If using a piezomagnetic coefficient ($c\partial S^{\mathrm{ms}}/\partial H$) to replace the magnetostrictive strain and still using Eq. (3) rather than Eq. (16), the micromechanics modeling and averaging method discussed in Section 3.1.3 could be still applicable for such Terfenol-D based composites.

4.1.2. *Equivalent-circuit modeling*

For the 2-2 laminate composites, Dong *et al.* [156–158] proposed an equivalent-circuit approach which is more convenient for modeling the ME coupling in the dynamic cases. This approach is also based on magnetostrictive and piezoelectric constitutive equations, where the magnetostrictive and piezoelectric layers are mutually coupled through elastic interaction, via an equation of motion that is excited by a magnetic field H. We suppose that the magnetostrictive/piezoelectric laminates are long-type configurations (Fig. 21), in which the piezoelectric layer is polarized along either its thickness or length directions and stressed by two magnetostrictive layers along their length (i.e., principal strain) direction. Accordingly, the piezoelectric constitutive equations for one-dimensional motion are

$$S_{1p} = S_{11}^{E} T_{1P} + d_{31,p} E_3, \qquad D_3 = d_{31,p} T_{1p} + \varepsilon_{33}^{T} E_3, \tag{19a}$$

for the thickness poling case, and

$$S_{3p} = S_{33}^{D} T_{3p} + g_{33p} D_3, \qquad D_3 = -g_{33p} T_{3p} + \beta_{33}^{T} D_3. \tag{19b}$$

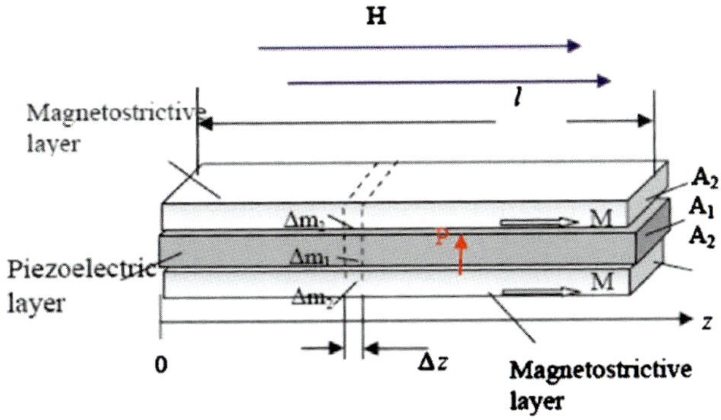

Figure 21. Long-type magnetostrictive/piezoelectric/magnetostrictive laminate [Ref. 156].

For the length poling case, where D_3 is the electric displacement, ε_{33}^T and β_{33}^T are the dielectric permittivity and impermeability under constant stress T, s_{33}^E and s_{33}^D are the elastic compliances of the piezoelectric material under constant electric field E or constant electric displacement $D, d_{31,p}$ and $g_{33,p}$ are the transverse piezoelectric constant and longitudinal piezoelectric voltage constants, and T_{1p}, T_{3p} and S_{1p}, S_{3p} are the stress and strain of the piezoelectric layer imposed by the magnetostrictive layers.

When **H** is applied parallel to the longitudinal axis of the laminate, a longitudinal strain is excited. The piezomagnetic constitutive equations for the longitudinal mode are

$$S_{3m} = S_{33}^H T_{3m} + d_{33,p} H_3, \quad B_3 = d_{33,m} T_{3m} + \mu_{33}^T H_3, \tag{20}$$

where B_3 is the magnetization along the length direction, μ_{33}^T is the permeability under constant stress, S_{33}^H is the elastic compliance of the magnetostrictive layer under constant **H**, $d_{33,m}$ is the longitudinal piezomagnetic constant, and T_{3m} and S_{3m} are the stress and strain in the longitudinal direction of the magnetostrictive layers imposed on the piezoelectric layer. These constitutive equations are linear relationships, which do not account for loss components. Significant nonlinearities in both piezoelectric and magnetostrictive materials are known to exist, especially under resonance drive. We will introduce a mechanical quality factor Q_m to include these losses later.

Assuming harmonic motion, along a given direction z, it will be supposed that three small mass units Δm_i in the laminate have the same displacement $u(z)$. This follows from Fig. 21, by assuming that the layers in the laminate act only in a coupled manner. Following Newton's second law, we then have an equation of motion to couple the piezoelectric and piezomagnetic constitutive equations [Eqs. (19a) and (19b)] as

$$\overline{\rho}\frac{\partial^2 u(z)}{\partial t^2} = n\frac{\partial T_{3,m}}{\partial z} + (1-n)\frac{\partial T_{i,p}}{\partial z}\ (0 < n < 1, i = 1\,\text{or}\ 3), \tag{21}$$

where $\overline{\rho} = (\rho_p A_p + \rho_m A_m)/A_{\text{lam}}$ is the average mass density of the laminate, and $n = A_m/A_{\text{lam}} = t_m/t_{\text{lam}}$ is a geometric factor, $t_m = t_{m1} + t_{m2}$ is the total thickness of the magnetic phase layers, A_p and $A_m(=A_{m1} + A_{m2})$ are the cross-sectional areas of the magnetic phase and piezoelectric phase layers, respectively, ρ_p and ρ_m are the mass densities of the piezoelectric and magnetostrictive layers. For a given laminate width w_{lam} and thickness t_{lam}, the total cross-sectional area of the laminate is $A_{\text{lam}} = t_{\text{lam}}w_{\text{lam}}$, and the total thickness is the sum of the layer thicknesses $t_{\text{lam}} = t_p + t_m$, where t_p is the thickness of the piezoelectric layer.

By combining piezoelectric and piezomagnetic constitutive equations [Eqs. (19a) and (19b)], solutions to the equation of motion [Eq. 21)] can be derived for longitudinally magnetized $\boldsymbol{M}$ and transversely or longitudinally poled $\boldsymbol{P}$ magnetoelectric modes, respectively, named as *L–T* or *L–L*. Correspondingly, the two magneto-(elasto)-electric (or ME) equivalent circuits for *L–T* and *L–L* modes under free boundary conditions can be derived, as given in Fig. 22 [156, 158], where an applied magnetic field $\boldsymbol{H}$ acts as a magnetic induced "mechanical voltage" ($\varphi_m H_3$), and then induces a "mechanical current" ($\dot{u}_1$ and $\dot{u}_2$) via the magnetoelastic effect with a coupling factor φ_m. In turn, $\varphi_m H_3$ results in an electrical voltage V, and $\dot{u}_1$ and $\dot{u}_2$ result in a current I_p across the piezoelectric layer due to electromechanical coupling. A transformer with a turn ratio of φ_p can then be used to represent the electromechanical coupling. In the circuits of Fig. 22, Z_1 and Z_2 are the characteristic mechanical

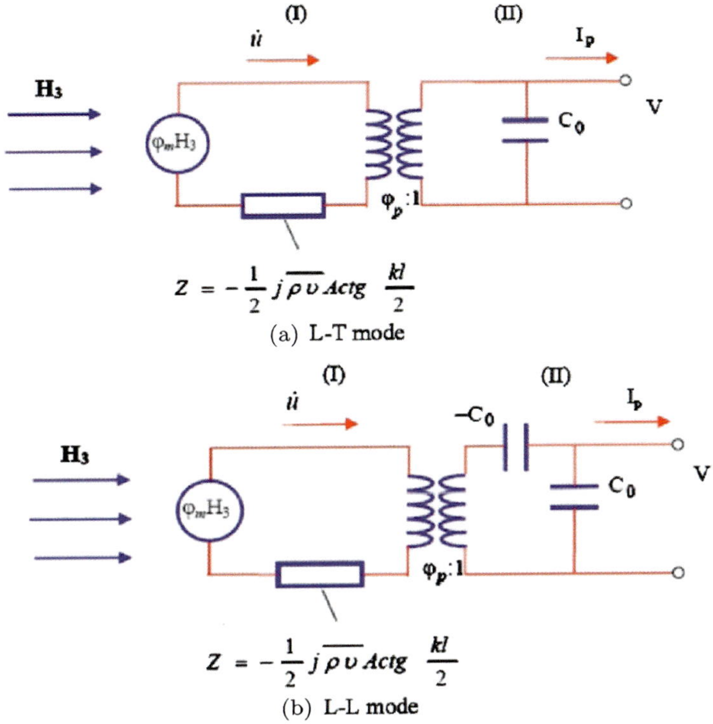

Figure 22. Magneto-elastic-electric bioeffect equivalent circuits for (a) *L–T* mode and (b) *L–L* mode [Refs. 156 and 158].

impedances of the composite, and C_0 is the clamped capacitance of the piezoelectric plate.

Following Fig. 22, open-circuit conditions have been supposed from above, where the current I_p from the piezoelectric layer is zero. Thus, the capacitive load C_0 (and $-C_0$) can be moved to the main circuit loop. Applying Ohm's law to the mechanical loop, the following ME coefficients at low frequency for *L–T* and *L–L* modes can be directly derived as

$$\left|\frac{dE}{dH_3}\right|_{(L-T)} = \frac{nd_{33,m}g_{31,p}}{ns_{11}^E(1-k_{31,p}^2)+(1-n)s_{33}^H}, \tag{22a}$$

$$\left|\frac{dE}{dH_3}\right|_{(L-L)} = \frac{nd_{33,m}g_{33,p}}{ns_{33}^E(1-k_{33}^2)+(1-n)s_{33}^H}, \tag{22b}$$

Thus the ME coefficients are also proportional to the piezomagnetic constant $d_{33,m}$, the piezoelectric voltage constants ($g_{31,p}$ or $g_{33,p}$), and the thickness ratio n of the Terfenol-D layers.

In order to derive the ME coefficients at resonance frequency, we assume the long-type ME laminate composites to be a L–L $\lambda/2$-resonator, operating in a length extensional mode, the series angular resonance frequency is $\omega_s = \pi\overline{v}/l$, where l is the length of the laminate and $\overline{v}$ is mean acoustic velocity. Under resonant drive [158], the mechanical quality factor Q_m of the laminate is finite due to both mechanical and electric dissipations. This limitation of the vibration amplitude must also be included in order to predict the resonant response. Finite values of Q_{mech} result in an effective motional mechanical resistance of $R_{\text{mech}} = \pi Z_0/8Q_{\text{mech}}$; accordingly, the equivalent circuit of the laminates for the L–L mode under resonance drive is given in Fig. 23. At electromechanical resonance ($\omega = \omega_s$), dV/dH of the L–L mode reaches a maximum value of

$$\left(\frac{dV}{dH}\right)_{\omega s} = \frac{4Q_m\varphi_m\varphi_p}{\pi Z_0\omega_s C_0}, \tag{23}$$

where Q_m is the effective mechanical quality factor of the laminate composite including contributions from the Terfenol-D and piezoelectric layers, and also from the bonding between layers. Analysis has shown that dV/dH at the resonance frequency is $\sim Q_m$ times higher than that at subresonant frequencies. Using a similar approach, it is ready to obtain the resonance equivalent circuit for the L–T mode.

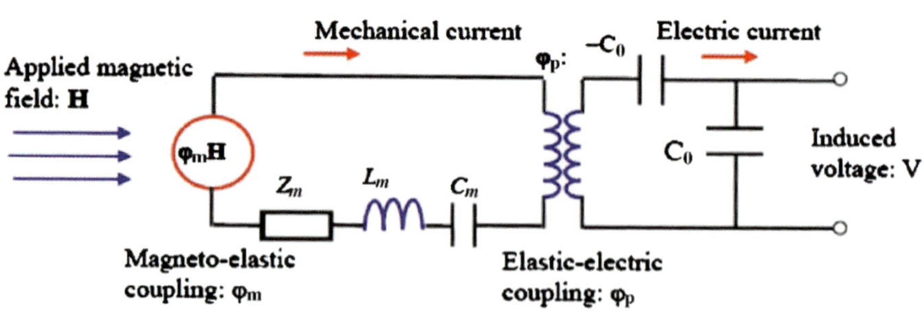

Figure 23. Magneto-elastic-electric equivalent circuits for L–L at resonance [Refs. 156 and 157].

The 2-2 laminate structures can also be modeled numerically by using finite-element method [159, 160].

4.2. *Experiments*

The GME effect predicted was reported first in the Terfenol-D/PZT laminate by Ryu *et al.* [47, 48] and in the Terfenol-D/PVDF laminate by Mori and Wuttig [116]. Since then, in particular, Dong *et al.* [49–66, 156–158] have reported the GME effect in a number of laminate composites of Terfenol-D and various piezoelectric materials including PZT ceramics, $Pb(Mg_{1/3}Nb_{2/3}O_3)$–$PbTiO_3$ (PMN-PT) or $Pb(Zn_{1/3}Nb_{2/3}O)_3$–TiO_3 (PZN-PT) single crystal, or electroactive PVDF copolymers.

The laminate composites are generally fabricated by bonding magnetostrictive and piezoelectric layers (Table 2) using an epoxy resin, followed by annealing at a modest temperature of 80–100°C. Typically, the laminate composites are made of two magnetostrictive layers and a single piezoelectric one, where the piezoelectric layer is sandwiched between the two magnetostrictive ones. However, ME laminates can be made in many different configurations [47–66, 156–166] including disk, rectangular, and ring shapes. These various configurations can be operated in numerous working modes including *T–T* (transverse magnetization and transverse polarization), *L–T*, *L–L*, symmetric *L–L* (push–pull) longitudinal vibrations; *L–T* unimorph and bimorph bending; *T–T* radial and thickness vibrations multilayer; and *C–C* (circumferential magnetization and circumferential polarization) vibration modes, as shown in Figs. 24–27.

Table 2. Various constituent phases used for ME laminate composites.

Piezoelectric Phase	Magnetic Phase
PZT	Terfenol-D
PMN-PT, PZN-PT	Ni, Permendur, Fe–Ga, Ni2MnGa
PVDF	Metglas

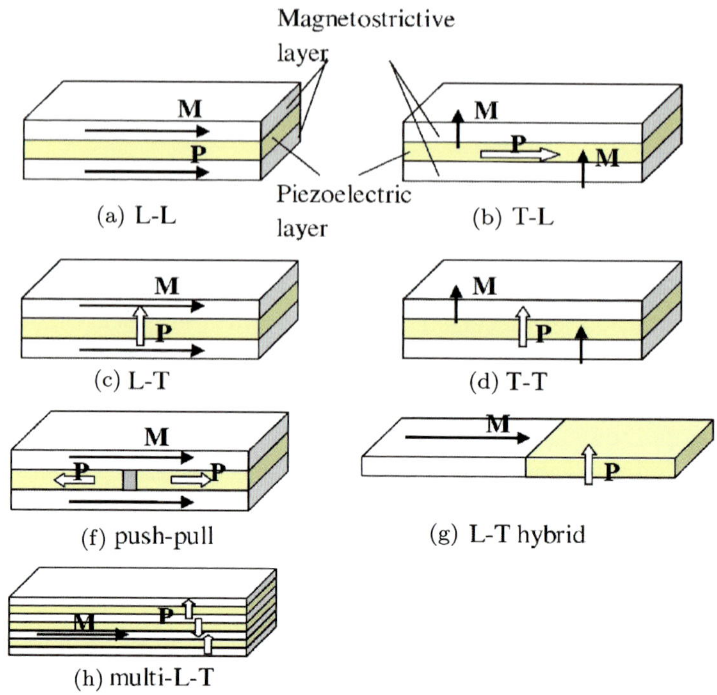

Figure 24. Various rectangular laminates operated in longitudinal vibration mode.

4.2.1. *T–T Terfenol-D/PZT laminate*

The first studies of Terfenol-D/PZT laminates [48] were performed on disk-shape three layer configurations that were operated in a transverse magnetization and transverse polarization (T–T) mode. Relatively large ME voltage coefficients of $\alpha_E = 4.8\,\text{V/cm Oe}$ was reported by Ryu *et al.* [48] under dc magnetic field bias of $H_{\text{dc}} \geq 4000\,\text{Oe}$; although later, other investigators repeated the actual value to be $\alpha_E = 1.3\,\text{V/cm Oe}$ [164, 165]. Figure 28 shows these later (repeated) experimental results for a T–T laminate, where a maximum ME voltage of 66 mV/Oe (or 1.32 V/cm Oe) was observed under $H_{\text{dc}} \approx 4000\,\text{Oe}$. This discrepancy might be due to the interfacial bonding [155], as calculated in Fig. 20.

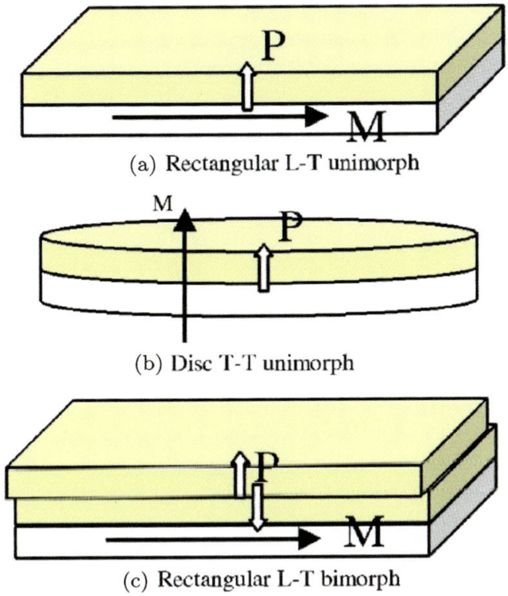

Figure 25. Unimorph and bimorph bending vibration modes.

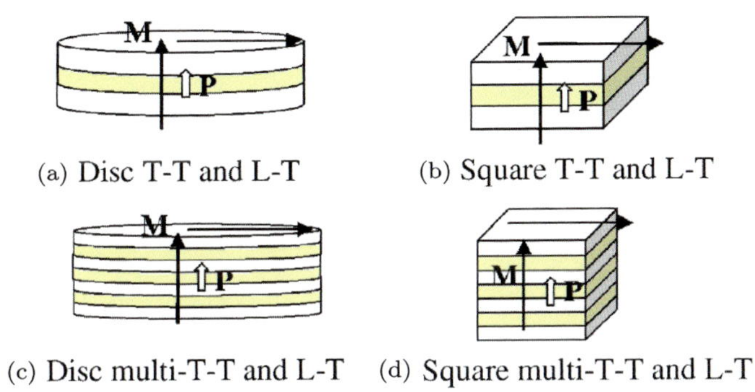

Figure 26. Radial and thickness vibration modes of disk- and square-type ME laminates.

The main problem with the T–T mode laminates is that a quite high dc magnetic bias H_{dc} is required to obtain a maximum value of α_E. This high H_{dc} is caused by a large demagnetization factor in transversely magnetized Terfenol-D layers.

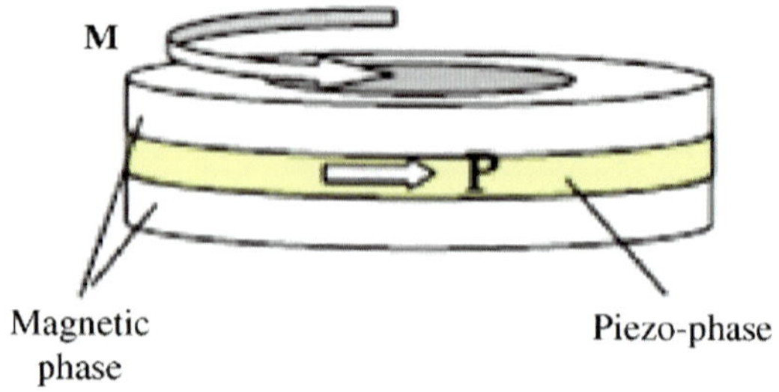

Figure 27. *C–C* mode of ring ME laminate [Ref 53].

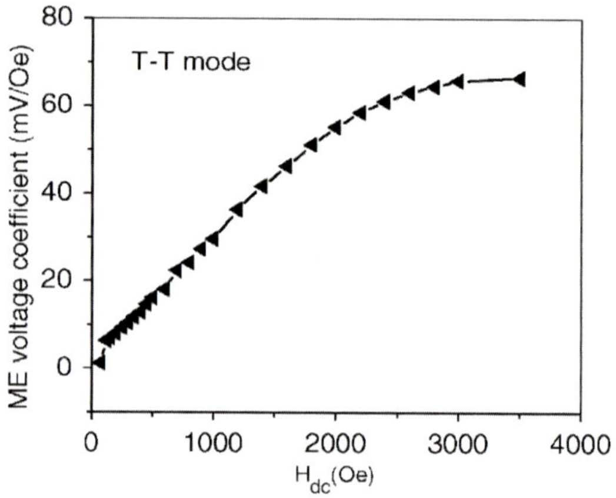

Figure 28. ME voltage coefficient as a function of the applied d_c magnetic field bias H_{dc} [Ref. 165].

4.2.2. *L–T Terfenol-D/PZT and PMN-PT laminates*

To reduce the demagnetization factor effect, a long-type configuration (Fig. 21) that uses a longitudinal magnetization was designed [49–51]. This dramatic decrease in the demagnetization factor results in a large reduction in the H_{dc}, required to achieve the maximum ME coefficient [51]. Long rectangular-shaped Terfenol-D/PZT/Terfenol-D and Terfenol-D/PMN-PT/Terfenol-D three layer laminates with a longitudinal magnetization and transverse polarization (*L–T*) were then reported [49–51] based on this consideration. Experimental results confirmed that at low magnetic biases of $H_{\mathrm{dc}} < 500\,\mathrm{Oe}$ much large values of α_E could be obtained for *L–T*

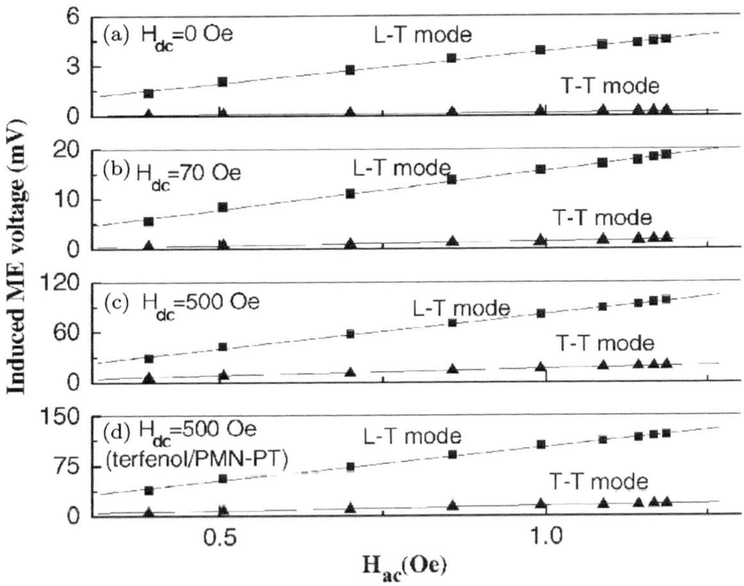

Figure 29. Induced ME voltage as a function of H_{ac} for a laminate of Terfenol-D and PZT-5 ceramic: (a) $H_{dc} = 0\,\text{Oe}$, (b) $H_{dc} = 70\,\text{Oe}$, and (c) $H_{dc} = 500\,\text{Oe}$; and (d) for a laminate of Terfenol-D and a ⟨001⟩-oriented PMN-PT crystal. The measurement frequency was 10^3 Hz [Ref. 50].

laminates relative to T–T ones. For example, Fig. 29 shows these measurements taken under an ac magnetic excitation of $H_{ac} = 1\,\text{Oe}$ at $f = 1\,\text{kHz}$. It can be seen that the induced ME voltage for the L–T Terfenol-D/PZT laminates under $H_{dc} = 500\,\text{Oe}$ was 0.085 V/Oe (or $\alpha_E = 1.7\,\text{V/cm Oe}$) and for L–T Terfenol-D/PMN-PT ones ~0.11V/Oe ($\alpha_E = 2.2\,\text{V/cm Oe}$); whereas that for the T–T mode of Terfenol-D/PZT laminate under $H_{dc} = 500\,\text{Oe}$ was only ~0.015 V/Oe ($\alpha_E = 0.3\,\text{V/cm Oe}$). Clearly, long type L–T laminates have significantly higher ME voltage coefficients than T–T ones under modest magnetic biases.

4.2.3. *L–L and push–push terfenol-D/PZT and PMN-PT laminates*

To achieve a high output voltage, a L–L mode of Terfenol-D/PZT/Terfenol-D or Terfenol-D/PMN-PT/Terfenol-D laminate is a good choice due to its large dielectric displacement along the length (longitudinal) direction [52, 156, 165]. For a long-type ME laminate, the length of the piezoelectric layer is much larger than its thickness, in addition, the longitudinal electromechanical coupling coefficient $k_{33,p}$ and piezoelectric voltage constant $g_{33,p}$ are higher than the corresponding transverse $k_{31,p}$ and $g_{31,p}$. Following Eq. (23), a Terfenol-D/PZT laminate operated in an L–L mode should have a much higher induced voltage VME under magnetic field excitation. For example, Fig. 30 shows measurements for an L–L mode terfenol-D/PZT laminate, which exhibits a maximum $V_{\mathrm{ME}} = 3.5\,\mathrm{V/Oe}$ at 1 kHz (or $\alpha_E = 2.4\,\mathrm{V/cm\ Oe}$) at $H_{\mathrm{dc}} = 500\,\mathrm{Oe}$. Although L–L ME laminates have the highest induced voltage (in V/Oe), its ME charge coefficient (in C/Oe) is quite low. This is because L–L ME laminates have a low capacitance. For a ME laminate with small capacitance, direct measurements of the induced ME voltages using a lock-in amplifier may result in big error [156]. In this case, a charge measurement method may be more reliable for obtaining the correct values.

A double (or symmetrically poled) L–L configuration has been designated as push–pull mode [Fig. 23(f)] [59, 158, 166]. The push–pull configuration is a compromised design, which takes advantages

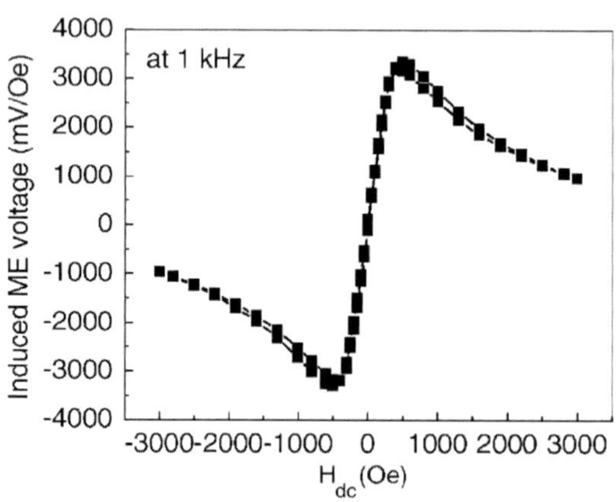

Figure 30. Induced ME voltage in the L–L Terfenol-D/PZT/Terfenol-D laminate [Ref. 156].

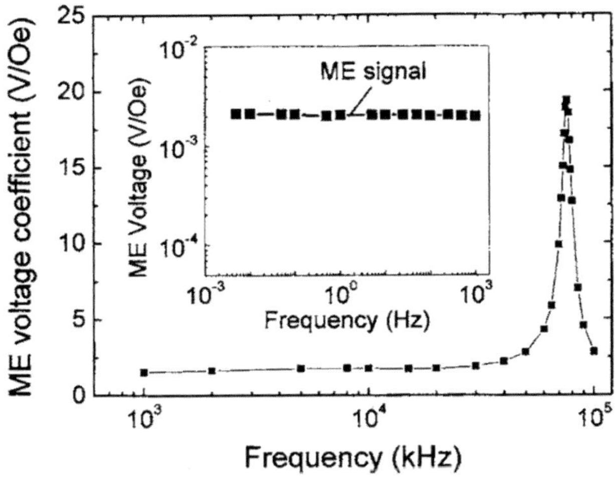

Figure 31. ME voltage coefficients as a function of ac magnetic field frequency [Ref. 166].

of the high induced voltage of L–L mode and the large ME charge of the L–T one. Figure 31 shows measurements of the ME voltage coefficients for a push–pull laminate, which had a maximum ME induced voltage of 1.6 V/Oe (or $\alpha_E = 2.5$ V/cm Oe) at 1 kHz and ~20 V/Oe (~31 V/cm Oe) at 75 kHz. Although this value is similar to an L–L mode, its ME charge coefficient was two times higher than that in L–L one [166].

4.2.4. *L–T bending mode of Terfenol-D/PZT laminates*

A single longitudinally magnetized Terfenol-D layer laminated together with one or two transversely poled piezoelectric PZT or PMN-PT layers is an L–T bending mode (Fig. 25) [164]. Under an external magnetic excitation, the stress applied to the piezoelectric layer by the Terfenol-D one is asymmetric, resulting in a bending motions rather than a longitudinal one for the L–T mode. In general, the ME voltage coefficients of the bending mode are smaller (~1.2 V/cm Oe) than those of the L–T or L–L ones, simply because the bending mode contains only a single Terfenol-D layer. However, the bending mode does have better low-frequency response to

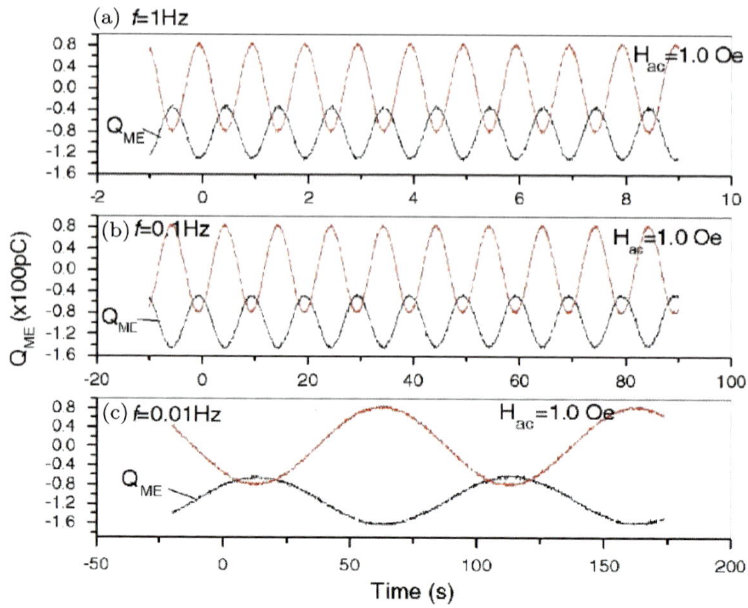

Figure 32. The ME response of a bimorph-type ME laminate [Ref. 164]

magnetic fields than either the *L–T* or *L–L*, making it suitable for low-frequency magnetic sensor applications. Figure 32 shows the induced charge response of the bimorph ME laminate [see Fig. 25(c)] to low-frequency magnetic field variations. The ME charge coefficient of this bimorph was ~800 pC/Oe at 10^{-2} Hz.

4.2.5. *C–C Terfenol-D/PZT and PZN-PT laminates*

In many situations, magnetic fields are excited by electric currents. In this case, the excited magnetic fields are vortexes. A ME ring-type laminate operated in circumferential magnetization and circumferential polarization has been designated as the *C–C* mode (see Fig. 27). Experimental investigations have shown that Terfenol-D/PMN-PT/Terfenol-D three layer laminate ring has a very high ME coefficient with maximum values of up to 5.5 V/cm Oe at 1 kHz in response to a vortex magnetic field [58]. High ME coupling in the *C–C* mode is due to the magnetic loop of the ring ME-type configuration, and which is suitable for capturing a vortex-type field. We will see later that this *C–C* ME laminate ring has potential for electric current sensor applications.

4.2.6. *ME laminates based on non-Terfenol-D materials*

Although Terfenol-D has the highest magnetostriction amongst all known magnetostrictive materials, this rare-earth iron alloy is quite costly, and also very brittle. Thus alternatively, there are other magnetostrictive metallic materials (see Table 2) — such as Permendur, Ni, Metglas, and other magnetic alloys, to be used [59, 60, 167–171]. Dong *et al.* developed new ME laminate composites of magnetostrictive Fe 20at.% Ga alloys with PZT ceramics and piezoelectric single crystals, which exhibited a large ME coupling [59, 60]. The maximum ME voltages observed for *L–T* Fe–Ga/PMN-PT laminates were ~1 V/cm Oe at 1 kHz and 70 V/cm Oe at resonance, as shown in Fig. 33. Although the low-frequency ME performance of the Fe–Ga/PMN-PT laminates is not as good as that of Terfenol-D/PZT laminates, its resonance ME performance is better than that of Terfenol-D/PZT laminates, due to a higher *Q*-factor.

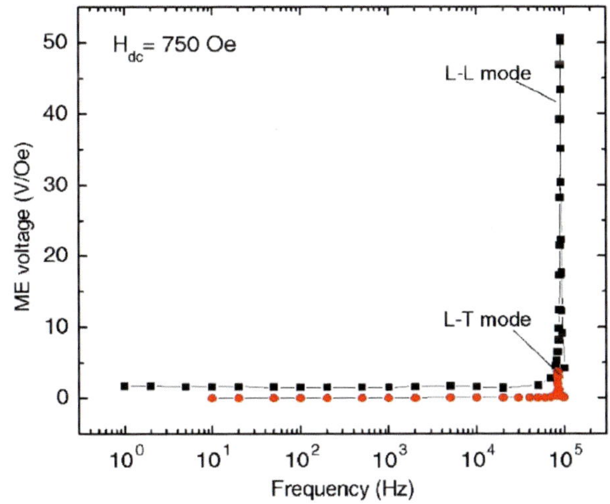

Figure 33. ME voltage coefficients of Fe–Ga/PMN-PT laminates as a function of magnetic field frequency [Ref. 60].

Recently, a thin ($<100\,\mu$m) and flexible laminate of Metglas and PVDF piezopolymer films has been reported [167]. The results showed that in both unimorph and three-layer configurations very low dc magnetic biases (as low as 8 Oe) can induce GME effect with ME voltage coefficients of 7.2 V/cm Oe at low frequency, and up to 310 V/cm Oe at resonance. Zhao *et al.* [168] reported a Ni_2MnGa/PZT laminate and observed a distinct peak of the ME coefficient near the martensitic transformation temperature (ca. 30°C) of the Ni_2MnGa alloy. This enhancement results from the martensitic transition in the magnetic Ni_2MnGa alloy.

5. Three-phase composites

5.1. *Quasi-0-3-type particulate composites*

In the two-phase Terfenol-D based laminates, the eddy current loss in bulk Terfenol-D is quite large at high frequency, which limits the applications of this laminate structure, and the GME response of the laminated composites is strongly influenced by the interfacial binders [155], as discussed above. In addition, Terfenol-D thin layers are very brittle. In order to overcome these difficulties in two-phase Terfenol-D/PZT laminate composites, a three-phase composite of Terfenol-D, PZT, and a polymer has been developed [69]. Such three-phase composites can be easily fabricated by a conventional low-temperature processing, and especially the three-phase particulate composites can be fabricated into a variety of forms such as thin sheets and molded shapes, and they exhibit greatly improved mechanical properties.

The simplest three-phase ME composite is quasi-0-3-type particulate composites where Terfenol-D grains are randomly oriented in the mixture matrix of PZT and polymer [69, 71]. The conductive Terfenol-D grains are well dispersed in the PZT/polymer matrix to keep the composite insulating, as schematized in Fig. 34(a), thus forming a 0-3-3 connectivity of phases. Figure 34(b) shows a micrograph of the three-phase particulate composite. In the three-phase ME composites, the inactive polymer is just used as a binder. Such a three-phase particulate composite shown in Fig. 34 can be

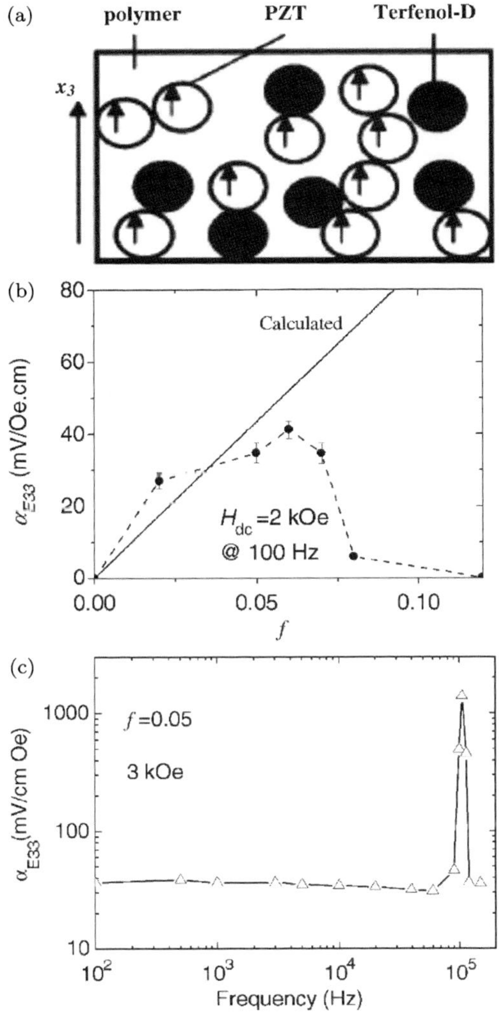

Figure 34. (a) Schematic representation of the particulate Terfenol-D/PZT/polymer composites. All grains are randomly embedded in the polymer matrix. The polarization in PZT particles (denoted by the arrows within the open circles) is assumed to be parallel to the x_3 direction. (b) Calculated (solid line) and measured (dots) α_{E33} at high field saturation for the f Terfenol-D/$(0.7 - f)$PZT/0.3 polymer composites. (c) Measured frequency-dependent α_{E33} for the 0.05Terfenol-D/0.65PZT/0.3polymer composite [Refs. 71 and 172].

denoted as f Terfenol-D/(1-f_m-f)PZT/f_m polymer, where f and f_m are the volume fractions of Terfenol-D and polymer, respectively. The ME response of the 0-3-3 composites can also be calculated by using Green's function technique as discussed above [172]. The solid line in Fig. 34(b) shows calculated α_{E33} (under the boundary condition of completely mechanically clamped in the x_3 direction but free in the transverse direction as in experimental measurement) of the f Terfenol-D/(0.7-f)PZT/0.3 polymer composite. As shown, in the low volume fraction range of $f < 0.1$, the α_{E33} values at high-magnetic-field saturation increase approximately linearly with increasing f. This increase in α_{E33} is attributed to the increase in magnetostrictively induced strain of the composites with f. Comparison in Fig. 34(b) shows that there is a good agreement between the experimental and calculated and measured α_{E33} as $f < 0.07$. However, as $f > 0.07$, the measured α_{E33} values decline with f, which is due to formation of a Terfenol-D grain percolation path through the composites, though the magnetostrictively induced strain of the composites still increases with f. The maximum α_{E33} value of the composites at 2 kOe is about 42 mV/cm Oe at $f = 0.06$, which is comparable with those measured for the ferrite/PZT ceramic particulate composites (see Section 3.2). At high resonance frequency, α_{E33} is as high as over 1500 mV/cm Oe [Fig. 34(c)]. The peak ME response is attributed to enhanced coupling elastic interaction at the resonance.

As observed, the ME effect in the Terfenol-D/PZT/PVDF particulate composites is mainly limited by the concentration threshold of the Terfenol-D grains allowed in the composites due to the low resistance of Terfenol-D, which is dependent on the composite microstructure. If f allowed in the composites is higher, the α_{E33} values will increase. The results imply that the ME response of the three-phase composites could be further improved by carefully controlling processing to achieve homogeneity and higher loading of the Terfenol-D grains well isolated in the PZT/polymer mixture.

5.2. *Quasi-2-2-type laminate composites*

In order to eliminate this limitation on the low volume fraction of Terfenol-D in the quasi-0-3-type particulate composites, 2-2-type laminate composites containing Terfenol-D/polymer and PZT/polymer layers have been made by laminating the Terfenol-D/polymer (denoted as T-layer) and the PZT/polymer (denoted as P-layer) particulate composite layers and then simply hot molding these layers together [72–75]. Such a simple hot-pressing procedure ensures a good interfacial bonding between composite layers, since the polymer used in the T-layer and P-layer is the same just as a matrix binder. Different laminate structures can be obtained, e.g., sandwich structure of P-layer/T-layer/P-layer (abbreviated as the P–T–P composite), or T-layer/P-layer/T-layer (abbreviated as the T–P–T composite), or transverse structure of P-layer and T-layer, as shown in Fig. 35.

In comparison with the quasi-0-3-type particulate composites, the quasi-2-2-type laminate composites exhibit strong anisotropy, and thus different longitudinal ME coefficient α_{E33} and transverse ME

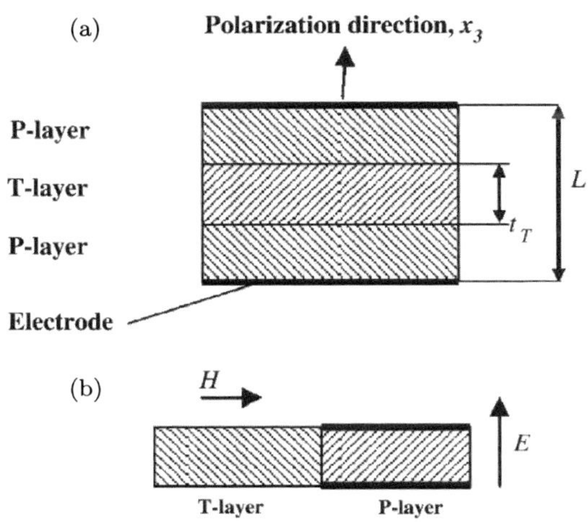

Figure 35. Schematic illustrations of (a) a three-phase laminate P–T–P composite and (b) a three-phase transverse P–T composite.

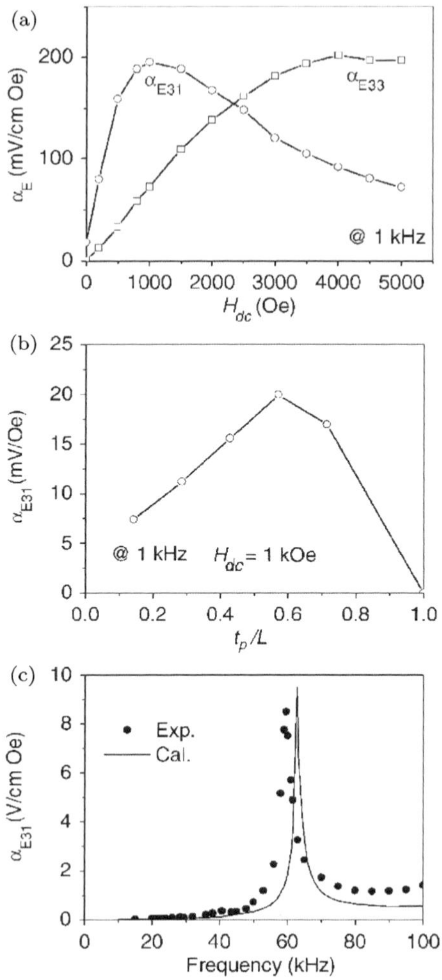

Figure 36. The ME coefficients measured at 1 kHz for the laminate T–P–T composites as a function of (a) the bias magnetic field and (b) t_p/L [Ref. 75]. (c) Comparison of frequency dependencies of the ME coefficient measured (solid dots) and calculated (solid line) for the transverse structure of P-layer and T-layer [Ref. 173].

coefficient α_{E31} [see, e.g., Fig. 36(a)]. α_{E33} increases first rapidly with the bias field and slowly changes with the magnetic field in the high bias field range where the effective magnetostriction becomes to approach its saturation. α_{E31} nonmonotonically depends on the bias field with a peak at about 1.5 kOe. Such a nonlinear ME response of the composites is closely related to the magnetostrictive behavior of the Terfenol-D. The difference between α_{E33} and α_{E31} is due to different magnetostrictions arising from the out-of-plane bias and in-plane bias in the anisotropic laminate composite samples. At high magnetic fields the magnetostriction gets saturated more faster under the in-plane bias than the out-of-plane bias producing a nearly constant electric field in the PZT, thereby decreasing α_{E31} with increasing bias field.

The ME effect of these composites is strongly dependent on the thickness ratio t_p/L and volume fraction of the polymer matrix. For example, Fig. 36(b) shows that both α_{E33} and α_{E31} of the P–T–P composites nonmonotonically depend on t_p/L with a maximum ME effect in the middle range of t_p/L. Both α_{E33} and α_{E31} of the composites first increase with t_p/L, which is attributed to the increase in the effective piezoelectric constant the composites with t_p/L. However, with further increasing t_p/L, the ME coefficient declines after a maximum value, which is due to the reduction in magnetostrictively induced strain of the laminated composites with increasing t_p/L, since much thinner T-layer than the P-layers leads to much lower magnetostrictively induced strain and thereby lower ME coefficient of the composites.

These 2-2 laminate composites exhibit similar frequency dependence [Fig.36(c)] of the ME coefficient to those 0-3 particulate composites, i.e., their ME coefficients keep at stable values in the low-frequency range but get greatly enhanced at resonance frequency. For example, the maximum ME coefficient values of these 2-2 laminate composites are about 100 mV/cm Oe order of magnitude in the low-frequency range, which are comparable to those for the PZT/ferrite ceramic composites; but reach up to several V/cm Oe at resonance.

A similar dependence of the ME effect of the composites on the volume fraction f_P of the polymer is also observed. $f_P < 0.3$, the low quality of the samples due to low concentration of the polymer binder leads to the large loss of stress/strain transfer from the T-layer to the P-layer and thus low ME response. As $f_P > 0.5$, the ME coefficient also declines with the increase in f_P. This decrease is attributed to the decrease in the concentration of the piezoelectric phase and magnetostrictive phase, which lead to low magnetostrictively induced strain and low piezoelectricity of the composites.

In comparison with the quasi-0-3-type particulate composites, these quasi-2-2-type laminate composites exhibit higher the ME effect. The ME response of the three-phase polymer-based laminate composites could be further improved by optimization of their structures, and processing; and would be an important smart multiferroic material for magnetic-electric devices.

Recently, it has been reported that a simple bilayered composite containing thin PZT ceramic chip and Terfenol-D/epoxy layer has large capacitance and large ME charge output of about 300 pC/Oe at the resonance frequency [78]. Three resonance modes including the first-order flexural mode, second-order flexural mode, and radial mode, have been observed in this simple three-phase composite bilayer. The measured phase spectra showed a significant phase change around each resonance frequency. The thinner PZT chip is benefit for higher flexural resonance ME effect. The ME effects measured under open and short electric boundary conditions show different behaviors.

5.3. *Quasi-1-3-type rod-array composites*

Besides 0-3 and 2-2 structures, the 1-3 structure, i.e., a fiber (or rod) reinforced composite, is another important structure. As for the 1-3 piezoelectric composites [18], such 1-3 multiferroic ME composite [76, 77] can be prepared via a dice-and-fill process. According to the designed volume fraction of PZT rods, a PZT bulk is diced to get a PZT rod array, and then the gap of the PZT rod array is filled with a mixture of Terfenol-D particles and epoxy. When the epoxy hardens, the pseudo-1-3-type multiferroic composite is obtained [see the inset of Fig. 37(a)].

The dependence of the ME coefficients on the volume fraction f_{PZT} of PZT rods is presented in Fig. 37(a). The calculations and experiments all show that the ME coefficient increases with increasing f_{PZT} due to the increasing piezoelectric effect. After reaching a maximum, the ME coefficient decreases when $f_{\mathrm{PZT}} > 0.1$.

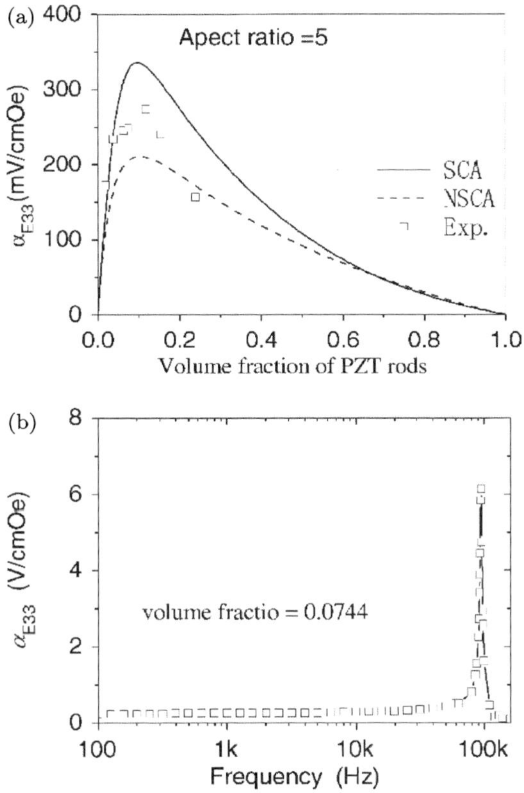

Figure 37. Variation in the ME coefficient of the pseudo-1-3-type multiferroic composites with (a) the volume fraction of PZT rods at low frequency and (b) frequency, at 2 kOe bias field. Solid and dashed lines in (a) are calculated results by SCA and NSCA [Ref. 76].

This decrease is attributed to the decrease in the volume fraction of the magnetostrictive Terfenol-D/epoxy matrix. The maximum α_{E33} reaches about 300 mV/cm Oe at 2 kOe, which is comparable with that for the pseudo-2-2-laminate ME composites and larger than that for the 0-3 particulate ME composites.

Figure 37(b) is a typical frequency dependence of the ME coefficient of the pseudo-1-3-type multiferroic composite with $f_{\text{PZT}} = 0.0744$. A GME coefficient of about 6.2 V/cm Oe at 90 kHz is observed. This GME effect at high frequency is also attributed to the piezoelectric resonance which significantly enhances the magnetic-mechanical-electric coupling between matrix and rods. The transverse ME coefficient of such 1-3–type ME composites is lower than the longitudinal one in the test frequency range, resulting from the large anisotropy of the rod reinforced composites [76, 77].

5.4. *Other three-phase composites*

Other three-phase ME composites have also been available. In the Terfenol-D based composites above, a high magnetic field is needed to achieve a maximum ME coefficients, which is due to the low magnetic permeability μ_r of the Terfenol-D alloys. Typically, μ_r in magnetostrictive Terfenol-D or Fe–Ga is as low as 3–10. However, by incorporating a third phase ferromagnetic layers with a high-permeability into Terfenol-D/PZT laminates, the effective permeability of the resultant three-phase ME composites can be dramatically increased, which in turn results in a larger effective piezomagnetic coefficient and thus a stronger ME coupling at lower magnetic field. For example, Fig. 38(a) illustrates a configuration of three-phase $MnZnFe_2O_4$/Terfenol-D/PZT laminate [174]. The comparison in Fig. 38(b) clearly shows the largely enhanced ME effect in the three-phase composite due to the incorporation of high-permeability $MnZnFe_2O_4$ layers.

6. Nanostructured composite thin films

Recently, motivated by a pioneer work of Zheng *et al.* [79], multiferroic nanocomposite thin films of ferroelectric [e.g., BTO, PTO, PZT, and $BiFeO_3$ (BFO)] and magnetic oxides (e.g., CFO, NFO, and LCMO) prepared via physical deposition techniques [e.g., pulsed laser deposition (PLD)] and chemical solution processing (e.g., sol–gel spin-coating method) have become new routes to multiferroic ME composites [94]. In comparison to bulk multiferroic ME composites, the nanostructured thin films provide more degrees of freedom, such as lattice strain or interlayer interaction, to modify the ME behavior. It also offers a way to investigate the physical mechanism of ME effect in nanoscale. The coupling interaction between two oxides in

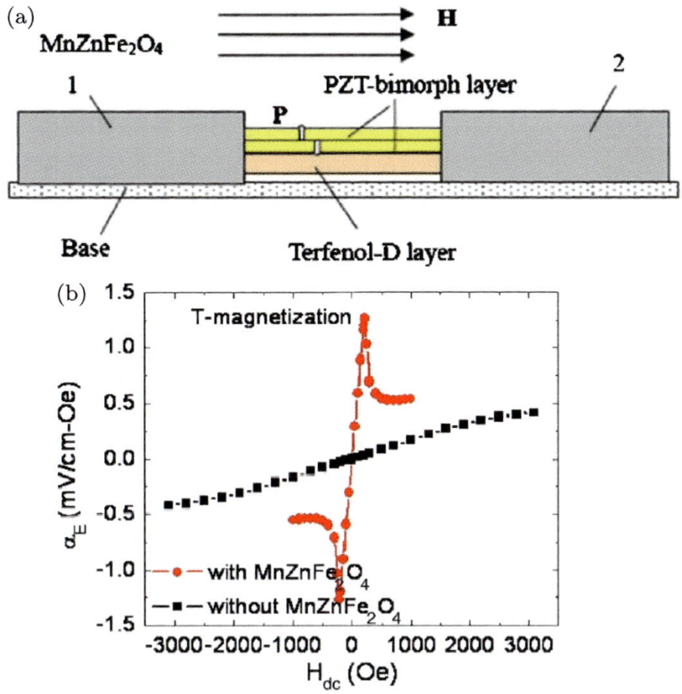

Figure 38. (a) Configuration of three-phase $MnZnFe_2O_4$/Terfenol-D/PZT laminate; (b) the effect of high-μ phase on its ME voltage coefficient [Ref. 174].

the multiferroic nanostructures is still due to elastic interaction as was the case in bulk composites. However, the mechanical constraint arising from the film on substrate and the good bonding between the two phases in the nanostructured composite films could significantly affect the ME coupling interactions.

Similarly to bulk ceramic composites shown in Fig. 1, according to the microstructure of the nanostructured composite films, there are also three kinds of nanostructured composite films, i.e., (1) 0-3–type structures with magnetic spinel nanoparticles (e.g., CFO and NFO) embedded in the ferroelectric films (e.g., PZT), (2) 1-3–type heterostructures (vertical heterostructures) consisting of magnetic spinel pillars (e.g., CFO) vertically embedded into the ferroelectric films (e.g., BTO, PTO, or BFO), and (3) 2-2–type heterostructures (horizontal nanostructures) consisting of alternating layers of a ferroelectric perovskite and magnetic oxide.

Few works were reported on the 0-3–type nanostructure [82, 90]. Wan *et al.* [82] prepared a PZT-CFO composite thin film using a sol–gel process and spin-coating technique. The phase separation of the CFO and PZT phases in the films was verified by X-ray diffraction. The films exhibited both good magnetic and ferroelectric properties, and the ME effect of these films was found to be strongly dependent on the magnetic bias and magnetic field frequency. In comparison, more publications were reported on 1-3 and 2-2 heterostructures.

6.1. *1-3–type vertical heterostructures*

The prototypical 1-3 vertical heterostructure consists of a magnetic spinel phase epitaxially embedded into the ferroelectric matrix. The first example was reported by Zheng *et al.* [79], where arrays of magnetic CFO nanopillars with diameters of 20–30 nm were embedded in a ferroelectric BTO matrix films (Fig. 39). Other different combinations of $PbTiO_3$–$CoFe_2O_4$ and $BiFeO_3$–$CoFe_2O_4$ have also been grown on $SrTiO_3$ single-crystal substrates. The morphologies of the component phases in such vertical composite films containing either magnetic or ferroelectric nanopillars vary markedly with the substrate orientation and phase fractions [87, 88, 175], as was inferred from their microscopy observations. However, the design and control of such heterostructures remain a challenge.

Recent work has demonstrated substantial ME coupling in such nanostructures, through switching of the magnetization on reversal of the ferroelectric polarization [81, 176], which was attributed to (1) a reduced clamping effect by the substrate in the vertical architecture, and (2) efficient strain coupling resulting from a larger interfacial surface area and intrinsically heteroepitaxial in three dimensions. Detailed studies suggest that the switching is mediated by strong mechanical coupling between the two lattices, which leads to a time-dependent modulation of the magnetic anisotropy in the nanopillars. But controllable switching of the magnetization can only be achieved if an additional weak magnetic field is superimposed to lift the time-reversal symmetry.

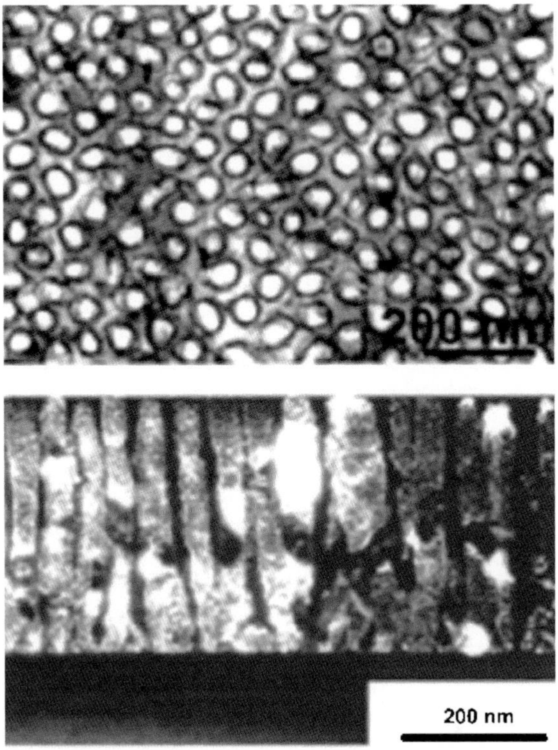

Figure 39. Plan view TEM image and cross-sectional dark field TEM image of $CoFe_2O_4$ nanopillars embedded in the $BaTiO_3$ matrix [Ref. 80].

Although the ME coupling has been observed via microscopy in such vertical nanostructures [81, 176], through switching of the magnetization on reversal of the ferroelectric polarization, the ME coefficients could not be directly measured in the 1-3–type vertical heterostructures because of leakage problem resulting from low resistance of the magnetic pillars penetrating through the films or the magnetic matrix. The leakage problem would erase the promising direct ME effect in the vertical nanostructures.

6.2. *2-2–type horizontal heterostructures*

In comparison with the vertical nanostructures, the 2-2–type horizontal heterostructures exhibit weak ME effects due to large in-plane constraint from substrates [117]. But the horizontal nanostructures are more easier to be fabricated, and rule out the leakage problem due to the ferroelectric layers shutting off circuit, which could lead to visible ME effect.

Murugavel *et al.* [83] and Singh *et al.* [84] reported $BaTiO_3$–$Pr_{0.7}Ca_{0.3}MnO_3$ perovskite superlattices grown on $SrTiO_3$ via PLD. Their studies found that the impedance and capacitance of the films varied with the applied magnetic field due to the ME coupling in the nanostructures. A negative magnetocapacitance value of 3% T at 1 kHz and 100 K was demonstrated.

Different 2-2–type combinations of ferroelectric perovskites (e.g., BTO, PZT) and magnetic oxides (e.g., $CoFe_2O_4$, $NiFe_2O_4$, and $La_{1.2}Sr_{1.8}Mn_2O_7$) have also been grown via PLD and sol–gel spin-coating method [85, 86, 92, 93]. These layered films showed good coexistence of ferroelectric and ferromagnetic behaviors. For example, Fig. 40 shows a simple bilayer BTO–NFO nanocomposite film, and the measurements illustrate obvious ME response of the heterostructured film [177]. As seen from Fig. 40, when an in-plane alternating magnetic field signal δH is applied, the single-phase BTO and NFO films do not produce any ME output δV, by comparison, the heterostructured NFO/BTO film clearly yields the ME output following the switch of the magnetic excitation signal δH. The observation of the ME coupling in the heterostructures can only be achieved as a weak ac magnetic field δH is superimposed to lift the time-reversal symmetry.

In particular, a special horizontal heterostructure has been recently constructed by growing an epitaxial ferroelectric PZT film on a magnetostrictive layered manganite single crystal of composition LSMO [89], and inversely, an epitaxial magnetic $LaCa(Sr)MnO_3$ film on a ferroelectric single crystals (e.g., $BaTiO_3$ and PMN-PT) [91, 178]. The efficient mechanical coupling at the interface results in a remarkable ME effect in such special horizontal heterostructures.

For example, in this PZT-film-on-LSMO structure [89], the manganite crystal serves as the substrate for the PZT film growth, as the magnetostrictive element and as the bottom electrode in the device. Without the constraint from conventional substrates,

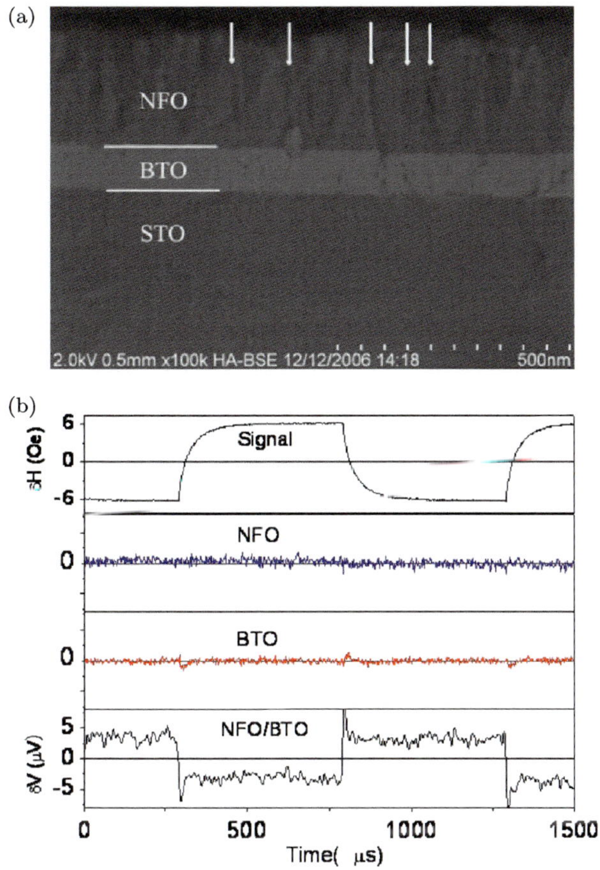

Figure 40. (a) A cross-sectional SEM image of the heterostructured NFO/BTO composite film grown in STO substrate; and (b) the ME response (induced ME voltage δV) of this film when the in-plane magnetic field is applied (ac magnetic field δH at 1 kHz with a bias magnetic field of 100 Oe). For comparison, the results for pure NFO and BTO films are also shown [Ref. 177].

such a geometry maximizes the efficiency of in-plane transfer of the magnetostrictive strain from LSMO to the PZT film. A large ME effect is observed at T_C (about 105 K) of LSMO where it exhibits the highest magnetostrictive sensitivity (0.13%) to an applied magnetic field, i.e., magnetically induced an abrupt 7.3% increase in switchable polarization of the PZT film [89]. This modulation of polarization originates from the contraction of the in-plane lattice parameters of

LSMO at T_C. Most recently, Eerenstein *et al.* [91] have reported an epitaxial magnetic LSMO film on BTO single crystal and achieved sharp and persistent converse ME effect. In the LSMO-film-on-BTO structure, an applied electric field modifies the ferroelectric domain structure of the BTO substrate, and then the switching of BTO domains alters in LSMO the local strain, magnetic anisotropy, and thus the magnetization.

6.3. *Theoretical modeling*

In theoretical analyses, some characteristics of films, e.g., giant residual stress/strain resulting from the lattice misfit between the film and substrate and spontaneous polarization in epitaxial films, have been considered to understand the ME response in nanostructured films [117]. As a result, the constitutive equations for the coupling magnetic–mechanical–electric interactions in the nanostructured films can be expressed as

$$\begin{aligned} \boldsymbol{\sigma} &= \boldsymbol{c\varepsilon} - \boldsymbol{e}^T\boldsymbol{E} - \boldsymbol{c\varepsilon}^{\mathrm{ms}} - \boldsymbol{\sigma}_s, \\ \mathbf{D} &= \mathbf{e}\boldsymbol{\varepsilon} + \boldsymbol{\kappa E} + \boldsymbol{\alpha H} + \boldsymbol{P}_s, \\ \boldsymbol{B} &= \boldsymbol{\mu}(\boldsymbol{\varepsilon}, \boldsymbol{E}, \boldsymbol{H})\boldsymbol{H} + \boldsymbol{M}_s. \end{aligned} \tag{24}$$

In comparison to the constitutive equations for bulk composites, the residual stress $\boldsymbol{\sigma}_s$ (or residual strain $\boldsymbol{\varepsilon}_s$), spontaneous polarization $\boldsymbol{P}_s$ and magnetization $\boldsymbol{M}_s$ are incorporated for the multiferroic composite films. Based on the modified constitutive equations, the ME coupling interaction and magnetically induced polarization for nanostructured composite films were firstly calculated using Green's function technique [117]. It was revealed that the 1-3–type vertical heterostructures could exhibit large ME response which is even larger than that in their bulk counterparts if there is no leakage problem, while the 2-2–type horizontal heterostructures show much weaker ME coupling due to large in-plane constraint effect. Most recently, the calculations for the ME effects in nanobilayers, nanopillars, and nanowires of nickel ferrite and PZT on MgO substrates or templates have also shown that the ME coupling decreases with increasing substrate clamping [179].

Because the magnetostriction of the magnetic phase can dynamically change the strain constraint in the nanostructured composite films, the magnetically induced polarization could also be calculated from the Landau–Ginsburg–Devonshire phenomenological thermodynamic theory [118, 121], where the boundary conditions are related to the magnetic field. This method was found to yield some similar results to those using Green's function technique [118], i.e., the calculation results showed that the 1-3–type nanostructured composite films exhibit a large ME coefficient, but the 2-2–type films exhibit much weaker ME coefficient due to large in-plane constraint.

A powerful approach, phase-field model, in which the elastic energy in the constrained thin film was incorporated including the effect of free film surface and the constraint from the substrate, has been recently developed for studying the ME coupling effect in the 1-3 nanocomposite thin films [120]. The phase-field calculations of the magnetic-field-induced electric polarization in 1-3–type $BaTiO_3$–$CoFe_2O_4$ nanocomposite film give similar results to Green's function technique. The phase-field simulation illustrates that the magnetic-field-induced electric polarization is highly dependent on the film thickness, morphology of the nanocomposite, and substrate constraint, which provide a number of degrees of freedom in controlling coupling in nanocomposite films, as shown in Fig. 41.

Recently, by first-principles density functional calculations of a model Fe/$BaTiO_3$ horizontal superlattice, Duan *et al.* [119] have demonstrated a ME effect in such a multiferroic heterostructure that arises from a purely electronic hybridization between Ti and Fe atoms, not mediated by strain. Thus this ME effect is driven by the coupling between ferroelectricity and magnetism through interface bonding. The displacement of atoms at the interface caused by ferroelectric instability alters the overlap between atomic orbitals at the interface, which in turn affects the interface magnetization, resulting in a ME effect. In this case, the ME effect is confined to the interface and represents a change of the interface magnetic moment at the coercive field of the ferroelectric. Such a ME effect, they argue, should be possible in both vertical and horizontal heterostructures. However, this remains in need of experimental verification.

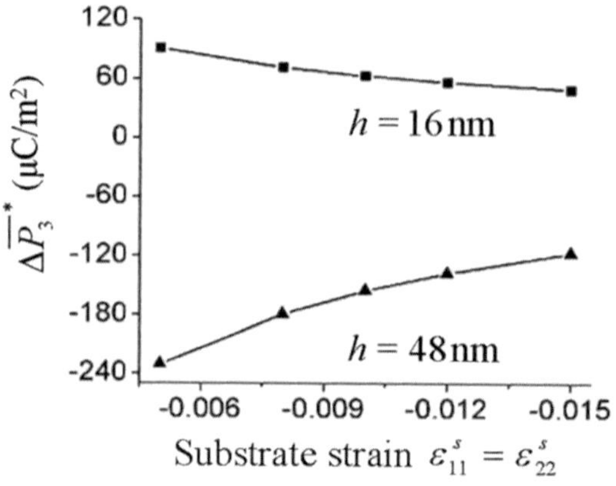

Figure 41. Dependence of magnetic-field-induced electric polarization ΔP_3 on the substrate strains (h is the thickness of the film, the distance between neighboring magnetic pillars is $d = 64$ nm, and the volume fraction of the magnetic pillars is 0.35) [Ref. 120].

7. Applications

So far only the bulk ME composites discussed above exhibit large ME effects above room temperature, which can exceed the intrinsic ME effect in single-phase compounds by many orders of magnitude. Thus the ME composites are ready for technological applications. Promising applications include magnetic field sensors (thus complementing Hall sensors and current measurement probes), transducer, filters, oscillators, phase shifters, memory devices, and so on. Since the magnetic field input is required to have two components, a dc bias and an ac probe, either of the two can be detected by providing the other component. The ME composite can thus be used as a magnetic probe for detecting ac or dc fields. At the resonance frequency the ME composite can be used as a transducer which converts the microwave magnetic field into a microwave electric field. Because of the shift in the resonance frequency in a static magnetic or electric bias field the composite materials hold promise in electrically tunable microwave applications such as filters, oscillators, and phase shifters. Due to the hysteretic nature of the ME effect, the composites may find applications in memory devices [3]. The linear ME effect has a positive or a negative sign, depending on the annealing conditions (parallel or antiparallel magnetic and electric fields). Thus the coupling could, in principle, permit data to be written electrically and read magnetically in memory technologies.

7.1. *Magnetic sensors*

The working principle of magnetic sensing in the ME composites is simple and direct [180]. When probing a magnetic field, the magnetic phase in the ME composites strains, producing a proportional charge in the piezoelectric phase. Highly sensitive magnetic field sensors can be obtained using the ME composites with high ME coefficients. The ME composites can be used as a magnetic probe for detecting ac or dc fields.

7.1.1. *Ac magnetic field sensors*

Figure 42 shows the voltages induced across the two ends of a PMN-PT layer in a push–pull ME Terfenol-D/PMN-PT/Terfenol-D three-layer laminate [see Fig. 24(f)] as a function of ac magnetic field (H_{ac}) at drive frequencies of (i) $f = 1\,\text{kHz}$ and (ii) $f = 77.5\,\text{kHz}$

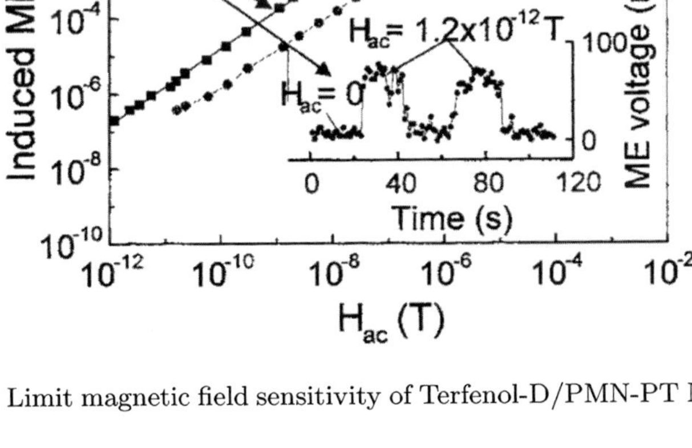

Figure 42. Limit magnetic field sensitivity of Terfenol-D/PMN-PT ME laminate [Ref. 166].

under a given H_{dc}. In this figure, the induced ME voltage can be seen to have a linear response to H_{ac} over a wide range of fields from $10^{-11} < H_{ac} < 10^{-3}$ T [49, 164, 166]. When the laminates were operated under resonance drive, an enhancement in sensitivity to small magnetic field variations was observed. The sensitivity limit of the ME laminates at ambient conditions was 1.2×10^{-12} T [166]. These results unambiguously demonstrate that ME laminates have an ultrahigh sensitivity to small ac magnetic field variations.

Apart from a bimorph [164], a multilayer configuration [57] of ME laminates has been reported that enables ultralow frequency detection of magnetic field variations. This multilayer ME laminate is illustrated earlier in Fig. 24(h). This configuration can greatly improve the low-frequency capability because of its high ME charge coupling and large capacitance. Figure 43 shows the magnetic field induced voltages for multilayer Terfenol-D/PMN-PT ME laminates at frequencies of 100, 1, and 0.01 Hz. At an extremely low frequency

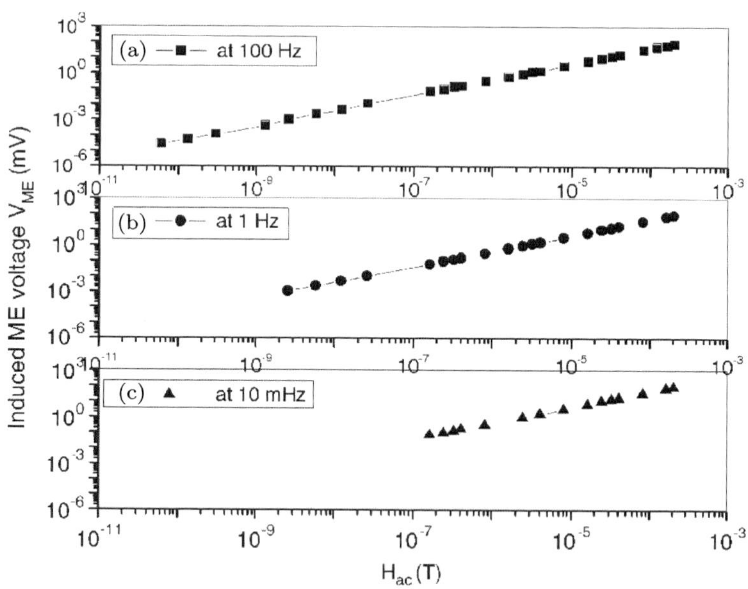

Figure 43. Low-frequency response of a multi-L–T Terfenol-D/PMN-PT laminate [Ref. 57].

of $f = 10\,\mathrm{MHz}$, the multilayer ME laminates can still detect a small magnetic field variation as low as 10^{-7} T.

7.1.2. *Dc magnetic field sensors*

It has also been reported that small dc magnetic field variations can be detected using ME laminates based on magnetic bias effect [61] or inverse ME effect [181, 182]. In fact, small long-type ME laminates of Terfenol-D and PZT are quite sensitive to small H_{dc} variations, when driven under a constant H_{ac}. The sensitivity limit is about 10^{-7} T using a constant amplitude low frequencies drive, which can be enhanced to 10^{-8} T under resonant drive. Figure 44 shows the sensitivity limit of an L–T ME laminate to small dc magnetic field variations, while under resonant drive. It can be seen that dc magnetic field changes as small as 10^{-8} T were readily detected.

7.1.3. *ME current sensors*

The ME composites have also been shown to be good candidates for electric current sensors. A straight wire containing ac or dc current I will excite an ac or dc vortex magnetic field H_{vor} around this wire: $H_{vor} = I/\pi r$, where r is radius of the vortex magnetic field. Accordingly, ring-type ME laminates [53, 54, 58] are ideal configurations for vortex magnetic field detection, or current I

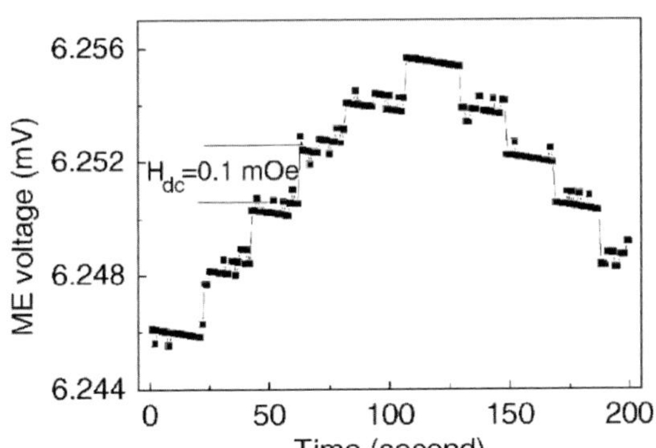

Figure 44. Sensitivity limit of an L–T ME laminate to small dc magnetic field variations under resonant drive [Ref. 61].

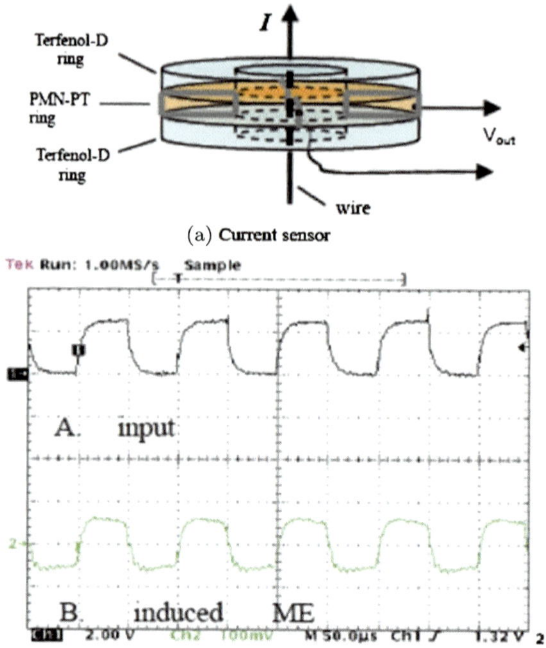

(a) Current sensor

(b) square-wave current detection

Figure 45. ME current sensor [Ref. 58].

detection. Previously, a ME ring-type laminate (see Fig. 27) has been made of a circumferentially magnetized Terfenol-D and a circumferentially poled piezoelectric PZT (or PMN-PT), which was shown to have a high ME sensitivity of up to 5.5 V/cm Oe at $f =$ 1 kHz to a vortex magnetic field. Figure 45(a) illustrates this ring-type ME laminate used as an electric current sensor, and Fig. 45(b) shows the ME voltage response to a square-wave current passing through the wire. Detection using a toroidal type variable reluctance coil (100 turns) exhibited a much smaller induced voltage (by a factor of 0.01 times) than that of this ME ring sensor.

7.2. *Transformers and gyrators*

ME transformers or gyrators have important applications as voltage gain devices, current sensors, and other power conversion devices. An extremely high voltage gain effect under resonance drive has been reported in long-type ME laminates consisting of Terfenol-D and PZT layers [55, 56]. A solenoid with n turns around the laminate that carries a current of I_{in} was used to excite a H_{ac}. The input ac voltage applied to the coils was V_{in}. When the frequency of H_{ac} was equal to the resonance frequency of the laminate, the magnetoelectric voltage coefficient was strongly increased, and correspondingly the output ME voltage (V_{out}) induced in the piezoelectric layer was much higher than V_{in}. Thus, under resonant drive, ME laminates exhibit a strong voltage gain, offering potential for high-voltage miniature transformer applications.

Figure 46 shows the measured voltage gain V_{out}/V_{in} as a function of the drive frequency for a ME transformer consisting of Terfenol-D layers of 40 mm in length and a piezoelectric layer of 80 mm in length. A maximum voltage gain of ~260 was found at a resonance frequency of 21.3 kHz. In addition, at the resonance state, the maximum voltage gain of the ME transformer was strongly dependent on an applied H_{dc}, which was due to the fact that Terfenol-D has a large effective piezomagnetic coefficient only under a suitable H_{dc}.

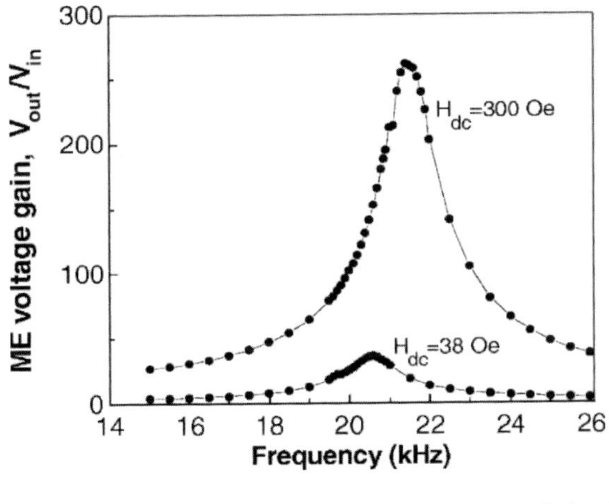

Figure 46. Voltage gain of ME transformer as a function of the drive frequency [Ref. 55].

Other reports have shown that a ME laminate with a coil carrying current I_{in} has a unique current-to-voltage (I–V) conversion capability. ME laminates actually act as a I–V gyrator, with a high I–V gyration coefficient G [65, 66]. Figure 49(a) illustrates the ME gyrator's configuration and Fig. 47(b) shows its ME gyration equivalent circuit. At electromechanical resonance, the ME gyrator shows a strong I–V conversion of ~2500 V/A, as shown in Fig. 48. We also observed (i) reverse gyration: an input current to the piezoelectric section induced a voltage output across coils, and (ii) impedance inversion: a resistor R_i connected in parallel to the primary terminals of the gyrator resulted in an impedance G^2/R_i in series with the secondary terminals.

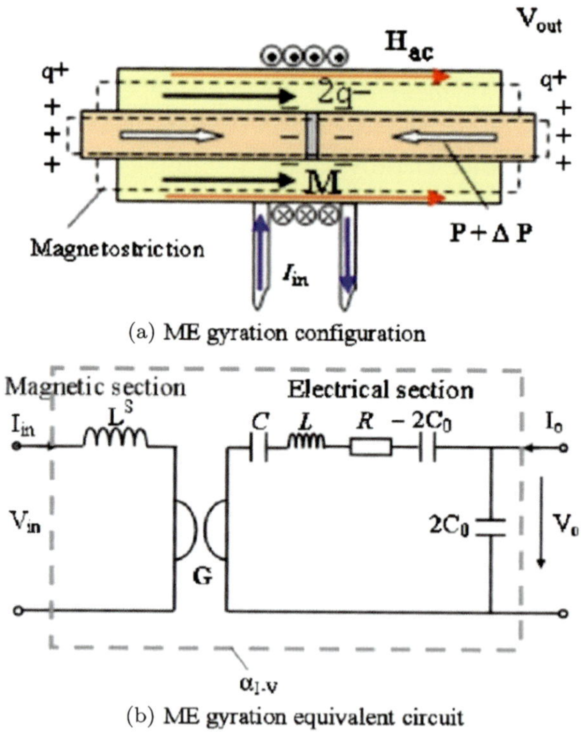

Figure 47. ME gyrator: (a) ME gyration configuration, and (b) ME gyration equivalent circuit [Ref. 65].

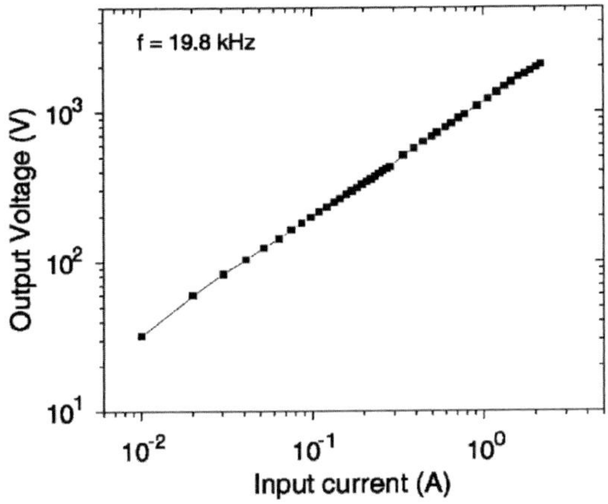

Figure 48. *I–V* gyration of the ME gyrator [Ref. 65].

7.3. *Microwave devices*

Since the composite ME effect is a "dynamic" effect, i.e., a pronounced linear ME response is only observed for an ac field oscillating in the presence of a stronger dc bias field, the composite ME effect is thus predetermined for microwave applications. Different frequency ranges are accessed by using the enhancement from the electromechanical resonance (~100 k Hz), the ferromagnetic resonance (~10 GHz), or the antiferromagnetic resonance (~100 GHz). Among the ME ceramic composites (e.g., ferrite-piezoelectric ceramic composites) represent a promising new approach to build a new class of fast electric field tunable low power microwave devices. An electric field E applied to the piezoelectric transducer produces a mechanical deformation that couples to the ferrite and produces a shift in the resonance field [127, 183–185]. Unlike the situation when magnetic fields are used for such tuning, the process is fast because there are no inductors, and the power budget is small because the biasing voltages involve minimal currents. This process has been demonstrated with ferromagnetic resonance and spin wave delay lines [153, 186–195]. The critical goal for the future is in the development of a wide class of efficient wide band and low loss electrically tunable magnetic film devices for battlefield radar, signal processing, and secure and experimental evaluation of characteristics. The anticipated advantages of ME devices are yet to be exploited.

7.3.1. *Tunable devices*

In the microwave region of the electromagnetic spectrum, the ME effect can be observed in the form of a shift in FMR profile in an external electric field E[127, 183, 184]. The response can be realized with a device similar to the one shown in Fig. 49, i.e., a device concept based ferrite-piezoelectric microwave resonator that is dual electric and magnetic field tunable. Mechanical stress in the PZT layer due to the electric field is coupled to the ferrite, and this leads to a shift δH_E in the FMR resonance field [189].

Figure 50 shows the data on the resonance field shift as a function of E for a bilayer of epitaxial YIG film and PMN-PT single crystal. The field shift δH_E converts to an equivalent frequency shift of about 3 MHz/Oe. This ME shift is determined by piezoelectric coupling, taken as d, and the magnetostriction λ. A large d and a large λ lead to a strong electric field induced shift. The data in Fig. 50 also provide the basis for a wide variety of *electric field tunable* FMR-based and spin wave based YIG signal processing devices.

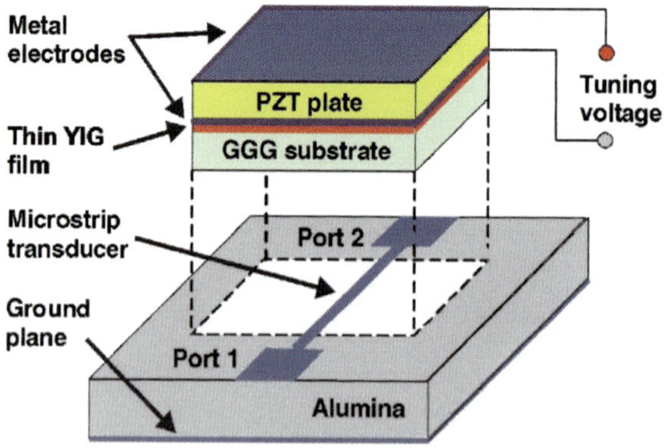

Figure 49. Example of a magnetic YIG film grown on gadolinium gallium garnet substrate and PZT electric field tunable microwave magnetic strip line device [Ref. 189].

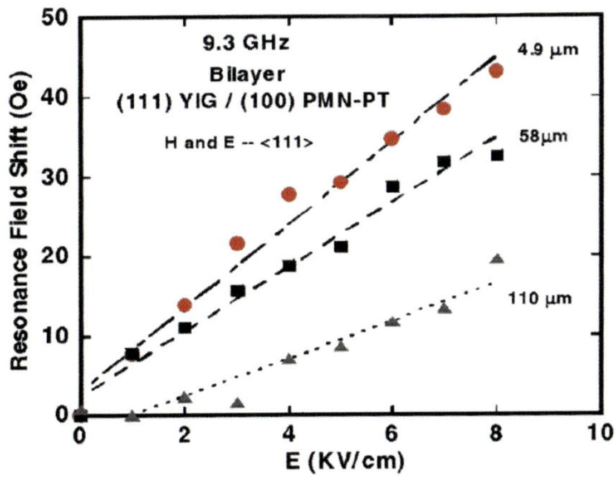

Figure 50. Data on FMR field shift vs electric field E for a YIG/PMN-PT bilayer [Ref. 186].

7.3.2. *Resonators*

The device in Fig. 49 is a basic YIG/piezoelectric resonator. Such devices are useful in frequency selective microwave devices such as bandpass or band-stop filters [186, 190]. The addition of the piezoelectric gives this important class of devices electric field tunability as shown above. Much work is needed if structures such as the one in Fig. 49 are to become viable components in real systems. This device can also serve as an ideal "test bed" for the selection of materials and device configurations for the full range of electric field tunable FMR and spin wave devices enumerated above. There are, for example, several issues.

(1) Options for different magnetic film components. Substituted YIG films can be used with higher magnetizations to cover different frequency ranges. Different substitutions also can give more or less of a ME interaction and better or worse tuning. Zinc-substituted spinel ferrite can be used to achieve very high magnetizations for the low millimeter wave frequency range, and hexagonal ferrite [153] films can be used for frequencies up to 100 GHz or so.

(2) Options for different piezoelectric and ferroelectric layer components. There are many other options for the electric layers, including single crystal PMT-PT, PZT, and barium strontium titanate.
(3) Interplay of magnetic/electric/interface property loss and high power capabilities. Electric field tunability of FMR and spin wave devices may come at a price. It is clear that there will be interface modes. In the case of layered devices of the sort in Fig. 49, these effects can be controlled to some extent through the use of different bonding methods and materials.
(4) Tuning: It is well known that for basic magnetic film FMR and spin wave based microwave devices, one can also tune the frequency of operation by changing the direction of the field relative to the film plane or the direction of spin wave propagation. The field for the data shown above was applied perpendicular to the YIG film in Fig. 49. It is known that the magnetic loss can depend on the field orientation. The effect of field orientation on the ME coupling and the tunability remain to be determined.
(5) Speed of tuning: The obvious advantage of electric field tunability is speed. What are the limits? How can they be achieved in a useful device? What if the electric field tunability could be as fast as the microwave signal itself? This could lead to an entire class of completely new devices.

Figure 51 shows representative FMR data for YIG/PMN-PT with an electric field across PMN-PT. These data were with the static magnetic field perpendicular to the YIG film. The field shift in this case is negative as the electric field is applied. If the static field is in plane, it turns out that the shift is positive. The tuning shown in Fig. 51, with a δH_E of about 50 MHz, is obtained for an electric field of 10 kV/cm. This tuning is more than five times the width of FMR linewidth in the YIG film and is more than adequate for a variety of device applications. With optimized choices for the electric layer and the electric field, one should be able to obtain a tuning range of 0.5–1 GHz and an insertion loss of a few decibels at most.

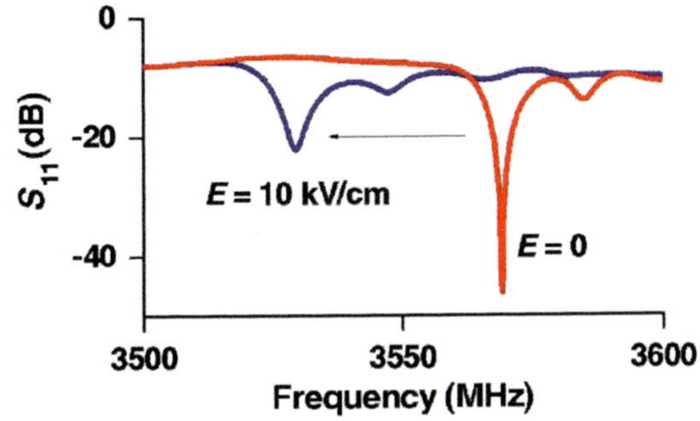

Figure 51. Electric field tunable 13 μm YIG/PMN-PT resonator with perpendicular magnetization (H = 3010 Oe). The frequency shift is -50 MHz for E = 10 kV/cm [Ref. 190].

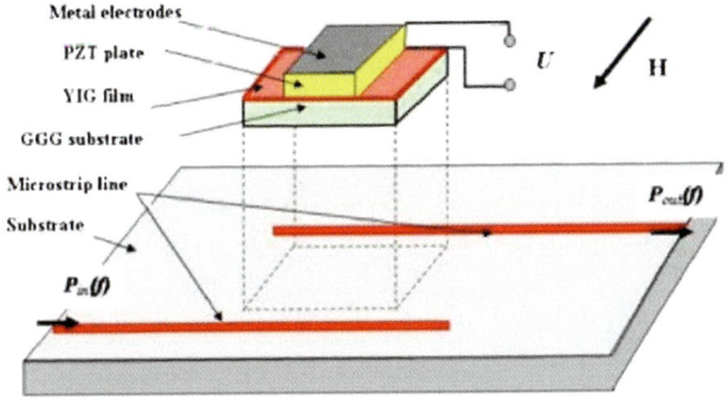

Figure 52. Diagram showing a microwave ME filter [Ref. 190].

7.3.3. *Filters*

The single-cavity ME filter [190], shown in Fig. 52, consists of a 1 mm thick dielectric ground plane (permittivity of 10), input and output microstrips of nonresonance lengths, and a ME element. Power is coupled from input to output under FMR in the ME element. The ME element consisted of epitaxial YIG film bonded to PZT. A 110 μm thick YIG film grown by liquid-phase epitaxy on a (111) gadolinium gallium garnet was used. A PZT plate with the dimensions $4 \times 1 \times 0.5\,\text{mm}^3$ was initially poled by heating up to 150°C and cooling back to room temperature in an electric field of 10 kV/cm perpendicular to the sample plane. The layered structure was made by bonding the YIG film surface to PZT with 0.08 mm thick layer of ethyl cyanoacrylate, a fast-dry epoxy. The layered structure was placed between the transducers as in Fig. 52 and was subjected to a field H parallel to the sample plane and perpendicular to the microstrips.

An input continuous-wave signal $P_{in}(f) = 1$ mW was applied to the filter. The frequency dependence of the insertion loss L, i.e., the transmitted power through the ME element, was measured at 4–10 GHz as a function of H and E applied across the PZT layer. Representative results are shown in Fig. 53. Consider first the profile for $E = 0$. The maximum input–output coupling is observed at $f_r = 6.77$ GHz that corresponds to FMR in YIG for an in-plane H $=$ 1.7 k Oe, $4\pi M = 1.75$ kG, and gyromagnetic ratio $\gamma = 2.8$ GHz/kOe. The loss increases sharply below f_r and the off-resonance isolation is 20–25 dB. Above f_r, secondary maxima due to magnetostatic modes are seen in Fig. 53. A significant modification of frequency dependent L profile is observed (Fig. 53) when $E = 1$ kV/cm is applied across PZT. f_r is downshifted by 28 MHz. The shift arises due to strain at YIG–PZT interface caused by the piezoelectric deformation in PZT. Further increase in E results in increase in the magnitude of the downshift, as shown in Fig. 53. An upshift in f_r can be observed when

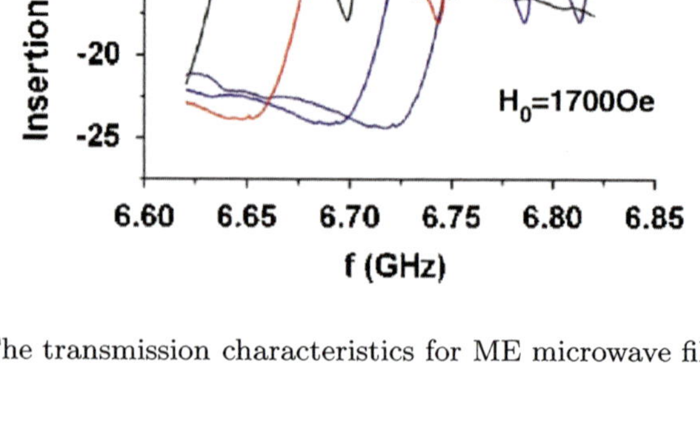

Figure 53. The transmission characteristics for ME microwave filter [Ref. 190].

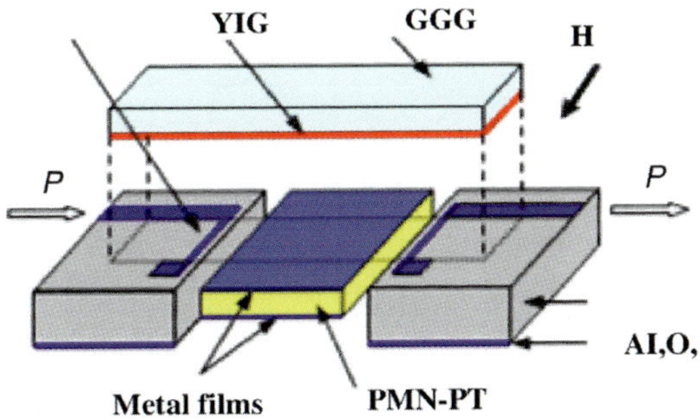

Figure 54. Stripline structure for phase shift and delay time measurements in YIG/PMN-PT [Ref. 187].

the direction of E was reversed by reversing the polarity of applied voltage, and is attributed to a switch from compressive to tensile strain in YIG. Thus, a microwave ME filter based on YIG/PZT layered structure has been designed and characterized. This filter can be tuned by 2% of the central frequency with a nominal electric field of 3 kV/cm.

7.3.4. *Phase shifters and delay lines*

A useful microstrip line structure for phase shift and delay time measurements is shown in Fig. 54 [187]. We used YIG/piezoelectric elements and measured their phase shift, delay time, and E-field tunability. The bias field H is applied parallel to the transducer to excite magnetostatic surface wave (MSSW). Typical results shown in Fig. 55 demonstrate the feasibility for realizing passive, electric field tunable phase shifter, and delay lines.

8. Future directions

8.1. *Bulk ceramic composites*

Although bulk ME ceramic composites were considered to exhibit promising larger linear ME effects, the ME effect so far observed in cofired ceramic composites is 10 times or more lower than that predicted, due mainly to their inherent preparation problems, such as atomic interfacial diffusion and/or reaction problems between two

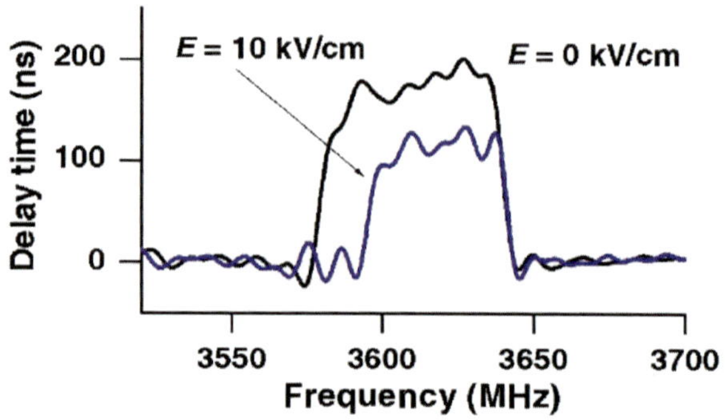

Figure 55. Delay time characteristics for MSSW excitations in YIG/PMN-PT for $H = 701\text{Oe}$ [Ref. 187].

ceramic phases during high-temperature preparation. A large number of combinations of piezoelectric oxides and magnetic oxides (mainly ferrites) would be attained. However, several important issues remain to be solved, including, the following.

(1) For the particulate ceramic composites, the challenge remains in achieving good dispersion of high concentration of ferrite phase in the piezoelectric ceramic matrix, favorable coherent interfaces, and sufficient bulk density while avoiding possible reaction and interfacial diffusion between the two ceramic phases to occur. Recently, chemical solution processing and novel sintering techniques such as SPS and hot pressing have been employed to fabricate the particulate ceramic composites and some improvement has been achieved. But obviously, more work still remains.
(2) For the laminate ceramics, the control of well cofiring the ferrite and piezoelectric layers with good interfacial contact but without interfacial diffusion/reaction is a key issue, which is like the cofiring of multilayer ceramic devices [196]. For the ME applications, strong ME coupling between these layers are desired. On the other hand, the laminate ceramics with high values of permittivity and permeability but minimum ME interaction are also highly desired for the integrated capacitor–inductor applications. The

cofiring especially at low temperature of laminate ferroelectric and ferrite ceramics has been of technological importance.

(3) There have been a lot of theoretical works devoted to the constitutive problem of the ME ceramic composites, especially, many simulations from the micromechanics field. However, no work is done on the fatigue of the ME ceramic composites under bias fields, which is particularly of technological importance.

(4) The ME ceramic composites are ready for practical ME device applications due to their large ME effect above room temperature. Although some prototypes of microwave devices (see Section 7) based on the ME ceramic laminates have been proposed, much work still remains for their applications in real systems.

8.2. *Magnetic alloy based composites*

Among the ME composites, the magnetostrictive alloy (such as Terfenol-D) based composites have shown strongest ME coupling over wide frequency range. For example, the highest subresonant ME voltage coefficient of 5.5 V/cm Oe was observed for a *C–C* mode Terfenol-D/PZN-PT/Terfenol-D three-layer ring-type laminate. For such magnetic alloy based composites, the future directions include the following:

(1) Further optimizations for such ME laminate composites still remain, such as multilayer long or ring-type configurations for structure optimization, incorporation of even higher-μ third phases, and engineering controlled connectivity of the magnetic–piezoelectric phases. Actually, there are many possibilities that have not been tested.

Recently, Metglas ribbons have been used as the magnetostrictive layers [62]. Metglas has an extremely high magnetic permeability ($\mu_r \sim 30000$), but low magnetostriction. However, because of its ultrahigh μ_r, the effective piezomagnetic coefficient is high under small dc magnetic bias, making it an ideal candidate for incorporation into ME composites. Accordingly, a two-dimension Metglas layer connected with a one-dimension piezoelectric PZT fiber layer (i.e., a 2–1 type connectivity of Metglas/PZT fiber laminate, as shown in Fig. 56) has recently been shown to have a largely enhanced ME coupling. Figure 57 shows the ME voltage coefficient as a function of dc magnetic bias

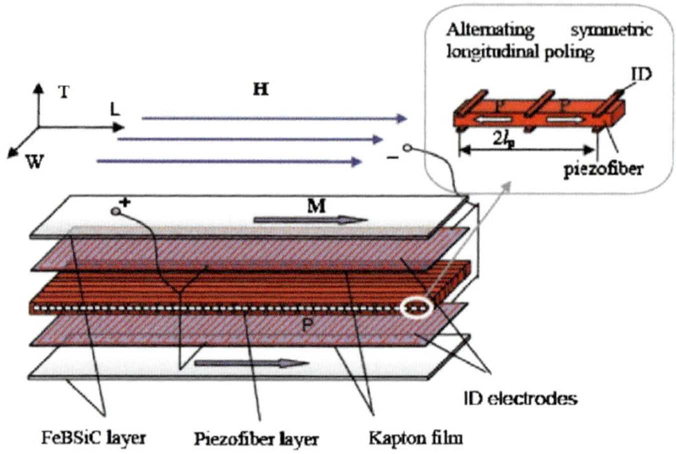

Figure 56. 2-1–type connectivity Metglas/PZT fiber laminate [Ref. 62].

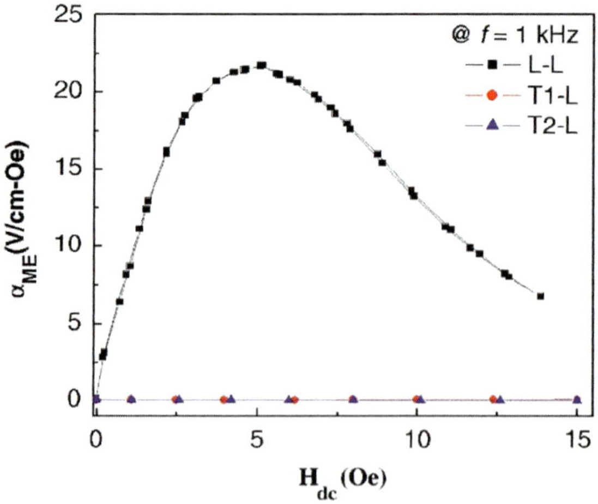

Figure 57. ME field coefficient as a function of dc magnetic bias for longitudinal (L), width (T_1), and thickness (T_2) magnetization [Ref. 62].

for longitudinal (L), width (T_1), and thickness (T_2) magnetizations. From this figure, it can be seen that the maximum dE/dH for the L–L mode is 22 V/cm Oe under a low H_{dc} of only 5 Oe. This is almost an order higher than that of corresponding L–L mode of Terfenol-D based ME laminates. Furthermore, under resonance drive, this value can be enhanced to ~500V/cm Oe. Clearly, Metglas/PZT fibers in such 2–1 connectivity are an important direction for new ME laminate developments.

Most recently, a small three-phase ME composite rod with a single PZT ceramic rod (see the inset in Fig. 58) embedded in Terfenol-D/epoxy (TDE) matrix has been reported [197]. The experimental results demonstrated that the coupling interaction between the PZT rod and TDE matrix can generate much larger ME response than other three-phase polymer-based composites. At high frequency where the electromechanical resonance appears, this ME rod showed a giant ME effect. This simple ME rod presented a size-dependent ME response, which implies that much smaller-sized ME rods (e.g., micro-ME rods) with large

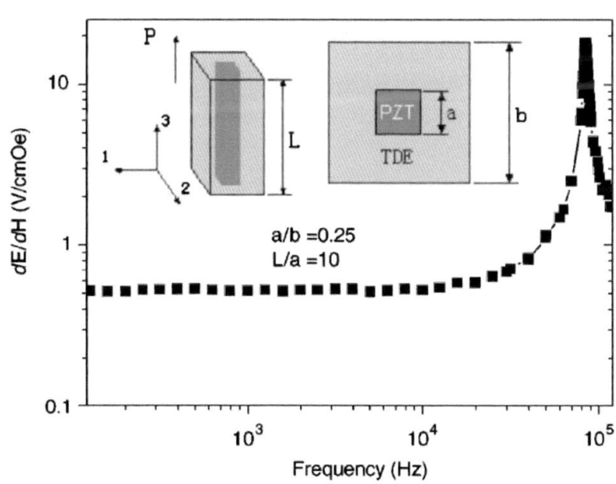

Figure 58. Frequency dependence of the ME coefficients for the small ME composite rod (see the inset) with $a/b = 0.25$ and $L/a = 10$ ($a = 1$ mm) [Ref. 197].

ME response can be obtained by using a single PZT fiber and promises future micro-ME devices.

(2) In the magnetic alloy based (two-phase or three-phase) composites, piezoelectric ceramics and magnetic alloys are bonded together by using polymer binders. The lifetimes of such ME composites are determined by the polymer binders. How do the polymer binders affect the ME response and the fatigue of such ME composites?

(3) Like the ME ceramic composites, these two-phase or three-phase magnetic alloy based ME composites are ready for practical ME device applications due to their giant ME effect above room temperature. Although some prototypes of devices based on these composites have been proposed, much work also remains for their real applications.

8.3. *Nanostructures*

ME nanostructures have become an important topic of ever-increasing interest in last few years, since they, especially ME thin films, are easy to on-chip integration, which is a prerequisite for incorporation into microelectronic devices. But ME nanostructures are just in infant stage. The potential of ME nanostructures and their devices is almost limitless, but we must first overcome two fundamental challenges at least.

(1) The first is physical: how do we control the identity, placement, and function of every important atom in a ME nanostructure, in ways that are practical to apply to real-world ME materials and devices? In terms of nanostructure assembly, controlling the growth of desired ME nanostructures with precisely controlled composition, atomic arrangements, and interfaces is the most interesting but difficult problem. Indeed, this has been an ongoing quest in the field of nanomaterials and nanodevices, with recent progress also driven by directed assembly processes. For example, the formation of self-assembled, vertical nanostructures with long-range ordering will undoubtedly have great impact, not only in the field of multiferroics, but in a broad range of photonic

applications [198]. High-quality horizontal heterostructures such as ME superlattices [199] hold particular promise for the rational design of new multiferroics and their devices.

(2) The second is conceptual: how do we understand systems that are too large to be handled by bruteforce calculation, but too small to be tackled by statistical methods? Many open questions regarding the ME coupling in nanostructures remain [94]: what is the timescale of the coupling process? How does the time dependence of the magnetism relate to the ferroelectric switching? What is the smallest value of magnetic field that can lead to full switching of the magnetization? How does the behavior depend on the chemistry of the nanostructures? Is there a critical dimension below which the coupling will disappear or change in nature?

(3) Like bulk ME composites, sensors, transducers, and heterogeneous read/write devices are among the suggested technical implementations of the composite ME effect in ME nanostructures.

Acknowledgments

C.W.N. was supported by Ministry of Science and Technology of China (Grant Nos. 2002CB613303 and 2006AA03Z101) and the NSF of China (Grant Nos. 50621201 and 10574078). B.M.I. was supported by the Russian Foundation for Basic Research (Project Nos. 06-08-00896-a, No. 06-02-08071-ofi, and No. 05-02-39002-GFEN-a) and Program of Russian Ministry of Education and Science. S.D. and D.V. were supported by Office of Naval Research, U.S. Department of Energy, and DARPA.

References

[1] Schmid, H. (1994). Multi-ferroic magnetoelectrics, *Ferroelectrics*, 162, p. 317.

[2] Fiebig, M. (2005). Revival of the magnetoelectric effect, *J. Phys. D*, 38, pp. R123–R152.

[3] Eerenstein, W., Mathur, N. D. and Scott, J. F. (2006). Multiferroic and magnetoelectric materials, *Nature*, 442, pp. 759–765.

[4] Magnetoelectric Interaction Phenomena in Crystals, Proceedings of the MEIPIC-2, edited by Schmid, H. (Ascona, Switzerland, 13–18 September 1993), [*Ferroelectrics*, vol. 161–162 (1994)].
[5] Proceedings of the MEIPIC-3, edited by Bichurin, M. (Novgorod, Russia, 16–20 September 1996) [*Ferroelectrics*, vol. 204 (1997)].
[6] Proceedings of the MEIPIC-4, edited by Bichurin, M. (Novgorod, Russia, 16–19 October 2001) [*Ferroelectrics*, vol. 279–280 (2002)].
[7] Magnetoelectric Interaction Phenomena in Crystals, edited by Fiebig, M., Eremenko, V. V. and Chupis, I. E. (Kluwer, Dordrecht, 2004); Proceedings of the MEIPIC-5, Sudak, Ukraine, 21–24 September 2003.
[8] Hill, N. A. (2000). Why are there so few magnetic ferroelectrics? *J. Phys. Chem. B*, 104, pp. 6694–6709.
[9] Kimura, T., Goto, T., Shintani, H., Ishizaka, K., Arima, T. and Tokura, Y. (2003). Magnetic control of ferroelectric polarization, *Nature*, 426, pp. 55–58.
[10] Wang, J., Neaton, J. B., Zheng, H., Nagarajan, V., Ogale, S. B., Liu, B., Viehland, D., Vaithyanathan, V., Schlom, D. G., Waghmare, U. V., Spaldin, N. A., Rabe, K. M., Wuttig, M. and Ramesh, R. (2003). Epitaxial $BiFeO_3$ multiferroic thin film heterostructures, *Science*, 299, pp. 1719–1722.
[11] Goto, T., Kimura, T., Lawes, G., Ramirez, A. P. and Tokura, Y. (2004). Ferroelectricity and giant magnetocapacitance in perovskite rare-earth manganites, *Phys. Rev Lett.*, 92, pp. 257201.
[12] Lottermoser, T., Lonkai, T., Amann, U., Hohlwein, D., Ihringer, J. and Fiebig, M. (2004). Magnetic phase control by an electric field, *Nature*, 430, pp. 541–544.
[13] Hur, N., Park, S., Sharma, P. A., Ahn, J. S., Guha, S. and Cheong, S. W. (2004). Electric polarization reversal and memory in a multiferroic material induced by magnetic fields, *Nature*, 429, pp. 392–395.
[14] Prellier, W., Singh, M. P. and Murugavel, P. (2005). The single-phase multiferroic oxides: from bulk to thin film, *J. Phys. Condens. Matter*, 17, pp. R803–R832.
[15] Cheong, S.-W. and Mostovoy, M. (2007). Multiferroics: a magnetic twist for ferroelectricity, *Nat. Mater.*, 6, pp. 13–20.
[16] van Suchtelen, J. (1972). Product properties: new application of composite-materials, *Philips Res. Rep.*, 27, p. 28.
[17] Nan, C. W. (1994). Magnetoelectric effect in composites of piezoelectric and piezomagnetic phases, *Phys. Rev. B*, 50, pp. 6082–6088.
[18] Newnham, R. E., Skinner, D. P. and Cross, L. E. (1978). Connectivity and piezoelectric-pyroelectric composites, *Mater. Res. Bull.*, 13, pp. 525–536.
[19] Boomgaard, J., Terrell, D. R., Born, R. A. J. and Giller, H. F. J. I. (1974). An *in situ* grown eutectic magnetoelectric composite material, *J. Mater. Sci.*, 9, pp. 1705–1709.
[20] Run, A. M. J. G., Terrell, D. R. and Scholing, J. H. (1974). An *in situ* grown eutectic magnetoelectric composite material, *J. Mater. Sci.*, 9, pp. 1710–1714.

[21] Boomgard, J., Run, A. M. J. G. and Suchtelen, J. (1976). Magnetoelectricity in piezoelectric-magnetostrictive composites, *Ferroelectrics*, 10, pp. 295–298.

[22] Boomgard, J. and Born, R. A. J. (1978). A sintered magnetoelectric composite material $BaTiO_3$-Ni(Co, Mn)Fe_2O_4, *J. Mater. Sci.*, pp. 1538–1548.

[23] Harshe, G., Dougherty, J. P. and Newnham, R. E. (1993). Theoretical modelling of multilayer magnetoelectric composites, *Int. J. Appl. Electromagn. Mater.*, 4, pp. 145–159.

[24] Lopatin, S., Lopatina, I. and Lisnevskaya, I. (1994). Magnetoelectric PZT/ferrite composite materials, *Ferroelectrics*, 162, pp. 63–68.

[25] Lupeiko, T. G., Lisnevskaya, I. V., Chkheidze, M. D. and Zvyagintsev, B. I. (1995). Laminated magnetoelectric composites based on nickel ferrite and PZT materials, *Inorg. Mater.*, 31, pp. 1139–1142.

[26] Bichurin, M. I., Kornev, I. A., Petrov, V. M. and Lisnevskaya, I. V. (1997). Investigation of magnetoelectric interaction in composite, *Ferroelectrics*, 204, pp. 289–297.

[27] Patankar, K. K., Patil, S. A., Sivakumar, K. V., Mahajan, R. P., Kolekar, Y. D. and Kothale, M. B. (2000). AC conductivity and magnetoelectric effect in $CuFe_{1.6}Cr_{0.4}O_4$-$BaTiO_3$ composite ceramics, *Mater. Chem. Phys.*, 65, pp. 97–102.

[28] Ryu, J., Carazo, A. V., Uchino, K. and Kim, H. E. (2001). Piezoelectric and magnetoelectric properties of Lead Zirconate Titanate/Ni-Ferrite particulate composites, *J. Electroceram.*, 7, pp. 17–24.

[29] Patankar, K. K., Nipankar, R. P., Mathe, V. L., Mahajan, R. P. and Patil, S. A. (2001). Role of sintering on magneto-electric effect in $CuFe_{1.8}Cr_{0.2}O_4$-$Ba_{0.8}Pb_{0.2}Ti_{0.8}Zr_{0.2}O_3$ composite ceramics, *Ceram. Int.*, 27, pp. 853–858.

[30] Mathe, V. L., Patankar, K. K., Jadhav, U. V., Patil, A. N., Lotake, S. D. and Patil, S. A. (2001). Studies on structural, dielectric and magnetoelectric properties in $CuFe_{1.8}Cr_{0.2}O_4$-$Pb(Mg_{1/3}V_{2/3})O_3$ composites, *Ceram. Int.*, 27, pp. 531–535.

[31] Srinivasan, G., Rasmussen, E. T., Gallegos, J., Srinivasan, R., Bokhan, Y. I. and Laletin, V. M. (2001). Magnetoelectric bilayer and multilayer structures of magnetostrictive and piezoelectric oxides, *Phys. Rev. B*, 64, pp. 214408.

[32] Srinivasan, G., Rasmussen, E. T., Levin, B. J. and Hayes, R. (2002). Magnetoelectric effects in bilayers and multilayers of magnetostrictive and piezoelectric perovskite oxides, *Phys. Rev. B*, 65, pp. 134402.

[33] Bichurin, M. I., Petrov, V. M., Petrov, R. V., Kiliba, Y. V., Bukashev, F. I., Smirnov, A. Y. and Eliseev, D. N. (2002). Magnetoelectric sensor of magnetic field, *Ferroelectrics*, 280, pp. 365–368.

[34] Kothale, M. B., Patankar, K. K., Kadam, S. L., Mathe, V. L., Rao, A. V. and Chougule, B. K. (2003). Dielectric behaviour and magnetoelectric effect in copper-cobalt ferrite plus barium lead titanate composites, *Mater. Chem. Phys.*, 77, pp. 691–696.

[35] Srinivasan, G., Hayes, R. and Bichurin, M. I. (2003). Low frequency and microwave magnetoelectric effects in thick film heterostructures of lithium zinc ferrite and lead zirconate titanate, *Solid State Commun.*, 128, pp. 261–266.

[36] Srinivasan, G., Rasmussen, E. T. and Hayes, R. (2003). Magnetoelectric effects in ferrite-lead zirconate titanate layered composites: The influence of zinc substitution in ferrites, *Phys. Rev. B*, 67, pp. 014418.

[37] Kadam, S. L., Patankar, K. K., Mathe, V. L., Kothale, M. B., Kale, R. B. and Chougule, B. K. (2003). Electrical properties and magnetoelectric effect in $Ni_{0.75}Co_{0.25}Fe_2O_4+Ba_{0.8}Pb_{0.2}TiO_3$ composites, *Mater. Chem. Phys.*, 78, pp. 684–690.

[38] Mazumder, S. and Bhattacharyya, G. S. (2004). Synthesis and characterization of in situ grown magnetoelectric composites in the BaO-TiO-FeO-CoO system, *Ceram. Int.*, 30, pp. 389–392.

[39] Zhai, J. Y., Cai, N., Shi, Z., Lin, Y. H. and Nan, C. W. (2004). Magnetic-dielectric properties of $NiFe_2O_4$/PZT particulate composites, *J. Phys. D*, 37, pp. 823–827.

[40] Zhai, J. Y., Cai, N., Shi, Z., Lin, Y. H. and Nan, C. W. (2004). Coupled magnetodielectric properties of laminated $PbZr_{0.53}Ti_{0.47}O_3/NiFe_2O_4$ ceramics, *J. Appl. Phys.*, 95, pp. 5685–5690.

[41] Zeng, M., Wan, J. G., Wang, Y., Yu, H., Liu, J. M., Jiang, X. P. and Nan, C. W. (2004). Resonance magnetoelectric effect in bulk composites of lead zirconate titanate and nickel ferrite, *J. Appl. Phys.*, 95, pp. 8069–8073.

[42] Devan, R. S., Lokare, S. A., Patil, D. R., Chougule, S. S., Kolekar, Y. D. and Chougule, B. K. (2006). Electrical conduction and magnetoelectric effect of $(x)BaTiO_3+(1-x)Ni_{0.92}Co_{0.03}Cu_{0.05}Fe_2O_4$ composites in ferroelectric rich region, *J. Phys. Chem. Solids*, 67, pp. 1524–1530.

[43] Li, Y. J., Chen, X. M., Lin, Y. Q. and Tang, Y. H. (2006). Magnetoelectric effect of $Ni_{0.8}Zn_{0.2}Fe_2O_4/Sr_{0.5}Ba_{0.5}Nb_2O_6$ composites, *J. Eur. Ceram. Soc.*, 26, pp. 2839–2844.

[44] Zhou, J.-p., He, H.-c., Shi, Z., Liu, G. and Nan, C. W. (2006). Dielectric, magnetic, and magnetoelectric properties of laminated $PbZr_{0.52}Ti_{0.48}O_3$/ $CoFe_2O_4$ composite ceramics, *J. Appl. Phys.*, 100, pp. 094106.

[45] Devan, R. S., Deshpande, S. B. and Chougule, B. K. (2007). Ferroelectric and ferromagnetic properties of $(x)BaTiO_3+(1-x)Ni_{0.94}Co_{0.01}Cu_{0.05}Fe_2O_4$ composite, *J. Phys. D*, 40, pp. 1864–1868.

[46] Chougule, S. S. and Chougule, B. K. (2007). Studies on electrical properties and the magnetoelectric effect on ferroelectric-rich $(x)Ni_{0.8}Zn_{0.2}Fe_2O_4+(1-x)$PZT ME composites, *Smart Mater. Struct.*, 16, pp. 493–497.

[47] Ryu, J. H., Priya, S., Carazo, A. V., Uchino, K. and Kim, H. E. (2001). Effect of the magnetostrictive layer on magnetoelectric properties in lead zirconate titanate/terfenol-D laminate composites, *J. Am. Ceram. Soc.*, 84, pp. 2905–2908.

[48] Ryu, J., Carazo, A. V., Uchino, K. and Kim, H. E. (2001). Magnetoelectric properties in piezoelectric and magnetostrictive laminate composites, *Jpn. J. Appl. Phys.*, Part 1, 40, pp. 4948–4951.

[49] Dong, S. X., Li, J. F. and Viehland, D. (2003). Ultrahigh magnetic field sensitivity in laminates of TERFENOL-D and $Pb(Mg_{1/3}Nb_{2/3})O_3$-$PbTiO_3$ crystals, *Appl. Phys. Lett.*, 83, pp. 2265–2267.

[50] Dong, S. X., Li, J. F. and Viehland, D. (2003). Giant magneto-electric effect in laminate composites, *IEEE Trans. Ultrason. Ferroelectr. Freq. Control*, 50, pp. 1236–1239.

[51] Dong, S. X., Li, J. F. and Viehland, D. (2004). Characterization of magnetoelectric laminate composites operated in longitudinal-transverse and transverse-transverse modes, *J. Appl. Phys.*, 95, pp. 2625–2630.

[52] Dong, S. X., Li, J. F. and Viehland, D. (2004). Vortex magnetic field sensor based on ring-type magnetoelectric laminate, *Appl. Phys. Lett.*, 85, pp. 2307–2309.

[53] Dong, S. X., Li, J. F. and Viehland, D. (2004). Circumferentially magnetized and circumferentially polarized magnetostrictive/piezoelectric laminated rings, *J. Appl. Phys.*, 96, pp. 3382–3387.

[54] Dong, S. X., Li, J. F. and Viehland, D. (2004). Vortex magnetic field sensor based on ring-type magnetoelectric laminate, *Appl. Phys. Lett.*, 85, pp. 2307–2309.

[55] Dong, S. X., Li, J. F., Viehland, D., Cheng, J. and Cross, L. E. (2004). A strong magnetoelectric voltage gain effect in magnetostrictive-piezoelectric composite, *Appl. Phys. Lett.*, 85, pp. 3534–3536.

[56] Dong, S. X., Li, J. F. and Viehland, D. (2004). Voltage gain effect in a ring-type magnetoelectric laminate, *Appl. Phys. Lett.*, 84, pp. 4188–4190.

[57] Dong, S. X., Zhai, J. Y., Xing, Z. P., Li, J. F. and Viehland, D. (2005). Extremely low frequency response of magnetoelectric multilayer composites, *Appl. Phys. Lett.*, 86, pp. 102901.

[58] Dong, S. X., Bai, J. G., Zhai, J. Y., Li, J. F., Lu, G. Q., Viehland, D., Zhang, S. J. and Shrout, T. R. (2005). Circumferential-mode, quasi-ring-type, manetoelectric laminate composite-a highly sensitive electric current and/or vortex magnetic field sensor, *Appl. Phys. Lett.*, 86, pp. 182506.

[59] Dong, S. X., Zhai, J. Y., Bai, F. M., Li, J. F., Viehland, D. and Lograsso, T. A. (2005). Magnetostrictive and magnetoelectric behavior of Fe-20 at. % Ga/Pb(Zr, Ti)O_3 laminates, *J. Appl. Phys.*, 97, pp. 103902.

[60] Dong, S. X., Zhai, J. Y., Wang, N. G., Bai, F. M., Li, J. F., Viehland, D. and Lograsso, T. A. (2005). Fe-Ga/$Pb(Mg_{1/3}Nb_{2/3})O_3$-$PbTiO_3$ magnetoelectric laminate composites, *Appl. Phys. Lett.*, 87, pp. 222504.

[61] Dong, S. X., Zhai, J. Y., Li, J. F. and Viehland, D. (2006). Small dc magnetic field response of magnetoelectric laminate composites, *Appl. Phys. Lett.*, 88, pp. 082907.

[62] Dong, S., Zhai, J., Li, J.-F. and Viehland, D. (2006). Magnetoelectric effect in Terfenol-D/Pb(Zr, Ti$O)_3$/mu-metal laminate composites, *Appl. Phys. Lett.*, 89, pp. 252904.

[63] Dong, S., Zhai, J., Li, J.-F. and Viehland, D. (2006). Magnetoelectric effect in Terfenol-D/Pb(Zr, $TiO)_3$/mu-metal laminate composites, *Appl. Phys. Lett.*, 89, pp. 122903.

[64] Dong, S., Zhai, J., Li, J. and Viehland, D. (2006). Enhanced magnetoelectric effect in three-phase $MnZnFe_2O_4/Tb_{1-x}Dy_xFe_{2-y}$/Pb (Zr, $Ti)O_3$ composites, *J. Appl. Phys.*, 100, pp. 124108.

[65] Dong, S., Zhai, J., Li, J. F., Viehland, D. and Bichurin, M. I. (2006). Magnetoelectric gyration effect in $Tb_{1-x}Dy_xFe_{2-y}$/Pb(Zr, $Ti)O_3$ laminated composites at the electromechanical resonance, *Appl. Phys. Lett.*, 89, pp. 243512.

[66] Zhai, J., Li, J., Dong, S., Viehland, D. and Bichurin, M. I. (2006). A quasi(unidirectional) Tellegen gyrator, *J. Appl. Phys.*, 100, pp. 124509.

[67] Li, P., Wen, Y. and Bian, L. (2007). Enhanced magnetoelectric effects in composite of piezoelectric ceramics, rare-earth iron alloys, and ultrasonic horn, *Appl. Phys. Lett.*, 90, pp. 022503.

[68] Yin, X. M., Zhang, N. and Bao, J. C. (2007). Single magnetic field-driven magnetoelectric effect in $Tb_{1-x}Dy_xFe_{2-y}$Pb(Zr, $Ti)O_3$ laminate composites, *Phys. Lett. A*, 361, pp. 434–436.

[69] Nan, C. W., Liu, L., Cai, N., Zhai, J., Ye, Y., Lin, Y. H., Dong, L. J. and Xiong, C. X. (2002). A three-phase magnetoelectric composite of piezoelectric ceramics, rare-earth iron alloys, and polymer, *Appl. Phys. Lett.*, 81, pp. 3831–3833.

[70] Wan, J. G., Liu, J. M., Chand, H. L. W., Choy, C. L., Wang, G. H. and Nan, C. W. (2003). Giant magnetoelectric effect of a hybrid of magnetostrictive and piezoelectric composites, *J. Appl. Phys.*, 93, pp. 9916–9919.

[71] Nan, C. W., Cai, N., Liu, L., Zhai, J., Ye, Y. and Lin, Y. (2003). Coupled magnetic-electric properties and critical behavior in multiferroic particulate composites, *J. Appl. Phys.*, 94, pp. 5930–5936.

[72] Cai, N., Zhai, J., Nan, C. W., Lin, Y. and Shi, Z. (2003). Dielectric, ferroelectric, magnetic, and magnetoelectric properties of multiferroic laminated composites, *Phys. Rev. B*, 68, pp. 224103.

[73] Cai, N., Nan, C. W., Zhai, J. Y. and Lin, Y. H. (2004). Large high-frequency magnetoelectric response in laminated composites of piezoelectric ceramics, rare-earth iron alloys and polymer, *Appl. Phys. Lett.*, 84, pp. 3516–3518.

[74] Lin, Y. H., Cai, N., Zhai, J. Y., Liu, G. and Nan, C. W. (2005). Giant magnetoelectric effect in multiferroic laminated composites, *Phys. Rev. B*, 72, pp. 012405.

[75] Nan, C. W., Cai, N., Shi, Z., Zhai, J., Liu, G. and Lin, Y. (2005). Large magnetoelectric response in multiferroic polymer-based composites, *Phys. Rev. B*, 71, pp. 014102.

[76] Shi, Z., Nan, C. W., Zhang, J., Cai, N. and Li, J. F. (2005). Magnetoelectric effect of Pb(Zr, $Ti)O_3$ rod arrays in a (Tb, $Dy)Fe_2$/epoxy medium, *Appl. Phys. Lett.*, 87, pp. 012503.

[77] Shi, Z., Nan, C.-W., Zhang, J., Ma, J. and Li, J.-F. (2006). Magnetoelectric properties of multiferroic composites with pseudo-1-3-type structure, *J. Appl. Phys.*, 99, pp. 124108.
[78] Shi, Z., Ma, J., Lin, Y. and Nan, C.-W. (2007). Magnetoelectric resonance behavior of simple bilayered Pb(Zr, Ti)O_3-(Tb, Dy)Fe_2/epoxy composites, *J. Appl. Phys.*, 101, pp. 043902.
[79] Zheng, H., Wang, J., Lofland, S. E., Ma, Z., Mohaddes-Ardabili, L., Zhao, T., Salamanca-Riba, L., Shinde, S. R., Ogale, S. B., Bai, F., Viehland, D., Jia, Y., Schlom, D. G., Wuttig, M., Roytburd, A. and Ramesh, R. (2004). Multiferroic $BaTiO_3$-$CoFe_2O_4$ nanostructures, *Science*, 303, pp. 661–663.
[80] Zheng, H., Wang, J., Mohaddes-Ardabili, L., Wuttig, M., Salamanca-Riba, L., Schlom, D. G. and Ramesh, R. (2004). Three-dimensional heteroepitaxy in self-assembled $BaTiO_3$-$CoFe_2O_4$ nanostructures, *Appl. Phys. Lett.*, 85, pp. 2035–2037.
[81] Zavaliche, F., Zheng, H., Mohaddes-Ardabili, L., Yang, S. Y., Zhan, Q., Shafer, P., Reilly, E., Chopdekar, R., Jia, Y., Wright, P., Schlom, D. G., Suzuki, Y. and Ramesh, R. (2005). Electric field-induced magnetization switching in epitaxial columnar nanostructures, *Nano Lett.*, 5, pp. 1793–1796.
[82] Wan, J. G., Wang, X. W., Wu, Y. J., Zeng, M., Wang, Y., Jiang, H., Zhou, W. Q., Wang, G. H. and Liu, J. M. (2005). Magnetoelectric $CoFe_2O_4$-Pb(Zr, Ti)O_3 composite thin films derived by a sol–gel process, *Appl. Phys. Lett.*, 86, pp. 122501.
[83] Murugavel, P., Saurel, D., Prellier, W., Simon, C. and Raveau, B. (2004). Tailoring of ferromagnetic $Pr_{0.85}Ca_{0.15}MnO_3$/ferroelectric $Ba_{0.6}Sr_{0.4}TiO_3$ superlattices for multiferroic properties, *Appl. Phys. Lett.*, 85, pp. 4424–4426.
[84] Singh, M. P., Prellier, W., Mechin, L., Simon, C. and Raveau, B. (2006). Correlation between structure and properties in multiferroic $La_{0.7}Ca_{0.3}MnO_3$/$BaTiO_3$ superlattices, *J. Appl. Phys.*, 99, pp. 024105.
[85] Zhou, J. P., He, H. C., Shi, Z. and Nan, C. W. (2006). Magnetoelectric $CoFe_2O_4$/$Pb(Zr_{0.52}Ti_{0.48})O_3$ double-layer thin film prepared by pulsed-laser deposition, *Appl. Phys. Lett.*, 88, pp. 013111.
[86] He, H. C., Zhou, J. P., Wang, J. and Nan, C. W. (2006). Multiferroic $Pb(Zr_{0.52}Ti_{0.48})O_3$–$Co_{0.9}Zn_{0.1}Fe_2O_4$ bilayer thin films via a solution processing, *Appl. Phys. Lett.*, 89, pp. 052904.
[87] Levin, I., Li, J., Slutsker, J. and Roytburd, A. L. (2006). Design of self-assembled multiferroic nanostructures in epitaxial films, *Adv. Mater.*, 18, pp. 2044–2047.
[88] Zheng, H., Straub, F., Zhan, Q., Yang, P.-L., Hsieh, W.-K., Zavaliche, F., Chu, Y.-H., Dahmen, U. and Ramesh, R. (2006). Self-assembled growth of $BiFeO_3$-$CoFe_2O_4$ nanostructures, *Adv. Mater.*, 18, pp. 2747–2752.

[89] Wu, T., Zurbuchen, M. A., Saha, S., Wang, R. V., Streiffer, S. K. and Mitchell, J. F. (2006). Observation of magnetoelectric effect in epitaxial ferroelectric film/manganite crystal heterostructures, *Phys. Rev. B*, 73, pp. 134416.
[90] Ortega, N., Bhattacharya, P., Katiyar, R. S., Dutta, P., Manivannan, A., Seehra, M. S., Takeuchi, I. and Majumder, S. B. (2006). Multiferroic properties of Pb(Zr, Ti)O_3/$CoFe_2O_4$ composite thin films, *J. Appl. Phys.*, 100, pp. 126105.
[91] Eerenstein, W., Wiora, M., Prieto, J. L., Scott, J. F. and Mathur, N. D. (2007). Giant sharp and persistent converse magnetoelectric effects in multiferroic epitaxial heterostructures, *Nat. Mater.*, 6, pp. 348–351.
[92] Ma, Y. G., Cheng, W. N., Ning, M. and Ong, C. K. (2007). Magnetoelectric effect in epitaxial $Pb(Zr_{0.52}Ti_{0.48})O_3/La_{0.7}Sr_{0.3}MnO_3$ composite thin film, *Appl. Phys. Lett.*, 90, pp. 152911.
[93] He, H.-C., Wang, J., Zhou, B.-P. and Nan, C.-W. (2007). Ferroelectric and ferromagnetic behavior of $Pb(Zr_{0.52}Ti_{0.48})O_3$-$Co_{0.9}Zn_{0.1}Fe_2O_4$ multi-layered thin films prepared via solution processing, *Adv. Funct. Mater.*, 17, pp. 1333–1338.
[94] Ramesh, R. and Spaldin, N. A. (2007). Multiferroics: progress and prospects in thin films, *Nat. Mater.*, 6, pp. 21–29.
[95] Nan, C. W. (1993). Physics of inhomogeneous inorganic materials, *Prog. Mater. Sci.*, 37, pp. 1–116.
[96] Nan, C. W. and Clarke, D. R. (1997). Effective properties of ferroelectric and/or ferromagnetic composites: A unified approach and its application, *J. Am. Ceram. Soc.*, 80, pp. 1333–1340.
[97] Benveniste, Y. (1995). Magnetoelectric effect in fibrous composites with piezoelectric and piezomagnetic phases, *Phys. Rev B*, 51, pp. 16424–16427.
[98] Getman, I. (1994). Magnetoelectric composite materials: theoretical approach to determine their properties, *Ferroelectrics*, 162, pp. 45–50.
[99] Chen, T. and Chiang, S. C. (1997). Electroelastic fields and effective moduli of a medium containing cavities or rigid inclusions of arbitrary shape under anti-plane mechanical and in-plane electric fields, *Acta Mech.*, 121, pp. 79–96.
[100] Lee, J. S., Boyd, J. G. and Lagoudas, D. C. (2005). Effective properties of three-phase electro-magneto-elastic composites, *Int. J. Eng. Sci.*, 43, pp. 790–825.
[101] Li, J. N. Y. and Dunn, M. L. (1998). Anisotropic coupled-field inclusion and inhomogeneity problems, *Philos. Mag. A*, 77, pp. 1341–1350.
[102] Li, J. Y. (2003). Uniqueness and reciprocity theorems for linear thermo-electro-magneto-elasticity, Q. *J. Mech. Appl. Math.*, 56, pp. 35–43.
[103] Srinivas, S. and Li, J. Y. (2005). The effective magnetoelectric coefficients of polycrystalline multiferroic composites, *Acta Mater.*, 53, pp. 4135–4142.
[104] Srinivas, S., Li, J. Y., Zhou, Y. C. and Soh, A. K. (2006). The effective magnetoelectroelastic moduli of matrix-based multiferroic composites, *J. Appl. Phys.*, 99, pp. 043905.

[105] Li, L. J. and Li, J. Y. (2006). Electromagnetic fields induced in a uniaxial multiferroic material by a point source or an ellipsoidal inclusion, *Phys. Rev. B*, 73, pp. 184416.
[106] Huang, J. H. and Kuo, W. S. (1997). The analysis of piezoelectric/piezomagnetic composite materials containing ellipsoidal inclusions, *J. Appl. Phys.*, 81, pp. 1378–1386.
[107] Huang, J. H. (1998). Analytical predictions for the magnetoelectric coupling in piezomagnetic materials reinforced by piezoelectric ellipsoidal inclusions, *Phys. Rev. B*, 58, pp. 12–15.
[108] Wu, T. L. and Huang, J. H. (2000). Closed-form solutions for the magnetoelectric coupling coefficients in fibrous composites with piezoelectric and piezomagnetic phases, *Int. J. Solids Struct.*, 37, pp. 2981–3009.
[109] Pan, E. (2002). Three-dimensional Green's functions in anisotropic magneto-electro-elastic bimaterials, *Z. Angew. Math. Phys.*, 53, pp. 815–838.
[110] Filippov, D. A. (2004). Theory of magnetoelectric effect in ferrite-piezoelectric hybrid composites, *Tech. Phys. Lett.*, 30, pp. 351–353.
[111] Huang, H. T. and Zhou, L. M. (2004). Micromechanics approach to the magnetoelectric properties of laminate and fibrous piezoelectric/magnetostrictive composites, *J. Phys. D*, 37, pp. 3361–3366.
[112] Bichurin, M. I., Petrov, V. M., Ryabkov, O. V., Averkin, S. V. and Srinivasan, G. (2005). Theory of magnetoelectric effects at magnetoacoustic resonance in single-crystal ferromagnetic-ferroelectric heterostructures, *Phys. Rev. B*, 72, pp. 060408.
[113] Zhou, Y. and Shin, F. G. (2006). Magnetoelectric effect of mildly conducting magnetostrictive/piezoelectric particulate composites, *J. Appl. Phys.*, 100, pp. 043910.
[114] Nan, C. W., Li, M. and Huang, J. H. (2001). Calculations of giant magnetoelectric effects in ferroic composites of rare-earth-iron alloys and ferroelectric polymers, *Phys. Rev. B*, 63, pp. 144415.
[115] Nan, C. W., Li, M., Feng, X. Q. and Yu, S. W. (2001). Possible giant magnetoelectric effect of ferromagnetic rare-earth-iron-alloys-filled ferroelectric polymers, *Appl. Phys. Lett.*, 78, pp. 2527–2529.
[116] Mori, K. and Wuttig, M. (2002). Magnetoelectric coupling in terfenol-D/polyvinylidenedifluoride composites, *Appl. Phys. Lett.*, 81, pp. 100–101.
[117] Nan, C. W., Liu, G., Lin, Y. H. and Chen, H. D. (2005). Magnetic-field-induced electric polarization in multiferroic nanostructures, *Phys. Rev. Lett.*, 94, pp. 197203.
[118] Liu, G., Nan, C. W., Xu, Z. K. and Chen, H. D. (2005). Coupling interaction in multiferroic $BaTiO_3$-$CoFe_2O_4$ nanostructures, *J. Phys. D*, 38, pp. 2321–2326.
[119] Duan, C.-G., Jaswal, S. S. and Tsymbal, E. Y. (2006). Predicted magnetoelectric effect in Fe/$BaTiO_3$ multilayers: Ferroelectric control of magnetism, *Phys. Rev. Lett.*, 97, pp. 047201.

[120] Zhang, J. X., Li, Y. L., Schlom, D. G., Chen, L. Q., Zavaliche, F., Ramesh, R. and Jia, Q. X. (2007). Phase-field model for epitaxial ferroelectric and magnetic nanocomposite thin films, *Appl. Phys. Lett.*, 90, pp. 052909.
[121] Lu, X., Wang, B., Zheng, Y. and Ryba, E. (2007). Coupling interaction in 1-3-type multiferroic composite thin films, *Appl. Phys. Lett.*, 90, pp. 133124.
[122] Nan, C. W. (1997). Comment on "The analysis of piezoelectric/piezomagnetic composite materials containing ellipsoidal inclusions" [*J. Appl. Phys.* 81, 1378 (1997)], *J. Appl. Phys.*, 82, pp. 5268–5269.
[123] Hill, R. (1964). Theory of mechanical properties of fibre-strengthened materials .1. elastic behaviour, *J. Mech. Phys. Solids*, 12, pp. 199–212.
[124] Avellaneda, M. and Harshe, G. (1994). Magnetoelectric effect in piezoelectric magnetostrictive multilayer (2-2) composites, *J. Intell. Mater. Syst. Struct.*, 5, pp. 501–513.
[125] Bichurin, M. I., Petrov, V. M. and Srinivasan, G. (2003). Theory of low-frequency magnetoelectric coupling in magnetostrictive-piezoelectric bilayers, *Phys. Rev. B*, 68, pp. 054402.
[126] Bichurin, M. I. and Petrov, V. M. (1989). Magnetic resonance in layered ferrite-ferroelectric structures, *Sov. Phys. Tech. Phys.*, 33, p. 1389.
[127] Bichurin, M. I., Petrov, V. M. Kiliba, Y. V. and Srinivasan, G. (2002). Magnetic and magnetoelectric susceptibilities of a ferroelectric/ferromagnetic composite at microwave frequencies, *Phys. Rev. B*, 66, pp. 134404.
[128] Bichurin, M. I., Petrov, V. M., Kornev, I. A., Tatarenko, A. S., Kiliba, Y. V., Konstantinov, N. A. and Srinivasan, G. (2002). Resonance magnetoelectric effect in multilayer composites, *Ferroelectrics*, 280, pp. 353–363.
[129] Bichurin, M. I., Filippov, D. A., Petrov, V. M., Laletsin, V. M., Paddubnaya, N. and Srinivasan, G. (2003). Resonance magnetoelectric effects in layered magnetostrictive-piezoelectric composites, *Phys. Rev. B*, 68, pp. 132408.
[130] Filippov, D. A., Bichurin, M. I., Petrov, V. M., Laletin, V. M., Poddubnaya, N. N. and Srinivasan, G. (2004). Giant magnetoelectric effect in composite materials in the region of electromechanical resonance, *Tech. Phys. Lett.*, 30, pp. 6–8.
[131] Bichurin, M. I. and Petrov, V. M. (1994). Composite magnetoelectrics: their microwave properties, *Ferroelectrics*, 162, pp. 33–35.
[132] Chen, X. M., Tang, Y. H., Chen, I. W., Xu, Z. C. and Wu, S. Y. (2004). Dielectric and magnetoelectric characterization of $CoFe_2O_4$/$Sr_{0.5}Ba_{0.5}Nb_2O_6$ composites, *J. Appl. Phys.*, 96, pp. 6520–6522.
[133] Tang, Y. H., Chen, X. M., Li, Y. J. and Zheng, X. H. (2005). Dielectric and magnetoelectric characterization of $CoFe_2O_4/Ba_{0.55}Sr_{0.25}Ca_{0.2}Nb_2O_6$ composites, *Mater. Sci. Eng., B*, 116, pp. 150–155.
[134] Patankar, K. K., Mathe, V. L., Patil, A. N., Patil, S. A., Lotke, S. D., Kolekar, Y. D. and Joshi, P. B. (2001). Electrical conduction and magnetoelectric effect in $CuFe_{1.8}Cr_{0.2}O_4$-$Ba_{0.8}Pb_{0.2}TiO_3$ composites, *J. Electroceram.*, 6, pp. 115–122.

[135] Patankar, K. K., Mathe, V. L., Mahajan, R. P., Patil, S. A., Reddy, R. M. and Siva Kumar, K. V. (2001). Dielectric behaviour and magnetoelectric effect in $CuFe_2O_4$-$Ba_{0.8}Pb_{0.2}TiO_3$ composites, *Mater. Chem. Phys.*, 72, pp. 23–29.

[136] Kothale, M. B., Patankar, K. K., Kadam, S. L., Mathe, V. L., Rao, A. V. and Chougule, B. K. (2003). Dielectric behaviour and magnetoelectric effect in copper-cobalt ferrite plus barium lead titanate composites, *Mater. Chem. Phys.*, 77, pp. 691–696.

[137] Kulkarni, S. R., Kanamadi, C. M., Patankar, K. K. and Chougule, B. K. (2005). Magnetic properties and magnetoelectric effect in $Ni_{0.8}Co_{0.1}Cu_{0.1}Fe_2O_4$+$PbZr_{0.2}Ti_{0.8}O_3$ composites, *J. Mater. Sci.*, 40, pp. 5691–5694.

[138] Dai, Y. R., Bao, P., Zhu, J. S., Shen, H. M., Liu, J. M. and Wang, Y. N. (2003). Internal friction and dielectric of $CuFe_2O_4/PbZr_{0.53}Ti_{0.47}O_3$ composites, *J. Appl. Phys.*, 39, pp. 1209–1211.

[139] Botello-Zubiate, M. E., Bueno-Baques, D., De Frutos Vaquerizo, J., Fuentes Cobas, L. E. and Matutes-Aquino, J. A. (2006). Synthesis and magnetoelectric characterization of cobalt ferrite-barium titanate composites using a new pulsed magnetic field method, Integr. *Ferroelectr.*, 83, pp. 33–40.

[140] Kanamadi, C. M., Pujari, L. B. and Chougule, B. K. (2005). Dielectric behaviour and magnetoelectric effect in $(x)Ni_{0.8}Cu_{0.2}Fe_2O_4$+$(1-x)Ba_{0.9}Pb_{0.1}Ti_{0.9}Zr_{0.1}O_3$ ME composites, *J. Magn. Magn. Mater.*, 295, pp. 139–144.

[141] Zeng, M., Wan, J. G., Wang, Y., Yu, H., Liu, J. M. and Jiang, X. P. (2004). Resonance magnetoelectric effect in bulk composites of lead zirconate titanate and nickel ferrite, *J. Appl. Phys.*, 95, pp. 8069.

[142] Islam, R. A. and Priya, S. (2006). Annealing and aging effect in 0.95 $Pb(Zr_{0.52}Ti_{0.48})O_3$-$_{0.05}NiFe_{1.9}Mn_{0.1}O_4$ particulate magnetoelectric composites, *Jpn. J. Appl. Phys.*, Part 2, 45, pp. L128–L131.

[143] Petrov, V. M., Srinivasan, G., Bichurin, M. I. and Gupta, A. (2007). Theory of magnetoelectric effects in ferrite piezoelectric nanocomposites, *Phys. Rev. B*, 75, pp. 174442.

[144] Ren, S. Q., Weng, L. Q., Song, S. H., Li, F., Wan, J. G. and Zeng, M. (2005). $BaTiO_3/CoFe_2O_4$ particulate composites with large high frequency magnetoelectric response, *J. Mater. Sci.*, 40, pp. 4375–4378.

[145] Srinivasan, G., DeVreugd, C. P., Flattery, C. S., Laletsin, V. M. and Paddubnaya, N. (2004). Magnetoelectric interactions in hot-pressed nickel zinc ferrite and lead zircon ante titanate composites, *Appl. Phys. Lett.*, 85, pp. 2550–2552.

[146] Jiang, Q. H., Shen, Z. J., Zhou, J. P., Shi, Z. and Nan, C.-W. (2007). Magnetoelectric composites of nickel ferrite and lead zirconnate titanate prepared by spark plasma sintering, *J. Eur. Ceram. Soc.*, 27, pp. 279–284.

[147] Shen, Z. and Nygren, M. (2005). Microstructural prototyping of ceramics by kinetic engineering: Applications of spark plasma sintering, *Chem. Rec.*, 5, pp. 173–184.

[148] Weng, L., Fu, Y., Song, S., Tang, J. and Li, J. (2007). Synthesis of lead zirconate titanate-cobalt ferrite magnetoelectric particulate composites via an ethylenediaminetetraacetic acid-citrate gel process, *Scr. Mater.*, 56, pp. 465–468.
[149] Wan, J.-G., Zhang, H., Wang, X., Pan, D., Liu, J.-M. and Wang, G. (2006). Magnetoelectric $CoFe_2O_4$-lead zirconate titanate thick films prepared by a polyvinylpyrrolidone-assisted sol–gel method, *Appl. Phys. Lett.*, 89, pp. 122914.
[150] Corral-Flores, V., Bueno-Baques, D., Carrillo-Flores, D. and Matutes-Aquino, J. A. (2006). Enhanced magnetoelectric effect in core-shell particulate composites, *J. Appl. Phys.*, 99, pp. 08J503.
[151] Islam, R. A. and Priya, S. (2006). Magnetoelectric properties of the lead–free cofired BaTiO3–(Ni0.8Zn0.2)Fe2O4 bilayer composite, *Appl. Phys. Lett.*, 89, pp. 152911; Zhang, N., Ke, W., Schneider, T. and Srinivasan, G. (2006). Dependence of the magnetoelectric coupling in NZFO-PZT laminate composites on ferrite compactness, *J. Phys.: Condens. Matter*, 18, pp. 11013–11019.
[152] Lisnevskaya, I. V., Bobrova, I. A., Bikyashev, E. A. and Lupeiko, T. G. (2006). Interfacial reactions and properties of $Y_3Fe_5O_{12}/Ba_1\text{-}xPbxTiO_3$ composites, *Inorg. Mater.*, 42, pp. 1147–1151.
[153] Srinivasan, G., Zavislyak, I. V. and Tatarenko, A. S. (2006). Millimeter-wave magnetoelectric effects in bilayers of barium hexaferrite and lead zirconate titanate, *Appl. Phys. Lett.*, 89, pp. 152508.
[154] Nan, C. W., Lin, Y. H. and Huang, J. H. (2002). Magnetoelectricity of multiferroic composites, *Ferroelectrics*, 280, pp. 319–329.
[155] Nan, C. W., Liu, G. and Lin, Y. H. (2003). Influence of interfacial bonding on giant magnetoelectric response of multiferroic laminated composites of $Tb_{1-x}Dy_xFe_2$ and $PbZr_xTi_{1-x}O_3$, *Appl. Phys. Lett.*, 83, pp. 4366–4368.
[156] Dong, S. X., Li, J. F. and Viehland, D. (2003). Longitudinal and transverse magnetoelectric voltage coefficients of magnetostrictive/piezoelectric laminate composite: Theory, *IEEE Trans. Ultrason. Ferroelectr. Freq. Control*, 50, pp. 1253–1261.
[157] Dong, S. X., Li, J. F. and Viehland, D. (2006). Magnetoelectric coupling, efficiency, and voltage gain effect in piezoelectric-piezomagnetic laminate composites, *J. Mater. Sci.*, 41, pp. 97–106.
[158] Dong, S. X., Cheng, J. R., Li, J. F. and Viehland, D. (2003). Enhanced magnetoelectric effects in laminate composites of Terfenol-D/$Pb(Zr, Ti)O_3$ under resonant drive, *Appl. Phys. Lett.*, 83, pp. 4812–4814.
[159] Liu, Y. X., Wan, J. G., Liu, J. M. and Nan, C. W. (2003). Effect of magnetic bias field on magnetoelectric coupling in magnetoelectric composites, *J. Appl. Phys.*, 94, pp. 5118–5122.
[160] Liu, G., Nan, C. W., Cai, N. and Lin, Y. H. (2004). Calculations of giant magnetoelectric effect in multiferroic composites of rare-earth-iron alloys and PZT by finite element method, *Int. J. Solids Struct.*, 41, pp. 4423–4434.

[161] Jia, Y. M., Or, S. W., Chan, H. L. W., Zhao, X. Y. and Luo, H. S. (2006). Converse magnetoelectric effect in laminated composites of PMN-PT single crystal and Terfenol-D alloy, *Appl. Phys. Lett.*, 88, pp. 242902.

[162] Yu, H., Zeng, M. and Wang, Y. (2005). Magnetoelectric resonance-bandwidth broadening of Terfenol-D/epoxy-Pb(Zr,Ti)O_3 bilayers in parallel and series connections, *Appl. Phys. Lett.*, 86, pp. 032508; Guo, S. S., Lu, S. G., Xu, Z., Zhao, X. Z. and Or, S. W. (2006). Enhanced magnetoelectric effect in Terfenol-D and flextensional cymbal laminates, *Appl. Phys. Lett.*, 88, pp. 182906.

[163] Wan, J. G., Li, Z. Y., Wang, Y., Zeng, M., Wang, G. H. and Liu, J. M. (2005). Strong flexural resonant magnetoelectric effect in Terfenol-D/epoxy-Pb(Zr,Ti)O_3bilayer, *Appl. Phys. Lett.*, 86, pp. 202504.

[164] Zhai, J. Y., Xing, Z. P., Dong, S. X., Li, J. F. and Viehland, D. (2006). Detection of pico-tesla magnetic fields using magneto-electric sensors at room temperature, *Appl. Phys. Lett*, 88, pp. 062510.

[165] Dong, S. X., Li, J. F. and Viehland, D. (2004). Longitudinal and transverse magnetoelectric voltage coefficients of magnetostrictive/piezoelectric laminate composite: Experiments, *IEEE Trans. Ultrason. Ferroelectr. Freq. Control*, 51, pp. 794–799.

[166] Dong, S., Zhai, J., Bai, F., Li, J. F. and Viehland, D. (2005). Push-pull mode magnetostrictive/piezoelectric laminate composite with an enhanced magnetoelectric voltage coefficient, *Appl. Phys. Lett.*, 87, pp. 062502.

[167] Zhai, J., Dong, S., Xing, Z., Li, J. and Viehland, D. (2006). Giant magnetoelectric effect in Metglas/polyvinylidene-fluoride laminates, *Appl. Phys. Lett.*, 89, pp. 083507.

[168] Zhao, K., Chen, K., Dai, Y. R., Wan, J. G. and Zhu, J. S. (2005). Effect of martensitic transformation on magnetoelectric properties of $Ni_2MnGa/PbZr_{0.52}Ti_{0.48}O_3$ composite, *Appl. Phys. Lett.*, 87, pp. 162901.

[169] Laletin, V. M., Paddubnaya, N., Srinivasan, G., De Vreugd, C. P., Bichurin, M. I., Petrov, V. M. and Filippov, D. A. (2005). Frequency and field dependence of magnetoelectric interactions in layered ferromagnetic transition metal-piezoelectric lead zirconate titanate, *Appl. Phys. Lett.*, 87, pp. 222507.

[170] Babu, S. N., Bhimasankaram, T. and Suryanarayana, S. V. (2005). Magnetoelectric effect in metal-PZT laminates, *Bull. Mater. Sci.*, 28, pp. 419–422.

[171] Jia, Y. M., Or, S. W., Lam, K. H., Chan, H. L. W., Tang, Y. X., Zhao, X. Y. and Luo, H. S. (2007). Magnetoelectric effect in composites of magnet, metal-cap, and piezoceramic, *Appl. Phys. A*, 86, pp. 525–528.

[172] Shi, Z., Nan, C. W., Liu, J. M., Filippov, D. A. and Bichurin, M. I. (2004). Influence of mechanical boundary conditions and microstructural features on magnetoelectric behavior in a three-phase multiferroic particulate composite, *Phys. Rev. B*, 70, pp. 134417.

[173] Filippov, D. A., Bichurin, M. I., Nan, C. W. and Liu, J. M. (2005). Magnetoelectric effect in hybrid magnetostrictive-piezoelectric composites in the electromechanical resonance region, *J. Appl. Phys.*, 97, pp. 113910.

[174] Dong, S., Zhai, J., Li, J. and Viehland, D. (2006). Enhanced magnetoelectric effect in three-phase $MnZnFe_2O_4/Tb_{1-x}Dy_xFe_{2-y}/Pb(Zr,Ti)O_3$ composites, *J. Appl. Phys*, 100, pp. 124108.
[175] Slutsker, J., Levin, I., Li, J., Artemev, A. and Roytburd, A. L. (2006). Effect of elastic interactions on the self-assembly of multiferroic nanostructures in epitaxial films, *Phys. Rev. B*, 73, pp. 184127.
[176] Zavaliche, F., Zhao, T., Zheng, H., Straub, F., Cruz, M. P., Yang, P. L., Hao, D. and Ramesh, R. (2007). Electrically assisted magnetic recording in multiferroic nanostructures, *Nano Lett.*, 7, pp. 1586–1590.
[177] Deng, C., Zhang, Y., Ma, J., Lin, Y. and Nan, C.-W. (2007). Magnetic-electric properties of epitaxial multiferroic $NiFe_2O_4$-$BaTiO_3$ heterostructure, *J. Appl. Phys.*, 102, pp. 074114.
[178] Thiele, C., Doerr, K., Bilani, O., Roedel, J. and Schultz, L. (2007). Influence of strain on the magnetization and magnetoelectric effect in $La_{0.7}A_{0.3}MnO_3$/PMN-PT(001) (A=Sr, Ca), *Phys. Rev. B*, 75, pp. 054408.
[179] Chopdekar, R. V. and Suzuki, Y. (2006). Magnetoelectric coupling in epitaxial $CoFe_2O_4$ on $BaTiO_3$, *Appl. Phys. Lett.*, 89, pp. 182506.
[180] Petrov, V. M., Srinivasan, G., Bichurin, M. I. and Gupta, A. (2007). Theory of magnetoelectric effects in ferrite piezoelectric nanocomposites, *Phys. Rev. B*, 75, pp. 224407.
[181] Bergs, R., Islam, R. A., Vickers, M., Stephanou, H. and Priya, S. (2007). Magnetic field anomaly detector using magnetoelectric composites, *J. Appl. Phys.*, 101, pp. 024108.
[182] Wan, J. G., Liu, J. M., Wang, G. H. and Nan, C. W. (2006). Electric-field-induced magnetization in $Pb(Zr,Ti)O_3$/Terfenol-D composite structures, *Appl. Phys. Lett.*, 88, pp. 182505.
[183] Huang, Z. (2006). Theoretical modeling on the magnetization by electric field through product property, *J. Appl. Phys.*, 100, pp. 114104.
[184] Bichurin, M. I., Kornev, I. A., Petrov, V. M., Tatarenko, A. S., Kiliba, Y. V. and Srinivasan, G. (2001). Theory of magnetoelectric effects at microwave frequencies in a piezoelectric/magnetostrictive multilayer composite, *Phys. Rev. B*, 64, pp. 094409.
[185] Shastry, S., Srinivasan, G., Bichurin, M. I., Petrov, V. M. and Tatarenko, A. S. (2004). Microwave magnetoelectric effects in single crystal bilayers of yttrium iron garnet and lead magnesium niobate-lead titanate, *Phys. Rev. B*, 70, pp. 064416.
[186] Bichurin, M. I., Petrov, R. V. and Kiliba, Y. V. (1997). Magnetoelectric microwave phase shifters, *Ferroelectrics*, 204, pp. 311–319.
[187] Srinivasan, G., Tatarenko, A. S. and Bichurin, M. I. (2005). Electrically tunable microwave filters based on ferromagnetic resonance in ferrite-ferroelectric bilayers, *Electronics Letters*, 41, pp. 596–598.
[188] Fetisov, Y. K. and Srinivasan, G. (2005). Ferrite/piezoelectric microwave phase shifter: Studies on electric field tunability, *Electron. Lett.*, 41, pp. 1066–1067.

[189] Semenov, A. A., Karmanenko, S. F., Demidov, V. E., Kalinikos, B. A., Srinivasan, G., Slavin, A. N. and Mantese, J. V. (2006). Ferrite-ferroelectric layered structures for electrically and magnetically tunable microwave resonators, *Appl. Phys. Lett.*, 88, pp. 033503.

[190] Fetisov, Y. K. and Srinivasan, G. (2006). Electric field tuning characteristics of a ferrite-piezoelectric microwave resonator, *Appl. Phys. Lett.*, 88, pp. 143503.

[191] Tatarenko, A. S., Gheevarughese, V. and Srinivasan, G. (2006). Magnetoelectric microwave bandpass filter, *Electron. Lett.*, 42, pp. 540–541.

[192] Semenov, A. A., Kamanenko, S. F., Kalinikos, B. A., Srinivasan, G., Slavin, A. N. and Mantese, J. V. (2006). Dual-tunable hybrid wave ferrite-ferroelectric microwave resonator, *Electron. Lett.*, 42, pp. 641–642.

[193] Srinivasan, G. and Fetisov, Y. K. (2006). Microwave magnetoelectric effects and signal processing devices, *Integr. Ferroelectr.*, 83, pp. 89–98.

[194] Ustinov, A. B., Tiberkevich, V. S., Srinivasan, G., Slavin, A. N., Semenov, A. A., Karmanenko, S. F., Kalinikos, B. A., Mantese, J. V. and Ramer, R. (2006). Electric field tunable ferrite-ferroelectric hybrid wave microwave resonators: Experiment and theory, *J. Appl. Phys.*, 100, pp. 093905.

[195] Srinivasan, G. and Fetisov, Y. K. (2006). Ferrite-piezoelectric layered structures: Microwave magnetoelectric effects and electric field tunable devices, *Ferroelectrics*, 342, pp. 65–71.

[196] Setter, N. and Waser, R. (2000). Electroceramic materials, *Acta Mater.*, 48, pp. 151–178.

[197] Ma, J., Shi, Z. and Nan, C.-W. (2007). Magnetoelectric properties of composites of single Pb(Zr,Ti)O_3 rods and Terfenol-D/epoxy with a single-period of 1-3-type structure, *Adv. Mater.*, 19, pp. 2571–2573.

[198] Figotin, A. and Vitebskiy, I. (2001). Nonreciprocal magnetic photonic crystals, *Phys. Rev. E*, 63, pp. 066609; Figotin, A. and Vitebskiy, I. (2003). Oblique frozen modes in periodic layered media, *Phys. Rev. E*, 68, pp. 036609.

[199] Singh, M. P. and Prellier, W. (2007). Oxide superlattices for multiferroics: Opportunities, issues, and challenges, *Philos. Mag. Lett.*, 87, pp. 211–222.

Chapter 6

Lead-free Piezoelectric Ceramics: Alternatives for PZT?*

Thomas R. Shrout† and Shujun J. Zhang

Materials Research Institute
Pennsylvania State University, University Park, PA 16802, USA
†*trs16@psu.edu*

Investigations in the development of lead-free piezoelectric ceramics have recently claimed comparable properties to the lead-based ferroelectric perovskites, represented by $Pb(Zr,Ti)O_3$, or PZT. In this work, the scientific and technical impact of these materials is contrasted with the various families of "soft" and "hard" PZTs. On the scientific front, the intrinsic nature of the dielectric and piezoelectric properties are presented in relation to their respective Curie temperatures (T_C) and the existence of a morphotropic phase boundary (MPB). Analogous to PZT, enhanced properties are noted for MPB compositions in the $(Na,Bi)TiO_3$-$BaTiO_3$ and ternary system with $(K,Bi)TiO_3$, but offer properties significantly lower. The consequences of a ferroelectric to antiferroelectric transition well below T_C further limits their usefulness. Though comparable with respect to T_C, the high levels of piezoelectricity reported in the $(K,Na)NbO_3$ family are the result of enhanced polarizability associated with the orthorhombic–tetragonal polymorphic phase transition being compositionally shifted downward. As expected, the properties are strongly temperature dependent, while degradation occurs through the thermal cycling between the two distinct ferroelectric domain states. Extrinsic contributions arising from domains and domain wall mobility were determined using high field strain and polarization measurements. The concept of "soft" and "hard" lead-free piezoelectrics were discussed in relation to donor and acceptor modified

*This article was firstly published as "Shrout and Zhang (2007). Lead-free piezoelectric ceramics: Alternatives for PZT? *J. Electroceram.*, 19, pp. 111–124". Reprinted with the permission of Springer Science + Business Media, LLC.

PZTs, respectively. Technologically, the lead-free materials are discussed in relation to general applications, including sensors, actuators, and ultrasound transducers.

Keywords: Piezoelectric; perovskites; lead-free ceramic PZT-KNN.

1. Introduction

In the past, innovations in actuators, sensors, and ultrasonic transducers have been the driving force for new developments in piezoelectric ceramics. Currently, the driving force is the result of environmental regulations[1]; hence, an enormous insurgence in the compositional development of lead-free piezoelectric materials.

$Pb(Zr_{1-x}Ti_x)O_3$ (PZT) and related perovskite compositions have been the mainstay for high performance actuators and transducers, owing to their superior dielectric, piezoelectric, and electromechanical coupling coefficients. Compositionally, PZT ceramics lie near a morphotropic phase boundary (MPB) separating tetragonal and rhombohedral phases, at $\sim x = 0.48$ PT, shown in Fig. 1. MPB

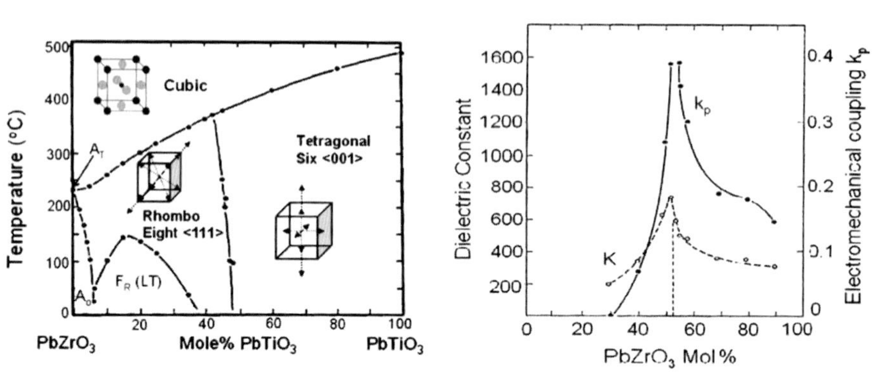

Figure 1. Morphotropic phase boundary (*left*) and enhanced dielectric and piezoelectric properties (*right*) in the $Pb(Zr,Ti)O_3$ system. (after [1, 2]): MPB is defined as an abrupt structural change within a solid solution with composition. *Note*: The MPB is nearly temperature independent.

[1]For example: the legislation will be enforced in the EU as the draft directives on waste from electrical and electronic equipment (WEEE), restriction of hazardous substances (RoHS), and end-of life vehicles (ELV).

compositions have anomalously high dielectric and piezoelectric properties as a result of enhanced polarizability, arising from the coupling between two equivalent energy states, that is, the tetragonal and rhombohedral phases, allowing optimum domain reorientation during the poling process. Further modifications using acceptor and donor dopants give us the wide range of piezoelectric compositions we have today. Excellent reviews on piezoelectric ceramics, including both modified and undoped PZT ceramics, are given by Jaffe *et al.* [1, 2] and Herbert *et al.* [3, 4].

Recently, lead-free based piezoelectrics have been reported to offer comparable properties to that of PZT ceramics, making them viable candidates for various applications [5–40]. In this chapter, we present an overview of current developments in lead-free piezoelectric ceramics and contrast them to lead-based perovskite materials. Specifically, we discuss the "origin" of the enhanced properties reported in relation to their underlying crystallographic structure relationships, that is, polymorphic phase transitions and morphotropic phase boundaries.

2. Background on piezoelectric ceramics

Among the important material parameters for piezoelectric ceramics, electromechanical coupling (k_{ij}), dielectric constant (K), or relative permittivity ε_r, and associated piezoelectric coefficients (d_{ij}) are key parameters. In general, the piezoelectric properties of a ferroelectric ceramic can be expressed using the simplistic term,

$$d_{ij} \sim 2Q_{ij}K\varepsilon_0 P_i, \tag{1}$$

where d_{ij} is the piezoelectric coefficient, P_i the remnant polarization on poling, K the dielectric constant, ε_0 the permittivity of free space, and Q_{ij} the electrostriction coefficient. Since both Q_{ij} and P_i exhibit little dependence on composition or temperature below T_C in perovskite ferroelectric ceramics such as PZT, the piezoelectric coefficient d_{ij} and dielectric constant K are interrelated (i.e., a ceramic with high piezoelectric coefficient also exhibits a large dielectric constant). To achieve a high dielectric constant or piezoelectric coefficient, MPB-based ceramics are further engineered by compositionally adjusting the Curie temperature (T_C) downward relative to room temperature. A reduction in T_C, however, results in more temperature dependent properties and less polarization stability (i.e., aging and loss of piezoelectric activity). As a general rule of thumb, piezoelectric materials can be safely used to approximately 1/2 T_C without significant reduction in piezoelectric activity. Therefore, not only must the dielectric and piezoelectric properties of a transducer material be considered, but also the T_C and any polymorphic phase transition that may occur.

3. Lead-free piezoelectrics

Recent reports on lead-free piezoelectrics can be categorized into two main perovskite families: (1) $K_{0.5}Na_{0.5}NbO_3$ (KNN) and (2) $Na_{0.5}Bi_{0.5}TiO_3$ (NBT) [5–39]. Room temperature values of the dielectric, piezoelectric, and electromechanical properties have been compiled and tabulated in Table 1. In addition to the KNN and NBT families, properties for $BaTiO_3$-based piezoelectrics are also included [40–42]. Relaxor-based piezoelectrics of (Ba)$Sr_{0.7}Bi_{0.2}TiO_3$-$K_{0.5}Bi_{0.5}TiO_3$ [43] and tungsten bronzes [44], though included, will not be discussed owing to their relatively low properties.

For comparison with PZT materials, the room temperature values of K, d_{ij}, and coupling are plotted as a function of T_C and shown in Figs. 2, 3, and 4, respectively. As stated above, the dielectric and piezoelectric properties are intrinsically enhanced relative to a decrease in Curie temperature (T_C) for PZT-based materials, with only a modest (if any) enhancement noted for the three families of lead-free perovskites. In terms of the electromechanical coupling factors (k_p and k_{33}), no corresponding relationship with T_C for PZT or lead-free systems is observed.

Though $BaTiO_3$ was the first polycrystalline piezoelectric ceramic, its inherently low T_C and numerous polymorphic phase transitions limits its overall usefulness.

From the overview of the properties presented above, several questions arise. For the case of the KNN family, are the properties

Table 1. Dielectric and piezoelectric properties of lead-free perovskite piezos "past and present".

Material	$\varepsilon_r/\varepsilon_0$	Loss	d_{33}(pC/N)	k_p	k_{33}	T_C(°C)	T_{O-T}/T_d(°C)	Ref.
$BaTiO_3$	1,700	0.01	190	0.36	0.5	115	0	[40, 41]
$BaTiO_3$-$CaTiO_3$-Co	1,420	0.005	150	0.31	0.46	105	−45	[40, 42]
$(K_{0.5}Na_{0.5})NbO_3$ (HP)	500	0.02	127	0.46	0.6	420	/	[45, 46]
$(K_{0.5}Na_{0.5})NbO_3$	290	0.04	80	0.35	0.51	420	195	[47]
KNN-Li (7%)	950	0.084	240	0.45	0.64	460	~20	[6]
KNN-Li3%;Ta20%(LF3)	920–1,256	0.024–0.02	190–230	0.46–0.505	0.62	310–323	50–70	[6, 7]
KNN-LF4a	1,570	/	410	0.61	/	253	25	[5, 13]
KNN-$SrTiO_3$ (5%)	950	/	200	0.37	/	277	27	[8, 9]
KNN-$LiTaO_3$ (5%)	570	0.04	200	0.36	/	430	55	[10]
KNN-$LiNbO_3$ (6%)	500	0.04	235	0.42	0.61	460	70	[11]
KNN-$LiSbO_3$ (5%)	1,288	0.019	283	0.50	/	392	45	[12]
NBT-KBT-LBT	1,550	0.034	216	0.401	/	350	160	[21]
NBT-KBT-BT	820	0.03	145	0.612	0.519	302	224	[43]
NBT-KBT-BT (MPB)	730	0.02	173	0.33	0.59	290	162	[43]
NBT-KBT-BT	770	0.034	183	0.367	0.619	290	100	[26]
NBT-BT	665	0.028	64(d_{31})	0.28(k_{31})	/		/	[28]
NN-BT10%	1,000	0.015	147	/	/	235	/	[29]
NBT-KBT50%	25	0.03	150	0.22	/	320	210	[31]
SBT-KBT90	870	0.04	110	0.15	0.507	296	/	[43]
SBT-KBT85	1,000	0.05	120	0.16	0.491	250	/	[43]
BBT-KBT90	837	0.05	140	0.23	0.538	297	144	[43]
BBT-KBT80	630	0.04	95	0.15	0.361	290	238	[43]
$Sr_2NaNb_5O_{15}^a$	1,100	/	120	/	/	280	/	[44]

Notes: [a]Textured ceramics; T_{O-T}: Orthorhombic to tetragonal phase transition; HP: Hot Pressed; NN: $NaNbO_3$; NBT: $(Na_{0.5}Bi_{0.5})TiO_3$; KBT: $(K_{0.5}\ Bi_{0.5})TiO_3$; LBT: $(Li_{0.5}Bi_{0.5})TiO_3$; BT: $BaTiO_3$; SBT: $(Sr_{0.7}Bi_{0.2})TiO_3$; BBT: $(Ba_{0.7}Bi_{0.2})TiO_3$; KNN: $(K_{0.5}Na_{0.5})NbO_3$; T_d: depolarization temperature.

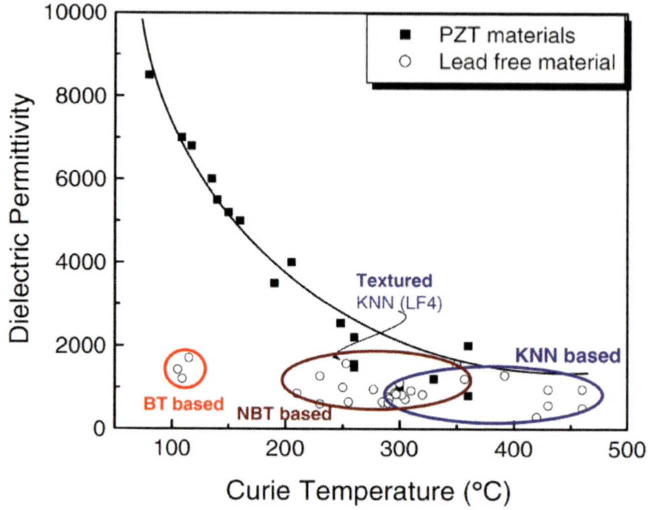

Figure 2. Room temperature values of dielectric permittivity as a function of T_C for various piezoceramics.

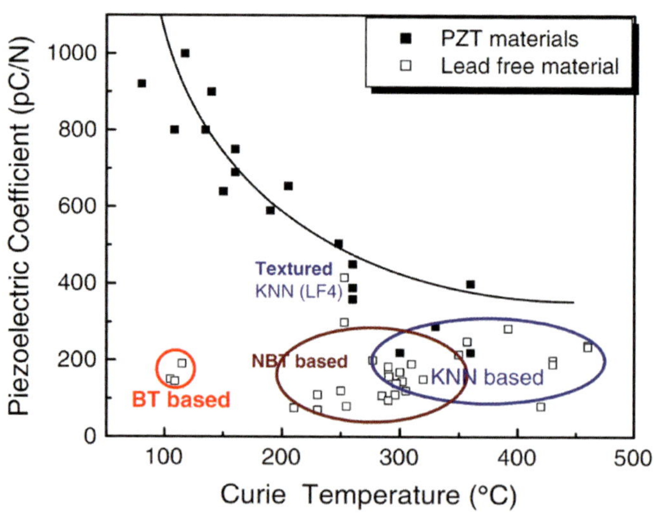

Figure 3. Room temperature values of d_{33} as a function of T_C for various piezoceramics.

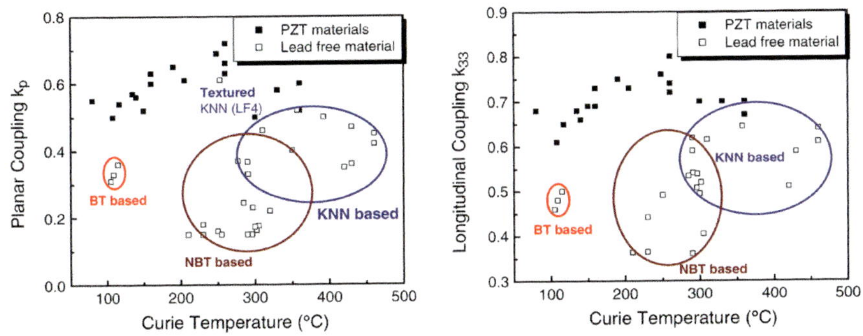

Figure 4. Room temperature electromechanical coupling k_p (*left*) and k_{33} (*right*) for various piezoelectric ceramics (The data in Figs. 2, 3, and 4 are obtained from [2, 5–54]).

reported currently superior to that of the past? What is the origin of the enhanced properties that have been widely reported? For the KNN and NBT families, are the T_C's reported valid in relation to temperature usage range? From a technological perspective, are there distinct families of "soft" and "hard" lead-free piezoelectrics? Finally, how the lead-free piezoelectrics "stack up" against PZT in various applications?

4. $(K_{0.5}Na_{0.5})NbO_3$ system: Past and present

The phase diagram for the solid solution $(1-x)KNbO_3-(x)NaNbO_3$ system is given in Fig. 5. Reported in 1955, a morphotropic phase boundary was identified at $x \sim 0.5$ separating two orthorhombic ferroelectric phases. Inherent to MPBs, a maximum in remnant polarization (P_r) and minimum in coercive field (E_c) were reported [45]. A broad maximum in the planar coupling coefficient was observed, however, with no enhancement in the dielectric properties for either air-fired or hot pressed materials, see Fig. 5(b). Table 2 presents dielectric and piezoelectric data for KNN with MPB compositions, reported prior to 1962 and for material recently processed using "optimal" processing and various fluxes. As presented, no significant differences in the piezoelectric properties are observed.

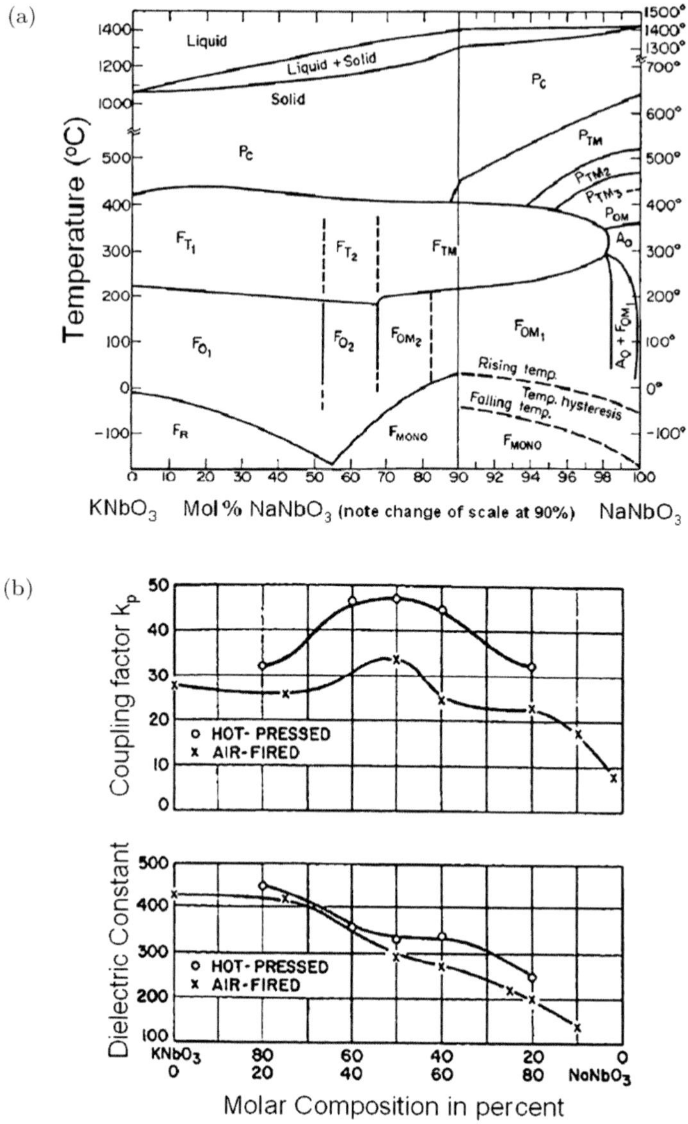

Figure 5. (a) Phase diagram for the $KNbO_3$-$NaNbO_3$ system (after [1]) and (b) dielectric and piezoelectric properties as a function of composition (after [45]). *Note*: The orthorhombic–orthorhombic MPB is at ~50 mol% $NaNbO_3$.

Table 2. Dielectric and piezoelectric properties of KNN ceramics: "past" and "present".

Properties	Air-fired 1959 [47]	Hot Pressed 1962 [45]	Standard 2005–2006 [16, 48, 52]	CuO Flux 2005 [16]	KCT/KCN Flux 2005 [17, 18]	ZnO Flux 2006 [52, 54]
ε_r	290	490	429–605	231	250	500–650
d_{33}(pC/N)[a]	80	127	98–107	86	90(High Field-190)	117–123
k_p	36%	∼46%	33.4–40%	37.8%	41%	0.4–0.44

Note: [a]Comment: Berlincourt d_{33} values obtained on soft piezoelectric ceramic disks are generally higher (∼10–20%) than values obtained using the IEEE Standards [55, 56].

It is important to note that KNN material prepared by spark plasma sintering reported significantly higher dielectric and piezoelectric properties than those given in Table 2, with $\varepsilon_r \sim 700$ and $d_{33} \sim 148$ pC/N [49, 50]. It is suggested that the enhancement is the result of extrinsic contributions to the polarizability associated with submicron grain sizes, similar to that found in fine grain $BaTiO_3$ [51], warranting further investigations.

5. Enhanced properties: Morphotropic phase boundary versus a polymorphic phase transition

Analogous to PZT, a morphotropic phase boundary separating ferroelectric tetragonal, and rhombohedral phases exists in the $Na_{0.5}Bi_{0.5}TiO_3$-$BaTiO_3$ and related ternary system with $K_{0.5}Bi_{0.5}TiO_3$ [32] (the former shown in Fig. 6(a)). Both the dielectric and piezoelectric properties are significantly enhanced, as evident in Fig. 6(b). Unlike that of the PZT system, the MPB is strongly curved, and prior to the prototypic cubic transformation, a phase transformation to an anti-ferroelectric phase occurs, as shown in Fig. 7. The consequence of this transformation is a loss of polarization and thus piezoelectric activity. Figure 8 shows pyroelectric data for BNBK79 and BNBK88 ceramics, reflecting that the temperature of depolarization (T_d) is significantly lower than T_C for compositions in

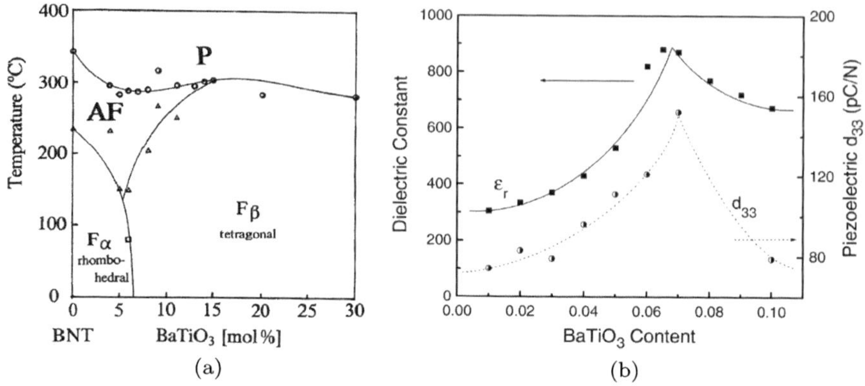

Figure 6. (a) Phase diagram for NBT-BaTiO3 (after [32]) and (b) temperature behavior (data obtained from [33]).

Note: AF: Anti-ferroelectric; P: Paraelectric Cubic Phase.

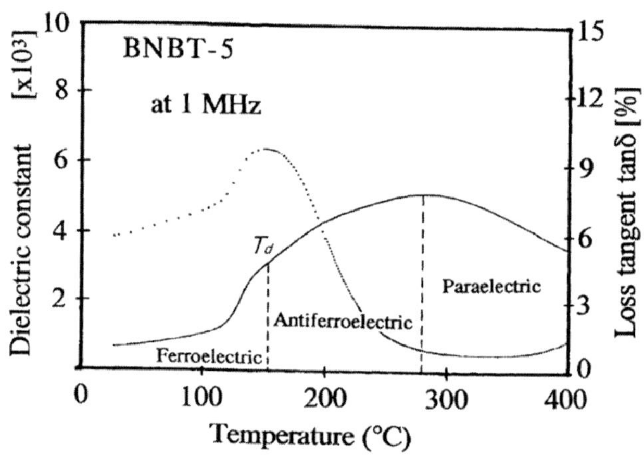

Figure 7. Phase transformation in the NBT-BT system (after [32]).

these systems. It is also confirmed by the electromechanical behavior of the BNBK system in Fig. 9, where the materials were found to be depoled at T_d. The T_d values for various BNBK compositions are listed in Table 1.

For the KNN family, reports of enhanced dielectric and piezoelectric properties were attributed to a morphotropic phase boundary

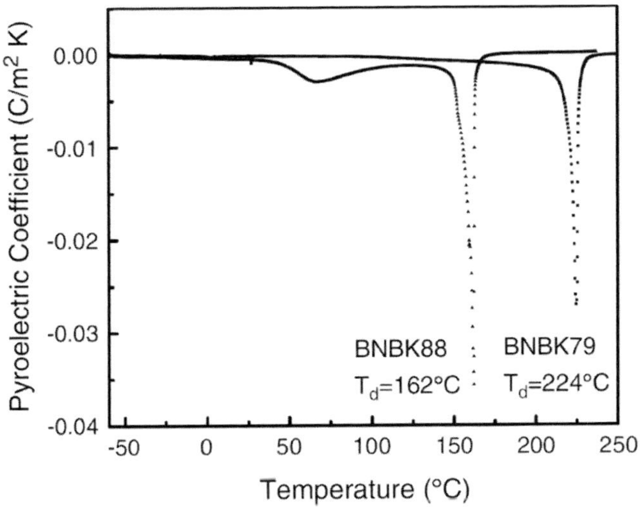

Figure 8. Depolarization (T_d) temperature of two compositions in the NBT-KBT-BT system, as determined by pyroelectric measurements (after [59]).

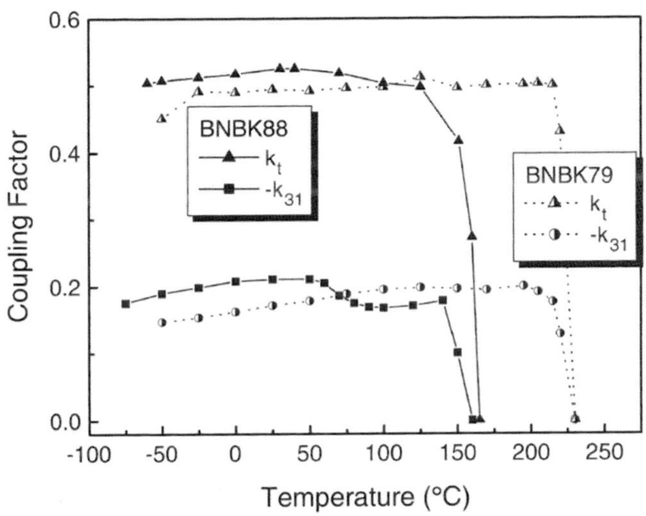

Figure 9. Temperature dependence of electromechanical coupling factors k_t and k_{31} for BNBK79 and BNBK88 ceramics (after [59]).

separating orthorhombic and tetragonal phases [5–11]. Though technically a MPB, the enhancement in properties is a result of compositionally shifting the $T_{\mathrm{O-T}}$ polymorphic phase transition downward to room temperature. As first reported in $BaTiO_3$ [1–58], and later in $KNbO_3$ [1, 57], the enhanced polarizability associated with a polymorphic phase transition leads to maxima in the dielectric and piezoelectric properties. This behavior is clearly shown for the case of Cd modified $NaNbO_3$ [1, 60] and pure $KNbO_3$ [1, 57] in Figs. 10 and 11, respectively. As shown, maxima occurred at all the ferroelectric–ferroelectric polymorphic phase transitions, with the highest levels of piezoelectric activity belonging to the orthorhombic–tetragonal transition.

Figure 12 gives the dielectric permittivity and dielectric loss for a modified KNN over the temperature range of −50–450°C. As shown, two dielectric anomalies are presented with the $T_C \sim 380$°C belonging to the tetragonal-cubic phase transition and the anomaly at room temperature being associated with the orthorhombic–tetragonal

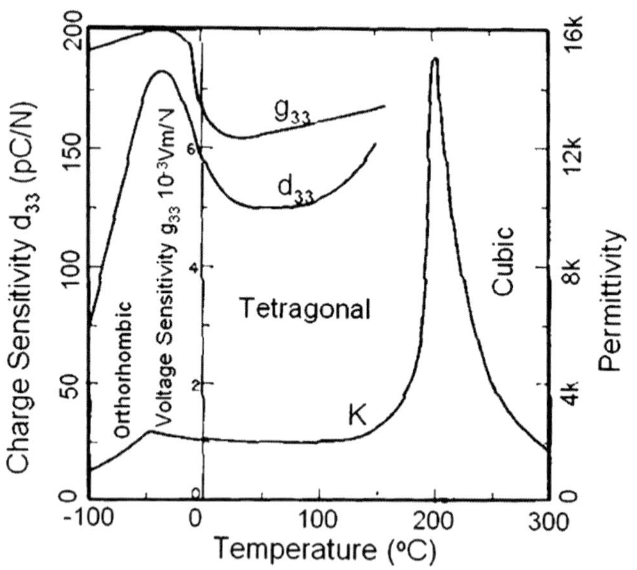

Figure 10. Dielectric constant and piezoelectric coefficient versus temperature for $Na_{0.75}Cd_{0.125}NbO_3$ (after [1]).

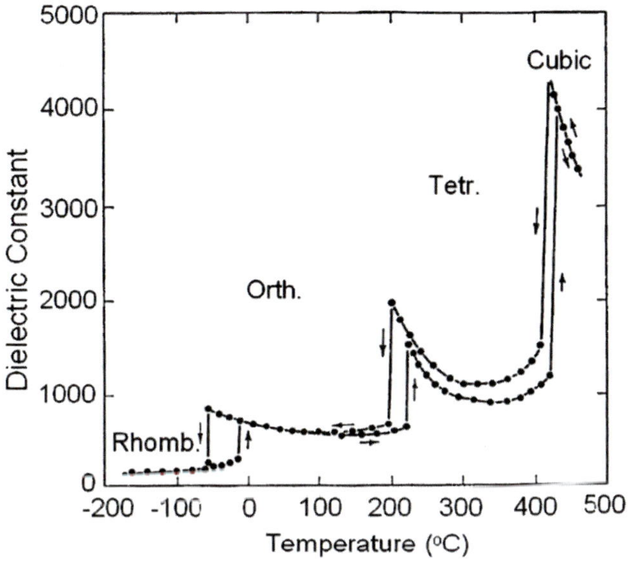

Figure 11. Dielectric constant versus temperature for $KNbO_3$, showing the various polymorphic phase transitions (after [1, 57]).

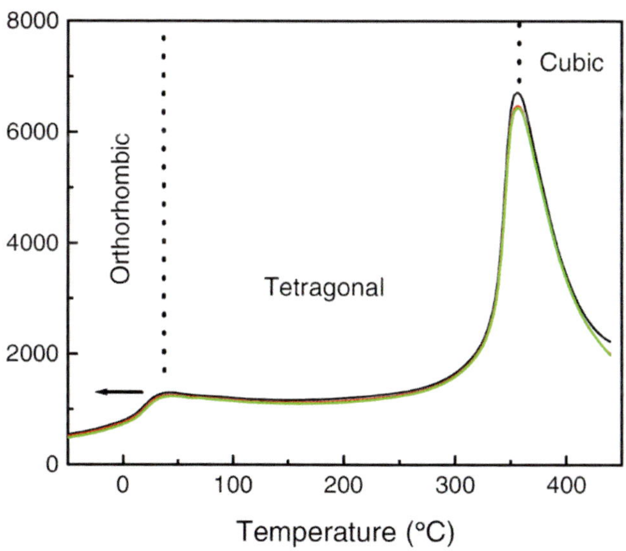

Figure 12. Dielectric permittivity as a function of temperature for a KNN-LS5% material (after [62]).

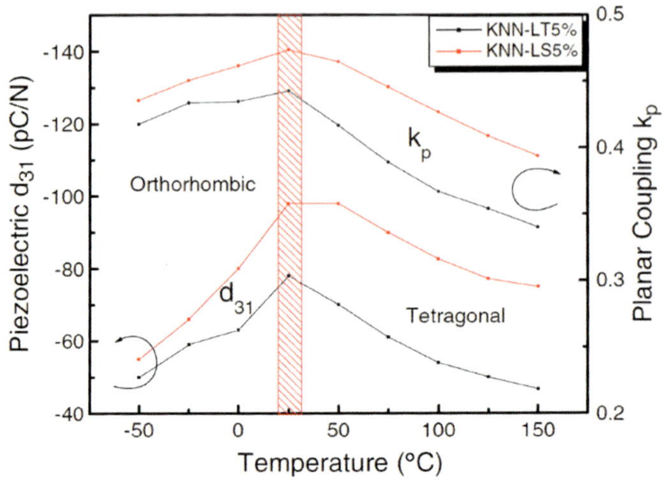

Figure 13. Piezoelectric coefficient and coupling factor as a function of temperature for $LiTaO_3$ (LT) and $LiSbO_3$ (LS) modified KNN systems (after [62]).

(T_{O-T}) polymorphic phase transition. The enhancement in piezoelectricity is clearly evident in Fig. 13, where maxima in both coupling coefficient and piezoelectric d_{31} are again found at the T_{O-T} for two representative KNN compositions.

Figure 14 presents the phase diagram of the KNN system as a function of compositional modifications, including $LiTaO_3$, $LiNbO_3$, and $SrTiO_3$ [8–11], in which T_{O-T} can be readily shifted from ~200°C for pure KNN to room temperature with ~5 mol% additions. Similar phase diagrams and their compositional temperature dependence on the various polymorphic phase transitions are well documented for $KNbO_3$-$KTaO_3$ [1, 61] and $BaTiO_3$ [1, 58], the latter being the underlying basis for the development of temperature stable ceramic capacitors.

From the above, it is evident that the temperature dependence of a polymorphic ferroelectric–ferroelectric phase transition mimics the compositional dependence of properties for a true MPB. In addition to strong temperature dependent behavior, thermal cycling between two distinct ferroelectric phases is expected to result in a reduction in polarization. Figure 15 gives the electromechanical coupling k_{31}

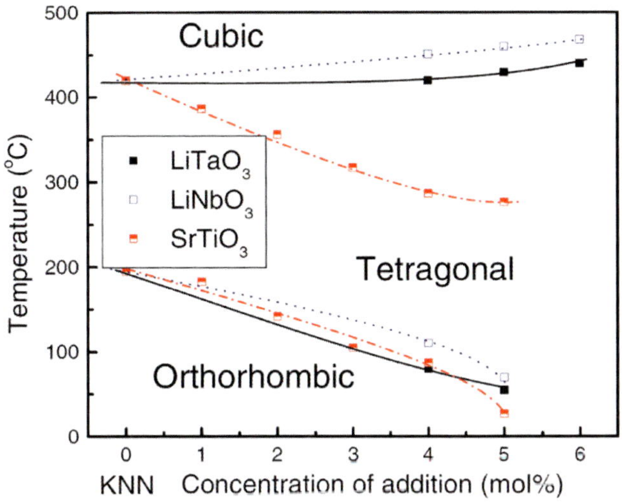

Figure 14. Phase diagram for modified KNN (data from [8–11]). *Note*: $BaTiO_3$ additions are analogous to $SrTiO_3$ [9].

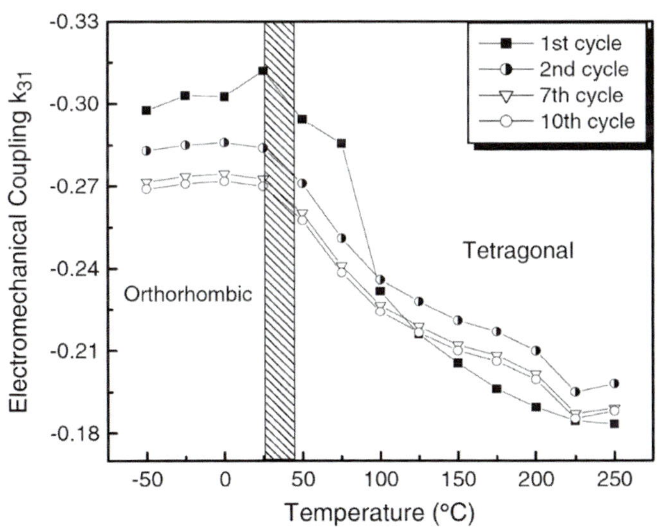

Figure 15. Electromechanical coupling factor k_{31} as a function of temperature for KNN-LS as a function of thermal cycling (after [63]).

of a KNN-LS system as a function of thermal cycling. Following the initial cycle, k_{31} decreased by nearly 10% at room temperature, ultimately stabilizing after several thermal cycles. The decrease in properties is clearly the result of cycling through regions of distinctly different domain states. As a point of future investigations, the degree of degradation may be correlated to the specific state at which the poling took place.

Presented in Fig. 16, a maximum in P_r is evident at the T_{O-T} polymorphic phase transition near room temperature, analogous to a MPB composition. This maximum in P_r is expected owing to the summation of possible crystallographic orientations, with 12 $\langle 110 \rangle$ spontaneous polarization directions for an orthorhombic phase and six $\langle 001 \rangle$ directions for a tetragonal phase. However, at this time, it is difficult to explain the maximum in E_C, which generally reaches a minimum near a MPB. The decrease of E_C with increasing temperature can be explained by the fact that the domain wall motion and polarization reversal becomes easier with increasing temperature, a thermally activated process. Additional aspects that

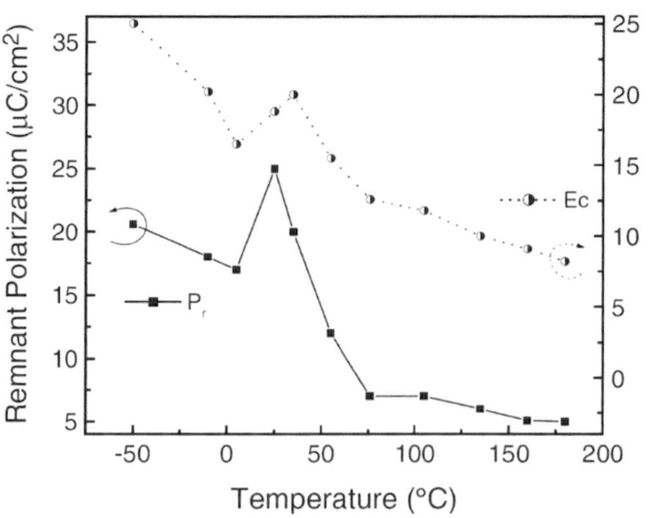

Figure 16. Remnant polarization and coercive field as a function of temperature for KNN-LS (after [63]).

distinguish a polymorphic based piezoelectric and/or based on MPB will be discussed in later sections.

6. "Soft" and "hard": PZT piezoelectrics

The diversity of piezoelectric applications has led to the development of an extensive array of modified PZT and related ceramics. Generally, based near the MPB, the substitution of small amounts of cations (1–2 mol%) modifies the dielectric and piezoelectric properties. Doping with cations with a low valence, e.g. Fe^{+3} replacing Ti^{+4}/Zr^{+4} or K^{+1} for Pb^{+2}, results in oxygen vacancies, subsequently resulting in re-orientable dipoles that align in the direction of the polarization vector within domains, creating internal fields that stabilize the domain configuration and reduce domain wall mobility. This hardening effect results in a reduction in dielectric constant and loss and piezoelectric activity while the coercive field (E_C) and mechanical Q (inverse of mechanical loss) increase. In addition, the electrical resistivity of acceptor doped PZTs is reduced. This more stable behavior makes them suitable for high power and high voltage applications. Typical values of the dielectric and piezoelectric properties for two classes of "hard" PZT, DoD Type I and III, are given in Table 3 [2, 64, 65]. A representative P/E (polarization versus electric field) behavior is shown in Fig. 17 (left), where the off-set in coercive field reflecting the level of internal bias is clearly shown [66]. It is this development of internal bias that effectively increases the coercive field of hard piezoelectrics.

Donor doping of PZT results in the opposite effect of acceptor modified piezoelectrics, with dielectric and piezoelectric activity increasing, while the coercive field and mechanical Q decrease. This change is the result of lead vacancies arising from the replacement of Pb^{+2} with cations such as La^{+3} and/or Nb^{+5} for Ti^{+4}/Zr^{+4}, enhancing the mobility of domain walls. Though the dielectric loss increases, the electrical resistivity significantly improves, allowing operation at relatively high temperatures. The donor doped materials are in general both electrically and mechanically "soft". This family of piezoelectrics is generally used for medical ultrasound transducers,

significantly lower than PZT, limiting their usefulness in today's applications.

For the KNN family, unmodified compositions exhibit "soft" like behavior, with high loss, low mechanical Q, and modest $E_C \sim 5-$ $-9\,\mathrm{kV/cm}$. Properties affected by modifications to date have been primarily the result of enhanced densification as with additions of ZnO, SnO_2, Sc_2O_3, acting as fluxes with minor changes in both T_C and T_{O-T} [52, 54]. The addition MnO_2, generally used as an acceptor dopant, resulted in reduced grain growth, however, no evidence of "hardening" was observed [15].

The addition of CuO (and/or KCuTaO, KCuNbO) flux, however, did result in a "hardening" effect, as reported in [16–18]. The incorporation of Cu^{+1} cations onto the B-site, replacing Nb^{+5}, would act as an acceptor dopant. From the properties summarized in Table 4, the dielectric and piezoelectric properties were reduced, while mechanical Q and dielectric loss increased and decreased, respectively.

For the orthorhombic–orthorhombic MPB based compositions of KNN modified with the T_{O-T} engineered to near room temperature, "soft" like behavior is noted. From Table 4, we see mechanical Q values are about 40–50, with dielectric loss on the order of a few percent, typical of a soft material. However, from Fig. 18, E_C values of $>15\,\mathrm{kV/cm}$ are more indicative of hard piezoelectrics, and interestingly significantly higher than pure and acceptor modified KNN. Attempts to introduce vacancies in modified KNN with Sr^{+2} or Ba^{+2} ions on the A-site and Ti^{+4} replacing Nb^{+5} on the B-site, resulted in little effect on the dielectric and piezoelectric properties [37]. As with $BaTiO_3$ based piezoelectrics, changes observed were the result of a shift in the polymorphic phase transition and grain size effects. It should also be noted that a "pinched" polarization hysteresis was observed for thermally cycled samples, as shown in the small insert in Fig. 18, analogous to a "hard" PZT material, with an internal bias of $5\,\mathrm{kV/cm}$. The "pinched" behavior could be eliminated when the sample was heated up to 100°C.

In the BNBK family (unmodified), the P/E loops, see Fig. 19, exhibit E_Cs significantly higher than that of the hardest MPB PZTs,

Table 4. Detailed properties for BT, KNN and BNBK based lead-free materials.

Material	T_C(°C)	T_{O-T}/T_d(°C)	ε_r	Loss	k_{33}	d_{33} (pC/N)	d_{33}^*	$g_{33}(10^{-3}$ Vm/N)	E_C (kV/cm)	Q
$BaTiO_3$	115	/	1,700	0.01	0.50	190	/	13	1–2	300
BT-CT-Co	109	/	1,420	0.005	0.46	150	190	12	7	800
KNN (HP)	420	195	496	0.02	0.61	127	/	29	∼5	240
KNN-Cu	414	180	231	0.003	0.38 (k_p)	86	85	42	/	2,280
KNN-KCT	385	190	250	0.004	0.42 (k_p)	90	180	40	9	1,300
KNN-LS	368	35	1,380	0.02	0.62	260	360	20	∼18	40
KNN-LT	323	60	1,256	0.016	0.51 (k_p)	230	320	21	/	73
BNBK79	280	224	650	0.04	0.54	135	180	24	37	110
BNBK88	262	162	810	0.02	0.60	170	210	24	29	150

Note: $x(Na_{0.5}Bi_{0.5})TiO_3 - y(K_{0.5}Bi_{0.5})TiO_3 - zBaTiO_3$ $[x + y + z = 1;\ y : z = 2 : 1]$ – BNBK100x.

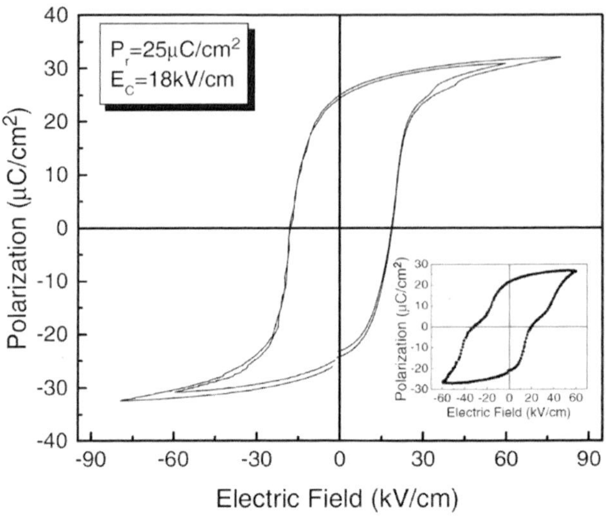

Figure 18. Polarization hysteresis for modified KNN ceramic (after [63]).

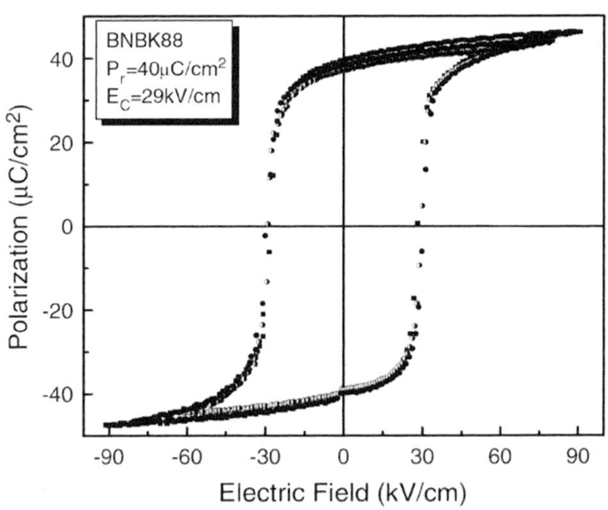

Figure 19. Polarization hysteresis for BNBK88 with MPB composition (after [59]).

with E_{C}s $\sim 30\,kV/cm$. However, they exhibit dielectric loss and mechanical Q values reflective of "soft" materials, as reported in Table 4. Attempts to modify the BNBK type materials with dopants have been limited. Isovalent additions including La^{+3} have reported only modest changes in properties [67–69], being associated with enhanced densification and a downward shift in the ferroelectric–antiferroelectric (T_d) phase transition. A decrease in E_C was associated with downward shift in T_d. A significant decrease in E_C and T_d was also reported for CeO_2 doped BNBK, however, with a modest "hardening" effect, with mechanical Qs increasing from ~40 to 110 [27].

8. Extrinsic response

In addition to the piezoelectric activity of the crystallographic contribution in ferroelectric materials, their domains react to an applied electrical and/or mechanical field, adding to the piezoelectric response. This extrinsic effect arises primarily from domain wall contributions, although additional contributions can occur, such as grain boundary effects and displacement of inter-phase boundaries [66]. This extrinsic effect can be clearly seen for a "soft" PZT in Fig. 20, where a highly non-linear–strain-electric field behavior is observed. As compared to a "hard" PZT, the strain–electric field behavior is accompanied by hysteresis (the strain hysteresis is calculated by $\Delta x/x$ in which Δx is the width of the widest part of the loop and x is the peak strain [70]). For the case of PZT Type II and VI piezoelectrics, strain hysteresis reaches levels on the order of 25% at fields of ~25 kV/cm, resulting in effective d^*_{33} 30–60% higher than the low field values, see Table 3. Interestingly, similar levels of hysteresis and enhanced d_{33} are found for the KNN and BNBK based lead-free materials. As shown in Fig. 21, the high field piezoelectric d^*_{33} is about 30–50% higher than the low-field values.

Figure 22 shows that the high levels of strain-E field hysteresis occurs over the entire usage temperature range for both modified KNN and BNBK. The strongly temperature dependence of d^*_{33} is

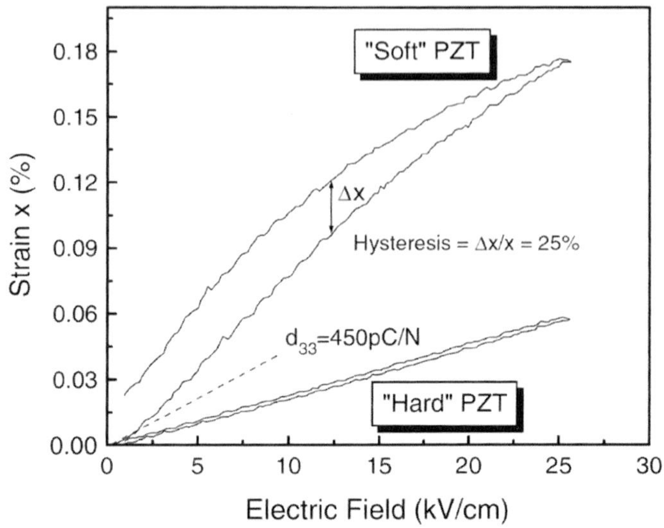

Figure 20. Strain versus electric field for a "soft" and "hard" PZT, showing respective levels of hysteresis.

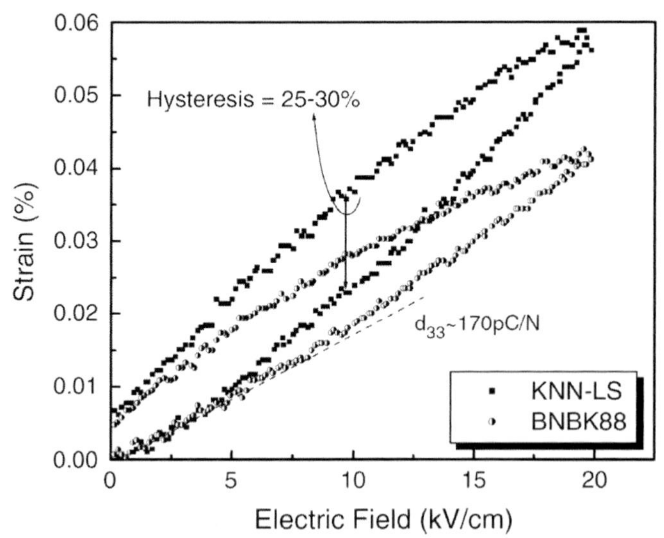

Figure 21. Unipolar strain versus electric field for lead-free piezoelectrics.

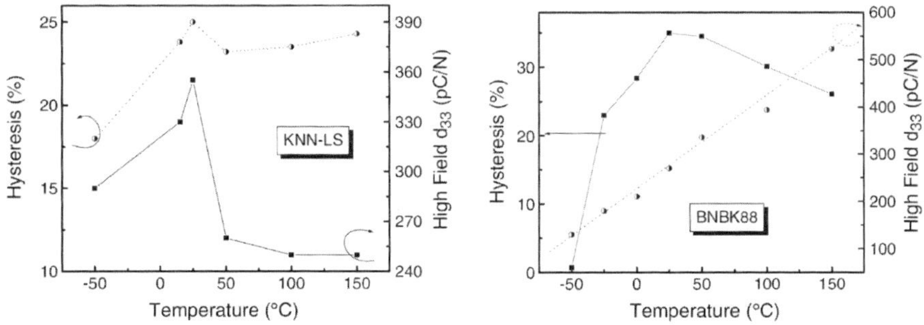

Figure 22. Strain hysteresis and high field d_{33}^* values as a function of temperature for KNN-LS (*left*) (after [63]) and BNBK88 (*right*) (after [59]).

noted near T_{O-T} in the KNN system with a strongly linear increasing in d_{33}^* obtained in the BNBK.

9. Summary

The dielectric and piezoelectric properties of lead-free perovskites, both "past" and "present" were tabulated and compared to the family of lead base perovskites, represented by PZT. The various lead-free materials were grouped into three distinct families: (1) $BaTiO_3$; (2) (Na,Bi)Ti-$BaTiO_3$-(K,Bi)TiO_3 (BNBK) and (3) (K,Na)NbO_3 (KNN).

Initially, the dielectric permittivity (ε_r) and piezoelectric strain coefficient d_{33} were contrasted as a function of transition temperature (T_C), whereby little intrinsic dependency was observed. The inherently low T_C of $BaTiO_3$ (T_C less than 120°C), directed further details on the BNBK and KNN families. Analogous to PZT, a morphotropic phase boundary separating ferroelectric rhombohedral and tetragonal phases is observed in the BNBK system(s), leading to enhanced properties. However, the onset of an anti-ferroelectric phase well below the ferroelectric–paraelectric transition (T_C) results in a loss of piezoelectric activity. Furthermore, the MPB is strongly curved as compared to the PZT system. Upon examination of the relatively high T_Cs of the KNN materials (T_Cs $>$ 400°C), they appear to be comparable to PZTs. Unfortunately, this family is plagued with the occurrence of polymorphic phase transitions as found in $BaTiO_3$ and related perovskite systems. These ferroelectric–ferroelectric phase transitions occur at temperatures far below T_C with the most piezoelectrically active transition occurring for the orthorhombic–tetragonal transition (T_{O-T}), being ~200°C for pure $(K,Na)NbO_3$. Though an orthorhombic–orthorhombic MPB boundary does occur in the $(1-x)KNbO_3 - xNaNbO_3$ ($x \sim 0.5$), the enhancement in piezoelectric properties is modest as compared to PZT with little or no enhancement in dielectric properties. Optimized processing of these materials has yet to yield significant piezoelectric performance.

Piezoelectric properties comparable to PZTs reported for KNN materials are noted for systems where the polymorphic phase transition T_{O-T} is shifted downward near room temperature. As expected, the properties are strongly temperature dependent and thermal cycling through the two distinct domain states results in the degradation of properties.

Analogous to donor and acceptor doped PZTs, the lead-free materials were further categorized in terms of "soft" and "hard" piezoelectrics. For the BNBK family, their relatively high coercive fields ($E_C > 30\,kV/cm$) were indicative of hard piezoelectrics, but their high dielectric loss, low mechanical Qs (~100) reflected soft behavior. The addition of various dopants had little effect on the properties, with the underlying changes being related to a downward shift in the T_d and grain size effects. The orthorhombic–orthorhombic MPB based KNN materials also exhibited "soft" like behavior with high dielectric loss, and low coercive fields with $E_C \sim 5\,kV/cm$ and low to intermediate Q values. The effect of various additives such as SnO_2, ZnO improved the properties, acting as a sintering aid, improving densification and/or affecting grain size. The addition of CuO, however, was found to act as an acceptor dopant, significantly increasing Q and decreasing dielectric loss.

The KNN materials, compositionally engineered with their polymorphic phase transition T_{O-T} shifted to near room temperature, all exhibited soft-like behavior, with low Qs. Interestingly, the coercive field of these materials was more indicative of a hard piezoelectric with E_Cs $> 15\,kV/cm$, being also higher than unmodified KNN.

Further evidence of soft behavior was found for both the BNBK and KNN materials with extrinsic contributions exemplified by strain — E field hysteresis to the levels similar to soft PZTs.

In addition to the levels of piezoelectric activity, secondary issues of property temperature dependence, temperature usage range and extrinsic response to various boundary conditions are necessary to technologically contrast the lead-free materials as they relate to various generic applications. For the case of sensors, the piezoelectric charge coefficient d_{33} and voltage coefficient g_{33} are the principle figure of merits with values for the lead-free materials being only one half to one third of "soft" PZTs, the material typically used in accelerometers and related pressure sensing devices. The need for temperature independent properties underscores the significance of the MPB based PZT compositions and its correspondingly high $T_{\rm C}$s, allowing operation up to temperatures above 200°C. In contrast, the highly curved MPB of the BNBK and its low $T_{\rm d}$ and polymorphic phase transition of the KNN are clearly inadvantageous. Applications using the converse piezoelectric effect, as for high strain actuators, require a high d_{33} and broad temperature usage range as reflected for the development of multilayer actuators used in fuel injection systems, being operational from −50 to 150°C with high strain d_{33}s greater than 600 pC/N desired [71]. It is further evident that such devices must be stable as a function of thermal cycling.

Applications that have less constraint on transition temperature include ultrasonic transducers used in medical diagnostics. In these applications, the figure of merit in principally the longitudinal coupling square (k_{33}^2), which translate into device performance in terms of bandwidth and sensitivity. As such, the optimum lead based materials offer values nearly 50% higher. Equally important is the need for high clamped permittivity ε_r^s for electrical impedance matching of small elements in high frequency arrays, where PZT Type VI offers nearly three times the value of lead-free materials.

Finally, manufacturability plays a very important role in the implementation of materials into actual devices. For KNN, processing in an aqueous environment generally leads to issues of high dielectric loss, though recent work has shown that moisture tolerant

Figure 23. SEM images for PZT (*left*) and KNN-LT (*right*) lead-free materials, showing different microstructure.

materials can be achieved by careful processing that ensure full densification [7].

The presence of a polymorphic phase transitions and the relatively low T_d in the KNN and BNBK families, respectively, can lead to significant degradation in the piezoelectric properties upon machining of the ceramics due to both temperature and stress effects. In contrast to PZTs, the microstructures shown in Fig. 23, reflect contrasting morphologies, with nearly cubic like grains for the lead-free materials (right) as compared to the nearly dodecahedron grains of PZTs (left). The effect of the distinct microstructures on machining and mechanical properties warrants further investigations.

10. Future work

In this section, we leave the reader with a few questions:

(1) For morphotropic phase boundary systems in the BNBK family, can we eliminate or increase T_d, providing better temperature dependence and wider usage range?
(2) For the KNN family, can we shift the polymorphic phase transitions to improve temperature stability while maintaining high piezoelectric properties?
(3) For the above systems, can acceptor and donor dopants be identified for the engineering of "soft" and "hard" classes of lead-free piezoelectrics?

(4) Fundamentally, what are the role of domains and domain boundaries on the extrinsic response in these materials? Can "domain engineering" be used to enhance the properties?

For lead-free materials to be successfully implemented, investigations of manufacturability as it pertains to specific applications must be made. Examples of this are issues of electrode-ceramic interactions in co-fired multilayer actuators, depolarization during soldering, etc.

Acknowledgment

This work supported by Office of Naval Research (ONR) and National Institutes of Health (NIH) under grant number P41-RR11795. The authors also thank Prof. Tadashi Takenaka and Dr. Hajime Nagata for the BNBK samples.

References

[1] Jaffe, B. C., Cook, W. and Jaffe, H. (1971). *Piezoelectric Ceramics* (Academic Press, New York) p. 92.
[2] Berlincourt, D. (1971). *Ultrasonic Transducer Materials: Piezoelectric Crystals and Ceramics*, ed. by Mattiat, O. E. (Plenum, London), Chapter 2.
[3] Moulson, A. H. and Herbert, J. M. (1990). *Electroceramics: Materials, Properties, Applications* (Chapman & Hall, London).
[4] Herbert, J. M. (1982). *Ferroelectric Transducers and Sensors* (Gordon & Breach Science, New York).
[5] Saito, Y., Takao, H., Tani, T., Nonoyama, T., Takatori, K., Homma, T., Nagaya, T. and Nakamura, M. (2004). Lead-free piezoceramics, *Nature*, 432, pp. 84–87.
[6] Hollenstein, E., Davis, M., Damjanovic, D. and Setter, N. (2005). Piezoelectric properties of Li- and Ta-modified $(K_{0.5}Na_{0.5})NbO_3$ ceramics, *Appl. Phys. Lett.*, 87, pp. 182905.
[7] Saito, Y. and Takao, H. (2006). High performance lead-free piezoelectric ceramics in the $(K,Na)NbO_3$-$LiTaO_3$ solid solution system, *Ferroelectrics*, 338, pp. 1433–1448.
[8] Wang, R., Xie, R., Hanada, K., Matsusaki, K., Bandoand, H. and Itoh, M. (2005). Phase diagram and enhanced piezoelectricity in the strontium titanate doped potassium-sodium niobate solid solution, *Phys. Status Solidi*, A, 202, pp. R57–R59.
[9] Wang, R., Xie, R.-J., Hanada, K., Matsusaki, K., Bando, H., Sekiya, T. and Itoh, M. (2006). Phase diagram of the $(Na_{0.5}K_{0.5})NbO_3$-$ATiO_3$ solid solution, *Ferroelectrics*, 336, pp. 39–46.

[10] Guo, Y. P., Kakimoto, K. and Ohsato, H. (2005). $(Na_{0.5}K_{0.5})NbO_3$-$LiTaO_3$ lead-free piezoelectric ceramics, *Mater. Lett.*, 59, pp. 241–244.

[11] Guo, Y. P., Kakimoto, K. and Ohsato, H. (2004). Phase transitional behavior and piezoelectric properties of $(Na_{0.5}K_{0.5})NbO_3$-$LiNbO_3$ ceramics, *Appl. Phys. Lett.*, 85, pp. 4121–4123.

[12] Zang, G. Z., Wang, J. F., Chen, H. C., Su, W. B., Wang, C. M., Qi, P., Ming, B. Q., Du, J. and Zheng, L. M. (2006). Perovskite $(Na_{0.5}K_{0.5})_{1-x}(LiSb)_xNb_{1-x}O_3$ lead-free piezoceramics, *Appl. Phys. Lett.*, 88, pp. 212908.

[13] Lang, S. B., Zhu, W. Y. and Cross, L. E. (2006). Piezoelectric and pyroelectric properties of $(K_{0.5}Na_{0.5})_{1-x}(Nb_{1-y}Ta_y)O_3$ ceramics, *Ferroelectrics*, 336, pp. 15–21.

[14] Takao, H., Saito, Y., Aoki, Y. and Horibuchi, K. (2006). Microstructural evolution of crystalline-oriented $(K_{0.5}Na_{0.5})NbO_3$ piezoelectric ceramics with a sintering aid of CuO, *J. Am. Ceram. Soc.*, 89, pp. 1951–1956.

[15] Tashiro, S. and Ishii, K. (2006). Grain size and piezoelectric properties of $(Ba,K,Na)NbO_3$ lead-free ceramics, *J. Ceram. Soc. Jpn.*, 114, pp. 386–391.

[16] Saito, Y. and Takao, H. (2005). *12th US–Japan Seminar on Dielectric and Piezoelectric Ceramics*, ed. by Wu, C., Kishi, H., Randall, C., Pinceloup, P. and Funakubo, H. (Maryland) pp. 103–107.

[17] Matsubara, M., Kikuta, K. and Hirano, S. (2005). Piezoelectric properties of $(K_{0.5}Na_{0.5})(Nb_{1-x}Ta_x)O_3$-$K_{5.4}CuTa_{10}O_{29}$ ceramics, *J. Appl. Phys.*, 97, pp. 114–105.

[18] Matsubara, M., Yamaguchi, T., Sakamoto, W., Kikuta, K., Yogo, T. and Hirano, S. (2005). Processing and piezoelectric properties of lead-free $(K,Na)(Nb,Ta)O_3$ ceramics, *J. Am. Ceram. Soc.*, 88, pp. 1190–1196.

[19] Yoo, J., Hong, J., Lee, H., Jeong, Y., Lee, B., Song, H. and Kwon, J. (2006). Piezoelectric and dielectric properties of La_2O_3 added Bi(Na, K)TiO_3-$SrTiO_3$ ceramics for pressure sensor application, *Sens. Actuators A*, 126, pp. 41–47.

[20] Choy, S. H., Wang, X. X., Chan, H. L. W. and Choy, C. L. (2006). Study of compressive type accelerometer based on lead-free BNKBT piezoceramics, *Appl. Phys. A*, 82, pp. 715–718.

[21] Yuan, Y., Zhang, S. R., Zhou, X. H. and Liu, J. S. (2006). Phase transition and temperature dependences of electrical properties of $Bi_{0.5}(Na_{1-x-y}K_xLi_y)_{0.5}TiO_3$ ceramics, *Jpn. J. Appl. Phys.*, 45, pp. 831–834.

[22] Lin, D. M., Xiao, D. Q., Zhu, J. G. and Yu, P. (2006). Piezoelectric and ferroelectric properties of $Bi_{0.5}(Na_{1-x-y}K_xLi_y)_{0.5}TiO_3$ lead-free piezoelectric ceramics, *Appl. Phys. Lett.*, 88, pp. 062901.

[23] Wang, X. X., Tang, X. G. and Chan, H. L. W. (2004). Electromechanical and ferroelectric properties of $(Bi_{1/2}Na_{1/2})TiO_3$-$(Bi_{1/2}K_{1/2})TiO_3$-$BaTiO_3$ lead-free piezoelectric ceramics, *Appl. Phys. Lett.*, 85, pp. 91–93.

[24] Hiruma, Y., Aoyagi, R., Nagata, H. and Takenaka, T. (2004). Piezoelectric properties of $BaTiO_3$-$(Bi_{1/2}K_{1/2})TiO_3$ ferroelectric ceramics, *Jpn. J. Appl. Phys.*, 43, pp. 7556–7559.

[25] Takenaka, T. and Nagata, H. (1999). Present status of non-lead-based piezoelectric ceramics, *Key Eng. Mater.*, 157–158, pp. 57–64.

[26] Takenaka, T. and Nagata, H. (2005). Current status and prospects of lead-free piezoelectric ceramics, *J. Eur. Ceram. Soc.*, 25, pp. 2693–2700.

[27] Choy, S. H., Wang, X. X., Chan, H. L. W. and Choy, C. L. (2006). $0.75(Bi_{1/2}Na_{1/2})TiO_3$-$0.20(Bi_{1/2}K_{1/2})TiO_3$-$0.05BaTiO_3$ lead-free ceramics with addition of CeO_2, *Ferroelectrics*, 336, pp. 69–79.

[28] Song, J. S., Jeong, S. J., Kim, I. S., Lee, D. S. and Park, E. C. (2006). Piezoelectric and dielectric properties in grain oriented $(Bi_{0.5}Na_{0.5})TiO_3$-$BaTiO_3$ ceramics, *Ferroelectrics*, 338, pp. 1419–1424.

[29] Zeng, J. T., Kwok, K. W. and Chan, H. L. W. (2006). Ferroelectric and piezoelectric properties of $Na_{1-x}Ba_xNb_{1-x}Ti_xO_3$ ceramics, *J. Am. Ceram. Soc.*, 89, pp. 2828–2832.

[30] Wang, X. X., Chan, H. L. W. and Choy, C. L. (2003). Piezoelectric and dielectric properties of CeO_2-added $(Bi_{0.5}Na_{0.5})0.94Ba_{0.06}TiO_3$ lead-free ceramics, *Solid State Commun.*, 125, pp. 395–399.

[31] Zhao, S., Li, G., Ding, A., Wang, T. and Yin, Q. (2006). Ferroelectric and piezoelectric properties of $(Na,K)_{0.5}Bi_{0.5}TiO_3$ lead free ceramics, *J. Phys. D: App. Phys.*, 39, pp. 2277–2281.

[32] Takenaka, T., Maruyama, K. and Sakata, K. (1991). $(Bi_{1/2}Na_{1/2})TiO_3$-$BaTiO_3$ system for lead-free piezoelectric ceramics, *Jpn. J. Appl. Phys.*, 30, pp. 2236–2239.

[33] Nagata, H., Yoshida, M., Makiuchi, Y. and Takenaka, T. (2003). Large piezoelectric constant and high Curie temperature of lead-free piezoelectric ceramic ternary system based on bismuth sodium titanate-bismuth potassium titanate-barium titanate near the morphotropic phase boundary, *Jpn. J. Appl. Phys.*, 42, pp. 7401–7403.

[34] Hiruma, Y., M. Y., Aoyagi, R., Nagata, H. and Takenaka, T. (2006). Lead-free piezoelectric ceramic based on $(Bi_{1/2}N_{a1/2})TiO_3$-$(Bi_{1/2}K_{1/2})TiO_3$-$BaTiO_3$ solid solution, *Ceram. Trans.*, 174, pp. 139–146.

[35] Wu, L., Xiao, D. Q., Lin, D. M., Zhu, J. G. and Yu, P. (2005). Synthesis and properties of Bi-$0.5(Na_{1-x}Ag_x)_{0.5(1-y)}Ba_yTiO_3$ piezoelectric ceramics, *Jpn. J. Appl. Phys.*, 44, pp. 8515–8518.

[36] Lin, D. M., Xiao, D. Q., Zhu, J. G. and Yu, P. (2005). Electrical properties of $Bi_{1-z}(Na_{1-x-y-z}K_xLi_y)_{0.5}Ba_zTiO_3$ multi-component lead-free piezoelectric ceramics, *Phys. Status Solidi, A Appl. Res.*, 202, pp. R89–R91.

[37] Wang, R., Tachibana, N., Miura, N., Hanada, K., Matsusaki, K., Bando, H. and Itoh, M. (2006). Effects of vacancies on the dielectric and piezoelectric properties of $(Na_{0.5}K_{0.5})NbO_3$-$SrTiO_3$ solid solution, *Ferroelectrics*, 331, pp. 135–139.

[38] Maeder, M. D., Damjanovic, D. and Setter, N. (2004). Lead free piezoelectric materials, *J. Electroceram.*, 13, pp. 385–392.

[39] Chen, W., Li, Y. M., Xu, Q. and Zhou, J. (2005). Electromechanical properties and morphotropic phase boundary of $Na_{0.5}Bi_{0.5}TiO_3$-$K_{0.5}Bi_{0.5}TiO_3$-$BaTiO_3$ lead-free piezoelectric ceramics, *J. Electroceram.*, 15, pp. 229–235.

[40] Berlincourt, D. (1971). *Ultrasonic Transducer Materials: Piezoelectric Crystals and Ceramics*, ed. By Mattiat, O. E. (Plenum, London, 1971), Chapter 2.
[41] Jaffe, H. (1958). Piezoelectric ceramics, *J. Am. Ceram. Soc.*, 41, pp. 494–498.
[42] Schofield, D. and Brown, R. F. (1957). An investigation of some barium titanate compositions for transducer applications, *Can. J. Phys.*, 35, pp. 594–607.
[43] Zhang, S. J., Xia, R., Shrout, T. R., Zang, G. Z. and Wang, J. F. (2007). Characterization of lead free $(K_{0.5}Na_{0.5})NbO_3$-$LiSbO_3$ piezoceramic, *Solid State Commun.*, 141, pp. 675–679.
[44] Doshida, Y., Kishi, H., Hattori, Y., Makiya, A., Tanaka, S., Uematsu, K. and Kimura, T. (2005). *12th US–Japan Seminar on Dielectric and Piezoelectric Ceramics*, ed. by Wu, C., Kishi, H., Randall, C., Pinceloup, P. and Funakubo, H. (Maryland) pp. 123–127.
[45] Jaeger, R. E. and Egerton, L. (1962). Hot pressing of potassium-sodium niobates, *J. Am. Ceram. Soc.*, 45, pp. 209–213.
[46] Haertling, G. H. (1967). Properties of hot-pressed ferroelectric alkali niobate ceramics, *J. Am. Ceram. Soc.*, 50, pp. 329–330.
[47] Egerton, L. and Dillon, D. M. (1959). Piezoelectric and dielectric properties of ceramics in the system potassium sodium niobate, *J. Am. Ceram. Soc.*, 42, pp. 438–442.
[48] Du, H., Li, Z., Tang, F., Qu, S., Pei, Z. and Zhou, W. (2006). Preparation and piezoelectric properties of $(K_{0.5}Na_{0.5})NbO_3$ lead-free piezoelectric ceramics with pressure-less sintering, *Mater. Sci. Eng. B*, 131, pp. 83–87.
[49] Li, J. F., Wang, K., Zhang, B. P. and Zhang, L. M. (2006). Ferroelectric and piezoelectric properties of fine-grained $Na_{0.5}K_{0.5}NbO_3$ lead-free piezoelectric ceramics prepared by spark plasma sintering, *J. Am. Ceram. Soc.*, 89, pp. 706–709.
[50] Zhang, B. P., Li, J. F., Wang, K. and Zhang, H. L. (2006). Compositional dependence of piezoelectric properties in $Na_xK_{1-x}NbO_3$ lead-free ceramics prepared by spark plasma sintering, *J. Am. Ceram. Soc.*, 89, pp. 1605–1609.
[51] Arlt, G., Hennings, D. and Dewith, G. (1985). Dielectric-properties of fine-grained barium-titanate ceramics, *J. Appl. Phys.*, 58, pp. 1619–1625.
[52] Zuo, R., Roedel, J., Chen, R. and Li, L. (2006). Sintering and electrical properties of lead-free $Na_{0.5}K_{0.5}NbO_3$ piezoelectric ceramics, *J. Am. Ceram. Soc.*, 89, pp. 2010–2015.
[53] Chu, S. Y., Water, W., Juang, Y. D. and Liaw, J. T. (2003). Properties of $(Na, K)NbO_3$ and $(Li, Na, K)NbO_3$ ceramic mixed systems, *Ferroelectrics*, 287, pp. 23–33.
[54] Park, S. H., Ahn, C. W., Nahm, S. and Song, J. S. (2004). Microstructure and piezoelectric properties of ZnO-added $(Na_{0.5}K_{0.5})NbO_3$ ceramics, *Jpn. J. Appl. Phys.*, 43, pp. L1072–L1074.
[55] Zhang, S. J., Alberta, E. F., Eitel, R. E., Randall, C. A. and Shrout, T. R. (2005). Elastic, piezoelectric, and dielectric characterization of modified $BiScO_3$-$PbTiO_3$ ceramics, *IEEE Trans. Ultrason, Ferroelectr. Freq. Control*, 52, pp. 2131–2139.

[56] ANSI/IEE Standard 176 (1987). *IEEE Standards on Piezoelecricity.*
[57] Shirane, G., Danner, H., Pavlovic, A. and Pepinsky, R. (1954). Phase transitions in ferroelectric $KNbO_3$, *Phys. Rev.*, 93, pp. 672–673.
[58] Berlincourt, D. (1956). Recent developments in ferroelectric transducer materials, *IRE Trans. Ultrason. Ferroelectr. Freq.*, PGUE-4, pp. 53–65.
[59] Zhang, S. J., Shrout, T. R., Nagata, H., Hiruma, Y. and Takenaka, T. (2007). Piezoelectric properties in $(K_{0.5}Bi_{0.5})TiO_3$-$(Na_{0.5}Bi_{0.5})TiO_3$-$BaTiO_3$ lead-free ceramics, *IEEE Trans. Ultrason. Ferroelectr. Freq. Control*, 54, pp. 910–917.
[60] Lewis, B. and L. A. T., (1960). Electronics telecommnus, *Proc. Internat. Conf. Solid State Phys.*, pp. 883–890.
[61] Triebwasser, S. (1959). Study of ferroelectric transitions of solid-solution single crystals of $KNbO_3$-$KTaO_3$, *Phys. Rev.*, 114, pp. 63–70.
[62] Zhang, S. J., Xia, R. and Shrout, T. R. (2007). Lead-free piezoelectric ceramics vs. PZT? *J. Electroceram.*, 19, pp. 251–257.
[63] Zhang, S., Xia, R., Shrout, T. R., Zang, G. and Wang, J. (2006). Piezoelectric properties in perovskite $0.948(K_{0.5}Na_{0.5})NbO_3$-$0.052LiSbO_3$ lead-free ceramics, *J. Appl. Phys.*, 100, pp. 104–108.
[64] Setter, N. (2002). ABC of piezoelectricity and piezoelectric materials, in *Piezoelectric Materials in Devices*, ed. By Setter, N. (Lausanne, Switzerland) pp. 1–28.
[65] R. C. Pohanka, P. L. S. (1987). *Electronic Ceramics: Properties, Devices, and Applications* (CRC Press, New York).
[66] Carl, K. and Hardtl, K. H. (1978). Electrical aftereffects in $Pb(Ti,Zr)O_3$ ceramics, *Ferroelectrics*, 17, pp. 473–486.
[67] Yuan, Y., Zhang, S. R., Zhou, X. H. and Liu, J. S. (2006). Phase transitions and electrical properties in La3+-substituted $Bi_{0.5}(Na_{0.75}K_{0.15}Li_{0.10})_{0.5}$ TiO_3 ceramics, *J. Mater. Sci.*, 41, pp. 565–567.
[68] Son, Y. J., Choi, Y. G., Kwon, J. C., Cho, K. W., Kim, Y. M., Kweon, S. Y., Hong, T. H., Lee, Y. G., Ryu, S. L., Yoon, M. S. and Ur, S. C. (2006). Effects of La_2O_3 on the piezoelectric properties of lead-free $(Bi_{0.5}Na_{0.5})_{0.94}Ba_{0.06}TiO_3$ piezoelectric ceramics, *Mater. Sci. Forum*, 510, pp. 538–541.
[69] Jiang, X. P., Li, L. Z., Zeng, M. and Chan, H. L. W. (2006). Dielectric properties of Mn-doped $(Na_{0.8}K_{0.2})_{0.5}Bi_{0.5}TiO_3$ ceramics, *Mater. Lett.*, 60, pp. 1786–1790.
[70] Zhang, S. J., Luo, J., Xia, R., Rehrig, P. W., Randall, C. A. and Shrout, T. R. (2006). Field-induced piezoelectric response in $Pb(Mg_{1/3}Nb_{2/3})O_3$-$PbTiO_3$ single crystals, *Solid State Commun.*, 137, pp. 16–20.
[71] K. Lubitz, Schuh, C., Steinkopf, T. and Wilff, A. (2000). Active Materials: Behavior and Mechanics, in *Proc. 7th International Conference on New Actuators*, ed. By H. Borgmann (Messe Bremen GmbH, Bremen) pp. 58–61.

Chapter 7

Pyroelectric and Electrocaloric Materials*

Xinyu Li†, Sheng-Guo Lu†, Xiang-Zhong Chen†,‡, Haiming Gu†, Xiao-shi Qian† and Q. M. Zhang†,§

†*Department of Electrical Engineering and Materials Research Institute, The Pennsylvania State University, University Park, PA 16802, USA*

‡*Department of Polymer Science & Engineering and Key Laboratory of Mesoscopic Chemistry of MOE, School of Chemistry & Chemical Engineering, Nanjing University, Nanjing 210093, China*

§*qxz1@psu.edu*

Both the pyroelectric and electrocaloric effects originate from the cross-coupling between polarization and temperature in insulating dielectrics. Although both effects have been studied for many decades for various applications and large pyroelectric effect has been observed in many polar-dielectrics, it is only very recently that a large electrocaloric effect (ECE) was obtained in ferroelectric ceramic thin films and polymers, which revives the interest in the ECE. This review will summarize typical properties of pyroelectric and electrocaloric materials, present figures of merit for both phenomena, and examine the relationship between the pyroelectric effect and ECE. Moreover, we will also present theoretical works, experimental results, and material modifications to achieve large responses in electrocaloric materials.

*This article was firstly published as "Li, X., Lu, S.-G., Chen, X.-Z., Gu, H., Qian, X.-S. and Zhang, Q. M. (2013). Pyroelectric and electrocaloric materials, *J. Mater. Chem. C.*, 1, pp. 23–37". Reprinted with the permission of The Royal Society of Chemistry.

1. Introduction

The pyroelectric and electrocaloric effects originate from the cross-coupling between polarization and temperature in certain classes of insulating dielectrics. In simple terms, the pyroelectric effect refers to the polarization change caused by temperature change, while the electrocaloric effect (ECE) is the reversed process, i.e., a temperature change will be generated when there is a polarization change. These effects can be related by the Maxwell relations [1].

Pyroelectricity was first observed more than 2400 years ago [2], and has been studied for a long time [3–6]. Pyroelectric materials have found a wide range of applications [7], such as thermal imaging, laser detectors, radiometers, infrared sensors, pollution monitor and tools for gas analysis, fire alarms, and intruder alarms.

Although the study on ECE can be dated back to 1930, when a very weak ECE was measured in Rochelle salt [8], it was not until the 1960s that more studies on the electrocaloric materials were carried out. However, the small ECE observed in the second half of 20th century, where the adiabatic temperature change ΔT was less than 2°C, makes it not attractive for practical applications. The field has been revived since 2006, when Mischenko *et al.* reported a giant ECE in ceramic thin films in which ΔT of 12°C was observed at 226°C [9], and Neese *et al.* discovered a giant ECE near room temperature in a class of ferroelectric polymer, which exhibits both a large ΔT (>12°C) and isothermal entropy change ΔS (>50 J kg^{-1} K^{-1}) [10]. Electrocaloric materials with large ECE are promising to realize compact cooling devices for next-generation refrigeration, temperature regulation and air-conditioning, with high efficiency and environmentally friendliness.

This review will summarize main properties of typical pyroelectric and electrocaloric materials, including figures of merit for both effects. In contrast to the study of pyroelectric materials and their device applications, which is a relatively mature field, the understanding of electrocaloric materials is still at an early stage. Hence, this review will devote most of the volume to covering theoretical works and experimental results related to the ECE. The relationship between the pyroelectric and electrocaloric effect will be discussed, and more importantly, considerations for material modifications to achieve large ECE (with an emphasis on electrocaloric polymers) will be presented.

2. Pyroelectricity and pyroelectric materials

2.1. *Pyroelectricity*

Pyroelectricity is the electrical response of an insulating dielectric to a change in temperature. In most experimental studies, the pyroelectric coefficient p is,

$$p^X = \left(\frac{dD}{dT}\right)_{X,E}, \tag{1}$$

where D is the electric displacement (in most pyroelectric materials, D is approximately equal to the P, the polarization), T is the temperature, X is the stress, and E is the electric field. Subscript X means the measurement is made in stress free conditions.

Pyroelectric materials are usually polar materials in which the lattice is non-centrosymmetrical. Among the 32 point group symmetries of crystals, only 10 belong to the polar point group, i.e., non-centrosymmetrical. They are: triclinic, 1 (C_1); monoclinic, m and 2 (C_2); orthorhombic, 2 mm (C_{2V}); trigonal, 3 (C_3) and 3 m (C_{3V}); tetragonal, 4 (C_4) and 4 mm (C_{4V}); hexagonal, 6 (C_6) and 6 mm (C_{6V}). In addition, pyroelectricity is allowed in the two limiting Curie groups ∞ ($C\infty$) and ∞ mm ($C\infty V$), which represent symmetry of a textured material (e.g., fabric structure) and that of an electrically poled polycrystalline material, respectively.

Pyroelectricity was first observed more than 2400 years ago by the Greek philosopher Theophrastus [2], and has been studied for a long time [3–6]. To date, a large number of pyroelectric materials have been reported [7, 11–15]. Examples include barium strontium titanate (BST), triglycinesulphate (TGS), lithium niobate (LNB), lead zirconatetitanate (PZT), lead magnesium niobate-lead titanate

(PMN-PT), strontium barium niobate (SBN), and polyvinylidene fluoride (PVDF).

2.2. *Figure of merit (FOM) for pyroelectric materials*

Pyroelectric materials have found a number of applications in areas such as thermal imaging, laser detectors, radiometers, infrared sensors, pollution monitoring and gas analysis, fire alarms, and intruder alarms [7]. For these devices, three figures of merit are usually employed to evaluate a pyroelectric material and its interface with the electronic readout circuit, which can be a charge sensitive circuit or a voltage sensitivity circuit. They are current response FOM $F_{\mathrm{i}} = p$ for the charge sensitive readout, voltage response FOM $F_{\mathrm{V}} = p/\varepsilon'$ for the readout circuit in voltage mode, and for detectors, the signal to noise ratio (SNR) is a critical parameter and FOM $F_{\mathrm{D}} = p/\sqrt{\varepsilon''}$ measures the SNR of the sensor. Here, ε' and ε'' are the real and imaginary parts of permittivity, respectively. Some examples of these FOM will be presented in the next section.

2.3. *Secondary pyroelectric effect*

The primary pyroelectric coefficient is defined as the pyroelectric coefficient measured in a clamped condition p^x (strain $x =$ constant). As has been observed, pyroelectricity measured in a clamped condition can be quite different from that measured under stress free conditions ($X =$ constant) because of the piezoelectric contribution caused by the thermal expansion of the material. The pyroelectric coefficient caused by the thermal expansion is termed as the secondary pyroelectric coefficient. The total pyroelectric coefficient measured at constant stress p^X can be expressed as [16, 17]

$$p_i^X = p_i^x + \mathrm{d}_{ijk}^X \mathrm{c}_{jklm}^{X,E} \alpha_{lm}^E, \tag{2}$$

where d_{ijk}^X is the piezoelectric strain tensor under a constant stress, $\mathrm{c}_{jklm}^{X,E}$ the elastic stiffness tensor under constant stress and electric field, and α_{lm}^E the thermal expansion tensor under a constant electric field. Hence, the first term in Eq. (2) represents the primary pyroelectric effect and the second term is the secondary effect.

Table 1. Pyroelectric coefficients of several organic and inorganic materials [13].

Materials	Primary Coefficient (μC m^{-2} K^{-1})	Secondary Coefficient (μC m^{-2} K^{-1})	Measured Coefficient (μC m^{-2} K^{-1})
(A) Nonferroelectrics			
CdS (6 mm)	−3.0	−1.0	−4.0
CdSe (6 mm)	−2.94	−0.56	−3.50
ZnO (6 mm)	−6.9	−2.5	−9.4
BeO (6 mm)	−3.39	−0.01	−3.40
Tourmaline (3 m)	−0.48	−3.52	−4.0
$Li_2SO_4 \cdot 2H_2O$ (2)	+60.2	+20.1	+80.3
(B) Ferroelectrics			
$LiNbO_3$ (3 m)	−95.9	+12.9	−83
$LiTaO_3$ (3 m)	−178	+2.0	−176
$NaNO_2$ (2 mm)	−135	−5.0	−140
$Pb_5Ge_3O_{11}$ (3)	−110.5	+15.5	−95
$Sr_{0.5}Ba_{0.5}Nb_2O_6$ (4 mm)	−529	−21	−550
$Ba_2NaNb_5O_{15}$ (2 mm)	−141.8	+41.8	−100
TGS (2)	−330	+60	−270
PVDF (2 mm)	−14	−13	−27
$BaTiO_3$ (∞ m)	−260	+60	−200
$Pb(Zr_{0.95}Ti_{0.05})O_3$($\infty$ m)	−305.7	+37.7	−268
$Pb(Zr_{0.52}Ti_{0.48})O_3$($\infty$ m)	−110	+60	−50

As examples, Table 1 lists the primary and secondary pyroelectric coefficients of several organic and inorganic materials measured at room temperature, along with their symmetry groups. As can be seen, for some pyroelectric materials, the secondary effect can be comparable or even larger than that of the primary effect.

2.4. *Pyroelectric materials*

Large numbers of ferroelectrics have been investigated for pyroelectric applications, which involve organic and inorganic materials. On the organic material side, PVDF based polymers, and side-chain liquid crystal polymers have been studied. On the inorganic material side, single crystals, poled polycrystalline ceramics, and ceramic thin films have been explored. Near a continuous ferroelectric–paraelectric (FE–PE) phase transition, an extremely large pyroelectric coefficient may be obtained when under a DC bias electric field. In addition, a hybrid approach, ceramic–polymer composites, has also been studied.

PVDF as well as many PVDF based copolymers and terpolymers possess ferroelectricity. In most cases, these are semi-crystalline polymers and the crystalline phase is in a lamellar form. Polarization is the long range ordering of the dipoles formed from the C–H/C–F bonds, which are perpendicular to the backbone chains [18]. PVDF does not show a FE–PE phase transition before melting (~180°C) [19]. The copolymers of PVDF with trifluoroethylene (TrFE >20 mol%) display FE–PE phase transition before melting and hence at room temperature show stronger pyroelectric coefficient [20]. Table 2 lists the pyroelectric coefficients of the P(VDF-TrFE) copolymers at different compositions at room temperature [21, 22].

Liquid crystal side-chain polymers (LCP), where the polymer architecture consists of a polymer backbone, flexible spacers, and mesogenic groups based on a cyanobiphenyl moiety, also exhibit some pyroelectric response [23]. One example is the polymethacrylates separated with alkyl chains of 5–7 methylene groups from 4′-cyano-4-biphenyl mesogenic moieties.

There are many reports on the pyroelectric coefficients for bulk single crystals and ceramics, especially the $Pb(Mg_{1/3}Nb_{2/3})O_3$-$0.26PbTiO_3$(PMN-PT) and $Pb(In_{1/2}Nb_{1/2})O_3$–$Pb(Mg_{1/3}Nb_{2/3})O_3$–$PbTiO_3$(PIMNT) relaxor ferroelectric single crystals and SrBa Nb_2O_6(SBN) ceramics synthesized by the hot forging method [24–34].

Table 2. Room temperature pyroelectric properties of P(VDF-TrFE) copolymers and a LCP [21, 22].

Materials	$F_i = p$ (μC m^{-2} K^{-1})	ε'	$F_V = p/\varepsilon'$ (μC m^{-2} K^{-1})	$F_D = p/\sqrt{\varepsilon''}$ (μC m^{-2} K^{-1})
P(VDF-TrFE) 50/50	40	18	2	5.06
P(VDF-TrFE) 56/44	38	12	3	70
P(VDF-TrFE) 70/30	55	8	6.8	200
P(VDF-TrFE) 75/25	33	7.4	4	94
P(VDF-TrFE) 80/20	31	7	4.4	91
PVDF	27	11	2.46	33.24
LCP	0.98	6.7	0.15	3.46

Table 3. Pyroelectric properties of PMN-PT crystals and SBN ceramics at room temperature [24–34].

Materials	$F_i = p$ (μC m^{-2} K^{-1})	ε'	$F_V = p/\varepsilon'$ (μC m^{-2} K^{-1})	$F_D = p/\sqrt{\varepsilon''}$ (μC m^{-2} K^{-1})
PMN-0.13PT ⟨111⟩ poled	3260	3107	1.0	1018
PMN-0.21PT ⟨111⟩ poled	1790	961	1.8	1059
Fe doped PMN-0.38PT	568	310	1.8	394
PMN-0.29PT	1280	515	2.4	710
Mn doped PMN-0.38PT	1620	688	2.3	2790
Mn doped PMN-0.26PT	1720	660	2.6	2994
PIMNT (42/30/28)	900	702	1.28	760
SBN53/47 HF (⊥)[a]	510	980	0.52	120
SBN53/47 HF (‖)[a]	400	468	0.85	263

[a] ⊥: parallel to pressing axis; ‖: perpendicular to pressing axis.

Single crystals demonstrate large pyroelectric coefficients and FOMs as listed in Table 3 [24–34].

Thin films are attractive for pyroelectric applications because they can form texture or epitaxial structure, leading to properties approaching that of single crystals. Thin film devices are also more compatible with the semiconductor fabrication process, since many thin films can be integrated on Si substrates at temperatures as low as 600°C. In addition, it was also observed that the relaxor ferroelectric thin films exhibit much larger pyroelectric coefficients than conventional perovskite structured thin films [15]. Table 4 presents pyroelectric coefficients and FOMs of several thin films [35–45].

The ceramic–polymer composites combine polymeric properties such as mechanical flexibility, formability, and low cost with the ceramic properties such as large p and excellent mechanical strength. These hybrids exhibit better pyroelectric properties over the individual constituent phase while possessing excellent mechanical strength, formability, and robustness of the polymer, which may someday be useful for IR detectors without supporting substrates. Table 5 lists pyroelectric coefficients and FOMs of several ceramic–polymer composites [46–52].

Table 4. Pyroelectric coefficients and FOMs of thin films [35–45].

Materials	$F_i = p$ (μC m^{-2} K^{-1})	ε'	$F_V = p/\varepsilon'$ (μC m^{-2} K^{-1})	$F_D = p/\sqrt{\varepsilon''}$ (μC m^{-2} K^{-1})
$PbTiO_3$/Pt/Si	130–145	180–260	0.50–0.81	43–91
PZT 15/85/Pt/Si	160–220	200–230	0.70–1.10	86–156
PLT10-20/Pt/Si	200–576	153–550	2.3	55–466
Porous PLT/Pt/Si	220	90	2.4	232
$LiNbO_3$ (006)/Pt/Si	71	(30)	2.4	130
PLT5-15/Pt/MgO	400–1300	100–350	1.1–13	214–1678
PLZT 7.5/8/92-20/80	360–820	193–260	1.4–4.2	171–518
$PbSc_{1/2}Ta_{1/2}O_3$/ sapphire	6000	6500	0.92	430
$K_{0.89}Na_{0.11}Ta_{0.55}Nb_{0.45}O_3/KTaO_3$	5200	12000	0.43	336

Table 5. Pyroelectric coefficients and FOMs of ceramic–polymer composites [46–52].

Materials	$F_i = p$ (μC m^{-2} K^{-1})	ε'	$F_V = p/\varepsilon'$ (μC m^{-2} K^{-1})	$F_D = p/\sqrt{\varepsilon''}$ (μC m^{-2} K^{-1})
TGS-PVDF	90	27.3	3.3	—
PT-P(VDF-TrFE)	40.7	57.3	0.71	28
PCLT-P(VDF-TrFE)	56.5	15.1	3.74	113
PZT-PU	90	24	3.75	22
PZT-P(VDF-TrFE)	92	29	3.2	93
$LiTaO_3$/P(VDF-TrFE) @70°C	137.5	20.7	6.6	67

3. Electrocaloric effect and electrocaloric materials

3.1. *General consideration of the electrocaloric effect*

3.1.1. *Maxwell relations*

In general, the Gibbs free energy G for a dielectric material could be expressed as a function of temperature T, entropy S, stress X, strain x, electric field E, and electric displacement D in the form

$$G = U - TS - X_i x_i - E_j D_j, \tag{3}$$

where U is the internal energy of the system, the stress and field terms are written using Einstein notation where i runs from 1 to 6 and j is from 1 to 3. The differential form of Eq. (3) is

$$\mathrm{d}G = -S\mathrm{d}T - x_i\mathrm{d}X_i - D_j\mathrm{d}E_j. \tag{4}$$

Hence, the entropy S and electric displacement D_j are,

$$S = -\left(\frac{\partial G}{\partial T}\right)_{X,E}, \quad D_j = -\left(\frac{\partial G}{\partial E_j}\right)_{T,X}. \tag{5}$$

Equation (5) leads to the Maxwell relation, linking the ECE to the pyroelectric effect for a thermodynamically reversible system [1],

$$\left(\frac{\partial S}{\partial E_j}\right)_{T,X} = \left(\frac{\partial D_j}{\partial T}\right)_{E,X} = p_E, \tag{6}$$

and

$$-\left(\frac{\partial T}{\partial E}\right)_S = \frac{T}{c_E}\left(\frac{\partial D}{\partial T}\right)_E = \frac{Tp_E}{c_E}, \tag{7}$$

where c_E is the heat capacity. Under a constant stress X, the isothermal entropy change ΔS and adiabatic temperature change ΔT of an ECE material as the E is changes from E_1 to E_2 can be expressed as [1]

$$\Delta S = \int_{E_1}^{E_2}\left(\frac{\partial D}{\partial T}\right)_E dE, \tag{8}$$

$$\Delta T = -T\int_{E_1}^{E_2}\frac{1}{c_E}\left(\frac{\partial D}{\partial T}\right)_E dE. \tag{9}$$

Equations (6) through (9) indicate that in order to achieve a large ΔS and ΔT, the dielectric materials should possess a large pyroelectric coefficient over a broad electric field range. For ferroelectric materials, a large pyroelectric effect exists near the FE–PE phase transition. It is also noted that a large ΔT may be achieved even if ΔS is small when c_E of a dielectric material is small. However, for practical refrigeration applications, an ECE material should possess both a large ΔS and ΔT [53].

It is noted that in the temperature region including a first-order FE–PE transition, Eq. (8) should be modified to take into account the discontinuous change of the polarization ΔP (and ΔD) at the transition temperature, i.e.,

$$\Delta S = \int_0^E \left(\frac{\partial D}{\partial T}\right)_E dE - \Delta D\left(\frac{\partial E}{\partial T}\right). \tag{10}$$

It should be emphasized that one should be cautious in using the Maxwell relation to deduce the ECE near first order FE–PE transitions. The hysteresis associated with first order FE–PE transition means the process is not thermodynamically reversible and the Maxwell relation is derived based on reversible thermodynamic process.

Although a few studies on the ECE were conducted in which direct measurement of ΔT and/or ΔS was made [54, 55], most experimental studies were based on the Maxwell relation where the electric displacement D *versus* temperature T under different electric fields was characterized. ΔS and ΔT were deduced from Eqs. (8) and (9).

3.1.2. *ECE derived from the Landau–Devonshire phenomenological theory*

The Landau–Devonshire (L–D) phenomenological theory has been widely utilized to describe the macroscopic phenomena that occur in the polar materials, e.g., ferroelectric or ferromagnetic materials near their phase transitions. Here the L–D phenomenological theory is used to estimate the ECE of ferroelectrics. From the L–D theory, the Gibbs free energy of a ferroelectric material can be written as an expansion of the polarization P (for most ferroelectrics, $P \approx D$) as [1]

$$G = \frac{1}{2}\alpha P^2 + \frac{1}{4}\xi P^2 + \frac{1}{6}\zeta P^6 - EP, \tag{11}$$

where $\alpha = \beta(T - T_0)$, and β, ξ and ζ are temperature-independent phenomenological coefficients. From $\left(\frac{\partial G}{\partial T}\right)_E = -\Delta S$ and at temperatures above FE–PE transition where a single P value exists under an applied field E, one obtains,

$$\Delta S = -\frac{1}{2}\beta P^2, \tag{12}$$

and the adiabatic temperature change $\Delta T(= -T\Delta S/c_E)$ can be obtained, i.e.,

$$\Delta T = \frac{1}{2c_E}\beta T P^2, \tag{13}$$

Equation (12) indicates that the entropy will be reduced when the material increases its polarization or changes to a polar state from a non-polar state when an external action, e.g., temperature, electric field or stress, is applied. The entropy change and temperature change are associated with the phenomenological coefficient β and polarization, *viz.* proportional to β and P^2. Both parameters will affect the ECE values of the materials. A material with a large β and a large P will generate a large ECE entropy change and temperature change near the FE–PE phase transition temperature.

As an example, the L–D phenomenological theory predicts large ECE values in ferroelectric P(VDF-TrFE) polymers. P(VDF-TrFE) 65/35 mol% copolymer, with $\beta = 3.5 \times 10^7\,\mathrm{J\,m\,C^{-2}\,K^{-1}}$ and $D = 0.08\,\mathrm{C m^{-2}}$ [56], will exhibit a $\Delta S = 62\,\mathrm{J\,kg^{-1}\,K^{-1}}$. Making use of its specific heat capacity $c_E = 1.4 \times 10^3\,\mathrm{J\,kg^{-1}\,K^{-1}}$ [57] and Curie temperature $T_c = 102°\mathrm{C}$ [56], yields a $\Delta T = 16.6°\mathrm{C}$. The large ΔS and ΔT values suggest that a large ECE may be achieved in ferroelectric P(VDF-TrFE) copolymers.

In addition, the heat of FE–PE phase transition can also be used to assess the ECE ($Q = T\Delta S$) in a ferroelectric material at temperatures above the FE–PE transition. For example, P(VDF-TrFE) 68/32 mol% copolymer shows a heat of FE–PE transition larger than $2.1 \times 10^4\,\mathrm{J\,kg^{-1}}$ (or $\Delta S \approx 56.0\,\mathrm{J\,kg^{-1}\,K^{-1}}$), which is consistent with the prediction based on the phenomenological theory that a large ECE can be obtained in P(VDF-TrFE) ferroelectric polymers [10, 58, 59].

3.1.3. *Contribution of secondary pyroelectricity to the ECE*

For polymeric materials, the secondary pyroelectric effect can be significant. In order to derive the thermodynamic relations correctly, the secondary effect should be included in the considerations. In a real experimental situation, it is the force $\boldsymbol{F}$, displacement $\boldsymbol{R}$, voltage

V, charge Q, temperature T, and the total entropy S of the sample that are directly measured. The general form of the elastic Gibbs energy, hence, can be written as [60]

$$\mathrm{d}G = -S\mathrm{d}T - \boldsymbol{R}\cdot\mathrm{d}\boldsymbol{F} - Q\mathrm{d}V, \tag{14}$$

which yields a general form of the Maxwell relation,

$$\left(\frac{\partial Q}{\partial T}\right)_{V,F} = \left(\frac{\partial S}{\partial V}\right)_{T,F}. \tag{15}$$

When the dimensional change is small, Eq. (15) is reduced to Eq. (6).

Assuming the ECE film has a volume Ω, an area A_z in XY plane (perpendicular to the film thickness and the applied electric field), the electric field is along the z direction, and let $D = P$ for ferroelectric polymers, then $\mathrm{d}Q$ should be written as

$$\mathrm{d}Q = A_z\mathrm{d}P + P\mathrm{d}A_z, \tag{16}$$

if $\mathrm{d}A_z$, the dimensional change, is not negligibly small.

So the unit volume entropy $\Delta s = \Delta S/\Omega$ is

$$\Delta S = \int_{E_1}^{E_2} \left[\left(\frac{\partial Q}{\partial T}\right)_{E,X}\right] \Big/ A_Z dE. \tag{17}$$

On the device side, the total dipole moment M of the sample is related to the measured charge Q and polarization P as [60, 61]

$$Q = AP = M/d. \tag{18}$$

Making use of $\frac{1}{d}\left(\frac{\partial M}{\partial T}\right)_V = A\left(\frac{\partial P}{\partial T}\right)_V$ and Eq. (6), we have

$$\Delta S = -\left(\frac{1}{2}\beta P^2 + \int_{V_1}^{V_2}\left[\frac{P}{d^2}\left(\frac{\partial d}{\partial T}\right)_V\right] dV\right). \tag{19}$$

For freestanding films, the mechanical boundary conditions are $X_i = 0$ ($I = 1, 2,$ and 3). Then the entropy change per unit volume

Δs in Eq. (19) has the form [62]

$$\Delta S = -\left(\frac{1}{2}\beta P^2 + \int_{E_1}^{E_2}\left[\frac{P}{d}\left(\frac{\partial d}{\partial T}\right)_E\right]dE\right). \tag{20}$$

Using $(\partial d/d)/\partial T = \alpha$ and equation ($E = \beta(T - T_c)P + \gamma P^3$ for a second-order phase transition), Δs becomes

$$\Delta S = \frac{1}{2}\beta P^2 - \frac{1}{2}\alpha PE - \frac{1}{4}\alpha\gamma P^4. \tag{21}$$

Taking the typical data of $\beta = 2.4 \times 10^7\,\mathrm{J\,m\,C^{-2}\,K^{-1}}$ [58, 63], $\alpha = 2 \times 10^{-3}\,\mathrm{K^{-1}}$ (in the phase transition regime) [64], and $\gamma = 8.3 \times 10^{11}\,\mathrm{J\,m^5\,C^{-4}}$ [65], for a P(VDF-TrFE)55/45 mol% copolymer, it can be estimated that $\frac{1}{2}\beta P^2 = 5.88 \times 10^4\,\mathrm{J\,m^{-3}\,K^{-1}}$, the sum of the last two terms on the right side of Eq. (21) is $2.05 \times 10^4\,\mathrm{J\,m^{-3}\,K^{-1}}$, where $P = 0.07\,\mathrm{Cm^{-2}}$ and $E = 150\,\mathrm{MV\,m^{-1}}$ are used. The contribution of secondary pyroelectricity to the ECE is 26%. The results indicate that for polymers, the thermal expansion has a significant impact on the ECE entropy change through the interaction among the thermal expansion, electric field, and polarization.

Now if the polymer films are fixed to an inorganic substrate, the boundary conditions become $\alpha_1 = \alpha_2$, $x_1 = x_2 = 0$, $X_1 = X_2 \neq 0$ (the film surface is along the XY plane and Z is the film thickness direction). Using the electrostriction instead of piezoelectricity [66–68], near the phase transition temperature, the elastic Gibbs free energy can be written as

$$\begin{aligned} G = {} & \frac{1}{2}\beta(T - T_C)P^2 + \frac{1}{4}\gamma P^4 - EP - (X_1 + X_2)Q_{13}P^2 \\ & - \frac{1}{2}s_{11}(X_1^2 + X_2^2) - s_{12}X_1X_2, \end{aligned} \tag{22}$$

The equations of state could be expressed as [11]

$$\mathrm{d}x_1 = \mathrm{d}x_2 = \left(\mathrm{s}_{11}^{E,T} + \mathrm{s}_{12}^{E,T}\right)\mathrm{d}X_1 + 2\mathrm{Q}_{13}^{X,T}P_3\mathrm{d}P_3 = 0, \tag{23}$$

and it can be deduced that

$$-\Delta S = \frac{1}{2}\beta P^2 + \frac{2\alpha_1 S_1 + 2\alpha_1^2 T}{s_{11} + s_{12}}. \tag{24}$$

Equation (24) indicates that the clamping effect of the substrate can change the ECE of polymer films. Since for P(VDF-TrFE) based polymers, the strain $S_1 = Q_{13}P^2 > 0$ [69], the second term of the right-hand side of Eq. (24) will enhance the ECE entropy change in the clamped situation in addition to the contribution of the field induced phase transition [70].

3.2. *ECE measurement methods*

Generally, ECE measurement methods can be categorized into two groups: the indirect method, where entropy change ΔS and temperature change ΔT are deduced using Maxwell's relation; and the direct method, where ΔS and/or ΔT are acquired from specially designed calorimeters.

The indirect method is based on the Maxwell relation, shown in Eq. (6), from which the ΔS and ΔT of the ECE as the electric field changes from E_1 to E_2 can be deduced, i.e., Eqs. (8) and (9). The indirect method has been widely used to characterize the ECE since the 1960s [71], due to the fact that the apparatus measuring the polarization as a function of temperature is quite universal and relatively convenient to use. However, it should be noted that since the Maxwell relation is derived for ergodic systems, one needs to be cautious when using the indirect method to deducing the ECE in material systems that are not thermodynamically reversible such as the relaxor ferroelectric polymers [72].

The direct method (measuring entropy change (extensive) or temperature change (intensive)) can thus be divided into two subgroups. Since there is no standard commercial equipment for measuring the ECE, various calorimeters have been developed by several research groups to measure the electrocaloric effect under different electric fields over a wide temperature range. The commercial DSC system can be modified so that electric fields can be applied to the sample, and the heat absorbed or ejected by the sample is simultaneously recorded [28]. A calorimeter using a heat flux sensor was developed by Zhang's group at Penn State [73], as shown in Fig. 1, where the heat generated by the ECE of the sample is calibrated using the heat generated by a standard reference resistor R, from which

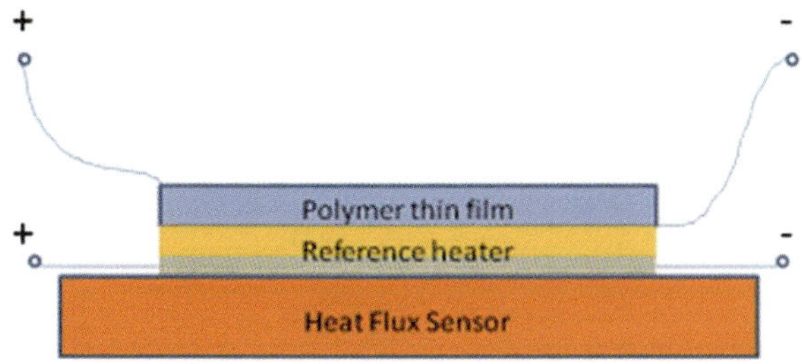

Figure 1. Schematic configuration of direct ECE measurement setup with heat flux sensor.

ΔS is determined. By substituting the heat flux sensor with an infrared sensor or thermal couple, the adiabatic temperature change can be obtained using the same calorimeter [63]. A high resolution calorimeter using a miniature thermistor and weak thermal link to the thermal bath was also developed for direct ECE measurement in thin films [63]. Additionally, other apparatuses using multiple thermometers to measure the temperature difference between the sample and background were also reported [55, 74]. Factors that may influence the accuracy of a direct ECE measurement over a certain temperature range include the sensitivity, signal-to-noise ratio and response time of the thermal sensor, the stability of temperature controller, and the ability of the system to provide good thermal contact during the measurement.

3.3. *Electrocaloric polymers*

Very large ECE has been observed in PVDF-based polymers. These materials can also be fabricated into thin films (to operate the cooling device in voltages normally used in commercial cooling devices) and large size devices, these polymers are probably the most likely ECE materials to be developed into commercial applications. This section will summarize the main advancements achieved in this field.

In PVDF-based polymers, it has been reported that there exist five crystalline phases. The most interesting one is the ferroelectric β phase, in which, polymer chains adopt all-trans (TTTT) conformation, where all the fluorine atoms are on one side of the chain, forming

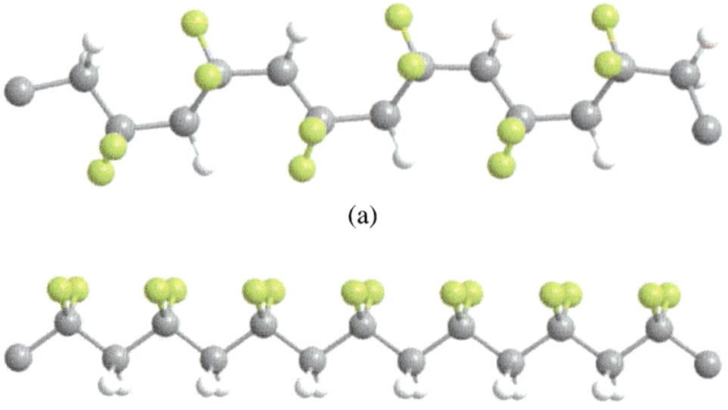

(a)

(b)

Figure 2. Schematic illustration of the two most common crystalline chain conformations in PVDF (a) tg+tg− conformation and (b) all-trans conformation.

dipoles perpendicular to the chain direction, as illustrated in Fig. 2 [18, 75, 76]. The ferroelectric β phase will undergo a phase transition and turn into a paraelectric phase, a phase consisting of a random sequence of *trans*-gauche (TG) bonds, such as TGTG′, and T_3GT_3G' isomers. Correspondingly, the dipoles in the crystallites also change from an ordered state to a disordered state. Additionally, with defect modification such as high energy irradiation or copolymerizing with a third bulky monomer, the normal ferroelectric polymers can be converted into relaxor ferroelectrics, in which randomly distributed nano-polar regions are embedded in a non-polar matrix. When an external electric field is applied, the random dipoles in either the paraelectric phase of normal ferroelectric polymers or in the relaxor ferroelectrics, will reorient along the electric field direction, causing a structure change in the polymer as well as a temperature change as compensation for the entropy change.

The PVDF-based polymers that have been studied include P(VDF-TrFE) copolymers, high-energy-electron-irradiated P(VDF-TrFE) copolymers, (P(VDF-TrFE-CFE)) terpolymers (CFE: chlorofluoroethylene), and P(VDF-TrFE) copolymer/P(VDF-TrFE-CFE) terpolymer blends.

3.3.1. *ECE in the normal ferroelectric P(VDF-TrFE) copolymer*

In a ferroelectric copolymer, the FE–PE transition corresponds to the order–disorder transition of the dipole states, thus a large ECE associated with dipole rearrangement would be generated near the FE–PE phase transition. P(VDF-TrFE) 55/45 mol% copolymer was studied as an electrocaloric material because its FE–PE phase transition is continuous.

Figure 3 presents ΔS and ΔT at several electric field levels deduced by indirect measurement [10]. As can be seen, the copolymer exhibits a ΔS of more than $55\,\mathrm{J\,kg^{-1}\,K^{-1}}$ and ΔT of more than

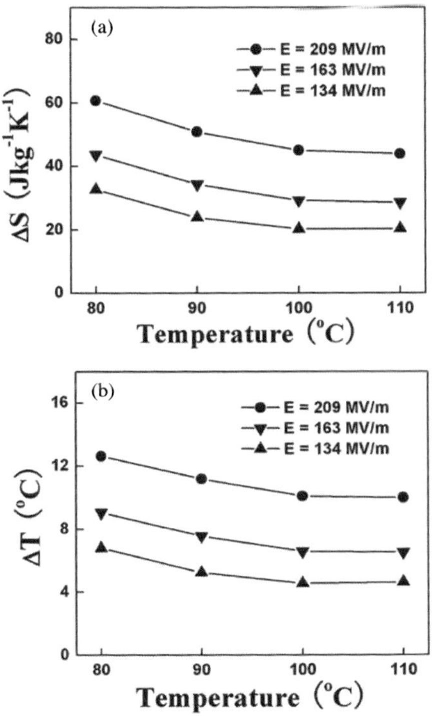

Figure 3. Isothermal entropy change DS and (b) adiabatic temperature change DT, measured by indirect method, as a function of ambient temperature at different electric fields for P(VDF-TrFE) 55/45 copolymer.

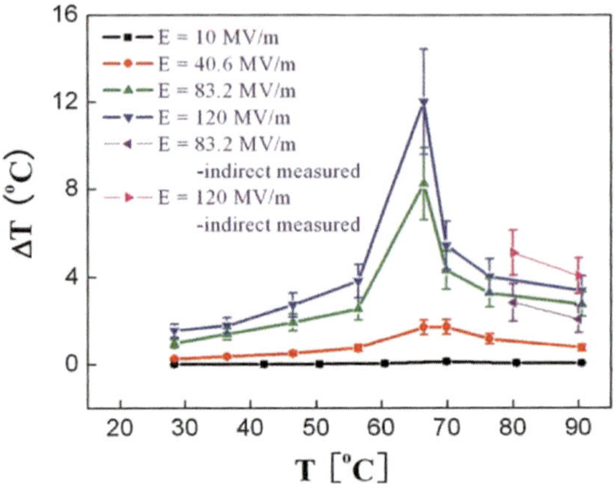

Figure 4. Directly measured DT as a function of temperature under several electric fields for non-stretched P(VDF-TrFE) 55/45 mol% copolymers and comparison with that indirectly measured.

12°C under an electric field of 209 MV m^{-1} and at temperatures around 80°C, which are the first experimental result demonstrating a giant ECE in polymers. Figure 4 presents ΔS and ΔT as a function of temperature measured under several electric fields in P(VDF-TrFE) 55/45 mol% copolymer using the direct method as well as the comparison between the ECE results directly measured and calculated from the indirect method [62]. The data show that the ECE reaches a maximum at the FE–PE transition, where a $\Delta T = 12$°C can be induced under a 120 MV m^{-1} electric field. At temperatures away from the FE–PE transition, ΔT drops rapidly, which is a characteristic feature for normal ferroelectric materials. The comparison result reveals that within the experimental error, the ECE deduced from the Maxwell relation is consistent with that directly measured, and also indicates that for a ferroelectric material at temperatures above FE–PE transition, the Maxwell relation is valid for deducing ECE.

3.3.2. *ECE in the relaxor polymers*

Defect modification, either through copolymerization with a bulky monomer such as CFE or CTFE (chlorotrifluoroethylene) to form a terpolymer or by direct high-energy-electron irradiation on copolymer films, can convert the normal ferroelectric P(VDF-TrFE) polymer into a ferroelectric relaxor, with a large room-temperature dielectric constant, slim P–E hysteresis loop and disorder dipole states at room temperature [66, 77]. Defect modification shifts the operating temperature of the ECE material to around room temperature, and relaxor polymers have the potential to generate an even larger ECE than the normal ferroelectric P(VDF-TrFE) copolymer.

The ECE in a dielectric is determined by the dipolar entropy change ΔS_{p} between polar and non-polar states, as shown in Eqs. (25) and (26),

$$\Delta T = \frac{T}{C_E}[S_P(0,T) - S_P(E,T)], \tag{25}$$

$$\Delta T_{\mathrm{sat}} = \frac{T \ln \Omega}{3\varepsilon_0 \Theta C_E} P_{\mathrm{sat}}^2, \tag{26}$$

where $S_{\mathrm{p}}(0,T)$ is the dipolar entropy when $E = 0$, and $S_{\mathrm{p}}(E,T)$ corresponds to the entropy of a dipole aligned state when external electric field E is applied, C_E is the heat capacity, Ω is the number of possible polar states, P_{sat} is the saturated polarization value, and Θ is the Curie constant in electrocaloric materials [78]. For a dipolar disordered state at $E = 0$, such as the relaxor at temperatures above the broad dielectric constant peak, $S_{\mathrm{p}}(0,T)$ is proportional to $P^2 \ln(\Omega)$. Thus, through defect modification to introduce local states and enhance the number of available polar states, relaxor ferroelectrics have the potential to realize a large ECE around room temperature.

3.3.2.1. *Giant electrocaloric effect in the irradiated P(VDF-TrFE) relaxor copolymers*

It has been shown that by employing high energy electron irradiation, the normal ferroelectric P(VDF-TrFE) copolymer can be converted to a relaxor ferroelectric polymer which displays high dielectric constant (~50 at 1 kHz), large reversible polarization change, and high

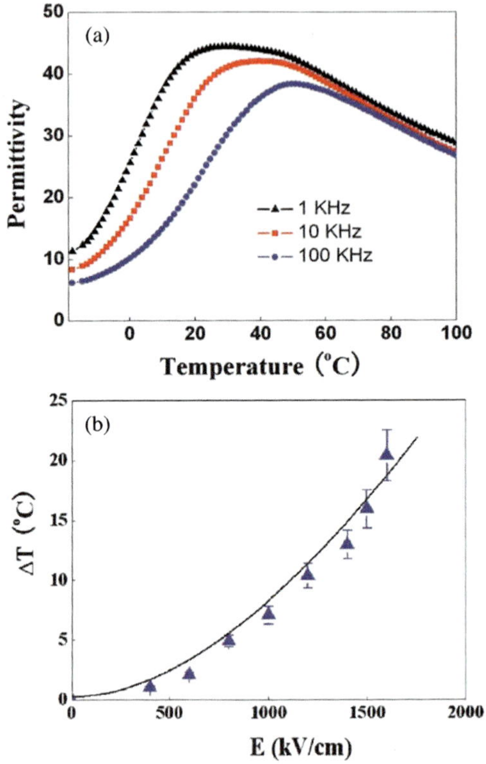

Figure 5. (a) Dielectric constant vs. temperature measured at different frequencies and (b) directly measured ECE for the high energy electron irradiated P(VDF-TrFE) 68/32% relaxor copolymer.

electrostriction at room temperature [66]. Figure 5(a) presents the dielectric constant of a high energy electron irradiated P(VDF-TrFE) 68/32 mol% copolymer, which has a broad dielectric constant peak around room temperature and that peak position moves progressively toward higher temperatures with frequency, a characteristic feature of relaxor ferroelectrics. The high energy electron irradiation breaks up long range polar-correlation in the polymer, which stabilizes dipolar disordered states around room temperature and generates local polar-states that may enhance the ECE, as discussed earlier. The ECE of the irradiated P(VDF-TrFE) 68/32 mol% copolymer measured near the broad dielectric peak (33°C) as a function of electric field is presented in Fig. 5(b) [63]. Under a field of $160\,\mathrm{MV\,m^{-1}}$, an adiabatic temperature change $\Delta T = 20°\mathrm{C}$ and isothermal entropy change $\Delta S = 95\,\mathrm{J\,kg^{-1}\,K^{-1}}$ can be obtained. Such a large ECE response shows promise for compact cooling applications with high efficiency and cooling power, and may open up a new direction for electrocaloric materials research.

3.3.2.2. *ECE in relaxor ferroelectric P(VDF-TrFE-CFE) terpolymers*

In the P(VDF-TrFE-CFE) relaxor ferroelectric polymers, large ECE has also been observed. ECE of the P(VDF-TrFE-CFE) 59.2/33.6/7.2 mol% terpolymer directly measured at 30°C, as presented in Fig. 6(a), showing a very large $\Delta T \sim 16°\mathrm{C}$ induced under $160\,\mathrm{MV\ m^{-1}}$ electric field; besides the large ECE, several relaxor ferroelectric polymers also display a nearly temperature independent ECE as presented in Fig. 6(b) [73]. For a P(VDF-TrFE-CFE) 59.2/33.6/7.2 mol% relaxor ferroelectric terpolymer, the ECE response is nearly temperature independent from 0°C to 45°C, which is in sharp contrast to that in normal ferroelectrics where the ECE peaks at the FE–PE transition and displays strong temperature dependence. Such a temperature independent ECE over a broad range is attractive for practical cooling device applications, especially for those requiring large temperature spans between cold and hot ends. These results also reveal the intricate roles played by the defects in tailoring the ferroelectric response and its polar nanostructures to generate a large ECE and its temperature response behavior. As has been shown here, the temperature dependence of the ECE of the terpolymer films depends critically on the film preparation conditions, while the uniaxially stretched terpolymer films show a pronounced temperature dependence on the ECE, the non-stretched films exhibit a nearly temperature independent ECE from 5°C to 45°C. At 5°C and 55°C, the ΔT of the uniaxially stretched films is more than 15% smaller, compared with that of the non-stretched films. Such a difference is likely to be caused by the changes in possible polar states and polar-correlation length due to film stretching.

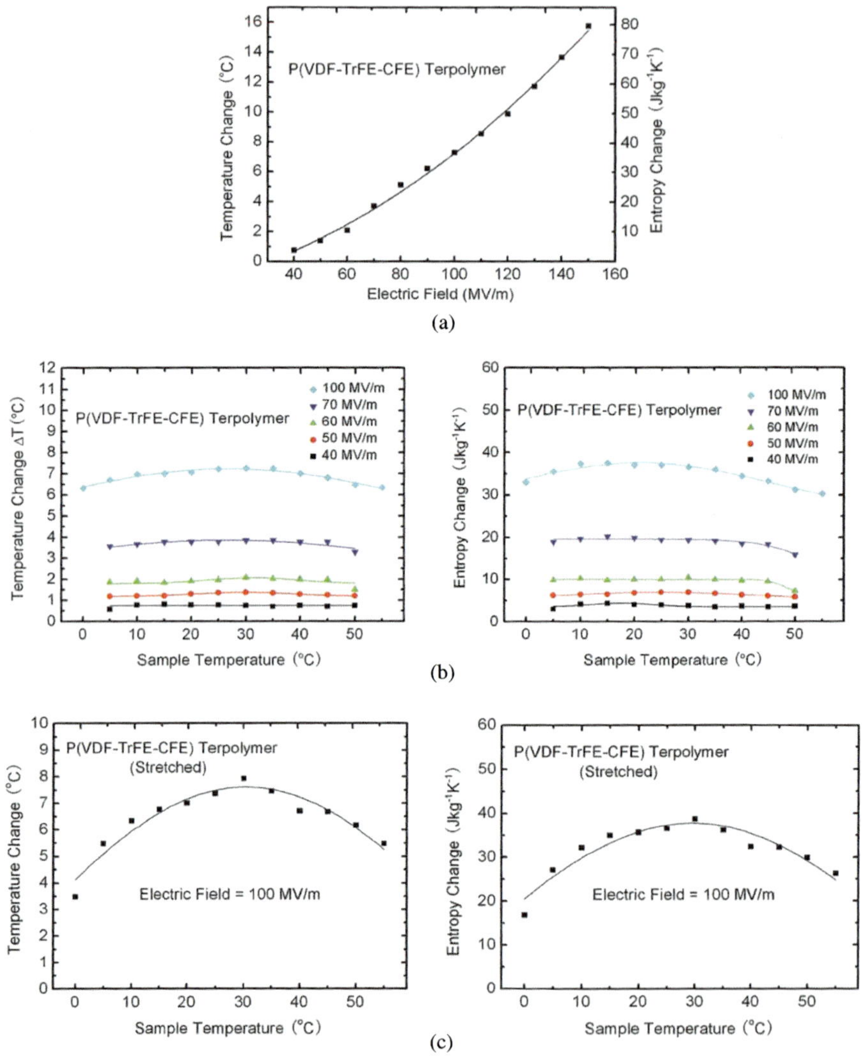

Figure 6. Directly measured ECE in non-stretched P(VDF-TrFE-CFE) 59.2/33.6/7.2 mol% terpolymer (a) adiabatic temperature change DT vs. applied electric field at 30°C and (b) adiabatic temperature change DT and isothermal entropy change DS vs. sample temperature under different applied fields, (c) adiabatic temperature change DT and isothermal entropy change DS vs. sample temperature under different applied fields in stretched P(VDF-TrFE-CFE) terpolymer under a constant electric field of $100\,\mathrm{MV\,m^{-1}}$.

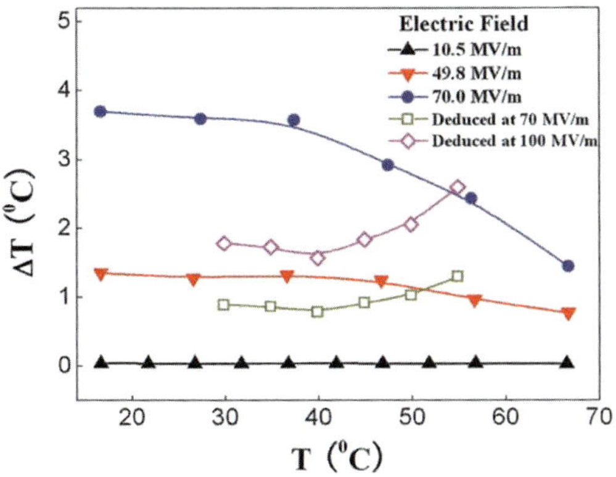

Figure 7. Temperature dependences of the directly measured DT of the terpolymer under different measuring electric fields.

It has also been observed that the ECE acquired in terpolymer by direct and indirect methods are very different, as presented in Fig. 7. The directly measured ECE from the relaxor terpolymer is much larger than that deduced from the Maxwell relation. Moreover, the directly measured ECE displays a much weaker temperature dependence at $E < 70\,\mathrm{MV\,m^{-1}}$. The results indicate that the Maxwell relation is not suitable to deduce ECE for the relaxor ferroelectric polymers even at temperatures above the broad dielectric constant maximum, since the Maxwell relations are valid only for systems in thermodynamic equilibrium (ergodic systems), while the relaxor ferroelectric polymers are non-ergodic systems even at temperatures above the dielectric constant maximum [72].

3.3.2.3. *ECE in the relaxor ferroelectric P(VDF-TrFE-CFE) terpolymers/P(VDF-TrFE) copolymer blends*

It is well known that nano-composites such as polymer blends, which exploit the merits of both the base polymer and the additive polymer, offer a great opportunity to enhance and tailor the material properties [19, 79–81]. It has been recently reported that the blends of P(VDF-TrFE-CFE) relaxor terpolymer with 10 wt% of P(VDF-TrFE) exhibit a 30% increase in the adiabatic temperature change compared with that of pure terpolymer, as shown in Fig. 8 [82]. Both

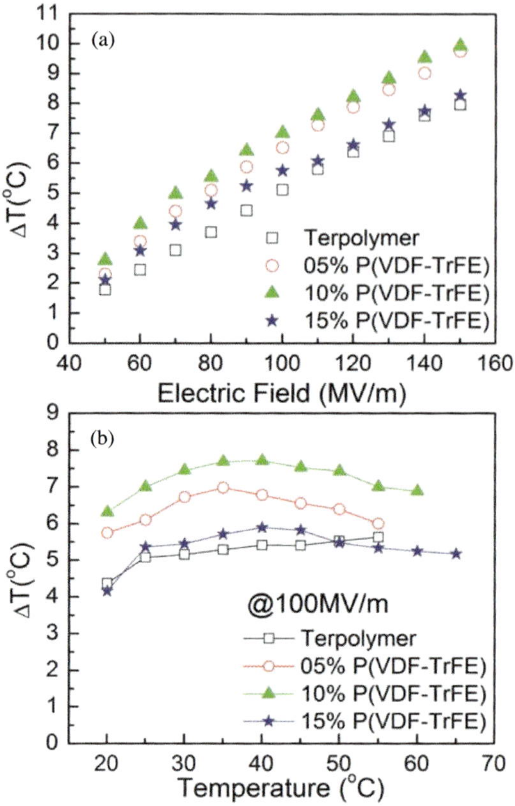

Figure 8. (a) Adiabatic temperature change as a function of electric field at room temperature and (b) Adiabatic temperature changes as a function of sample temperature at fixed electric field of 100 MV m^{-1}.

increased crystallinity in the blends and conversion of the copolymer from a normal ferroelectric to a relaxor lead to an increased relaxor polarization response and an enhanced ECE. Another advantage of the blends over pure terpolymer is that the Young's modulus of the blends is improved compared to the pure terpolymer.

3.4. *Electrocaloric ceramic materials*

3.4.1. *Bulk materials and thick films*

Many early ECE studies were limited to ceramics in the low temperature range of 4–15 K where the ceramics have low specific heat and hence may exhibit a large adiabatic temperature change [83]. The largest ECE effect was found in $SrTiO_3$ [84, 85], which is not ferroelectric but has a temperature-dependent dielectric constant. The maximum ΔT is 0.3°C from the ECE was observed at 10 K [85]. The ECE at low temperature was also studied in $KTaO_3$ single crystals [86].

After 1960s, the ECE in many perovskite ceramics were studied. It is well known that the properties of these ceramics can be tuned by doping, through which normal ferroelectrics (FE), anti-ferroelectrics (AFE), and ferroelectric relaxors can be obtained, providing very good models to understand the key factors that may influence the ECE.

Substitution of B-cations in PZT by Sn (PSZT) allows tuning of the temperature and composition ranges of the ferroelectric and antiferroelectric phases [71, 87, 88]. The largest ECE (i.e., the change of entropy at the phase transition) appears in compositions where the both the AFE and FE reach their stability limit at high temperature and a first order transition from a PE to FE can be induced electrically. On the other hand, in compositions where an electric field-induced AFE to FE transition is observed, the ECE is much smaller and changes slowly in a wide range of temperature [89]. The largest electrocaloric effect in bulk ceramics (2.6°C) was measured for PSZT annealed at high temperature. The magnitude of this effect strongly depends on microstructure. Large grain ceramics with small temperature and field ranges of AFE phase stability yield the largest electrocaloric effect [90]. La-doped PZT thick films were also examined and a maximum temperature change of 8.5°C was observed under a field of 7.5 MV m^{-1} [91].

In normal ferroelectrics, the sharp FE-PE transition usually leads to a narrow temperature range in which a large ECE exists. However, the ECE can be found in a wider temperature range in ferroelectric relaxors. For example, PMN ($Pb(MgNb)TiO_3$) exhibits a diffused dielectric phase transition around −13°C, which leads to a relatively large ECE near room temperature, i.e., maintaining a $\Delta T \sim 2.5$°C in temperature range from 16°C to 67°C under an electric field of 9 MV m^{-1} [92]. PMN-PT ($PbTiO_3$) with different compositions were also investigated in both ceramics and single crystals [93–98].

In lead scandium niobate (PSN) and its solid solution with lead scandium tantalate (PST), the ordering of perovskite heterovalent ions in sublattice B can be tailored either by substitution or thermal treatment [99, 100]. A high degree of long-range order leads to a high ECE maximum ($\Delta T = 1.7°C$) as well as a high phase transition temperature, while a decreased order causes a lower ECE (which occurs over a broader temperature range) and diffuse phase transitions.

Some lead-free ceramics such as $Ba_{0.73}Sr_{0.27}TiO_3$, $Na_{0.5}Bi_{0.5}TiO_3$, and $Ba_{0.3}Na_{0.7}Ti_{0.3}Nb_{0.7}O_3$ are also investigated for their ECE [101–103]. Although ECE is small, lead-free materials are attractive because of their environmentally friendliness.

The ECE reported in all the bulk ceramic samples is below 3°C, which makes them unimpressive for practical applications. The main reason is the low dielectric strength of bulk ceramics and the difficult in fabricating bulk ceramic to thin thickness (below 50 μm). By fabricating thick films using an interdigitated multilayer geometry, the film thickness of each layer can be reduced to $\sim$10 μm, which allows for application of high electric fields. Multilayered $PbSc_{0.5}Ta_{0.5}O_3$ [104], $BaTiO_3$ [105–108] were also studied.

3.4.2. *Thin films*

Thin ceramic films allow for the application of high electric fields ($>$10 MV m^{-1}) and thus a high ECE may be realized. Here we refer the thin films as to those thinner than 1 μm.

Mischenko *et al.* [9] first reported a temperature change of 12°C in 350 nm $PbZr_{0.95}Ti_{0.05}O_3$ thin film near Curie temperature of 222°C. A temperature change of about 11°C was also observed in a 700 nm $PbZrO_3$ thin film near its phase transition temperature at 235°C [109]. However, the phase transitions in these thin films are first-order transitions and the high ECE can only exist in a narrow temperature range. Additionally, the phase transition temperature is too high for applications near room temperature. To obtain a large ECE at a wide range of temperature near room temperature, La-doped PZT thin films were explored and a $\Delta T \sim 40°C$ was reported under the electric field of 120 MV m^{-1} at 45°C [63].

A ΔT of 9 K under 72.3 MV m^{-1} in 0.93PMN–0.07PT thin films was observed at depolarizing temperature 18°C instead of the temperature of 35°C, where the dielectric constant peak appeared, suggesting a dipolar glass–relaxor phase transition in this system [110]. For 0.9PMN–0.1PT, the maximum ΔT of 5°C was observed at 75°C, where a pseudocubic relaxor ferroelectric transforms to a cubic paraelectric phase [111]. The PMN-PT with 30%–35% PT are extremely interesting, because of the morphotropic phase boundary (MPB) can contribute additionally to the ECE [112–115].

Other thin films such as $PbSc_{0.5}Ta_{0.5}O_3$ relaxor with a wide temperature range of ECE were also investigated. A $\Delta T = 6.2$°C at 77.4 MV m^{-1} was reported [116]. $SrBiTa_2O_9$ [117], a lead-free perovskite-type bismuth layered oxide, was investigated where a ΔT of 4.93°C at around 290°C was reported.

Table 6 summarizes the ECE properties for organic and inorganic materials. For comparison, selected ECE results reported in the literature on inorganic materials, especially inorganic ferroelectric thin films where very high voltage can be applied, are also included. As can be seen, the relaxor ferroelectric polymers, because of their large ΔT and ΔS, broad operation temperature range and easy scaling up for various sized cooling devices, offer the most attractive ECE properties for practical cooling device applications.

4. Concluding remarks

Pyroelectric effects in major polar materials are summarized, including the effects in polymers, single crystals, ceramics, thin films, and liquid crystalline polymers. It is shown that the pyroelectric effect experimentally measured often includes both the primary effect and the secondary effect (dimensional effect). Three figures of merit are usually used to evaluate the properties of the pyroelectric materials, which are necessary since besides the material, a sensor system also consists of an electronic readout circuit, which can be in either the charge detection mode (FOM $F_{\mathrm{i}} = p$) or voltage detection mode (FOM $F_{\mathrm{V}} = p/\varepsilon'$). Also the signal-to-noise ratio is important in sensor designs which is related to the FOM $F_{\mathrm{D}} = p/\sqrt{\varepsilon''}$.

Table 6. ECE properties for some organic and inorganic materials investigated.

Material	Form	Experiment Temp. (°C)	ΔT(K)	ΔS ($J\,kg^{-1}K^{-1}$)	$\Delta T\ \Delta S$ ($J\,kg^{-1}$)	E ($MV\,m^{-1}$)	Measurement Method
P(VDF-TrFE) 55/45 [10]	Films	80	12.6	62	781.2	209	Indirect
P(VDF-TrFE) 55/45 [72]	Films	67	12			120	Direct
P(VDF-TrFE) 70/30 [118]	Films	117	21			300	Indirect
Irradiated P(VDF-TrFE) 68/32 [63]	Films	33	20	95	1900	160	Direct
P(VDF-TrFE-CFE) 59.2/33.6/7.2 [73]	Films	30	15.7	80	1256	150	Direct
$SrTiO_3$ [84]	Ceramics	−256	0.06			0.8	Direct
$SrTiO_3$ [85]	Single crystal	−262	0.3			0.7	Direct
$KTaO_3$ [86]	Single crystal	−260	0.25			1.56	Direct
$PbZr_{0.455}Sn_{0.455}Ti_{0.09}O_3$ [71]	Ceramics	44–55	~1.4			3	Direct
$PbZr_{0.75}Sn_{0.20}Ti_{0.05}O_3$ [90]	Ceramics	161	2.6			3	Direct
$Pb_{0.97}La_{0.02}(Zr_{0.95}Ti_{0.05})O_3$ [91]	Thick film	1.7	8.5			112(75)	Indirect
PMN [92]	Ceramics	16–67	~2.5	2.5	6.25	9	Direct
0.9PMN-0.1PT [93]	Ceramics	50	0.55			2.91	Direct
0.75PMN-0.25PT [94]	Ceramics	120–130	~0.5			1.35	DSC
0.87PMN-0.13PT [98]	Ceramics	18	~0.5			2	Direct
$Pb(Sc_{0.5}Nb_{0.5})O_3$ [99]	Ceramics	92	0.9			2	Direct
$Pb(Sc_{0.5}Ta_{0.5})O_3$–$Pb(Sc_{0.5}Nb_{0.5})O_3$ [100]	Ceramics	$-2 \sim 10$	1–1.7			2.5	Direct

(*Continued*)

Table 6. (*Continued*)

Material	Form	Experiment Temp. (°C)	ΔT(K)	ΔS ($J\,kg^{-1}K^{-1}$)	$\Delta T\ \Delta S$ ($J\,kg^{-1}$)	E ($MV\,m^{-1}$)	Measurement Method
$Pb(Mg_{0.5}W_{0.5})_{0.5}Ti_{0.5}O_3$ [119]	Ceramics	139	0.4			2.3	Indirect
$Na_{0.5}Bi_{0.5}TiO_3$ [102]	Ceramics	140	0.33	0.45	0.1485	5	Indirect
$Ba_{0.73}Sr_{0.27}TiO_3$ [101]	Ceramics	25	~1			2.4	DSC
$Ba_{0.3}Na_{0.7}Ti_{0.3}Nb_{0.7}O_3$ [103]	Ceramics	−13		0.026		1.5	DSC
PST($PbSc_{0.5}Ta_{0.5}O_3$) [104]	Multilayer		2.4 (3.5)			1.38	Direct
$BaTiO_3$ (commercial) [74]	Multilayer	80	0.55			30	Direct
$PbZr_{0.95}Ti_{0.05}O_3$ [9]	Thin films	226	12	7.9	94.8	77.6(50)	Indirect
$Pb_{0.88}La_{0.08}Zr_{0.65}Ti_{0.35}O_3$ [63]	Thin films	45	40	50	2000	120	Direct
$PbZr_{0.52}Ti_{0.48}O_3$ [120]	Thin films	387	11.1	6.17	68.49	57.7	Indirect
$PbZrO_3$ [109]	Thin films	235	11.4			51(40)	Indirect
0.9 PMN-0.1 PT [111]	Thin films	60	5			89.5	Indirect
0.93 PMN-0.07 PT [110]	Thin films	25	9			72.3	Indirect
0.67 PMN-0.33 PT [112]	Thin films	152	14.5			60	Indirect
0.65 PMN-0.35 PT [114]	Thin films	140	31			74.7	Indirect
$SrBiTa_2O_9$ [117]	Thin films	288	4.93			60	Indirect
$PbSc_{0.5}Ta_{0.5}O_3$ [116]	Thin films	68	6.2	6.3	39.06	77.4	Indirect

Compared with the pyroelectric materials, the study of ECE is still in its early stage. We note that in general the ECE and pyroelectric effect can be related through the Maxwell relation, and hence most ECE studies have been based on this relation (the indirect method). Since the Maxwell relation is valid only for systems and processes which are thermodynamically reversible and many ECE are studied near first order phase transitions, one should be cautious when using the indirect method to deduce ECE and interpret the experimental results.

Based on general thermodynamic considerations, a polar material with a large polarization change and short polar-correlation length (and hence a small Curie constant (or a large β coefficient) in the L–D theory) will exhibit a large ECE. In ferroelectrics, the large polarization change occurs near a FE–PE transition and hence the largest ECE in the material will occur near the FE–PE transition. Moreover, since the dipolar entropy is basically directly related to the degree of dipolar disorder, by increasing the number of polar states in the disordered phase can lead to a larger ECE. These considerations suggest that relaxor ferroelectrics may exhibit larger ECE than the normal ferroelectrics. Working with dielectrics near morphotropic phase boundary may also lead to higher ECE. All of these have indeed been experimentally observed. The advantage of working with the relaxor ferroelectrics is that the large ECE can occur over a broad temperature range compared with the normal ferroelectric in which the large ECE occurs over a narrower temperature range about the ferroelectric transition. Analogous to the pyroelectric effect, the secondary effect (the dimensional change of the EC material) can also play an important role.

The reports of large ECE in several materials systems create great opportunity to exploit these materials for solid-state cooling applications such as cooling of electronic components and on-chip cooling, refrigeration, and air-conditioning. Compared with ceramic thin films which are supported by a substrate and in which large ECEs are obtained, polymer thin films with large ECE are more attractive for practical cooling device applications since they can be fabricated into large size devices with high breakdown field.

Moreover, polymer/nanoparticle composites can be very promising for device applications, due to the enhancement in both ECE and thermal conductivity, which are equally important from cooling device application point of view. In developing materials with large ECE, one should also pay special attention to the possibility of these materials and phenomena for practical applications. With the advancement of fundamental understanding of ECE in insulating dielectrics, further experimental studies of ECE materials, as well as cooling device studies exploiting these ECE materials, we expect that ECE materials possessing better properties will be developed.

Acknowledgments

Xinyu Li, Xiao-shi Qian, and Q. M. Zhang were supported by the US DoE, Office of Basic Energy Sciences, Division of Materials Science and Engineering under Award No. DE-FG02-07ER46410. Sheng-Guo Lu and Xiang-Zhong Chen were supported by the Army Research Office under Grant No. W911NF-11-1-0534. Haiming Gu was supported by DOE SBIR Phase II, Contract No. DE-SC0003340 (subcontract from Strategic Polymers, Inc.) We thank Quinn Burlingame and Shan Wu for assistance in preparing the manuscript.

References

[1] Lines, M. E. and Glass, A. M. (1977). *Principles and Applications of Ferroelectrics and Related Materials* (Clarendon Press, Oxford Eng.).

[2] Lang, S. B. (2005). Pyroelectricity: From ancient curiosity to modern imaging tool, *Phys. Today*, 58, pp. 31–36.

[3] Whatmore, R. W. and Watton, R. (2001). Infrared detectors and emitters: Materials and devices, in *Infrared Detectors and Emitters: Materials and Devices*, edited by Capper, P. and Elliott, C. T. (Kluwer Academic, Boston).

[4] Lang, S. B. and Das-Gupta, D. K. (2000). Pyroelectricity: Fundamentals and applications, *Ferroelectr. Rev.*, 2, pp. 217–354.

[5] Muralt, P. (2001). Micromachined infrared detectors based on pyroelectric thin films, *Rep. Prog. Phys.*, 64, pp. 1339–1388.

[6] Lang, S. B. and Bhalla, A. S. (1991). Special issue on pyroelectricity, *Ferroelectrics*, 118, pp. R4–R5.

[7] Whatmore, R. W. (1986). Pyroelectric devices and materials, *Rep. Prog. Phys.*, 49, pp. 1335–1386.
[8] Kobeko, P. and Kurtschatov, J. (1930). Dielektrische Eigenschaften der Seignettesalzkristalle, *Z. Phys. A: Hadrons Nucl.*, 66, pp. 192–205.
[9] Mischenko, A. S., Zhang, Q., Scott, J. F., Whatmore, R. W. and Mathur, N. D. (2006). Giant electrocaloric effect in thin-film $PbZr_{0.95}Ti_{0.05}O_3$, *Science*, 311, pp. 1270–1271.
[10] Neese, B., Chu, B., Lu, S.-G., Wang, Y., Furman, E. and Zhang, Q. M. (2008). Large electrocaloric effect in ferroelectric polymers near room temperature, *Science*, 321, pp. 821–823.
[11] Liu, S. T. and Long, D. (1978). Pyroelectric detectors and materials, *Proc. IEEE*, 66, pp. 14–26.
[12] Kulwicki, B. M., Amin, A., Beratan, H. R. and Hanson, C. M. (1992). *ISAF 92: Proceedings of the Eighth IEEE International Symposium on Applications of Ferroelectrics*, pp. 1–10.
[13] Lee, M. H., Guo, R. and Bhalla, A. S. (1998). Pyroelectric sensors, *J. Electroceram.*, 2, pp. 229–242.
[14] Raevskaya, S. I., Zakharov, Y. N., Lutokhin, A. G., Emelyanov, A. S., Raevski, I. P., Panchelyuga, M. S., Titov, V. V. and Prosandeev, S. A. (2008). Critical nature of the giant field-induced pyroelectric response in $Pb(Mg_{1/3}Nb_{2/3})O_3$-$x$$PbTiO_3$ single crystals, *Appl. Phys. Lett.*, 93, pp. 042903.
[15] Alguero, M., Stewart, M., Cain, M. G., Ramos, P., Ricote, J. and Calzada, M. L. (2010). Properties of morphotropic phase boundary $Pb(Mg_{1/3}Nb_{2/3})O_3$-$PbTiO_3$ films with submicrometre range thickness on Si-based substrates, *J. Phys. D: Appl. Phys.*, 43, p. 205401.
[16] Nye, J. F. (1957). *Physical Properties of Crystals: Their Representation by Tensors and Matrices*, (Clarendon Press, Oxford).
[17] Newnham, R. E. (2005). *Properties of Materials: Anisotropy, Symmetry, Structure* (Oxford University Press, Oxford, New York).
[18] Lovinger, A. J. (1983). Ferroelectric polymers, *Science*, 220, pp. 1115–1121.
[19] Nalwa, H. S. (1995). *Ferroelectric Polymers: Chemistry, Physics, and Applications* (Marcel Dekker, New York).
[20] Zhang, Q. M. C. H., Xia, F. and Su, J. (2004). Electric EAP, in *Electroactive Polymer (EAP) Actuators as Artificial Muscles: Reality, Potential, and Challenges*, edited by Bar-Cohen, Y. (SPIE Press, Bellingham).
[21] Ploss, B. and Bauer, S. (1991). Characterization of materials for integrated pyroelectric sensors, *Sens. Actuators, A*, 26, pp. 407–411.
[22] Kim, J. H., Park, T. K., Lee, H. and Lee, D. J. (1995). Pyroelectricity in poly(vinylidene fluoride) and poly(vinylidene fluoride-co-tetrafluoroethylene) synthesized with various initiators, *Korea Polym. J.*, 3, pp. 101–105.
[23] Mehl, G. H., Nordmann, I., Lacey, D., Goodby, J. W., Hogg, J. H. C. and DasGupta, D. K. (1996). Pyroelectric and dielectric properties of side-chain liquid crystal polymers, *Polym. Eng. Sci.*, 36, pp. 1032–1037.

[24] Tang, Y., Luo, L., Jia, Y., Luo, H., Zhao, X., Xu, H., Lin, D., Sun, J., Meng, X., Zhu, J. and Es-Souni, M. (2006). Mn-doped $0.71Pb(Mg_{1/3}Nb_{2/3})O_3$-$0.29PbTiO_3$ pyroelectric crystals for uncooled infrared focal plane arrays applications, *Appl. Phys. Lett.*, 89, p. 162906.
[25] Wan, X. M., Tang, X. G., Wang, J., Chan, H. L. W., Choy, C. L. and Luo, H. S. (2004). Growth and pyroelectric property of 0.2 mol% Fe-doped $Pb(Mg_{1/3}Nb_{2/3})O_3$-$0.38PbTiO_3$ single crystals measured by a dynamic technique, *Appl. Phys. Lett.*, 84, pp. 4711–4713.
[26] Kumar, P., Sharma, S., Thakur, O. P., Prakash, C. and Goel, T. C. (2004). Dielectric, piezoelectric and pyroelectric properties of PMN-PT (68: 32) system, *Ceram. Int.*, 30, pp. 585–589.
[27] Davis, M., Damjanovic, D. and Setter, N. (2004). Pyroelectric properties of $(1-x)$ $Pb(Mg_{1/3}Nb_{2/3})O_3$-$xPbTiO_3$ and $(1-x)Pb(Zn_{1/3}Nb_{2/3})O_3$-$xPbTiO_3$ single crystals measured using a dynamic method, *J. Appl. Phys.*, 96, pp. 2811–2815.
[28] Sebald, G., Seveyrat, L., Guyomar, D., Lebrun, L., Guiffard, B. and Pruvost, S. (2006). Electrocaloric and pyroelectric properties of $0.75Pb(Mg_{1/3}Nb_{2/3})O_3$-$0.25PbTiO_3$ single crystals, *J. Appl. Phys.*, 100, p. 124112.
[29] Tang, Y. X., Wan, X. M., Zhao, X. Y., Pan, X. M., Lin, D., Luo, H. S., Sun, J. L., Meng, X. J. and Zhu, J. H. (2005). Large pyroelectric response in relaxor-based ferroelectric $(1-x)Pb(Mg_{1/3}Nb_{2/3})O_3$-$xPbTiO_3$ single crystals, *J. Appl. Phys.*, 98, p. 084104.
[30] Wang, J. T. and Zhang, C. (2005). Pyroelectric properties of $0.7PbMg_{1/3}(Nb_{0.9}Ta_{0.1})_{2/3}O_3$-$0.3PbTiO_3$ ceramics, *J. Appl. Phys.*, 98, p. 054103.
[31] Tang, Y. X., Zhao, X. Y., Wan, X. M., Feng, X. Q., Jin, W. Q. and Luo, H. S. (2005). Composition, dc bias and temperature dependence of pyroelectric properties of $\langle 111 \rangle$-oriented $(1-x)Pb(Mg_{1/3}Nb_{2/3})O_3$-$xPbTiO_3$ crystals, *Mater. Sci. Eng., B*, 119, pp. 71–74.
[32] Venet, M., Santos, I. A., Eiras, J. A. and Garcia, D. (2006). Potentiality of SBN textured ceramics for pyroelectric applications, *Solid State Ionics*, 177, pp. 589–593.
[33] Aleksandrov, S. E., Gavrilov, G. A., Kapralov, A. A., Smirnova, E. P., Sotnikova, G. Y. and Sotnikov, A. V. (2004). Relaxer ferroelectrics as promising materials for IR detectors, *Tech. Phys.*, 49, pp. 1176–1180.
[34] Liu, L., Wu, X., Zhao, X., Feng, X., Jing, W. and Luo, H. (2010). Pyroelectric performances of rhombohedral $0.42Pb(In_{1/2}Nb_{1/2})O_3$-$0.3Pb(Mg_{1/3}Nb_{2/3})O_3$-$0.28PbTiO_3$ single crystals, *IEEE Trans. Ultrason. Ferroelectr. Freq. Control*, 57, pp. 2154–2158.
[35] Seifert, A., Muralt, P. and Setter, N. (1998). High figure-of-merit porous $Pb_{1-x}Ca_xTiO_3$ thin films for pyroelectric applications, *Appl. Phys. Lett.*, 72, pp. 2409–2411.
[36] Zomorrodian, A. R., Lin, H., Wu, N. J., Huang, T. Q., Liu, D. and Ignatiev, A. (1996). High temperature polarization retention of a

$Pb(Zr_{0.52}Ti_{0.48})O_3/YBa_2Cu_3O_{7-x}$ memory cell, *Appl. Phys. Lett.*, 69, pp. 1789–1791.

[37] Sitar, Z., Gutmann, R., Pierhofer, H. and Gunter, P. (1995). Liquid phase epitaxy of $Na_{1-y}K_yTa_{1-x}Nb_xO_3$ for pyroelectric applications, *Mater. Res. Soc. Symp. Proc.*, pp. 589–594.

[38] Kohli, M. and Muralt, P. (1999). Poling of ferroelectric thin films, *Ferroelectrics*, 225, pp. 155–162.

[39] Watton, R. and Todd, M. A. (1991). Induced pyroelectricity in sputtered lead scandium tantalate films and their merit for IR detector arrays, *Ferroelectrics*, 118, pp. 279–295.

[40] Gebre, T., Batra, A. K., Guggilla, P., Aggarwal, M. D. and Lal, R. B. (2004). Pyroelectric properties of pure and doped lithium niobate crystals for infrared sensors, *Ferroelectr., Lett. Sect.*, 31, pp. 131–139.

[41] Bell, A., Huang, Y., Paul, O., Nemirovsky, Y. and Setter, N. (1995). A thin-film pyroelectric detector, *Integr. Ferroelectr.*, 6, pp. 231–240.

[42] Kobune, M., Fujii, S., Takayama, R. and Tomozawa, A. (1996). Preparation and pyroelectric properties of $(Pb,La)(Zr,Ti)O_3$ (PLZT) thin films, *Jpn. J. Appl. Phys.*, 35, pp. 4980–4983.

[43] Kobune, M., Ishito, H., Mineshige, A., Fujii, S., Takayama, R. and Tomozawa, A. (1998). Relationship between pyroelectric properties and electrode sizes in $(Pb,La)(Zr,Ti)O_3$ (PLZT) thin films, *Jpn. J. Appl. Phys.*, 37, pp. 5154–5157.

[44] Iijima, K., Tomita, Y., Takayama, R. and Ueda, I. (1986). Preparation of c-axis oriented pbtio3 thin-films and their crystallographic, dielectric, and pyroelectric properties, *J. Appl. Phys.*, 60, pp. 361–367.

[45] Liu, D. H. and Payne, D. A. (1995). Lower temperature crystallization and ordering in sol-gel derived $Pb(Sc_{0.5}Ta_{0.5})O_3$ powders and thin-layers, *J. Appl. Phys.*, 77, pp. 3361–3364.

[46] Wang, M., Fang, C. S. and Zhou, H. S. (1991). Study on the pyroelectric properties of TGS-PVDF composites, *Ferroelectrics*, 118, pp. 191–197.

[47] Chan, H. L. W., Chan, W. K., Zhang, Y. and Choy, C. L. (1998). Pyroelectric and piezoelectric properties of lead titanate/polyvinylidene fluoride-trifluoroethylene 0-3 composites, *IEEE Trans. Dielectr. Electr. Insul.*, 5, pp. 505–512.

[48] Dietze, M., Krause, J., Solterbeck, C. H. and Es-Souni, M. (2007). Thick film polymer-ceramic composites for pyroelectric applications, *J. Appl. Phys.*, 101, p. 054113.

[49] Lam, K. S., Wong, Y. W., Tai, L. S., Poon, Y. M. and Shin, F. G. (2004). Dielectric and pyroelectric properties of lead zirconate titanate/polyurethane composites, *J. Appl. Phys.*, 96, pp. 3896–3899.

[50] Fang, C. S., Wang, Q. W., Zhuo, H. S., Zhang, P. L. and Zhong, W. L. (1998). Preparation and pyroelectric properties of oriented composite ATGS-PVDF film, *J. Korean Phys. Soc.*, 32, pp. S1843–S1844.

[51] Batra, A. K., Corda, J., Guggilla, P., Aggarwal, M. D. and Edwards, M. E. (2009). Dielectric and pyroelectric properties of $LiTaO_3$: P(VDF-TrFE) composite films, *P. Soc. Photo-opt. Ins.*, 7213, p. 721313.
[52] Li, J. H., Yuan, N. G. and Chan, H. L. W. (2002). Preparation of PCLT/P(VDF-TrFE) pyroelectric sensor based on plastic film substrate, *Sens. Actuators. A*, 100, pp. 231–235.
[53] Wood, M. E. and Potter, W. H. (1985). General-analysis of magnetic refrigeration and its optimization using a new concept — maximization of refrigerant capacity, *Cryogenics*, 25, pp. 667–683.
[54] Sinyavsky, Y. V., Pashkov, N. D., Gorovoy, Y. M. and Lugansky, G. E. (1989). The optical ferroelectric ceramic as working body for electrocaloric refrigeration, *Ferroelectrics*, 90, pp. 213–217.
[55] Xiao, D. Q., Wang, Y. C., Zhang, R. L., Peng, S. Q., Zhu, J. G. and Yang, B. (1998). Electrocaloric properties of $(1-x)Pb(Mg_{1/3}Nb_{2/3})O_3-xPbTiO_3$ ferroelectric ceramics near room temperature, *Mater. Chem. Phys.*, 57, pp. 182–185.
[56] Furukawa, T. (1984). Phenomenological aspect of a ferroelectric vinylidene fluoride trifluoroethylene copolymer, *Ferroelectrics*, 57, pp. 63–72.
[57] Furukawa, T., Nakajima, T. and Takahashi, Y. (2006). Factors governing ferroelectric switching characteristics of thin VDF/TrFE copolymer films, *IEEE Trans. Dielectr. Electr. Insul.*, 13, pp. 1120–1131.
[58] Lu, S.-G. and Zhang, Q. (2009). Electrocaloric materials for solid-state refrigeration, *Adv. Mater.*, 21, pp. 1983–1987.
[59] Neese, B., Lu, S. G., Chu, B. and Zhang, Q. M. (2009). Electrocaloric effect of the relaxor ferroelectric poly(vinylidene fluoride-trifluoroethylene-chlorofluoroethylene) terpolymer, *Appl. Phys. Lett.*, 94, p. 042910.
[60] Dveyaharon, H. and Taylor, P. L. (1981). Thermodynamics of pyroelectricity and piezoelectricity in polymers, *Ferroelectrics*, 33, pp. 103–110.
[61] Nix, E. L., Nanayakkara, J., Davies, G. R. and Ward, I. M. (1988). Primary and secondary pyroelectricity in highly oriented polyvinylidene fluoride, *J. Polym. Sci., Part B: Polym. Phys.*, 26, pp. 127–140.
[62] Lu, S. G., Rozic, B., Zhang, Q. M., Kutnjak, Z. and Neese, B. (2011). Enhanced electrocaloric effect in ferroelectric poly(vinylidene-fluoride/trifluoroethylene) 55/45 mol% copolymer at ferroelectric–paraelectric transition, *Appl. Phys. Lett.*, 98, pp. 122906.
[63] Lu, S. G., Rozic, B., Zhang, Q. M., Kutnjak, Z., Li, X., Furman, E., Gorny, L. J., Lin, M., Malic, B., Kosec, M., Blinc, R. and Pirc, R. (2010). Organic and inorganic relaxor ferroelectrics with giant electrocaloric effect, *Appl. Phys. Lett.*, 97, p. 162904.
[64] Tashiro, K., Nishimura, S. and Kobayashi, M. (1990). Thermal contraction and ferroelectric phase-transition in vinylidene fluoride trifluoroethylene copolymers 2. An effect of tensile-stress applied in the direction perpendicular to the chain axis, *Macromolecules*, 23, pp. 2802–2806.

[65] Furukawa, T., Johnson, G. E., Bair, H. E., Tajitsu, Y., Chiba, A. and Fukada, E. (1981). Ferroelectric phase-transition in a co-polymer of vinylidene fluoride and trifluoroethylene, *Ferroelectrics*, 32, pp. 61–67.
[66] Zhang, Q. M., Bharti, V. and Zhao, X. (1998). Giant electrostriction and relaxor ferroelectric behavior in electron-irradiated poly(vinylidene fluoride-trifluoroethylene) copolymer, *Science*, 280, pp. 2101–2104.
[67] Cheng, Z. Y., Bharti, V., Xu, T. B., Wang, S. X., Zhang, Q. M., Ramotowski, T., Tito, F. and Ting, R. (1999). Transverse strain responses in electrostrictive poly(vinylidene fluoride-trifluoroethylene) films and development of a dilatometer for the measurement, *J. Appl. Phys.*, 86, pp. 2208–2214.
[68] Tajitsu, Y., Chiba, A., Furukawa, T., Date, M. and Fukada, E. (1980). Crystalline phase-transition in the co-polymer of vinylidenefluoride and trifluoroethylene, *Appl. Phys. Lett.*, 36, pp. 286–288.
[69] Cheng, Z. Y., Xu, T. B., Bharti, V., Wang, S. X. and Zhang, Q. M. (1999). Transverse strain responses in the electrostrictive poly(vinylidene fluoride-trifluorethylene) copolymer, *Appl. Phys. Lett.*, 74, pp. 1901–1903.
[70] Lu, S. G., Rozic, B., Zhang, Q. M., Kutnjak, Z. and Pirc, R. (2012). Electrocaloric effect in ferroelectric polymers, *Appl. Phys. A: Mater. Sci. Process.*, 107, pp. 559–566.
[71] Thacher, P. D. (1968). Electrocaloric effects in some ferroelectric and antiferroelectric Pb(Zr,Ti)O_3 compounds, *J. Appl. Phys.*, 39, pp. 1996–2002.
[72] Lu, S. G., Rozic, B., Zhang, Q. M., Kutnjak, Z., Pirc, R., Lin, M., Li, X. and Gorny, L. (2010). Comparison of directly and indirectly measured electrocaloric effect in relaxor ferroelectric polymers, *Appl. Phys. Lett.*, 97, p. 202901.
[73] Li, X., Qian, X. S., Lu, S. G., Cheng, J., Fang, Z. and Zhang, Q. M. (2011). Tunable temperature dependence of electrocaloric effect in ferroelectric relaxor poly(vinylidene fluoride-trifluoroethylene-chlorofluoroethylene terpolymer, *Appl. Phys. Lett.*, 99, p. 052907.
[74] Kar-Narayan, S. and Mathur, N. D. (2010). Direct and indirect electrocaloric measurements using multilayer capacitors, *J. Phys. D: App. Phys.*, 43, pp. 032002.
[75] Hasegawa, R., Takahashi, Y., Tadokoro, H. and Chatani, Y. (1972). Crystal-structures of three crystalline forms of poly(vinylidene fluoride), *Polym. J.*, 3, pp. 600–610.
[76] Bachmann, M. A. and Lando, J. B. (1981). A reexamination of the crystal-structure of phase-ii of poly(vinylidene fluoride), *Macromolecules*, 14, pp. 40–46.
[77] Xia, F., Cheng, Z. Y., Xu, H. S., Li, H. F., Zhang, Q. M., Kavarnos, G. J., Ting, R. Y., Abdul-Sedat, G. and Belfield, K. D. (2002). High electromechanical responses in a poly(vinylidene fluoride-trifluoroethylene-chlorofluoroethylene) terpolymer, *Adv. Mater.*, 14, pp. 1574–1577.

[78] Pirc, R., Kutnjak, Z., Blinc, R. and Zhang, Q. M. (2011). Upper bounds on the electrocaloric effect in polar solids, *Appl. Phys. Lett.*, 98, p. 021909.
[79] Chu, B., Lin, M., Neese, B., Zhou, X., Chen, Q. and Zhang, Q. M. (2007). Large enhancement in polarization response and energy density of poly(vinylidene fluoride-trifluoroethylene-chlorofluoroethylene) by interface effect in nanocomposites, *Appl. Phys. Lett.*, 91, pp. 122909.
[80] Li, J. J., Seok, S. I., Chu, B., Dogan, F., Zhang, Q. and Wang, Q. (2009). Nanocomposites of ferroelectric polymers with TiO_2 nanoparticles exhibiting significantly enhanced electrical energy density, *Adv. Mater.*, 21, pp. 217–221.
[81] Chu, B., Neese, B., Lin, M., Lu, S. G. and Zhang, Q. M. (2008). Enhancement of dielectric energy density in the poly(vinylidene fluoride)-based terpolymer/copolymer blends, *Appl. Phys. Lett.*, 93, p. 152903.
[82] Chen, X. Z., Qian, X. S., Li, X., Lu, S. G., Gu, H.-m., Lin, M., Shen, Q. D. and Zhang, Q. M. (2012). Enhanced electrocaloric effect in poly(vinylidene fluoride-trifluoroethylene)-based terpolymer/copolymer blends, *Appl. Phys. Lett.*, 100, p. 222902.
[83] Radebaugh, R., Lawless, W. N., Siegwarth, J. D. and Morrow, A. J. (1979). Feasibility of electrocaloric refrigeration for the 4-15-k temperature-range, *Cryogenics*, 19, pp. 187–208.
[84] Hegenbarth, E. (1961). Studies of the electrocaloric effect of ferroelectric-ceramics at low temperatures, *Cryogenics*, 1, pp. 242–243.
[85] Kikuchi, A. and Sawaguchi, E. (1964). Electrocaloric effect in $SrTiO_3$, *J. Phys. Soc. Jpn.*, 19, pp. 1497–1498.
[86] Lawless, W. N. (1977). Specific-heat and electro-caloric properties of $ktao_3$ at low-temperatures, *Phys. Rev. B: Soild State*, 16, pp. 433–439.
[87] Shvartsman, V. V., Aksenov, S. E. and Politova, E. D. (2001). Phase relations and ferroelectric properties of $Pb(Zr, Sn, Ti)O_3$ ceramics, *Ferroelectrics*, 258, pp. 13–20.
[88] Olsen, R. B., Butler, W. F., Payne, D. A., Tuttle, B. A. and Held, P. C. (1980). Observation of a polarocaloric (electrocaloric) effect of 2°C in lead zirconate modified with Sn^{4+} and Ti^{4+}, *Phys. Rev. Lett.*, 45, pp. 1436–1438.
[89] Sternberg, A., Birks, E., Shebanovs, L., Klotins, E., Ozolinsh, M., Tyunina, M., Zauls, V. and Kundzinsh, M. (1999). Phase transitions and properties of perovskite ferroelectric ceramics and films for certain applications, *Ferroelectrics*, 226, pp. 217–241.
[90] Tuttle, B. A. and Payne, D. A. (1981). The effects of microstructure on the electrocaloric properties of $Pb(Zr, Sn, Ti)O_3$ ceramics, *Ferroelectrics*, 37, pp. 603–606.
[91] Hao, X., Yue, Z., Xu, J., An, S. and Nan, C.-W. (2011). Energy-storage performance and electrocaloric effect in (100)-oriented $Pb_{0.97}La_{0.02}(Zr_{0.95}Ti_{0.05})O_3$ antiferroelectric thick films, *J. Appl. Phys.*, 110, p. 064109.
[92] Rozic, B., Malic, B., Ursic, H., Holc, J., Kosec, M. and Kutnjak, Z. (2011). Direct measurements of the electrocaloric effect in bulk $PbMg_{1/3}Nb_{2/3}O_3$ (PMN) ceramics, *Ferroelectrics*, 421, pp. 103–107.

[93] Sternberg, A., Shebanovs, L., Zauls, V. and Kundzins, K. (2003). Ceramics and thin films of some new $Pb(B^{3+},Nb)TiO_3$-$PbTiO_3$ systems, *Ferroelectrics*, 286, pp. 1049–1056.

[94] Guyomar, D., Sebald, G., Guiffard, B. and Seveyrat, L. (2006). Ferroelectric electrocaloric conversion in $0.75(PbMg_{1/3}Nb_{2/3}O_3)$-$0.25(PbTiO_3)$ ceramics, *J. Phys. D: Appl. Phys.*, 39, pp. 4491–4496.

[95] Sebald, G., Pruvost, S., Seveyrat, L., Lebrun, L., Guyomar, D. and Guiffard, B. (2007). Electrocaloric properties of high dielectric constant ferroelectric ceramics, *J. Eur. Ceram. Soc.*, 27, pp. 4021–4024.

[96] Sebald, G., Guiffard, B., Seveyrat, L., Benayad, A., Lebrun, L., Pruvost, S. and Guyomar, D. (2007). *2006 15th IEEE International Symposium on Applications of Ferroelectrics* (*IEEE*, New York), pp. 190–193.

[97] Guyomar, D. and Sebald, G. (2009). Pyroelectric/electrocaloric energy scavenging and cooling capabilities in ferroelectric materials, *Int. J. Appl. Electrom.*, 31, pp. 41–46.

[98] Hagberg, J., Uusimaki, A. and Jantunen, H. (2008). Electrocaloric characteristics in reactive sintered $0.87Pb(Mg_{1/3}Nb_{2/3})O_3$-$0.13PbTiO_3$, *Appl. Phys. Lett.*, 92, p. 132909.

[99] Birks, E. H. (1986). The electrocaloric effect in $Pb(Sc_{0.5}Nb_{0.5})O_3$ ceramic, *Phys. Status Solidi A*, 94, pp. 523–527.

[100] Shebanov, L. A., Birks, E. H. and Borman, K. J. (1989). X-ray studies of electrocaloric lead-scandium tantalate ordered solid-solutions, *Ferroelectrics*, 90, pp. 165–172.

[101] Lin, G. C., Xiong, X. M., Zhang, J. X. and Wei, Q. (2005). Latent heat study of phase transition in $Ba_{0.73}Sr_{0.27}TiO_3$ induced by electric field, *J. Therm. Anal. Calorim.*, 81, pp. 41–44.

[102] Bai, Y., Zheng, G.-P. and Shi, S.-Q. (2011). Abnormal electrocaloric effect of $Na_{0.5}Bi_{0.5}TiO_3$-$BaTiO_3$ lead-free ferroelectric ceramics above room temperature, *Mater. Res. Bull.*, 46, pp. 1866–1869.

[103] Zampiere, R. B., Dias, G. S., Cotica, L. F. and Santos, I. A. (2012). Enhanced ferroism in mechanically processed and environmentally friendly $Ba_{0.30}Na_{0.70}Ti_{0.30}Nb_{0.70}O_3$ ceramics, *Scr. Mater.*, 66, pp. 542–545.

[104] Shebanovs, L., Borman, K., Lawless, W. N. and Kalvane, A. (2002). Electrocaloric effect in some perovskite ferroelectric ceramics and multilayer capacitors, *Ferroelectrics*, 273, pp. 2515–2520.

[105] Karmanenko, S. F., Pakhomov, O. V., Prudan, A. M., Starkov, A. S. and Eskov, A. (2007). Layered ceramic structure based on the electrocaloric elements working as a solid state cooling line, *J. Eur. Ceram. Soc.*, 27, pp. 3109–3112.

[106] Bai, Y., Zheng, G. P. and Shi, S. Q. (2010). Kinetic electrocaloric effect and giant net cooling of lead-free ferroelectric refrigerants, *J. Appl. Phys.*, 108, p. 104102.

[107] Bai, Y., Zheng, G. and Shi, S. (2010). Direct measurement of giant electrocaloric effect in $BaTiO_3$ multilayer thick film structure beyond theoretical prediction, *Appl. Phys. Lett.*, 96, p. 192902.

[108] Bai, Y., Zheng, G. P., Ding, K., Qiao, L., Shi, S. Q. and Guo, D. (2011). The giant electrocaloric effect and high effective cooling power near room temperature for $BaTiO_3$ thick film, *J. Appl. Phys.*, 110, p. 094103.

[109] Parui, J. and Krupanidhi, S. B. (2008). Electrocaloric effect in antiferroelectric $PbZrO_3$ thin films, *Phys. Status Solidi A*, 2, pp. 230–232.

[110] Correia, T. M., Young, J. S., Whatmore, R. W., Scott, J. F., Mathur, N. D. and Zhang, Q. (2009). Investigation of the electrocaloric effect in a $PbMg_{2/3}Nb_{1/3}O_3$-$PbTiO_3$ relaxor thin film, *Appl. Phys. Lett.*, 95, p. 182904.

[111] Mischenko, A. S., Zhang, Q., Whatmore, R. W., Scott, J. F. and Mathur, N. D. (2006). Giant electrocaloric effect in the thin film relaxor ferroelectric $0.9PbMg_{1/3}Nb_{2/3}O_3$-$0.1PbTiO_3$ near room temperature, *Appl. Phys. Lett.*, 89, p. 242912.

[112] Feng, Z., Shi, D. and Dou, S. (2011). Large electrocaloric effect in highly (001)-oriented $0.67PbMg_{1/3}Nb_{2/3}O_3$-$0.33PbTiO_3$ thin films, *Solid State Commun.*, 151, pp. 123–126.

[113] Feng, Z., Shi, D., Zeng, R. and Dou, S. (2011). Large electrocaloric effect of highly (100)-oriented $0.68PbMg_{1/3}Nb_{2/3}O_3$-$0.32PbTiO_3$ thin films with a $Pb(Zr_{0.3}Ti_{0.7})O_3/PbO_x$ buffer layer, *Thin Solid Films*, 519, pp. 5433–5436.

[114] Saranya, D., Chaudhuri, A. R., Parui, J. and Krupanidhi, S. B. (2009). Electrocaloric effect of PMN-PT thin films near morphotropic phase boundary, *Bull. Mater. Sci.*, 32, pp. 259–262.

[115] He, Y., Li, X. M., Gao, X. D., Leng, X. and Wang, W. (2011). Enhanced electrocaloric properties of PMN-PT thin films with LSCO buffer layers, *Funct. Mater. Lett.*, 4, pp. 45–48.

[116] Correia, T. M., Kar-Narayan, S., Young, J. S., Scott, J. F., Mathur, N. D., Whatmore, R. W. and Zhang, Q. (2011). PST thin films for electrocaloric coolers, *J. Phys. D: Appl. Phys.*, 44, p. 165407.

[117] Chen, H., Ren, T. L., Wu, X. M., Yang, Y. and Liu, L. T. (2009). Giant electrocaloric effect in lead-free thin film of strontium bismuth tantalite, *Appl. Phys. Lett.*, 94, p. 182902.

[118] Liu, P. F., Wang, J. L., Meng, X. J., Yang, J., Dkhil, B. and Chu, J. H. (2010). Huge electrocaloric effect in Langmuir–Blodgett ferroelectric polymer thin films, *New J. Phys.*, 12, p. 023035.

[119] Shvartsman, V. V. and Politova, E. D. (2001). Thermal behaviour of polarization in relaxor ferroelectrics $Pb(Mg_{1/2}W_{1/2})_{(1-x)}Ti_xO_3$, *Ferroelectrics*, 257, pp. 75–84.

[120] Chukka, R., Cheah, J. W., Chen, Z., Yang, P., Shannigrahi, S., Wang, J. and Chen, L. (2011). Enhanced cooling capacities of ferroelectric materials at morphotropic phase boundaries, *Appl. Phys. Lett.*, 98, p. 242902.

Chapter 8

Development of Polymer-based 0–3 Composites with High Dielectric Constant*

Lin Zhang and Z.-Y. Cheng[†]

Materials Research and Education Center,
Auburn University, Auburn, AL 36849, USA
[†] *chengzh@eng.auburn.edu*

Composites have been extensively studied for dielectric and related applications. This is a review of polymer-based 0–3 composites that exhibit a high dielectric constant. These composites are classified into two types: dielectric–dielectric composite and conductor–dielectric composite. The physical principles and related models are presented with associated assumptions and approximations. In general, a dielectric–dielectric composite needs a higher concentration of the fillers to reach a high dielectric constant than a conductor–dielectric composite. The high dielectric constant observed in the conductor–dielectric composites is usually associated with a high dielectric loss and a low electric breakdown field. The experimental results are summarized to illustrate the principles for, and the achievements in, the development of these composites. The challenges facing the fundamental understanding and the further development of these composites for different applications are discussed.

Keywords: Dielectric property; dielectric constant; composite; 0–3 composite; percolation.

---

*This article was firstly published as "Zhang, L. and Cheng, Z.-Y. (2011). Development of polymer-based 0–3 composites with high dielectric constant, *J. Adv. Dielect.* 1, pp. 389–406". Reprinted with the permission of World Scientific Publishing Company.

1. Introduction

Dielectrics are widely used in electronics, electrical devices, and systems [1–6]. The performance of a dielectric material is characterized by its complex permittivity including dielectric constant (k or ε) and dielectric loss ($\tan\delta$) for the dielectrics used under weak electric field and by its polarization–electric field (P–E) relationship and its breakdown field (i.e., dielectric strength, E_b) for the dielectrics used under a high electric field. Based on the application, there are different needs for dielectric materials [6, 7]. For some applications, such as IC insulation materials, dielectrics with a low dielectric constant (i.e., low-k materials) are highly desirable, while for some other applications, such as gate dielectrics in field-effect transistors, dielectrics with a high dielectric constant are preferred [8]. Dielectrics used under a high electric field generally require a high breakdown field with a low loss at high field. Various dielectrics such as gases, liquids, and solids are widely used in the current industry for different applications based on their different and unique dielectric properties. For example, gas/air is used due to its low dielectric constant and minimal dielectric loss in addition to its extremely small temperature and frequency dependence [9–13]. With regard to solid dielectrics, both polymers and inorganic compounds, such as ceramics and crystals, are common materials. In general, polymers have attributes such as low process temperature, flexibility, the ability to withstand a high mechanical impact, and exhibit a high breakdown field and a small dielectric constant (mostly less than 5). Ceramics and crystals require a high process temperature, are brittle, and have a high dielectric constant with a low breakdown field.

To meet the different needs, composites have been widely studied in order to combine the unique properties of different materials [6, 13, 14]. For example, plastic foam has been studied and used as dielectric material in order to utilize the advantages of gas as dielectrics [15]. Dielectric materials, which are flexible with a low process temperature and possess a high dielectric constant and a high breakdown field are required for many applications such as capacitors and electronic packages. Therefore, a great deal of efforts has been given to the development of the polymer-based

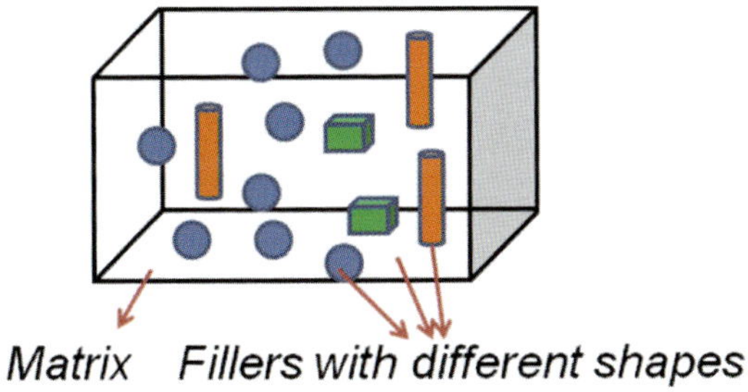

Figure 1. Schematic of 0–3 composites.

composites as dielectrics, especially 0–3 composites, as shown in Fig. 1, in which the flexible polymer matrix is filled with isolated powders/particles of inorganic compounds [16–20]. The composites with a high dielectric constant can be 0–3 composites, which have been thoroughly investigated, are flexible, easy to fabricate, and exhibit a low acoustic impedance. It should be pointed out that dielectric composites have been studied and utilized in the development of high performance electromechanical devices, such as transducer arrays. In this type of study, the electromechanical effect, such as piezoelectric and electrostriction [21, 22], is the key. In these composites, dielectrics with high electromechanical performance are used as its constitute(s). Electromechanical effect is harmful and therefore should be eliminated in the case of pure dielectric applications such as capacitors. This chapter is a review of high-dielectric-constant composites for dielectric applications. Research achievements and progresses in the development of electromechanical composites and/or devices are not included.

Based on the fillers used, the polymer-based 0–3 composites with a high dielectric constant can be classified into two types: One is dielectric–dielectric composites in which the fillers are dielectric materials; and the other is conductor–dielectric composites where the fillers are conducting materials [14, 23].

For dielectric–dielectric composites, it is extremely important to find fillers with a high dielectric constant. It is well known that most inorganic materials exhibit a dielectric constant ranging from a few to hundreds, while polar dielectrics, such as ferroelectric ceramics/crystals and relaxors, exhibit a dielectric constant as high as 10^3–10^4. Therefore, various ferroelectric-based ceramics have been widely used in the development of dielectric–dielectric composites [24–26]. It should be mentioned, however, that the dielectric constant of these polar dielectrics is very temperature dependent and these polar dielectrics exhibit a strong electromechanical effect, both of which are not desirable for many dielectric applications [27–29]. Ceramics with a giant dielectric constant reaching as high as 10^4 even 10^5, have been developed during last decade [30, 31]. These ceramics have a weak electromechanical effect and their dielectric constants exhibit a weak temperature dependence over a broad temperature range. These characteristics are necessary for the development of advanced dielectrics, including composites [32].

The conductor-dielectric composites are based on the percolation theory [23]. When the filler content is smaller than a certain value (percolation threshold), the composite is an insulator/dielectric and its dielectric constant increases with increasing filler content. As the filler content approaches the percolation threshold, the dielectric constant reaches a very high value. If the filler content is higher than the percolation threshold, the composite is a conductor. Therefore, the percolation threshold is very critical for the conductor–dielectric composites. The percolation threshold is dependent on the geometry, shape, size, and distribution of the conducting fillers [33, 34]. Conducting fillers with various shapes (spherical, core–shell, tube, and bar/wire-like) ranging in size from micrometers to nanometers have been studied [35–38]. Semiconductors have also been used as fillers [39–41].

Polymer matrices play a key role in the performance of composites. Various polymers have been used based on their processibility, flexibility, dielectric response, dielectric strength, melting temperature, and glass transition temperature. Examples of these are epoxy, poly(methyl methacrylate) (PMMA), polystyrene, pol(vinyl chloride) (PVC), polyetherimide, polyurethances (PU), poly(vinylidene fluoride) (PVDF), and PVDF-based co/ter-polymers. Among these polymers, PVDF and its co/ter-polymer have been widely used as polymer matrix since these polymers exhibit a high dielectric constant (>10) [21, 42–50].

Understanding the dielectric response of a composite (or heterogeneous dielectric) has been a topic of interest in fundamental research and applications for quite some time. Various models/formulas have been introduced simulating the dielectric constant of a composite using the dielectric property of its constituents. Based upon mathematical analogy, results are deemed valid for dielectric constant, electric conductivity, heat conductivity, and diffusivity of such materials. Many of the models/formulas presented below regarding the calculation of the dielectric property of a composite were originally developed based on the calculation of the conductivity or diffusivity of a mixture.

The achievements accomplished in the development of polymer-based 0–3 composites with a high dielectric constant are summarized here. The models/formulas introduced to describe the dielectric properties of composites are presented. Issues and challenges regarding the understanding of these composites and the development of high-performance composites are discussed.

2. Dielectric–Dielectric composites

2.1. *Model and simulation*

2.1.1. *Dielectric constant*

The simulation of the dielectric properties of a mixture of different dielectric materials is a problem of recurring interest and one of considerable theoretical and practical importance. Two well-known simplified models/formulas: Parallel and series models, shown in Fig. 2, where two dielectrics are used. The effective dielectric constant $\varepsilon_{\mathrm{eff}}$ of the mixture can be simply written as:

$$\varepsilon_{\mathrm{eff}}^{n} = v_1\varepsilon_1^{n} + v_2\varepsilon_2^{n}, \tag{1}$$

where v_1 and v_2 ($v_1 + v_2 = 1$) are the volume fraction of the dielectric-1 and dielectric-2, respectively, ε_1 and ε_2 are the dielectric constant of dielectric-1 and dielectric-2, respectively, and the n is

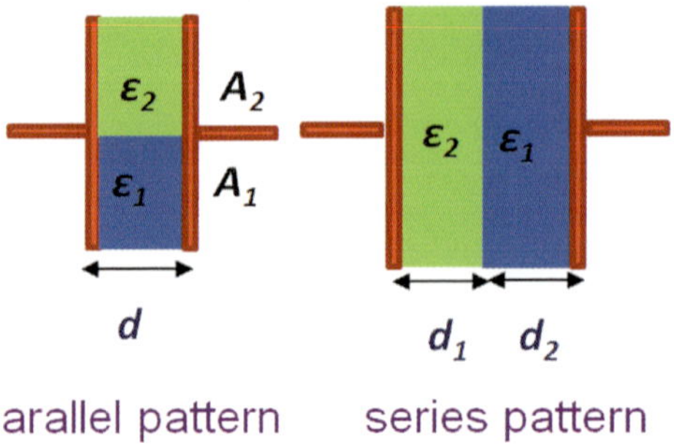

Figure 2. Schematic of structural patterns for parallel and series model.

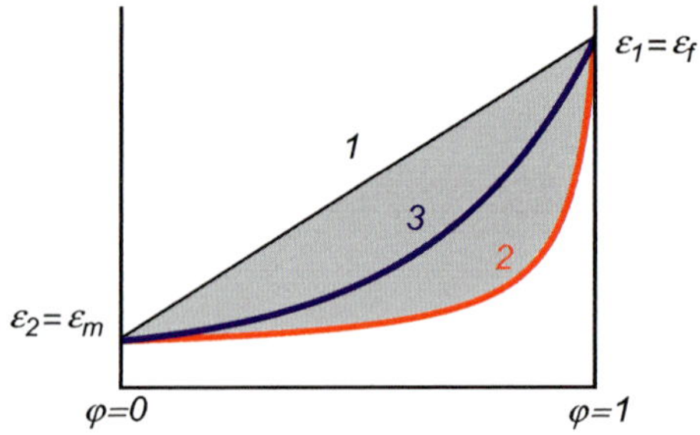

Figure 3. Schematic of effective dielectric constant of a mixture/composite versus the composition: 1-parallel connection; 2-series connection; 3-real composite (Lichtenecker's logarithmic mixing law).

either +1 for parallel case or −1 for series case. As shown in Fig. 3, the dielectric constant obtained for the parallel case is always higher than that obtained for series case.

Although the calculation using Eq. (1) does not represent the dielectric constant of a real composite, it is believed and widely accepted that the dielectric constant of a composite/mixture made of two dielectrics should be somewhere between the ε_{eff} of a composite made of two dielectrics should follow:

$$(v_1 \cdot \varepsilon_1^{-1} + v_2 \cdot \varepsilon_2^{-1})^{-1} \leq \varepsilon_{\text{eff}} \leq v_1 \cdot \varepsilon_1 + v_2 \cdot \varepsilon_2. \tag{2}$$

This is the Wiener limits introduced by Wiener in 1921.

Lichtenecker recognized the Wiener limits and proposed a formula for a composite [51]:

$$\varepsilon_{\text{eff}} = \frac{(v_1 \cdot \varepsilon_1 + v_2 \cdot \varepsilon_2)^u}{(v_1 \cdot \varepsilon_1^{-1} + v_2 \cdot \varepsilon_2^{-1})^{1-u}}, \tag{3}$$

where u is the relative weight of parallel connection and "$1 - u$" is the relative weight of series connection. The limits in the dielectric response of a mixture are important for the development of new dielectrics. Varies theorems have been established and applied to determine the limits in dielectric response of a mixture. Hashin and Shtrikman [52] given the following limits for a 0–3 composite:

$$\varepsilon_1 + v_2 \left(\frac{1}{\varepsilon_2 - \varepsilon_1} + \frac{v_1}{3\varepsilon_1} \right)^{-1} \leq \varepsilon_{\text{eff}} \leq \varepsilon_2 + v_1 \left(\frac{1}{\varepsilon_1 - \varepsilon_2} + \frac{v_2}{3\varepsilon_2} \right)^{-1}. \tag{4}$$

Lichtenecker treated composites using statistics and proposed [51]:

$$\log \varepsilon_{\text{eff}} = v_1 \log \varepsilon_1 + v_2 \log \varepsilon_1 \quad \text{or} \quad \varepsilon_{\text{eff}} = \varepsilon_1^{v_1} \cdot \varepsilon_2^{v_2}. \tag{5}$$

This is the Lichtenecker's logarithmic mixing law, which is a well-known equation for calculating the effective dielectric constant of a composite. It was also recognized by Lichtenecker that Eqs. (3) and (5) have the same result when $v_1 = v_2 = 1/2$ and $u = 1/2$.

Lichtenecker's logarithmic mixing law has been used in various literatures and generally considered as a quasi-empirical formula. Using Maxwell's equation, Eq. (5) was obtained through theory derivation recently [53]. The assumption used in derivation was that the particle shape and orientation of each constituent is random so that the charge density within a constituent can be replaced by the mean charge density of the mixture. In summary, Lichtenecker's logarithmic law can only be used for mixing in which the spatial distribution of shapes and orientations of the inclusions can be treated as statistically random. The success (or failure) of Lichtenecker's logarithmic law in predicting the effective dielectric constant of a dielectric–dielectric composite is dependent on how well this condition is satisfied [53].

In order to simulate the dielectric constant of a real composite, the polarization response to an electric field of a region that is filled with different dielectrics in some arbitrary way has to be calculated. This is an impossible task even if the detailed information about the dielectric constant and spatial distribution of each constitute is given [54]. Therefore, various models have been introduced to simulate the composites and many formulas have been proposed. Some of these formulas are purely empirical, while others are based on simplified structure/microstructure of a composite with some approximations in its derivation. In all of these formulas, the ε_{eff} of a composite is expressed as a function of composite's composition (i.e., the volume fraction of fillers, φ), dielectric constant ε_m of the matrix, the dielectric constant ε_f of the filler materials, and sometimes parameters related to the filler particles, such as shape and orientation. The following are the formulas introduced for 0–3 composites.

Model 1: Power Laws

Equation (1) has been extended as a general formula with the n as a variable related to the composite. Varies n values/expressions have been introduced. For example, n = 1/2 (refractive index model) was proposed based on the study of refractive index [55]. And n = 1/3 was derived using different approaches [56, 57]. Based on the results obtained using Monte Carlo simulation and finite element methods for spherical particles randomly dispersed in a matrix, it was suggested that $n = \varphi - \varphi_0$ [58]. Here φ_0 is the critical volume fraction at the point where the curves of the dielectric constant predicted by the new equation and the logarithmic mixing rule intercept each other. The simulation results also indicate that $\varphi_0 \approx 0.35$. Using the results obtained from the simulation based on Maxwell's equation, it was proposed $1.55\varphi + 0.25 \leq n \leq 1.65\varphi + 0.27$ for $\varphi < 0.25$ [59]. Another modification was proposed as [60]

$$\varepsilon_{\text{eff}}^n = z \cdot \varphi_1 \varepsilon_1^n + (1 - z \cdot \varphi_1)\varepsilon_2^n, \tag{6}$$

where z is a new correction factor, which is dependent on frequency, and $n = A\varphi + B$ with both A and B are constants and both dependent on frequency.

Model 2: Modified Logarithmic Law

Modifications were introduced soon after Lichtenecker's logarithmic mixing law was proposed. One of them is

$$\log \varepsilon_{\text{eff}} = (1-u\varphi)(1-\varphi)\log \varepsilon_m + (1+\text{u}(1-\varphi))\varphi \log \varepsilon_f$$
$$\text{or } \varepsilon_{\text{eff}} = \varepsilon_m^{(1-\varphi)(1-u\varphi)} \cdot \varepsilon_f^{\varphi(1+u(1-\varphi))}, \tag{7}$$

where u is a parameter introduced to modify the Lichtenecker's logarithmic mixing law. It was believed that the Lichtenecker's logarithmic mixing law Eq. (5) is good for the ratio of $\varepsilon_f/\varepsilon_m$ less than 4, while the modified mixture law Eq. (7) extends the useful range of the ratio of $\varepsilon_f/\varepsilon_m$ less than 10. However, the value of u is unknown. Later on, Bruggeman showed that the value of u is dependent on both ε_m and ε_f as $u = \frac{3}{2}\frac{\varepsilon_m - \varepsilon_m}{(2\sqrt{\varepsilon_m}+\sqrt{\varepsilon_m})(\sqrt{\varepsilon_f}+2\sqrt{\varepsilon_m})}$ [61].

Another one is

$$\log \varepsilon_{\text{eff}} = \log \varepsilon_m + (1-u)\varphi \log\left(\frac{\varepsilon_f}{\varepsilon_m}\right) \text{ or } \varepsilon_{\text{eff}} = \varepsilon_m^{1-\varphi} \cdot \varepsilon_f^{\varphi} \cdot \left(\frac{\varepsilon_m}{\varepsilon_f}\right)^{u\varphi}. \tag{8}$$

This is frequently used in literatures for 0–3 composite with the parameter u representing the shape of the fillers particles. It was experimentally found that the value of u is about 0.3 for most well dispersed ceramic–polymer composites. It should be mentioned that Eq. (8) is used for 0–3 composites, in which the volume fraction of the polymer matrix cannot be zero. Actually, it is apparent to find that Eq. (8) produces an unreasonable results for $\varphi = 1$.

Model 3: Rayleigh Models

Actually, the ε_{eff} of a 0–3 composite was first tackled by Rayleigh [62]. The composite was treated as a medium in which dielectric spheres in rectangular order are embedded into another dielectric. The calculation resulted in an infinite series, where each term represents an order of approximation. For the simplest case, the dielectric spheres in cube order are embedded in air ($\varepsilon_m = 1$), Rayleigh's calculation showed that the ε_{eff} for a moderate value of φ can be approximately calculated by

$$\varepsilon_{\text{eff}} = 1 + \frac{3\varphi}{\frac{\varepsilon_f+2}{\varepsilon_f-1} - \varphi - 1.65\frac{\varepsilon_f-1}{\varepsilon_f+4/3}\varphi^{10/3}}. \tag{9}$$

For a small value of φ, it becomes

$$\frac{\varepsilon_{\text{eff}} - 1}{\varepsilon_{\text{eff}} + 2} = \varphi \frac{\varepsilon_f - 1}{\varepsilon_f + 2}. \tag{10}$$

If the air is replaced with a dielectric with a dielectric constant ε_m, Eqs. (9) and (10) become:

$$\varepsilon_{\text{eff}} = 1 + \frac{3\varphi}{\frac{\varepsilon_f + 2\varepsilon_m}{\varepsilon_f - \varepsilon_m} - \varphi - 1.65 \frac{\varepsilon_f - \varepsilon_m}{\varepsilon_f + 4/3(\varepsilon_m)} \varphi^{10/3}}, \tag{11}$$

$$\frac{\varepsilon_{\text{eff}} - \varepsilon_m}{\varepsilon_{\text{eff}} + 2\varepsilon_m} = \varphi \frac{\varepsilon_f - \varepsilon_m}{\varepsilon_f + 2\varepsilon_m} (\varphi \ll 1). \tag{12}$$

If dielectric cylinders in a rectangular order are embedded in a dielectric, the ε_{eff} perpendicular to cylinder axes and along one side of the rectangular structure is approximately by Rayleigh as [62]

$$\varepsilon_{\text{eff}} = \varepsilon_m \left\{ 1 - 2\varphi \left[\frac{\varepsilon_m + \varepsilon_f}{\varepsilon_m - \varepsilon_f} + \varphi - 0.3058 \frac{\varepsilon_m - \varepsilon_f}{\varepsilon_m + \varepsilon_f} \varphi^4 \right]^{-1} \right\}. \tag{13}$$

For a small value of φ, this result would become

$$\varepsilon_{\text{eff}} - \varepsilon_m = 2\varphi\varepsilon_m \frac{\varepsilon_f - \varepsilon_m}{\varepsilon_f + \varepsilon_m} \text{ or } \varepsilon_{\text{eff}} + \varepsilon_m = 2\varepsilon_m \left[1 + \varphi \frac{\varepsilon_f - \varepsilon_m}{\varepsilon_f + \varepsilon_m} \right]. \tag{14}$$

A further approximation is obtained as

$$\frac{\varepsilon_{\text{eff}} - \varepsilon_m}{\varepsilon_{\text{eff}} + \varepsilon_m} = \varphi \frac{\varepsilon_f - \varepsilon_m}{\varepsilon_f + \varepsilon_m}. \tag{15}$$

This is the so-called Rayleigh cylinder formula for a small value of φ.

Considering Eqs. (10) and (15), Wiener proposed

$$\frac{\varepsilon_{\text{eff}} - \varepsilon_m}{\varepsilon_{\text{eff}} + u\varepsilon_m} = \varphi \frac{\varepsilon_f - \varepsilon_m}{\varepsilon_f + u\varepsilon_m}, \tag{16}$$

where $u(0 < u < \infty)$ is named as "form number." It was assumed that the u is dependent on the shape and size of the filler particles, but not the value of φ. It was also proposed that the value of u is also dependence on the ε_m and ε_f [63]. However, some experimental results suggested that the u increases with φ.

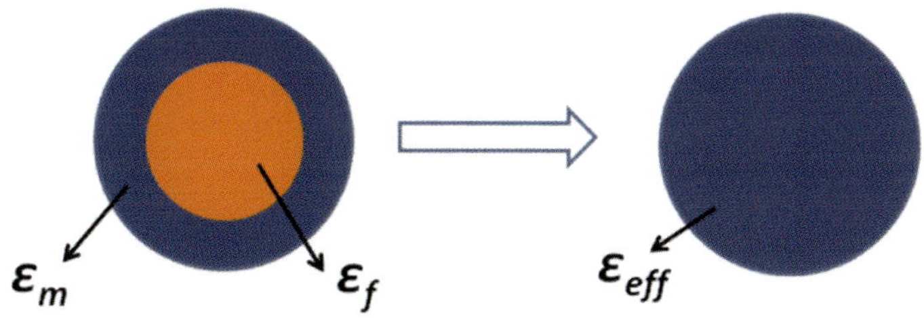

Figure 4. Schematic of Maxwell–Wagner equation.

Model 4: Maxwell–Wagner equation [54, 64, 65]

If filler particles are spherical in shape, the composite can be treated as a dielectric sphere (filler) surrounded by a concentric spherical shell (matrix) as shown in Fig. 4. The following equation is obtained:

$$\varepsilon_{\text{eff}} = \varepsilon_m \frac{2\varepsilon_m + \varepsilon_f + 2\varphi(\varepsilon_f - \varepsilon_m)}{2\varepsilon_m + \varepsilon_f - \varphi(\varepsilon_f - \varepsilon_m)} = \varepsilon_m \left[1 + \frac{3\varphi(\varepsilon_f - \varepsilon_m)}{(2+\varphi)\varepsilon_m + (1-\varphi)\varepsilon_f}\right]$$
$$= \varepsilon_m \left[1 + \frac{3\varphi(\varepsilon_f - \varepsilon_m)}{\varepsilon_m/(\varepsilon_f - \varepsilon_m) + (1-\varphi)/3}\right]. \tag{17}$$

This is the Maxwell–Wagner (also named as Maxwell–Garnett) mixing rule, which has been widely employed for calculating ε_{eff} of 0–3 composites. It is found that this rule is effective for infinite dilution of the dispersed phase. That is, the spherical filler particles are well separated by distances greater than their characteristic size. It is also argued that this mixing rule is an approximation of the Lichtenecker's logarithmic law [53]. Although Eq. (17) was originally derived under static field, this equation was also derived later on by Webman *et al.* [66] using effective medium theory for the dielectric behavior of microscopically inhomogeneous materials at optical and microwave frequency and by Skipetrov [67] for a random media in which the heterogeneities of the dielectric constant are much smaller than the wavelength of the electromagnetic wave.

Model 5: Bruggeman results [68]

The Rayleigh's calculation was further expended by Bruggeman to different cases. For a composite made of particles of two dielectrics (this is lightly different with 0–3 composites but more close to 0–0 composites), following results are obtained. In the derivation, it was assumed: (1) all particles are very small compared to dimension of the sample/composite, (2) the shape, size, and the locations of all particles are randomly distributed, (3) the number of particles is large, and (4) the all particles are closely packaged together so that there is no gap between particles. First of all, by using Rayleigh's result, Eq. (10), it was obtained as

$$\varphi\left(\frac{\varepsilon_f-\varepsilon_{\text{eff}}}{\varepsilon_f+2\varepsilon_{\text{eff}}}\right)+(1-\varphi)\left(\frac{\varepsilon_m-\varepsilon_{\text{eff}}}{\varepsilon_m+2\varepsilon_{\text{eff}}}\right)=0. \tag{18}$$

For a composite of two inter-dispersed materials. This is widely referred in literatures as Bruggeman mixing rule (this is also called as Polder–Van Santeen in literatures as it was also proposed in Ref [69]). This expression was also derived based on the study of conductivity for a random mixture in which each constituent (spherical particles) acts as if surrounded by a homogenous medium whose properties are those of the mixture [70]. Actually, this is a direct result of an effective medium approximation (EMA) or self-consistent approximation [66].

To fit the experimental results better, the Bruggeman mixing rule has been modified. For example, considering Eq. (18) and the modification of Eq. (15) into Eq. (16), it was proposed [71–73]:

$$\varphi\left(\frac{\varepsilon_f-\varepsilon_{\text{eff}}}{\varepsilon_f+u\varepsilon_{\text{eff}}}\right)+(1-\varphi)\left(\frac{\varepsilon_m-\varepsilon_{\text{eff}}}{\varepsilon_m+u\varepsilon_{\text{eff}}}\right)=0, \tag{19}$$

where u is a constant.

When the percolation phenomenon is considered, it was proposed that u equals to the critical volume fraction (φ_c) of the filler at which the filler particles first forms a continuous percolation path across the medium. For the study of the conductivity of a composite, McLachian proposed a general effective medium formula based on the effective media theory and consideration of percolation features in a composite [71–73]. This general formula for dielectric property of a composite is

$$\varphi\left(\frac{\varepsilon_f^{1/t}-\varepsilon_{\text{eff}}^{1/t}}{\varepsilon_f^{1/t}+u\varepsilon_{\text{eff}}^{1/t}}\right)+(1-\varphi)\left(\frac{\varepsilon_m^{1/t}-\varepsilon_{\text{eff}}^{1/t}}{\varepsilon_m^{1/t}+u\varepsilon_{\text{eff}}^{1/t}}\right)=0, \tag{20}$$

where $u = \varphi_c/(1-\varphi_c)$ is dependent on the critical volume fraction of filler particles, and t is parameter characterizing the microstructure (including size and shape) or distribution and interconnectivity of the components in a composite material.

Based on Rayleigh's results and assumptions, Bruggeman further introduced a change in the volume fraction of the filler and derived following equation for a random distribution of dielectric spheres in a matrix with a low concentration [61]:

$$1-\varphi = \left(\frac{\varepsilon_f - \varepsilon_{\text{eff}}}{\varepsilon_f - \varepsilon_m}\right)\left(\frac{\varepsilon_m}{\varepsilon_{\text{eff}}}\right)^{\frac{1}{3}}. \tag{21}$$

This is the Bruggeman equation. This is also named as Hanai equation since Hanai derived this equation for complex permittivity [74].

The Bruggeman equation has been modified based on different considerations to fit the experimental results. For example, the left of Eq. (21) was modified as [75]:

$$(1-\varphi)(u-(u-1)\varphi) = \left(\frac{\varepsilon_f - \varepsilon_{\text{eff}}}{\varepsilon_f - \varepsilon_m}\right)\left(\frac{\varepsilon_m}{\varepsilon_{\text{eff}}}\right)^{\frac{1}{3}}, \tag{22}$$

where u is a constant. It is experimentally found that the value of u can be positive and negative. Another widely used modification of Bruggeman equation is [76, 77]:

$$1-\varphi = \left(\frac{\varepsilon_f - \varepsilon_{\text{eff}}}{\varepsilon_f - \varepsilon_m}\right)\left(\frac{\varepsilon_m}{\varepsilon_{\text{eff}}}\right)^{\frac{1}{\alpha}}, \tag{23}$$

where α is an empirically determined value ($\geq$3) to fit the experimental data.

Model 6: Knott equation [15]

In the study of plastic foams, Knott introduced a model, in which the filler particles (i.e., air) are treated as small cubes that are surrounded by the matrix (i.e., foam) with the same thickness. Therefore, the effective capacitance (i.e., effective dielectric constant) of the foam is simulated using the capacitance of the cubes and the matrix. In the model, the capacitance of all these elements (i.e., cubes and slabs) is calculated using the parallel-plate mode. Based on these assumptions, it is obtained that

$$\varepsilon_{\text{eff}} = \varepsilon_m \left[1 - \frac{(\varepsilon_m - \varepsilon_f)\varphi}{\varepsilon_f + (\varepsilon_m - \varepsilon_f)\varphi^{1/3}}\right]. \tag{24}$$

Model 7: HS-CT equation [78]

Hashin–Shtrikman (HS) [52] developed an approach to calculate the effective magnetic permeability of multiphase materials. The HS approach was used by Cule and Torquato (CT) to calculate the effective dielectric constant [78]. By assuming that the composite is made up of composite cylinders consisting of a dielectric core (dielectric constant ε_f and radius a) surrounded by a concentric shell (dielectric constant ε_m and radius b) as shown in Fig. 5, it is obtained ($\varphi = (a/b)^2$):

$$\varepsilon_{\text{eff}} = \varepsilon_m \frac{\varepsilon_f + \varepsilon_m + (\varepsilon_f - \varepsilon_m)\varphi}{\varepsilon_f + \varepsilon_m - (\varepsilon_f - \varepsilon_m)\varphi}. \tag{25}$$

Model 8: Yamada equation [79]

In a study of dielectric and piezoelectric properties of 0–3 composite, Yamada *et al.* derived following formula using static field condition,

$$\varepsilon_{\text{eff}} = \varepsilon_m \left[1 + \frac{u\varphi(\varepsilon_f - \varepsilon_m)}{u\varepsilon_m + (1 - \varphi)(\varepsilon_f - \varepsilon_m)}\right], \tag{26}$$

where u is the parameter attributed to the shape and orientation of the ellipsoidal filler particles. Based on the experimental results, the value of u for ceramic particle is bigger than one (mostly about 10).

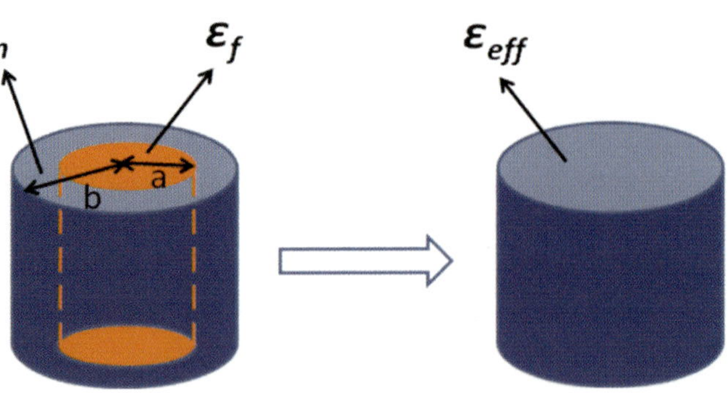

Figure 5. Schematic of assumption used for HS–CT equation.

Model 9: Nielsen equation [80]

Certainly the effective dielectric constant of a composite is dependent on the way the filler particles packing in the composite. Nielsen introduced a parameter Φ to present the way the filler particle packing and obtained the following equation for a 0–3 composite:

$$\frac{\varepsilon_{\text{eff}}}{\varepsilon_m} = \frac{1 + uB\varphi}{1 - B\Theta\varphi}, \tag{27}$$

where $u(u = 3.0$ for spheres) is a constant (the depolarization factor of the particle) depending on the shape of the filler particles, $B = \frac{\frac{\varepsilon_f}{\varepsilon_m} - 1}{\frac{\varepsilon_f}{\varepsilon_m} + u}$, $\Theta = 1 + \frac{1-\Phi}{\Phi^2}\varphi$, and Φ is the maximum packing fraction. For example, $\Phi = 0.637$ for random close packing and 0.524 for simple cubic packing of spheres.

Model 10: Jayasundere–Smith equation [81]

Based on Eq. (17) and an expression of effective dielectric constant that uses the average electric fields in each constituent [82], Jayasundere and Smith further included the interactions among the filler particles (spheres). When $\varepsilon_f \gg \varepsilon_m$, it was obtained:

$$\varepsilon_{\text{eff}} = \frac{\varepsilon_m(1-\varphi) + \varepsilon_f\varphi\left[\frac{3\varepsilon_m}{\varepsilon_f + 2\varepsilon_m}\right]\left[1 + \frac{3\varphi(\varepsilon_f - \varepsilon_m)}{\varepsilon_f + 2\varepsilon_m}\right]}{(1-\varphi) + \varphi\left[\frac{3\varepsilon_m}{\varepsilon_f + 2\varepsilon_m}\right]\left[1 + \frac{3\varphi(\varepsilon_f - \varepsilon_m)}{\varepsilon_f + 2\varepsilon_m}\right]}. \tag{28}$$

Model 11: Poon–Shin equation [83]

Poon and Shin noticed some inconsistent in the derivation of Eq. (23) and that Eq. (23) is not valid for $\varepsilon_f < \varepsilon_m$. They derived following formula for a 0–3 composite:

$$\frac{\varepsilon_{\text{eff}}}{\varepsilon_m} = 1 + \frac{\varphi\left(\frac{\varepsilon_f}{\varepsilon_m} - 1\right)}{\varphi + \frac{1}{3}(1-\varphi)\left[\frac{\varepsilon_f}{\varepsilon_m}(1-\varphi) + \varphi + 2\right]}. \tag{29}$$

This equation can be used for both $\varepsilon_f \geq \varepsilon_m$ and $\varepsilon_f \leq \varepsilon_m$.

Model 12: Reverse composite

Due to the assumptions and simplifications used in each of the above models, only a few formulas (i.e., Eqs. (5), (18), and (20)) are symmetrical in terms of its constituent, and all others are asymmetrical. The asymmetric result is more close to the reality of a 0–3 composite. For a real random 0–3 composite, it is likely that the filler becomes the matrix and the matrix may be separated into isolated particles when the volume fraction of the filler is high. Additionally, it is likely that the volume fraction of each constituent in a real composite is not a constant but changes with location. Considering these two possibilities, a methodology was proposed to fit the experimental data for a 0–3 composite over a broad composition range as [84]:

$$\varepsilon_{\text{eff}} = \int_0^{\varphi_c} [P(\varphi)\varepsilon_{\text{eff}}(\varepsilon_m, \varepsilon_f, \varphi)]\, d\varphi + \int_{\varphi_c}^{1} [P(\varphi)\varepsilon_{\text{eff}}^{\text{rev}}(\varepsilon_m, \varepsilon_f, \varphi)]\, d\varphi, \quad (30)$$

where $\varepsilon_{\text{eff}}(\varepsilon_m, \varepsilon_f, \varphi)$ is any equation mentioned above, $\varepsilon_{\text{eff}}^{\text{rev}}(\varepsilon_m, \varepsilon_f, \varphi)$ is the same equation but matrix and filler are exchanged, $P(\varphi)$ is the distribution function of the volume fraction so that $\int_0^1 P(x)dx = \varphi$. It was suggested that $P(\varphi)$ is approximately a Gaussian function, and φ_c is the volume fraction at which cluster size of the filler will form a channel through the sample. The reverse composite mode is illustrated in Fig. 6 using Maxwell–Wagner equation. Clearly, by exchanging the matrix and filler, the ε_{eff} changes a lot.

Model 13: Simulation

Due to the difficulty and complexity of a composite and the progress in the numerical simulation methodologies, numerical simulation has been used to determine the effective dielectric constant of a composite [60, 76, 77, 85–88].

Model 14: Interfacial layer

The existence of an interfacial layer between two constituents in a composite is obvious. Therefore, much research has focused on the influence of an interfacial layer on the dielectric property of a composite. Due to the difficulty in determining thickness and dielectric property of the interfacial layer, there are few quantified results. Whenever a phenomenon exists which is difficult to understand

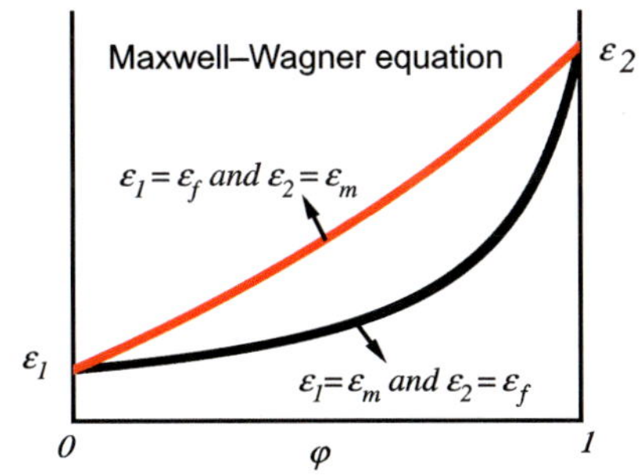

Figure 6. Schematic of reverse composite. Here Maxwell–Wagner equation is used. The bottom curve is the result obtained using dielectric-1 as matrix and dielectric-2 as filler, while the top curve is obtained using dielectric-1 as filler and dielectric-2 as matrix. Both curves are calculated using Maxwell–Wagner equation (i.e., Eq. (17)).

using two constituents, the interfacial layer is used to explain the phenomenon [89–93]. For example, it was experimentally found that the dielectric constant observed in a composite can be higher than the dielectric constant of two constituents [93]. This phenomenon is explained as the result of the appearance of the interfacial layer in the composite. This explanation would result in a conclusion that the interfacial layer (polymer) would have a dielectric constant much higher than the pure polymer [92, 93], which is hard to be accepted.

2.1.2. *Dielectric loss*

Most of the models/formulas given above were initially derived for the effective dielectric constant of a composite. Some of them were later on extended to complex permittivity. For example, Eq. (21) was obtained by Bruggeman for the ε_{eff} only but extended to complex permittivity by Hanai. Although the ε_{eff} is the primary focus for most of the research, the dielectric loss is also important. If the dielectric loss has to be simulated, the general practice is: (1) analyzing the ε_{eff} using the formulas above as only for dielectric constant; (2) then analyzing the dielectric loss using the formula as for complex permittivity. For example, when the Bruggeman equation, Eq. (21), is used for complex permittivity, the ε_{eff} is still simulated using

Eq. (21), while the imaginary part is calculated using

$$\varepsilon_{\text{eff}}^{''} = \frac{(\varepsilon_f - \varepsilon_{\text{eff}})(\varepsilon_f + 2\varepsilon_m)\varepsilon_{\text{eff}}}{(\varepsilon_f - \varepsilon_m)(\varepsilon_f + 2\varepsilon_{\text{eff}})\varepsilon_m}\varepsilon_m^{''} + \frac{3(\varepsilon_{\text{eff}} - \varepsilon_f)\varepsilon_{\text{eff}}}{(\varepsilon_f - \varepsilon_m)(\varepsilon_f + 2\varepsilon_{\text{eff}})}\varepsilon_f^{''}, \tag{31}$$

where ε_{eff} is from Eq. (21), $\varepsilon_{\text{eff}}^{''}$, $\varepsilon_m^{''}$, $\varepsilon_f^{''}$ are the imaginary parts of complex relative permittivity of the composite, the filler, and the matrix, respectively [94]. The volume fraction of the filler is represented by the ε_{eff}.

If the dielectric loss is considered, a composite can have a ε_{eff} higher than the dielectric constant of both constituents. This would break the Wiener limits, Eq. (2). However, we do believe that the dielectric loss of a composite is between the dielectric losses of two constituents. The above discussion is based on ideal dielectrics where the conductivity is zero. Actually, all dielectrics have a nonzero conductivity. It is well known that for a homogenous dielectric the conductivity (σ) contributes to the measured imaginary part of the complex relative permittivity as:

$$\varepsilon_{\text{con}}^{''} = \frac{\sigma}{\varepsilon_0 \omega}, \tag{32}$$

where ε_0 and ω are the permittivity of vacuum and the angular frequency. Therefore, the experimentally measured dielectric loss in a dielectric actually has two parts: one is the real dielectric loss (the imaginary part of the permittivity) and the other is due to the electrical conductivity. For a composite, the conductivity of its constituents would make the dielectric response of a composite much more complicated. For a composite under an electric field, the electric field is not uniform and its distribution is dependent on the dielectric property of each constituent. An electric current is induced in each constituent due to the electrical conductivity. The difference in the current density between two constituents would result in a charge accumulation around boundary between two constituents. The charge accumulation would redistribute the electric field in the composite, which would change the current density in each constituent. Therefore, an additional dielectric loss is expected in a composite due to the electrical conductivity of its constituents. This may be the reason that the dielectric loss obtained in many 0–3 composites is higher than the dielectric loss of both constituents. More research on this subject is needed.

2.2. *Polymer-based 0–3 composites*

Various composite systems have been investigated/developed for the development of high-dielectric-constant composites. In the research, ferroelectric and relaxor ferroelectrics, such as $BaTiO_3$ (BT), $PbTiO_3$ (PT), $Pb(Zr,Ti)O_3$ (PZT), $Ba_{0.65}Sr_{0.35}TiO_3$ (BST), $Pb(Mg_{1/3}Nb_{2/3})O_3$ (PMN), and $Pb(Mg_{1/3}Nb_{2/3})O_3$-$PbTiO_3$ (PMN-PT), have been used as fillers due to the fact that these ceramics exhibit a high dielectric constant. Table 1 shows some of 0–3 composites reported in literatures and the commercially available composites. BT is widely used as a filler since it is relatively cheap and has been successfully prepared in different sizes from nm to μm. There are commercially available composites ($\varepsilon_{\mathrm{eff}}$ is from 4 to 36) based on using BT as fillers [6]. For BT based composites, a dielectric constant about 40 was obtained in composite with 60 vol% BT using PMMA as matrix, in which it is also found that the effective dielectric constant is almost linearly dependent on the content of BT [44]; while a dielectric constant of 44 was obtained in composite with 40 vol% BT using epoxy as matrix [95]. Other polymers, such as PVDF and PVC, have also been used to prepare 0–3 composite with BT as fillers. A dielectric constant as high 100 was obtained [96–99]. It is well known that the dielectric constant of BT ceramics is strongly dependent on the grain size. That is, the dielectric constant of BT particles is strongly dependent on the size of crystal. This is one of the reasons that the dielectric property of the BT-based composite varies over a big range. PMN-PT exhibits a higher dielectric constant around room temperature and has been used as filler for 0–3 composites as shown in Table 1 [100, 101]. PZT is also a widely used filler material [102–104] Another widely used ceramic system is BST due to the fact that the BST can exhibit a high dielectric constant at temperature around room temperature [105]. It should be mentioned that some of these composites are developed for the

Table 1. The dielectric properties of ceramic–polymer composites at room temperature.

Dielectric Fillers	Polymers	Filler's Size	Composites				Ref.
			φ	ε_r	$\tan\delta$	Freq.	
$BaTiO_3$	PMMA	—	60%	40	0.012	1kHz	[44]
$BaTiO_3$	Epoxy	—	40%	44	0.2	1kHz	[95]
$BaTiO_3$	PVDF	0.1 μm	70%	152.3	<0.1	1kHz	[96]
$BaTiO_3$	Polystyrene	0.7 μm	70%	100	<1	1kHz	[97]
$BaTiO_3$	PVC	6.8–0.3 μm	40%	18	—	100kHz	[98]
$BaTiO_3$	polyamide (PA)	17–100 nm	70%	80	0.02	1MHz	[99]
$(Bi_{0.5}Na_{0.5})_{0.94}Ba_{0.06}TiO_3$	P(VDF-TrFE)70/30	—	30%	30	<0.02	1kHz	[94]
PMN	Epoxy	—	40%	24	0.05	100kHz	[95]
PMN–35PT	P(VDF-TrFE) 70/30	—	40%	50	<0.1	1kHz	[100]
PMN-PT	P(VDF-TrFE)	0.5 μm	50%	200	<0.1	10kHz	[101]
PZT	Polyurethane	—	30%	24	0.7	1kHz	[106]
PZT	PVDF	30 nm	32%	40	—	100Hz	[102]
PZT	Polyvinyl-butyral	100 μm	85%	155	0.05	1MHz	[103]
$(Ba_{0.65}Sr_{0.35})TiO_3$	P(VDF-TrFE)70/30	1 μm	50%	80	<0.2	1kHz	[105]
CCTO	P(VDF-TrFE) 55/45	10 μm	50%	610	<0.4	1kHz	[32]
CCTO	P(VDF-TrFE) 55/45	<500 nm	50%	60	0.05	1kHz	[107]
CCTO	Epoxy	0.3–0.5 μm	40%	50	—	100Hz	[108]
CCTO	PVDF	1–7 μm	55%	95	<0.24	100Hz	[109]
CCTO	Polystyrene	10–25 μm	64%	80	—	100Hz	[110]
CCTO	Polyimide	1–4 μm	40%	49	<0.2	100Hz	[111]
CCTO	polyethersulfone	—	50%	32.7	0.063	1kHz	[112]

electromechanical applications since these ceramics exhibit a strong electromechanical effect.

The dielectric behavior of the polymer matrix plays a critical role on the dielectric property of a composite. Unfortunately, most of polymers exhibit a very small dielectric constant (mostly less 5). It is well known that PVDF and P(VDF-TrFE) are ferroelectric and exhibit a high dielectric constant (>10) at room temperature. Therefore, a great deal of research has been done on the development of 0–3 composites using PVDF and P(VDF-TrFE) as matrix [94, 96, 100–102]. More recently it is found that the dielectric constant of P(VDF-TrFE) copolymer can be significantly increased by using high-energy electron irradiation and the irradiated copolymer exhibits a relaxor ferroelectric behavior [43, 46]. Based on this, high-energy electron irradiation was carried out on a composite system using PMN-PT ceramic powders as the filler and P(VDF-TrFE) copolymer as the matrix. The irradiated composite exhibits a much higher dielectric constant (~200) than other composites at room temperature [101]. The irradiated also made the BST P(VDF-TrFE) composites exhibit a high dielectric constant (~150) at room temperature [105]. Although PVDF based ferroelectric and relaxor ferroelectric polymers exhibit a high dielectric constant, their dielectric constant is strongly dependent on the temperature. In a search of electroactive polymer [50, 113, 114], it is found that P(VDF-CTFE) copolymer has a high E_b and exhibits a high dielectric constant (>10) that is almost independent of the temperature over a broad temperature range, which makes P(VDF-CTFE) copolymers a great candidate for the development of energy storage materials including composites. Actually, after this copolymer was passed to Penn State, a giant energy storage density was observed in it [115].

During the last decade, non-ferroelectric ceramics with high dielectric constant have been discovered [30, 31]. For example, it was reported that $CaCu_3Ti_4O_{12}$(CCTO) ceramics exhibit a dielectric constant as high as 10^5 and the dielectric constant is almost independent of temperature over a board temperature range [116]. More interestingly, these ceramics have a very weak electromechanical effect since it does not have piezoelectric effect. Therefore, the development of these giant-dielectric-constant ceramics opens a new avenue for the development of high-dielectric-constant composites [32]. Different composite systems have been studied/developed using CCTO as the filler [107–112]. For example, a dielectric constant more than 200 was obtained at room temperature in composites using micro-sized CCTO particles [32]. It is also interestedly found that the size of the CCTO particles has a strong influence on the dielectric behavior of the composites. For composites using nano-sized CCTO particles, a composite at room temperature exhibits a dielectric constant of 60 and loss about 0.05 at 1 kHz [107]. By combining nanosized CCTO particles with P(VDF-CTFE), a composite with a dielectric constant of 100 that is almost independent of temperature (up to 120 °C) was developed [113].

3. Conductor–polymer composites

3.1. *Percolation theory*

If the conductor particles are embedded in an insulator matrix, the composite is either an insulator/dielectric (i.e., the content of filler is less than the percolation threshold) or a conductor (i.e., the content of filler is higher than the percolation threshold) as shown in Fig. 7. The percolation threshold is the composition at which the conductor particles form channels through the composite [117]. Metal–polymer composites have been widely studied and employed for different applications, such as static or electromagnetic interference (EMI) shielding [118, 119]. The study of effective conductivity and dielectric constant of metal–polymer composites has been the subject of extensive studied in both theory and experiment. For the composites in insulator regime, the study has been focused on the dependence of the dielectric behavior of the composites on the filler concentration and on the percolation threshold of different composite systems. Similar to the dielectric–dielectric composites, the dielectric response of a conductor–dielectric composite would strongly dependent on the exact microstructure of the composite. For example, the percolation threshold is strongly dependent on the shape, size, and connectivity of the filler particles and the

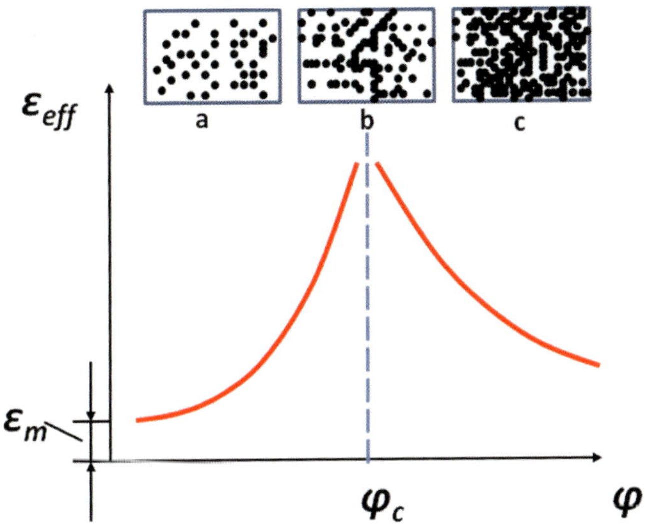

Figure 7. Schematic of the ε_{eff} versus composition of conductor dielectric composite, where $\varphi_c$is the percolation threshold. The insets show the geometric phase transition of fillers (denoted by dark spots) in the composites' microstructure near percolation.

morphology of the composites. Therefore, various results/models based on different assumptions and simplifications/approximation for a real composite have been introduced.

The dielectric response of a conductor–dielectric composite was first recognized, independently by Cavendish in 1773 and Faraday in 1837 when they noted that the dielectric in a capacitor does not attaint full polarization until certain time after the application of the charge to the plate [54]. This was misleadingly described as electric absorption. This phenomenon can be described by the combination of two components in series: one with a permittivity but no conductivity and the other with a conductivity but no permittivity. This results in a dielectric response similar to the Debye dielectric (in the calculation, Eq. (1) is used by replacing ε_f with Eq. (32)). Therefore, a dielectric dispersion, the frequency dependence of the dielectric property, is observed and this dispersion is named as Maxwell–Wagner effect.

For a real conductor–dielectric composite, the research has been focused on the percolation threshold and the composition dependence of the conductivity and dielectric constant. Based on random resistance network model [68], it was derived that the effective dielectric constant at low frequency can be approximately expressed as

$$\varepsilon_{\text{eff}} = \varepsilon_m \left[A(\varphi_c - \varphi)^\alpha + B(\varphi_c - \varphi)^\beta\right]^{-1}, \tag{33}$$

where φ_c is the percolation threshold and $\varphi < \varphi_c, A > 0, B > 0, \alpha > 0$, and $\beta > 0$. For a random binary system, it was derived as [66]

$$\varepsilon_{\text{eff}} \propto \varepsilon_m(\varphi_c - \varphi)^{-s}, \tag{34}$$

where $\varphi < \varphi_c$ and $s(>0)$ is a constant. The numerical simulations have suggested that s ≈ 0.7 [120]. Equation (34) is much simpler than Eq. (33). Therefore, this equation is widely used in the literatures to fit the experimental results [23, 33–36] In the literatures, Eq. (34) is normalized as

$$\varepsilon_{\text{eff}} = \varepsilon_m \left(\frac{\varphi_c - \varphi}{\varphi_c}\right)^{-s}, \tag{35}$$

since the dielectric constant of the composite should be ε_m when $\varphi = 0$.

It is also found that the effective dielectric constant of a conductor–dielectric composite is strongly dependent on the frequency. For example, if $\varphi \approx \varphi_c$, it was showed that [68]

$$\varepsilon_{\text{eff}} \propto \omega^{\gamma-1}, \tag{36}$$

where $\gamma(<1)$ is a constant and ω is the angular frequency. Based on numerical simulations, the value of γ has been proposed. For a random resistance networks, it was indicated that $\gamma \approx 0.73 \pm 0.05$ [68], while for a random binary medium, it was indicated that $\gamma = 0.75 \pm 0.05$ [121]. It should be mentioned that, in reality, the composite with composition close to φ_c may have a much higher conductivity than what has been obtained from theory research since quantum mechanical tunneling through the insulating barriers may become important.

Based on Eqs. (33) and (34), one would find that a high/giant dielectric constant can be achieved in a conductor–dielectric composites when the concentration of the conductive filler is approaching the percolation threshold from low concentration side. This approach has been widely studied to develop high-dielectric constant composites since 1980s [37, 38, 122–124]. For example, a giant dielectric constant ($\sim 10^4$) was reported for $IrCl_6$-dopoed polyacetylene in 1984 [124]. Equations (33) and (34) also indicate that the dielectric constant of the dielectric has a direct influence on the dielectric constant of the composites. It is highly desirable to use the dielectrics with high dielectric constant as the matrix. Conductor fillers with different shapes and sizes have been studied.

The polymer-based conductor–dielectric composites studied so far can be classified into three categories based on the filler form: particle, fiber, or core–shell. Metal particles are the most commonly used conductor fillers due to the fact that these particles can be easily prepared.

3.2. *Metal particles embedded in polymer*

Particles of different metals have been used in the development of high-dielectric-constant composites. The most commonly used metal particles are Ni, Ag, Cu, Fe, and Al, as shown in Table 2. As expected, the dielectric constant is enhanced by filling the polymer with conductor fillers. It appears that the intrinsic electrical property (i.e., conductivity) of a metal does not have a strong influence on the $\varepsilon_{\rm eff}$ of the composites, yet the size and shape of metal particles have a significant impact on the composite properties (i.e., percolation threshold and the $\varepsilon_{\rm eff}$). For example, a dielectric constant of 2000 is reported for an epoxy filled with 10 μm silver flakes [125], while a dielectric constant of 300 is reported for an epoxy filled with silver particles in diameter of 40 nm [126]. It is experimentally found that the composites with nano-sized metal particles have a lower dielectric constant with a low dielectric loss compared to those with micro-sized metal particles. For example, a dielectric constant of 400 with a loss of 0.18 is obtained in a PVDF filled with micro-sized nickel particles, while a dielectric constant of 70 with a loss of 0.1 is obtained in

Table 2. The dielectric properties of metal particle–polymer composites at room temperature.

Conducting Fillers	Polymers	Filler's Size	Composites				Ref.
			φ_c	$\varepsilon_{r\,\max}$	$\tan\delta$	Freq.	
Ag flake	epoxy	10 μm	11.43%	2000	0.24	10 kHz	[125]
Ag	epoxy	40 nm	15%	300	0.05	1 kHz	[126]
Ni	PVDF	5 μm	17%	400	0.18	1 kHz	[127]
Ni	PVDF	50 nm	20%	70	<0.1	1 kHz	[128]
Ni	PVDF	8–10 μm	28%	1273	2.04	1MHz	[130]
Al	epoxy	3 μm	80%	30	0.02	1 kHz	[129]
Al	polyethylene	102 nm	24 wt%	24	<0.04	100 kHz	[131]
Cu	polyethylene	20–25 μm	20%	50	<0.04	1 kHz	[132]
Fe	polyethylene	3.5 μm	22%	200	0.004	10 kHz	[133]
Fe	polyamide-12	3.5 μm	30%	150	0.07	1 kHz	[133]
Fe	polyoxymethylene	3.5 μm	25%	21	0.01	1 MHz	[133]

PVDF filled with nano-sized nickel particles [127, 128]. In Al-epoxy composites [129], a dielectric constant approximating 30 with a loss approximating 0.02 were also observed. This could be attributed to a tunneling network composed of the self-passivated Al particles with outside thick Al_2O_3 shells.

3.3. *Conductive fibers embedded in polymer*

Percolation thresholds can be significantly reduced by using a conductor fiber, as shown in Table 3. For example, the composites made of carbon fiber filling a PVDF resulted in a percolation threshold smaller than 10% [134, 135]. Additionally, fiber-based composites should have a better flexibility than the particle-based composites. Although different fibers have been studied, carbon fiber and carbon nanotube (CNT) have attracted a great deal of interest due to the fact that these fibers can be prepared with a larger length(L)/diameter(D) ratio and this aspect ratio can be tuned over a great range [136–139]. Graphene has also been studied as a conductor filler [140]. Carbon fillers have a higher intrinsic conductivity (10^5–10^8 S/m) and a significantly lower density than

Table 3. The dielectric properties of conducting fiber–polymer ccmposites at room temperature.

Conducting Fillers	Polymers	Filler's Size	Composites φ_c	$\varepsilon_{r\,\max}$	$\tan\delta$	Freq.	Ref.
Ag fiber	PVDF	D: 100 nm L: 20 μm	20%	379	0.15	1 kHz	[144]
Steel fiber	PVDF	D: 30 μm L: 500 μm	9.4%	427	0.3	50Hz	[145]
Carbon fiber	Polyethylene	D: 7–8 μm L: 100 μm	20%	5.5	—	2 kHz	[132]
Carbon fiber	Polyethylene	D: 2–8 μm L: 100 μm	30vol%	35	<1	200 kHz	[146]
Carbon fiber	PVDF	D: 8 μm L: 100 μm	7.4%	80	<0.2	1 kHz	[134]
Carbon fiber	Polypropylene (PP)	20–200 nm	5wt%	~300	<1	1 kHz	[147]
MWCNT	PVDF	—	1.6%	300	<0.8	1 kHz	[135]
MWCNT	PVDF	D: 10–20 nm L: 30 μm	3.8vol%	3600	<2	1 kHz	[141]
TFP-MWCNT	PVDF	—	8vol%	5000	<2	1 kHz	[142]
MWCNT	PVDF	D: 20–40 nm L: 15 μm	10.4vol%	2000	1	1 Hz	[143]
Graphite plate	PVDF	D: 0.5–25 μm T: 20–60 μm	1.01%	2700	0.48	100 Hz	[140]

metals. More importantly, carbon based fillers have a high compatibility with polymer matrices. CNT has a higher aspect ratio and a much higher conductivity and flexibility compared to carbon fibers. Therefore, a higher dielectric response is expected and obtained in CNT-based composites. For example, a dielectric constant of 10^3 has been reported in several composites [135, 141–143]. It is believed that there are other mechanisms responsible for high dielectric constant such as micro-capacitor networks in addition to the effect of size and high aspect ratio [34, 89].

3.4. *Core–shell particles embedded in polymer*

Although conductor–dielectric composites can exhibit a high dielectric constant, the data shown in Tables 2 and 3 indicate that a high dielectric constant observed in conductor–dielectric composites is usually associated with a high dielectric loss. This is undesirable for many dielectric applications. Since a high dielectric constant is obtained in a composite close to the percolation threshold, it is experimentally found that this composition window is very narrow and can be affected by many factors.

Therefore, the dielectric property of these composites would strongly depend on the process, possibly batch by batch, which would affect the repeatability and reproducibility of the composites. Furthermore, these composites have a very small E_b. Filler particles in core-shell forms have been recently studied based on these considerations. Some of the results are shown in Table 4 [148–152]. A core–shell filler particle has a conductor core and a dielectric/insulator

Table 4. The dielectric properties of core–shell composites at room temperature.

Polymer	Core		Shell		Composites				
Matrix	Material	Size	Material	Thickness	φ	$\varepsilon_{r\,max}$	$\tan\delta$	f	Ref.
Epoxy	Al	3 μm	Al_2O_3	40 nm	80%	105	0.2	10 kHz	[148]
Epoxy	Ag	80–90 nm	PVP	8–120 nm	20%	400	0.04	1 kHz	[149]
PVDF	$LiClO_4$	1–4 μm	PC	0.5 μm	20wt%	10^6	—	1 kHz	[150]
PI	$BaTiO_3$	100 nm	PAA	—	40%	18	0.01	1 kHz	[151]
PVDF	BST	50–100 nm	Ag	15 nm	55%	153	0.2	100 Hz	[152]

shell. It is expected that the dielectric layer coated on the conductor particles would increase the contact resistance of the filler particles and suppress the electron tunneling between adjacent filler particles and this, in turn, would significantly reduce the dielectric loss of the composite and increase the E_b. Compared to the data shown in Tables 2 and 3, the data in Table 4 indicates that the core–shell based composites exhibit a higher dielectric constant with a low dielectric loss. It is also recognized by the experimental results that an organic coating improves performance since the organic surface increases the compatibility between the filler particles and matrix. This promotes homogeneous dispersion of the filler particles in the matrix.

4. Challenges and improvement

As discussed above, simulating the dielectric constant of a real composite is still an open question. Some of the formulas are derived only for composites with a very small volume fraction of filler, while others are pure empirical. Although each formula can fit some experimental results well, it is difficult to predict formula(s) for a specific composite. There is no solid conclusion on simulating the percolation threshold, which is usually determined by the experiment results.

Dielectric loss is critical for many applications, however, there are few detailed studies on the dielectric loss of composites for both dielectric–dielectric and conductor–dielectric composites. Study of the physics related to dielectric loss in these composites is needed. Another topic that has to be exploited is the influence of the electrical conductivity of its constituents on the dielectric property of a composite. The dielectric performance of a composite can be enhanced by a proper design and usage of the conductivity of each constituent. It is important to determine the contributions from polarization and conductivity, respectively, in the measured dielectric loss in a composite. For example, for energy storage applications, the dielectric loss due to polarization is manageable, while the dielectric loss due to conductivity is unacceptable.

The existence of interfacial layer between two constituents of a composite is a widely accepted fact. Current studies refer to an unreasonably high dielectric constant for the interfacial layer. The role of the interfacial layer should be studied in a more detailed manner. New tools and/or methodologies have to be developed in order to characterize and observe this interfacial layer. If one wants to study the interfacial layer using the materials property, it has to be pointed out that different properties [91], rather than only dielectric property, have to be used. The dielectric loss should provide some unique information.

The interaction between constituents of a composite is important. Poor adhesion between two constituents of a 0–3 composite will result in the composite having a poor mechanical properties, a lower E_b, and a shorter lifetime. Coupling agents have been used to improve the adhesion. A general guideline on the selection of coupling agent for a composite system is needed. The coupling agent has to be uniformly coated onto the particle surface. The amount of coupling agent is also critical. Lesser amounts of the coupling agent would result in a poor adhesion, while additional amounts of the coupling agent would result in the composite having a high dielectric loss. The addition of coupling agent would also make simulating the dielectric property of a composite more different.

A conductor–dielectric composite requires a lower concentration of filler than a dielectric–dielectric composite in order to reach a high dielectric constant. The conductor–dielectric composites exhibit a high dielectric constant when the filler concentration is close to percolation threshold, which results in the composites having a high dielectric loss and a low E_b. Investigating approaches for increasing the E_b and reducing the dielectric loss in conductor–dielectric composites would be interesting. Core–shell particles show some promises, however, further research is needed. The thickness of the shell layer has to be optimized based on the desired dielectric constant and loss.

The percolation theory presented represents the dielectric property at a low frequency. If the frequency is such high that the size/dimension of the filler particles can be compared with the wavelength of the electromagnetic wave, the behavior of a composite would be governed by completely different mechanisms which can result in interesting phenomena, such as left-handed materials.

5. Summary

Dielectric composites are promising materials for many applications. Polymer-based 0–3 composites with a high dielectric constant represent an important category of dielectric composites. These composites are classified into dielectric–dielectric and conductor–dielectric composites. The physical principles and models for simulating dielectric property of a dielectric–dielectric and conductor–dielectric composite are presented with the associated assumptions and approximations. In general, simulating the dielectric property of a real composite is still an unresolved task. More research on the dielectric loss is needed. The achievements in the development of these composites are summarized.

In general, conductor–dielectric composites exhibit a higher dielectric constant, a higher dielectric loss, and a lower E_b than dielectric–dielectric composites. Some of the challenges facing the development of these composites are discussed.

References

[1] Borchardt, J., Alexander, J., Slenes, K. and Fuente, R. D. L. (2007). Ceramic–polymer composite for high energy density capacitors, *16th IEEE Int. Pulsed Power Conf.*, pp. 294–297.

[2] Tummala, R. T. and Laskar, J. (2004). Gigabit wireless: System-on-a-package technology, *Proc. IEEE*, 92, pp. 376–387.

[3] T. Osaka, M. D. (2001). *Energy Storage Systems for Electronics* (CRC Press, London).

[4] Karden, E., Ploumen, S., Fricke, B., Miller, T. and Snyder, K. (2007). Energy storage devices for future hybrid electric vehicles, *J. Power Sources*, 168, pp. 2–11.

[5] Fiedziuszko, S. J., Hunter, I. C., Itoh, T., Kobayashi, Y., Nishikawa, T., Stitzer, S. N. and Wakino, K. (2002). Dielectric materials, devices, and circuits, *IEEE Trans. Microwave Theory Tech.*, 50, pp. 706–720.

[6] Jillek, W. and Yung, W. K. C. (2005). Embedded components in printed circuit boards: A processing technology review, *Int. Adv. J. Manuf. Technol.*, 25, pp. 350–360.

[7] Nalwa, H. (1999). *Handbook of Low and High Dielectric Constant Materials and Their Applications* (Academic Press, London).
[8] Rao, Y., Ogitani, S., Kohl, P. and Wong, C. P. (2002). Novel polymer–ceramic nanocomposite based on high dielectric constant epoxy formula for embedded capacitor application, *J. Appl. Poly. Sci.*, 83, pp. 1084–1090.
[9] Rao, Y. and Wong, C. P. (2004). Material characterization of a high-dielectric-constant polymer–ceramic composite for embedded capacitor for RF applications, *J. Appl. Poly. Sci.*, 92, pp. 2228–2231.
[10] Dias, C. J. and Das-Gupta, D. K. (1996). Inorganic ceramic/polymer ferroelectric composite electrets, *IEEE Trans. Dielectr. Electr. Insul.*, 3, pp. 706–734.
[11] Reichmanis, E., Katz, H., Kloc, C. and Maliakal, A. (2005). Plastic electronic devices: From materials design to device applications, *Bell Labs Tech. J.*, 10, pp. 87–105.
[12] Yang, R., Jireh, Y. and Wong, C. P. (2001). High K polymer-ceramic nano-composite development, characterization, and modeling for embedded capacitor RF application, *Proc. 51st Electron. Comp. Technol. Conf.* (Cat. No.01CH37220), pp. 1408–1412.
[13] Yoon, J. R., Han, J. W., Lee, K. M. and Lee, H. Y. (2009). Dielectric properties of polymer-ceramic composites for embedded capacitors, *Trans. Electr. Electron. Mater.*, 10, pp. 116–120.
[14] Sebastian, M. T. and Jantunen, H. (2010). Polymer–ceramic composites of 0–3 connectivity for circuits in electronics: A review, *Int. J. Appl. Ceram. Technol.*, 7, pp. 415–434.
[15] Knott, E. F. (1993). Dielectric constant of plastic foams, *IEEE Trans. Antennas Propag.*, 41, pp. 1167–1171.
[16] Newnham, R. E., Skinner, D. P. and Cross, L. E. (1978). Connectivity and piezoelectric–pyroelectric composites, *Mater. Res. Bull.*, 13, pp. 525–536.
[17] Skinner, D. P., Newnham, R. E. and Cross, L. E. (1978). Flexible composite transducers, *Mater. Res. Bull.*, 13, pp. 599–607.
[18] Jayasundere, N., Smith, B. V. and Dunn, J. R. (1994). Piezoelectric constant for binary piezoelectric 0–3 connectivity composites and the effect of mixed connectivity, *J. Appl. Phys.*, 76, pp. 2993–2998.
[19] Wong, C. K., Poon, Y. M. and Shin, F. G. (2001). Explicit formulas for effective piezoelectric coefficients of ferroelectric 0–3 composites, *J. Appl. Phys.*, 90, pp. 4690–4700.
[20] Pilgrim, S. M. and Newnham, R. E. (1986). 3:0: A new composite connectivity, *Mater. Res. Bull.*, 21, pp. 1447–1454.
[21] Cheng, Z. and Zhang, Q. (2008). Field-activated electroactive polymers, *Mater. Res. Bull.*, 33, pp. 183–187.
[22] Safari, A. and Akdogan, E. K. (2008). *Piezoelectric and Acoustic Materials for Transducer Application* (Springer, New York).
[23] Nan, C. W., Shen, Y. and Ma, J. (2010). Physical properties of composites near percolation, *Annu. Rev. Mater. Res.*, 40, pp. 131–151.

[24] Newnham, R. E. (1986). Composite electroceramics, *Annu. Rev. Mater. Sci.*, 16, pp. 47–68.
[25] Fang, D.-N., Soh, A. K., Li, C.-Q. and Jiang, B. (2001). Nonlinear behavior of 0–3 type ferroelectric composites with polymer matrices, *J. Mater. Sci.*, 36, pp. 5281–5288.
[26] Komarneni, S. (1992). Feature article. Nanocomposites, *J. Mater. Chem.*, 2, pp. 1219–1230.
[27] Cheng, Z. Y., Katiyar, R. S., Yao, X. and Bhalla, A. S. (1998). Temperature dependence of the dielectric constant of relaxor ferroelectrics, *Phys. Rev. B*, 57, pp. 8166–8177.
[28] Cheng, Z. Y., Zhang, L. Y. and Yao, X. (1996). Investigation of glassy behavior of lead magnesium niobate relaxors, *J. Appl. Phys.*, 79, pp. 8615–8619.
[29] Lines, M. E. and A. M. G. (1977). *Principles and Applications of Ferroelectrics and Related Materials* (Clarendon Press, Oxford).
[30] Vasil'ev, A. N. and Volkova, O. S. (2007). New functional materials $AC_3B_4O_{12}$ (Review), *Low Temp. Phys.*, 33, pp. 895–914.
[31] Raevski, I. P., Prosandeev, S. A., Bogatin, A. S., Malitskaya, M. A. and Jastrabik, L. (2003). High dielectric permittivity in $AFe_{1/2}B_{1/2}O_3$ nonferroelectric perovskite ceramics (A=Ba, Sr, Ca; B=Nb, Ta, Sb), *J. Appl. Phys.*, 93, pp. 4130–4136.
[32] Arbatti, M., Shan, X. and Cheng, Z. Y. (2007). Ceramic–polymer composites with high dielectric constant, *Adv. Mater.*, 19, pp. 1369–1372.
[33] Stauffer, D. (1992). *Introduction to Percolation Theory* (Taylor & Francis, London).
[34] Garboczi, E. J., Snyder, K. A., Douglas, J. F. and Thorpe, M. F. (1995). Geometrical percolation threshold of overlapping ellipsoids, *Phys. Rev. E*, 52, pp. 819–828.
[35] Nan, C.-W. (1993). Physics of inhomogeneous inorganic materials, *Prog. Mater. Sci.*, 37, pp. 1–116.
[36] Yacubowicz, J. and Narkis, M. (1986). Dielectric behavior of carbon black filled polymer composites, *Polymer Eng. Sci.*, 26, pp. 1568–1573.
[37] Toker, D., Azulay, D., Shimoni, N., Balberg, I. and Millo, O. (2003). Tunneling and percolation in metal-insulator composite materials, *Phys. Rev. B*, 68, pp. 041403.
[38] Shen, Y., Yue, Z., Li, M. and Nan, C. W. (2005). Enhanced initial permeability and dielectric constant in a double-percolating $Ni_{0.3}Zn_{0.7}Fe_{1.95}O_4$-Ni-polymer composite, *Adv. Funct. Mater.*, 15, pp. 1100–1103.
[39] Wang, G., Deng, Y., Xiang, Y. and Guo, L. (2008). Fabrication of radial ZnO nanowire clusters and radial ZnO/PVDF composites with enhanced dielectric properties, *Adv. Funct. Mater.*, 18, pp. 2584–2592.
[40] Wang, G.-S., Wu, Y.-Y., Zhang, X.-J., Li, Y., Guo, L. and Cao, M.-S. (2014). Controllable synthesis of uniform ZnO nanorods and their enhanced dielectric and absorption properties, *J. Mater. Chem. A*, 2, pp. 8644–8651.

[41] Dang, Z.-M., Nan, C.-W., Xie, D., Zhang, Y.-H. and Tjong, S. C. (2004). Dielectric behavior and dependence of percolation threshold on the conductivity of fillers in polymer-semiconductor composites, *Appl. Phys. Lett.*, 85, pp. 97–99.
[42] Zhang, Z. and Chung, T. C. M. (2007). Study of VDF/TrFE/CTFE terpolymers for high pulsed capacitor with high energy density and low energy loss, *Macromolecules*, 40, pp. 783–785.
[43] Cheng, Z. Y., Zhang, Q. M. and Bateman, F. B. (2002). Dielectric relaxation behavior and its relation to microstructure in relaxor ferroelectric polymers: High-energy electron irradiated poly(vinylidene fluoride-trifluoroethylene) copolymers, *J. Appl. Phys.*, 92, pp. 6749–6755.
[44] Nalwa, H. S. (1995). *Ferroelectric Polymers: Chemistry, Physics, and Applications* (Marcel Dekker Inc., New York).
[45] Zhang, S., Chu, B., Neese, B., Ren, K., Zhou, X. and Zhang, Q. M. (2006). Direct spectroscopic evidence of field-induced solid-state chain conformation transformation in a ferroelectric relaxor polymer, *J. Appl. Phys.*, 99, pp. 044107.
[46] Zhang, Q. M., Bharti, V. and Zhao, X. (1998). Giant electrostriction and relaxor ferroelectric behavior in electron-irradiated poly(vinylidene fluoride-trifluoroethylene) copolymer, *Science*, 280, pp. 2101–2104.
[47] Li, Z.-M., Li, S.-Q. and Cheng, Z. Y. (2004). Crystalline structure and transition behavior of recrystallized-irradiated P(VDF-TrFE) 65/35 copolymer, *J. Appl. Phys.*, 97, pp. 014102.
[48] Zhou, X., Zhao, X., Suo, Z., Zou, C., Runt, J., Liu, S., Zhang, S. and Zhang, Q. M. (2009). Electrical breakdown and ultrahigh electrical energy density in poly(vinylidene fluoride-hexafluoropropylene) copolymer, *Appl. Phys. Lett.*, 94, pp. 162901.
[49] Wegener, M. and Arlt, K. (2008). PZT/P(VDF-HFP) 0–3 composites as solvent-cast thin films: preparation, structure and piezoelectric properties, *J. Phys. D: Appl. Phys.*, 41, pp. 165409.
[50] Li, Z., Wang, Y. and Cheng, Z. Y. (2006). Electromechanical properties of poly(vinylidene-fluoride-chlorotrifluoroethylene) copolymer, *Appl. Phys. Lett.*, 88, pp. 062904.
[51] Lichtenecker, K. (1929). Microbody therapy as feasibility problem, *Physik. Zeitschr.*, 30, pp. 805–809.
[52] Hashin, Z. and Shtrikman, S. (1962). A variational approach to the theory of the effective magnetic permeability of multiphase materials, *J. Appl. Phys.*, 33, pp. 3125–3131.
[53] Simpkin, R. (2010). Derivation of Lichtenecker's logarithmic mixture formula from Maxwell's equations, *IEEE Trans. Microwave Theory Tech.*, 58, pp. 545–550.
[54] Scape, B. K. P. (1989). *Principles of Dielectrics* (Clarendon Press, Oxford).
[55] Birchak, J. R., Gardner, C. G., Hipp, J. E. and Victor, J. M. (1974). High dielectric constant microwave probes for sensing soil moisture, *Proc. IEEE*, 62, pp. 93–98.

[56] Landau, L. D. and E. M. L. (1960). *Electrodynamics of Continuous Media* (London Pergamon Press, London and New York).
[57] Looyenga, H. (1965). Dielectric constants of heterogeneous mixtures, *Physica*, 31, pp. 401–406.
[58] Wakino, K. (1994). A new proposal on mixing rule of the dielectric constant of mixture, in *9th IEEE Int. Sym Appl. Ferro.*, p. 33.
[59] Stölzle, S., A. E. and Nimtz, G. (1992). Numerical simulation of random composite dielectrics, *J. Phys. I France*, 2, pp. 401–408.
[60] Kim, J.-B., Kim, T.-W. and Kim, C.-G. (2006). Simulation method for complex permittivities of carbon black/epoxy composites at microwave frequency band, *J. Appl. Polym. Sci.*, 100, pp. 2189–2195.
[61] Bruggeman, D. a. G. (1935). Berechnung verschiedener physikalischer Konstanten von heterogenen Substanzen. I. Dielektrizitätskonstanten und Leitfähigkeiten der Mischkörper aus isotropen Substanzen, *Ann. Phys.*, 416, pp. 636–664.
[62] Rayleigh, L. (1892). LVI. On the influence of obstacles arranged in rectangular order upon the properties of a medium, *Phil. Mag.*, 34, pp. 481–502.
[63] Burton, E. F. and Turnbull, L. G. (1937). Dielectric constants of solids at high frequencies and the influence of water of crystallization on dielectric constant, *Proc. R. Soc. Lond. A*, pp. 182–198.
[64] Maxwell, J. C. (1892). *Electricity and Magnetism* (Clarendon Press, Oxford).
[65] Wagner, K. W. (1924). *Die Isolierstoffe der Elektrotechnik*, Vol. 1, edited by Schering, H. (Springer, Berlin).
[66] Webman, I., Jortner, J. and Cohen, M. H. (1977). Theory of optical and microwave properties of microscopically inhomogeneous materials, *Phys. Rev. B*, 15, pp. 5712–5723.
[67] Skipetrov, S. E. (1999). Effective dielectric function of a random medium, *Phys. Rev. B*, 60, pp. 12705–12709.
[68] Bergman, D. J. and Imry, Y. (1977). Critical behavior of the complex dielectric constant near the percolation threshold of a heterogeneous material, *Phys. Rev. Lett.*, 39, pp. 1222–1225.
[69] Polder, D. and Van Santeen, J. H. (1946). The effective permeability of mixtures of solids, *Physica.*, 12, pp. 257–271.
[70] Landauer, R. (1952). The electrical resistance of binary metallic mixtures, *J. Appl. Phys.*, 23, pp. 779–784.
[71] Mclachlan, D. S. (1986). Equations for the conductivity of macroscopic mixtures, *J. Phys. C.*, 19, pp. 1339–1354.
[72] Mclachlan, D. S. (1988). Measurement and analysis of a model dual-conductivity medium using a generalised effective-medium theory, *J. Phys. C.*, 21, pp. 1521–1532.
[73] Mclachlan, D. S. (1987). An equation for the conductivity of binary mixtures with anisotropic grain structures, *J. Phys. C.*, 20, pp. 865–877.

[74] Hanai, T. (1960). Theory of the dielectric dispersion due to the interfacial polarization and its application to emulsions, *Kolloid, Z.*, 171, pp. 23–31.
[75] Puranik, S. M., Kumbharkhane, A. C. and Mehrotra, S. C. (1994). The static permittivity of binary mixtures using an improved bruggeman model, *J. Mol. Liq.*, 59, pp. 173–177.
[76] Tuncer, E., Gubański, S. M. and Nettelblad, B. (2001). Dielectric relaxation in dielectric mixtures: Application of the finite element method and its comparison with dielectric mixture formulas, *J. Appl. Phys.*, 89, pp. 8092–8100.
[77] Calame, J. P. (2008). Dielectric permittivity simulation of random irregularly shaped particle composites and approximation using modified dielectric mixing laws, *J. Appl. Phys.*, 104, pp. 114108.
[78] Cule, D. and Torquato, S. (1998). Electric-field distribution in composite media, *Phys. Rev. B*, 58, pp. R11829–R11832.
[79] Yamada, T., Ueda, T. and Kitayama, T. (1982). Piezoelectricity of a high-content lead zirconate titanate/polymer composite, *J. Appl. Phys.*, 53, pp. 4328–4332.
[80] Nielsen, L. E. (1974). The permittivity of suspensions and two-phase mixtures, *J. Phys. D: Appl. Phys.*, 7, pp. 1549–1554.
[81] Jayasundere, N. and Smith, B. V. (1993). Dielectric constant for binary piezoelectric 0–3 composites, *J. Appl. Phys.*, 73, pp. 2462–2466.
[82] Kerner, E. H. (1956). The electrical conductivity of composite media, *Proc. Phys. Soc., London, Sect. B*, 69, pp. 802–807.
[83] Poon, Y. M. and Shin, F. G. (2004). A simple explicit formula for the effective dielectric constant of binary 0–3 composites, *J. Mater. Sci.*, 39, pp. 1277–1281.
[84] Starke, T. K. H., Johnston, C., Hill, S., Dobson, P. and Grant, P. S. (2006). The effect of inhomogeneities in particle distribution on the dielectric properties of composite films, *J. Phys. D: Appl. Phys.*, 39, pp. 1305–1311.
[85] Karkkainen, K., Sihvola, A. and Nikoskinen, K. (2001). Analysis of a three-dimensional dielectric mixture with finite difference method, *IEEE Trans. Geosci. Remote Sensing*, 39, pp. 1013–1018.
[86] Sareni, B., Krähenbühl, L., Beroual, A. and Brosseau, C. (1997). Effective dielectric constant of random composite materials, *J. Appl. Phys.*, 81, pp. 2375–2383.
[87] Ang, C., Yu, Z., Guo, R. and Bhalla, A. S. (2003). Calculation of dielectric constant and loss of two-phase composites, *J. Appl. Phys.*, 93, pp. 3475–3480.
[88] Zhao, X., Wu, Y., Fan, Z. and Li, F. (2004). Three-dimensional simulations of the complex dielectric properties of random composites by finite element method, *J. Appl. Phys.*, 95, pp. 8110–8117.
[89] Tuncer, E., Serdyuk, Y. V. and Gubanski, S. M. (2002). Dielectric mixtures: Electrical properties and modeling, *IEEE Trans. Dielectrics Electr. Insul.*, 9, pp. 809–828.

[90] Pecharromán, C. and Moya, J. S. (2000). Experimental evidence of a giant capacitance in insulator–conductor composites at the percolation threshold, *Adv. Mater.*, 12, pp. 294–297.

[91] Li, Z.-M., Arbatti, M. D. and Cheng, Z. Y. (2004). Recrystallization study of high-energy electron-irradiated P(VDF-TrFE) 65/35 copolymer, *Macromolecules*, 37, pp. 79–85.

[92] Vo, H. T. and Shi, F. G. (2002). Towards model-based engineering of optoelectronic packaging materials: dielectric constant modeling, *Microelectron J.*, 33, pp. 409–415.

[93] Murugaraj, P., Mainwaring, D. and Mora-Huertas, N. (2005). Dielectric enhancement in polymer-nanoparticle composites through interphase polarizability, *J. Appl. Phys.*, 98, pp. 054304.

[94] Wang, X. X., Lam, K. H., Tang, X. G. and Chan, H. L. W. (2004). Dielectric characteristics and polarization response of lead-free ferroelectric $(Bi_{0.5}Na_{0.5})_{0.94}Ba_{0.06}TiO_3$-P(VDF-TrFE) 0–3 composites, *Solid State Commun.*, 130, pp. 695–699.

[95] Kuo, D.-H., Chang, C.-C., Su, T.-Y., Wang, W.-K. and Lin, B.-Y. (2004). Dielectric properties of three ceramic/epoxy composites, *Mat. Chem. Phys.*, 85, pp. 201–206.

[96] Dang, Z.-M., Wang, H.-Y., Peng, B. and Nan, C.-W. (2008). Effect of $BaTiO_3$ size on dielectric property of $BaTiO_3$/PVDF composites, *J. Electroceram.*, 21, pp. 381–384.

[97] Dang, Z.-M., Zheng, Y. and Xu, H.-P. (2008). Effect of the ceramic particle size on the microstructure and dielectric properties of barium titanate/polystyrene composites, *J. Appl. Polym. Sci.*, 110, pp. 3473–3479.

[98] Olszowy, M., Pawlaczyk, C., Markiewicz, E. and Kułek, J. (2005). Dielectric and pyroelectric properties of $BaTiO_3$–PVC composites, *Phys. Stat. Sol.* (a), 202, pp. 1848–1853.

[99] Kakimoto, M.-A., Takahashi, A., Tsurumi, T.-A., Hao, J., Li, L., Kikuchi, R., Miwa, T., Oono, T. and Yamada, S. (2006). Polymer–ceramic nanocomposites based on new concepts for embedded capacitor, *Mater. Sci. Eng. B*, 132, pp. 74–78.

[100] Lam, K. H., Chan, H. L. W., Luo, H. S., Yin, Q. R., Yin, Z. W. and Choy, C. L. (2003). Dielectric properties of 65PMN-35PT/P(VDF-TrFE) 0–3 composites, *Microelectron. Eng.*, 66, pp. 792–797.

[101] Bai, Y., Cheng, Z. Y., Bharti, V., Xu, H. S. and Zhang, Q. M. (2000). High-dielectric-constant ceramic-powder polymer composites, *Appl. Phys. Lett.*, 76, pp. 3804–3806.

[102] Hilczer, B., Kułek, J., Markiewicz, E., Kosec, M. and Malič, B. (2002). Dielectric relaxation in ferroelectric PZT–PVDF nanocomposites, *J. Non-Cryst. Sol.*, 305, pp. 167–173.

[103] Dong, L., Xiong, C., Quan, H. and Zhao, G. (2006). Polyvinyl-butyral/lead zirconate titanates composites with high dielectric constant and low dielectric loss, *Scr. Mater.*, 55, pp. 835–837.

[104] Lam, K. S., Wong, Y. W., Tai, L. S., Poon, Y. M. and Shin, F. G. (2004). Dielectric and pyroelectric properties of lead zirconate titanate/polyurethane composites, *J. Appl. Phys.*, 96, pp. 3896–3899.
[105] Adikary, S. U., Chan, H. L. W., Choy, C. L., Sundaravel, B. and Wilson, I. H. (2002). Characterisation of proton irradiated $Ba_{0.65}Sr_{0.35}TiO_3$/P(VDF-TrFE) ceramic–polymer composites, *Compos. Sci. Technol.*, 62, pp. 2161–2167.
[106] Lam, K. S., Zhou, Y., Wong, Y. W. and Shin, F. G. (2005). Electrostriction of lead zirconate titanate/polyurethane composites, *J. Appl. Phys.*, 97, pp. 104112.
[107] Zhang, L., Shan, X., Wu, P., Song, J. and Cheng, Z. Y. (2010). Microstructure and dielectric properties of CCTO-P(VDF-TrFE) nanocomposites, *Ferroelectrics*, 405, pp. 92–97.
[108] Shri Prakash, B. and Varma, K. B. R. (2007). Dielectric behavior of CCTO/epoxy and Al-CCTO/epoxy composites, *Compos. Sci. Technol.*, 67, pp. 2363–2368.
[109] Thomas, P., Varughese, K. T., Dwarakanath, K. and Varma, K. B. R. (2010). Dielectric properties of Poly(vinylidene fluoride)/$CaCu_3Ti_4O_{12}$ composites, *Compos. Sci. Technol.*, 70, pp. 539–545.
[110] Amaral, F., Rubinger, C. P. L., Henry, F., Costa, L. C., Valente, M. A. and Barros-Timmons, A. (2008). Dielectric properties of polystyrene-CCTO composite, *J. Non-Cryst. Sol.*, 354, pp. 5321–5322.
[111] Dang, Z.-M., Zhou, T., Yao, S.-H., Yuan, J.-K., Zha, J.-W., Song, H.-T., Li, J.-Y., Chen, Q., Yang, W.-T. and Bai, J. (2009). Advanced calcium copper titanate/polyimide functional hybrid films with high dielectric permittivity, *Adv. Mater.*, 21, pp. 2077–2082.
[112] Wang, F., Zhou, D. and Hu, Y. (2009). Preparation and dielectric properties of $CaCu_3Ti_4O_{12}$-polyethersulfone composites, *Phys. Status Solidi A*, 206, pp. 2632–2636.
[113] Li, Z. M. (2004). Novel electroactive poly(vinylidene fluoride)-based polymer systems and their applications (Ph. D. thesis, Auburn University).
[114] Shan, X. B. (2010). High dielectric constant 0–3 ceramic–polymer composites (Ph. D. thesis, Auburn University).
[115] Chu, B., Zhou, X., Ren, K., Neese, B., Lin, M., Wang, Q., Bauer, F. and Zhang, Q. M. (2006). A dielectric polymer with high electric energy density and fast discharge speed, *Science*, 313, pp. 334–336.
[116] Subramanian, M. A., Li, D., Duan, N., Reisner, B. A. and Sleight, A. W. (2000). High dielectric constant in $ACu_3Ti_4O_{12}$ and $ACu_3Ti_3FeO_{12}$ phases, *J. Solid State Chem.*, 151, pp. 323–325.
[117] Kirkpatrick, S. (1973). Percolation and conduction, *Rev. Mod. Phys.*, 45, pp. 574–588.
[118] Koskenmaki, D. C., Calhoun, C. D. and Huff, B. E. (1991) Metal/polymer composites, US Patent.

[119] Xue, Q. (2004). The influence of particle shape and size on electric conductivity of metal-polymer composites, *Eur. Polym. J.*, 40, pp. 323–327.
[120] Clarkson, M. T. (1988). Electrical conductivity and permittivity measurements near the percolation transition in a microemulsion. II. Interpretation, *Phys. Rev. A*, 37, pp. 2079–2090.
[121] Webman, I., Jortner, J. and Cohen, M. H. (1977). Critical exponents for percolation conductivity in resistor networks, *Phys. Rev. B*, 16, pp. 2593–2596.
[122] Berhan, L. and Sastry, A. M. (2007). Modeling percolation in high-aspect-ratio fiber systems. I. Soft-core versus hard-core models, *Phys. Rev. E*, 75, pp. 041120.
[123] Pecharromán, C., Esteban-Betegón, F., Bartolomé, J. F., López-Esteban, S. and Moya, J. S. (2001). New percolative $BaTiO_3$–Ni composites with a high and frequency-independent dielectric constant ($\epsilon_r \approx 80000$), *Adv. Mater.*, 13, pp. 1541–1544.
[124] Sichel, E. K., Rubner, M. F., Druy, M. A., Gittleman, J. I. and Bozowski, S. (1984). Giant dielectric constant of $IrCl_6$-doped polyacetylene, *Phys. Rev. B*, 29, pp. 6716–6721.
[125] Rao, Y. and C. P. W. (2002). *8th IEEE Proc. Electron. Compon. Technol. Conf.*, IEEE, Piscataway, NJ, p. 920.
[126] Qi, L., Lee, B. I., Chen, S., Samuels, W. D. and Exarhos, G. J. (2005). High-dielectric-constant silver–epoxy composites as embedded dielectrics, *Adv. Mater.*, 17, pp. 1777–1781.
[127] Dang, Z. M., Lin, Y. H. and Nan, C. W. (2003). Novel ferroelectric polymer composites with high dielectric constants, *Adv. Mater.*, 15, pp. 1625–1629.
[128] Xu, H.-P., Dang, Z.-M., Bing, N.-C., Wu, Y.-H. and Yang, D.-D. (2010). Temperature dependence of electric and dielectric behaviors of Ni/polyvinylidene fluoride composites, *J. Appl. Phys.*, 107, pp. 034105.
[129] Xu, J., Moon, K.-S., Tison, C. and Wong, C. P. (2006). A novel aluminum-filled composite dielectric for embedded passive applications, *IEEE Trans. Adv. Packaging*, 29, pp. 295–306.
[130] Panda, M., Srinivas, V. and Thakur, A. K. (2008). On the question of percolation threshold in polyvinylidene fluoride/nanocrystalline nickel composites, *Appl. Phys. Lett.*, 92, pp. 132905.
[131] Huang, X. Y., Jiang, P. K. and Kim, C. U. (2007). Electrical properties of polyethylene/aluminum nanocomposites, *J. Appl. Phys.*, 102, pp. 124103.
[132] Dang, Z.-M., Zhang, Y.-H. and Tjong, S. C. (2004). Dependence of dielectric behavior on the physical property of fillers in the polymer–matrix composites, *Synth. Met.*, 146, pp. 79–84.
[133] Zois, H., Apekis, L. and Mamunya, Y. P. (2003). Dielectric properties and morphology of polymer composites filled with dispersed iron, *J. Appl. Polym. Sci.*, 88, pp. 3013–3020.

[134] Dang, Z.-M., Wu, J.-P., Xu, H.-P., Yao, S.-H., Jiang, M.-J. and Bai, J. (2007). Dielectric properties of upright carbon fiber filled poly(vinylidene fluoride) composite with low percolation threshold and weak temperature dependence, *Appl. Phys. Lett.*, 91, pp. 072912.

[135] Wang, L. and Dang, Z.-M. (2005). Carbon nanotube composites with high dielectric constant at low percolation threshold, *Appl. Phys. Lett.*, 87, pp. 042903.

[136] Williams, C. D., Robles, R. O., Zhang, M., Li, S., Baughman, R. H. and Zakhidov, A. A. (2008). Multiwalled carbon nanotube sheets as transparent electrodes in high brightness organic light-emitting diodes, *Appl. Phys. Lett.*, 93, pp. 183506.

[137] Lin, K., Kumar, R. S., Cheng, L. S. and Jin, C. S. (2009). Performance improvement of carbon-nanotube-incorporated transparent conducting anode film for organic device application, *IEEE Trans. Electron Devices*, 56, pp. 31–37.

[138] Stankovich, S., Dikin, D. A., Dommett, G. H. B., Kohlhaas, K. M., Zimney, E. J., Stach, E. A., Piner, R. D., Nguyen, S. T. and Ruoff, R. S. (2006). Graphene-based composite materials, *Nature*, 442, pp. 282–286.

[139] Bryning, M. B., Islam, M. F., Kikkawa, J. M. and Yodh, A. G. (2005). Very low conductivity threshold in bulk isotropic single-walled carbon nanotube-epoxy composites, *Adv. Mater.*, 17, pp. 1186–1191.

[140] He, F., Lau, S., Chan, H. L. and Fan, J. (2009). High dielectric permittivity and low percolation threshold in nanocomposites based on poly(vinylidene fluoride) and exfoliated graphite nanoplates, *Adv. Mater.*, 21, pp. 710–715.

[141] Li, Q., Xue, Q., Hao, L., Gao, X. and Zheng, Q. (2008). Large dielectric constant of the chemically functionalized carbon nanotube/polymer composites, *Compos. Sci. Technol.*, 68, pp. 2290–2296.

[142] Dang, Z. M., Wang, L., Yin, Y., Zhang, Q. and Lei, Q. Q. (2007). Giant dielectric permittivities in functionalized carbon-nanotube/ electroactive-polymer nanocomposites, *Adv. Mater.*, 19, pp. 852–857.

[143] Yuan, J.-K., Yao, S.-H., Dang, Z.-M., Sylvestre, A., Genestoux, M. and Bai, J. (2011). Giant dielectric permittivity nanocomposites: realizing true potential of pristine carbon nanotubes in polyvinylidene fluoride matrix through an enhanced interfacial interaction, *J. Phys. Chem. C*, 115, pp. 5515–5521.

[144] Zheng, W., Lu, X., Wang, W., Wang, Z., Song, M., Wang, Y. and Wang, C. (2010). Fabrication of novel Ag nanowires/poly(vinylidene fluoride) nanocomposite film with high dielectric constant, *Phys. Stat. Sol.* (a), 207, pp. 1870–1873.

[145] Li, Y.-J., Xu, M., Feng, J.-Q. and Dang, Z.-M. (2006). Dielectric behavior of a metal–polymer composite with low percolation threshold, *Appl. Phys. Lett.*, 89, pp. 072902.

[146] Dang, Z., Shen, Y., Fan, L., Cai, N., Nan, C. and Zhao, S. (2003). Dielectric properties of carbon fiber filled low-density polyethylene, *J. Appl. Phys.*, 93, pp. 5543–5545.

[147] Sui, G., Jana, S., Zhong, W. H., Fuqua, M. A. and Ulven, C. A. (2008). Dielectric properties and conductivity of carbon nanofiber/semi-crystalline polymer composites, *Acta Mater.*, 56, pp. 2381–2388.

[148] Xu, J. and Wong, C. P. (2005). Low-loss percolative dielectric composite, *Appl. Phys. Lett.*, 87, p. 082907.

[149] Shen, Y., Lin, Y., Li, M. and Nan, C. W. (2007). High dielectric performance of polymer composite films induced by a percolating interparticle barrier layer, *Adv. Mater.*, 19, pp. 1418–1422.

[150] Dai, T., Chen, K., Qing, X., Lu, Y., Zhu, J. and Gao, F. (2010). Sequential polymer precipitation of core–shell microstructured composites with giant permittivity, *Macromol. Rapid Commun.*, 31, pp. 484–489.

[151] Dang, Z.-M., Lin, Y.-Q., Xu, H.-P., Shi, C.-Y., Li, S.-T. and Bai, J. (2008). Fabrication and dielectric characterization of advanced $BaTiO_3$/polyimide nanocomposite films with high thermal stability, *Adv. Funct. Mater.*, 18, pp. 1509–1517.

[152] Li, K., Wang, H., Xiang, F., Liu, W. and Yang, H. (2009). Surface functionalized $Ba_{0.6}Sr_{0.4}TiO_3$/poly(vinylidene fluoride) nanocomposites with significantly enhanced dielectric properties, *Appl. Phys. Lett.*, 95, pp. 202904.

Chapter 9

Flexible Nanodielectric Materials with High Permittivity for Power Energy Storage*

Zhi-Min Dang[†,‡,§,¶], Jin-Kai Yuan[†,‡,||], Sheng-Hong Yao[‡] and Rui-Jin Liao[§]

[†]*Department of Polymer Science and Engineering, School of Chemistry and Biological Engineering, University of Science and Technology Beijing, Beijing, 100083, P. R. China*

[‡]*State Key Laboratory of Chemical Resource Engineering Beijing University of Chemical Technology, Beijing, 100029, P. R. China*

[§]*State Key Laboratory of Power Transmission Equipment and System Security, Chongqing University, Chongqing, 400030, P. R. China*

[¶]*dangzm@ustb.edu.cn*

[||]*yuanjinkai@gmail.com*

Study of flexible nanodielectric materials (FNDMs) with high permittivity is one of the most active academic research areas in advanced functional materials. FNDMs with excellent dielectric properties are demonstrated to show great promise as energy-storage dielectric layers in high-performance capacitors. These materials, in common, consist of nanoscale particles dispersed into a flexible polymer matrix so that both the physical/chemical characteristics of the nanoparticles and the

*This article was firstly published as "Dang, Z.-M., Yuan, J.-K. Yao, S.-H. and Liao, R.-J. (2013), Flexible nanodielectric materials with high permittivity for power energy storage, *Adv. Mater.*, 25, pp. 6334–6365". Reprinted with the permission of John Wiley & Sons. Inc.

interaction between the nanoparticles and the polymers have crucial effects on the microstructures and final properties. This review first outlines the crucial issues in the nanodielectric field and then focuses on recent remarkable research developments in the fabrication of FNDMs with special constituents, molecular structures, and microstructures. Possible reasons for several persistent issues are analyzed and the general strategies to realize FNDMs with excellent integral properties are summarized. The review further highlights some exciting examples of these FNDMs for power-energy-storage applications.

1. Introduction

Ever-increasing energy requirements and the exhaustion of fossil fuels demand an improvement in the efficiency of energy usage, as well as the search for sustainable and renewable resources. As a result, there is an increase in the development of renewable energy production from solar and wind technologies. Due to the intermittency of such production and consumption requirements, it is of the utmost importance to develop new, low-cost, and environmentally friendly energy-conversion and storage systems, such as batteries [1], fuel cells [2], electrochemical capacitors [3], and dielectric-based capacitors [4]. Among these energy-storage technologies, dielectric capacitors possess an intrinsic high power density due to their very fast energy uptake and delivery, and thus hold great promise for the generation of high-performance power electronics used in hybrid electric vehicles, medical devices, and electrical weapon systems [5–8]. However, these potential applications are severely hindered by the low energy density of capacitors with current commercial dielectrics [9–13].

Generally, the energy density (U_{e}) of a dielectric material is the shaded area, as illustrated schematically in Fig. 1, which is equal to the integral:

$$U_e = \int_{D_{\max}}^{0} E dD, \tag{1}$$

where E is the electric field and $D_{\max}$ is the electric displacement (charge density) at the highest field. For a linear dielectric material, such as polymers and their composites, Eq. (1) becomes:

$$U_C = \frac{1}{2} DE = \frac{1}{2} \varepsilon_r \varepsilon_0 E^2, \tag{2}$$

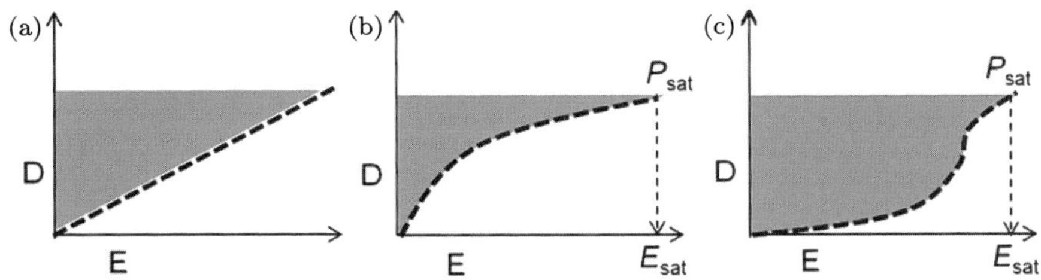

Figure 1. Electric energy density of: (a) linear, (b) ferroelectric, and (c) antiferroelectric dielectrics. The shaded area is related to the stored energy density.

where ε_r is the relative dielectric permittivity, and ε_0 is the permittivity of free space ($=8.85 \times 10^{-12}\,\mathrm{F\,m^{-1}}$). In linear dielectrics, the permittivity does not change with applied electric field. Energy storage is dominated by the field that can be applied to a capacitor, which is limited by the film's dielectric breakdown strength (E_b). Therefore, to increase the energy density of a capacitor, it is of tremendous importance to develop low-loss electroactive materials with high dielectric permittivity and high breakdown strength.

Conventional dielectric materials are ceramics with large dielectric permittivity, coupled with high stiffness and excellent thermal stability [7, 14–16]. However, their applicability for practical high-U_e capacitors is largely impeded by their small breakdown strength, high density, poor flexibility, and challenging processing conditions [4, 17]. As compared with conventional ceramics, polymer dielectric materials have a relatively high electric breakdown field [13, 18–21] and, importantly, offer processing advantages, including mechanical flexibility and the capability of being molded into various configurations for power electronic devices with reduced volume and weight. However, most polymers still fall significantly short of the rising demands for high-U_e capacitors because of their intrinsic low permittivity values (ca. $\varepsilon_r < 10$). This may cause the high energy density in the dielectric polymers to strongly rely on the high applied electric field (i.e., $E > 500\,\mathrm{MV/m}$) [13, 19, 22]. Thus, a key issue is to raise the dielectric permittivity of polymers substantially, while retaining their other excellent physical properties.

The introduction of inorganic particles into polymer matrices to form dielectric composites represents one of the most promising avenues to this target. The composite approach capitalizes upon the idea that the combination of the impressive permittivity of inorganic particles with good dielectric strength of polymers may result in a high energy-storage capacity. Previous investigations on dielectric composites concentrated mainly on the microsized particle filled systems, yet they are subsequently found to show impairment in the dielectric properties due to the surface defects and stress cracking after a long-serving time [20, 23–25]. The thickness of the microcomposite films is also often limited by the size of the microparticles [26, 27]. On the other hand, the continuous trend of structural miniaturization in modern electronics demands processing capacitor dielectrics into very thin films. As to maintaining the energy (W) stored constant in a capacitor with a given dielectric material, one has to decrease the thickness (d) of the dielectric film to counter the reduction in the capacitor volume ($\Delta\nu$), which can be clearly reflected by the following expression:

$$W = \frac{1}{2}CV^2 = \frac{1}{2}\varepsilon_r\varepsilon_0 V^2\left(\frac{\nabla v}{d^2}\right), \tag{3}$$

where C is capacitance of the parallel plate capacitor with a voltage V applied on it. The only way to avert the aforementioned conflict is replacing the microparticles with a nanometer-sized filler to prepare nanocomposite dielectrics. The advantages of nanocomposite dielectrics include low filler loading, nanometer sizes, and large internal surface areas [28, 29]. The low nanofiller loadings allow for composite formation without sacrificing some of the inherent polymeric properties, such as, density, flexibility, and facile processability. Furthermore, when the interfacial energy barriers came from the nanoparticle aggregation driven by the van der Waals forces are overcome, the interfiller distances in the nanocomposites can be in the range of nanometers, and the filler would interact chemically and physically with polymer matrices, resulting in the emergence of intermediate or mesoscopic properties. Such mesoscopic properties at interfaces always bring about unexpectedly excellent macroscopic properties of nanocomposites [30]. It has been demonstrated that, as the size of filler particles decreases to the nanometer scale, large interfacial areas in the composite between the polymer and nanoparticle would promote the exchange coupling effect through a dipolar interface layer, affording higher polarization levels, dielectric response, and breakdown strength [8, 26, 31–38]. All these features offer new opportunities for the design of an entirely different class of dielectrics, flexible nanodielectrics materials (FNDMs) with nanometric interfaces, that can offer higher energy storage or lower dielectric loss, and improved power dissipation [39]. The continuously increasing number (Fig. 2) of the scientific publications in this field indicates that FNDMs have been arousing enormous research attention and are gradually becoming a hot topic in the energy-storage field, especially in which high power uptake or delivery is essential.

To accelerate real applications of FNDMs in high-U_e capacitors, it is of utmost importance to be able to tailor their molecular structures and/or microstructures, and thus dielectric properties. Much remarkable progress has been achieved over the past decade, but many obstacles still remain, restricting further development

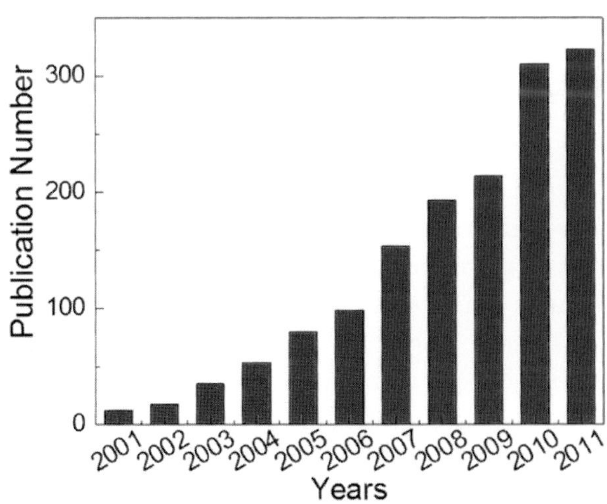

Figure 2. The number of publications in the field of FNDMs from 2001 to 2011. The data come from search results from Web of Knowledge, accessed November 5, 2012.

in this field. To clarify the emerging problems and to provide a comprehensive understanding of this field, we summarize the recent progress in the synthesis, structure, and dielectric properties of various FNDMs. Some long-standing problems and topics that warrant further investigation in the near future are also addressed.

2. Nanoparticle surface characteristics and interaction between nanoparticles and polymers

2.1. *Nanoparticle surface characteristics*

Nanoparticles are usually defined as microscopic particles with at least one dimension less than 100 nm, which can exhibit characteristics totally different from either atoms or bulk materials. As the particle shrinks in size from the micrometer to the nanometer scale, especially below 20 nm, the percentage of atoms at the surface of the particle becomes more significant, leading to a dramatic change in the physical properties, interfacial properties, and also the agglomeration behavior relative to bulk materials [40]. For instance, the dielectric properties of inorganic particles are largely dependent on their size [26, 41–46]. Taking $BaTiO_3$ as an example, the dielectric permittivity decreases dramatically from 5000 to hundreds as the grain size is reduced from 1 μm to 30 nm. Such a decrease in dielectric permittivity arises from the reduction of tetragonality with decreasing particle size and the transition to cubic structure (Fig. 3) when the grain size drops down to 30 nm [26, 43–46]. A parallel study has been performed on $PbZrO_3$ particles, and a ferroelectric to paraelectric phase transition is observed as the particles size becomes less than 100 nm, which also results in a decrease in the dielectric permittivity [41].

Another main concern along with the size effect in nanoparticles is their special surface characteristics associated with the large surface area to volume ratio. A demonstrative example of the size effect on the surface area is to fill a unit cube using micrometer and nanometer-sized cubes. The unit cube has a total edge length of about 1 cm, which would have a total surface area of about 6×10^{-4} m^2. Reducing the edge length of the filled cubes to 10 μm, the unit cube would

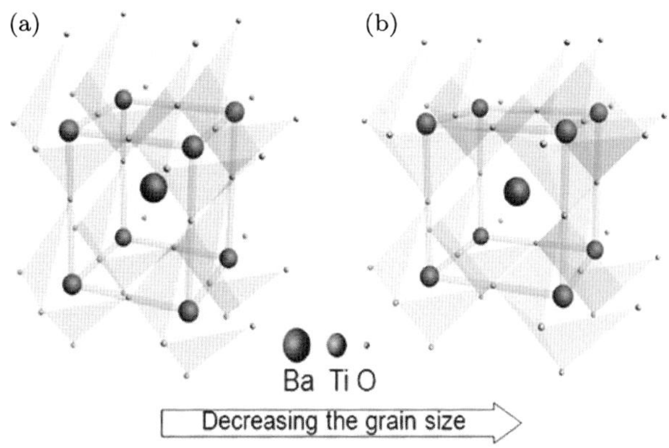

Figure 3. The crystal structure transition from: (a) tetragonal phase to (b) cubic phase in $BaTiO_3$ particle with decreasing the grain size from 1 μm to 30 nm.

consist of approximately 10^9 microcubes with a total surface area of 0.6 m^2. A further reduction to 5 nm would lead to 8×10^{18} nanocubes with a dramatically large surface area of 1200 m^2. This is schematically illustrated in Fig. 4(a). The great specific surface area would cause the nanoparticles to be more reactive with some other molecules. It has been well established that the dielectric behavior of nanoparticles would be significantly influenced by the properties of interfaces with their surroundings [34, 39, 47], since these interactions may give rise to a certain degree of polarization and charge separation at the interface [47], while the high surface energy of the nanoparticles encourages coalescence and largely impairs the uniformity of their nanocomposites, which is not desirable to realize the potential of nanoparticles in dielectric polymer nanocomposites.

In order to overcome the interfacial energy barriers due to a hydroxylated surface of fillers and aggregation of the particles driven by van der Waals forces, suitable organic modification of the particle surface is needed, aiming to alter the surface free energy and thus optimize their dispersibility in polymer matrices [27, 40, 48–52]. In an ideal case, assuming monomodal spherical nanoparticles are homogenously filled into a polymer matrix with aggregate free dispersion and a face-centered-cubic particle deposition as well, a maximum filling degree of 74 vol% can be obtained. However, the filling degree would

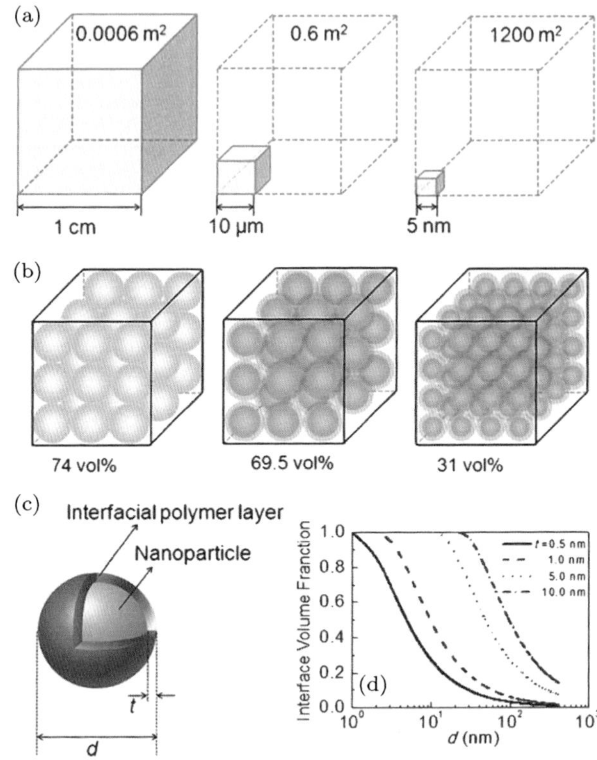

Figure 4. (a) Schematic representation of the increasing surface area as particle size decreases by filling a unit cube with micrometer-and nanometer-sized cubes. (b) Maximum filling degree of nanoparticles in a polymer matrix with assumption that spherical nanoparticles have a cubic face centered particle deposition. (c) A typical core–shell nanoparticle with a diameter of d and an interfacial polymer layer with thickness of t inside the diameter. (d) Interface volume percentage as a function of particle size, d, for different interface thickness 0.5, 1.0, 5, and 10.0 nm.

be significantly lower in a real nanocomposite. Due to the very high specific surface areas, an interfacial polymer layer would be attached to the nanoparticle surfaces to form a core–shell nanostructure [53]. Consequently, it is definitely core–shell nanoparticles rather than naked particles that are dispersed in real polymer nanocomposites. The presence of this shell would reduce the maximum filling degree of nanoparticles in a polymer matrix. Assuming the nanoparticles have a faced-centered-cubic arrangement and polymer shells with a constant thickness of 0.5 nm on them, the maximum filling degree drops from 69.5 vol% to 31 vol% on decreasing the particle size from 50 nm to 3 nm, as schematically shown in Fig. 4(b) [40]. This indicates that the interfacial polymer layer becomes increasingly prominent as the particle size is reduced. This situation can be well explained by the interfacial polymer volume fraction. Considering a system with spherical core–shell nanoparticles with a diameter d and an interface with thickness t inside the diameter of the core–shell nanoparticle, as shown in Fig. 4(c), the interface volume fraction f can be calculated by the following formula:

$$f = 3\frac{2t}{d}\left[1 - \left(\frac{2t}{d}\right) + \frac{1}{3}\left(\frac{2t}{d}\right)^2\right]. \tag{4}$$

If the interface is 1.0 nm in thickness, then the interfacial volume exceeds 50% of the core–shell particle volume when the particle diameter is less than 10 nm, as illustrated in Fig. 4(d). It should be emphasized that the physical properties of the mesoscopic interfacial region differ much from the nanoparticle or the bulk polymer matrix, depending on the local topographic features [40, 47, 54, 55]. Therefore, as the particle size decreases, the properties of the interfacial polymer layer of nanometer dimension would have an increasingly dominant role in determining the macroscopic properties of the nanocomposites. Clearly the properties of the interface can strongly influence the overall dielectric properties, and thus the energy density of nanocomposites. In this regard, it is necessary to consider how interfaces will respond to impressed electrical fields and contribute to the energy-storage capacity in polymer nanocomposites.

2.2. *Multi-layered core model*

In order to fully describe how the interfaces are formed chemically, physically, and electrically, Tanaka *et al.* [56] suggested and proposed a hypothetical multi-layered core model for interfaces in a nanocomposite based on a spherical inorganic particle embedded in a polymer matrix. The multicore structure involves a bonded layer,

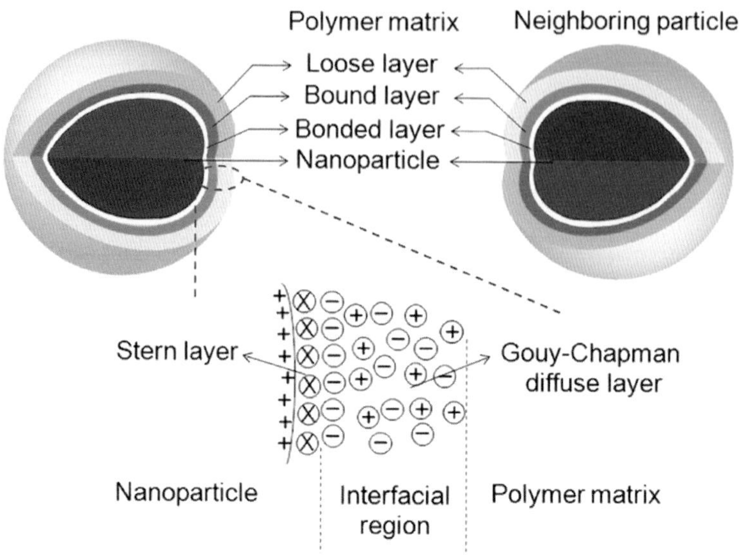

Figure 5. Multi-core model for interfaces between inorganic nanoparticles and polymer matrix in a polymer nanocomposite. It consists of a bonded layer, a bound layer, and a loose layer. As the nanoparticle is positively charged, a diffuse electrical double layer would form in the interfacial region and overlap the three layers above.

a bound layer and a loose layer, as shown in Fig. 5. The bonded layer corresponds to a transition layer with thickness of about 1 nm, which is tightly bonded to both inorganic and organic substances. Such bonding arises from either ionic, covalent or hydrogen bonds, or van der Waals force. The bound layer is a layer of polymer chains strongly bound and/or interacted to the bonded layer and the surface of the inorganic particle. This layer is heavily involved in altering the polymer chain conformation and/or mobility, the degree of chain folding, and other stereographic structures. Its thickness is in the range of 2–9 nm. The morphology and thickness of this layer are strongly dependent on the interfacial interaction strength in the bonded layer. It is postulated that tight binding such as ionic or covalent bonding will create a thick, bound layer with a highly ordered structure around the bonded layer. The loose layer is a region loosely coupling to the bound layer; nevertheless, in that layer chain conformation and mobility, and even free volume or crystallinity differ from the polymer matrix. It has a larger thickness, of several tens of nanometers, than the inner layers. Remembering that the interfiller spacing is of the same order of magnitude as the particle size in a nanocomposite loaded with several wt% of nanofillers, the interface of a nanoparticle is then prone to be overlapped with that of its nearest nanoparticle. In this case, the neighboring core layers are mutually superimposed, giving rise to a zone quite different from the filler and polymer matrix. It is anticipated that this interfacial overlapping phenomenon causes some collaborative effect among neighboring particles (far-field effect) so as to intensify the effect of interfaces on the macroscopic properties of the nanocomposites.

In fact, the morphology of interfacial layers is not only affected chemically by the surface characteristics of nanoparticles, but also electrically by the Gouy–Chapman diffusion layer formed when charges are injected [56], as shown in Fig. 5. There is an interfacial region (charge distribution) between the nanoparticle and the polymer matrix in this model. The nanoparticle becomes charged, possibly as a consequence of the equalization of the Fermi level or chemical potential, which causes ionization of surface groups as well as adsorption of ions from the polymer matrix. The polymer localized at the interfacial region then responds by establishing a screening diffuse double layer with excess counter charges [34]. Two mechanisms have been evoked for this response. The first is concerned with the polymer polarization involving both electronic polarization (dipole motion) and the orientation of permanent dipoles (molecular motion). The second is associated with a diffuse electrical double layer consisting of positive and negative ions if the polymer is solid electrolyte and contains mobile charges. In this layer, charge decays exponentially with distance according to the Born approximation [57]. Debye shielding length of the Gouy–Chapman diffusion layer can be calculated as several tens to 100 nm, indicating that the nanoparticles may also interact electrically with each other's nearest neighbors to result in a possible collaborative effect [56]. Therefore, it is plausible that the energy-storage capacity of the dielectric polymer nanocomposites largely drives from the interfacial polarization, which is mainly associated with the bonded layer. Meanwhile the bound and loose layers also influence the energy density, either by a local morphology change as a result of adhesion with bonded layer or through the far-field effect between the neighboring particles.

According to the multicore model, the dielectric properties of nanocomposites can be different from material to material due to relative differences in the interfacial properties. In detail, the binding strength in the bonded layer not only affects the overall thickness of interfacial region and thus determines the occurrence of the far-field effect, but also influences a lot the local morphology and crystallite structure of the inner bound layer, even the outer loose layer. It is widely accepted that the inner bound layer acts to impair the motion of dipoles originating from polar radicals, and the outer loose layer contributes to the reduction in free volume, both of which lead to a reduction in the dielectric permittivity [30, 56]. In this case, seeking an appropriate coupling agent to establish an optimal binding strength between the nanoparticle and the polymer matrix is crucial to enhance the interfacial polarization, meanwhile minimizing the negative effects, and consequently reaching a high energy density. Other important parameters are the shape and size of nanofiller, which determine the interfiller distances together with the filler content. The fillers with large aspect (length to diameter) ratio are considered to easily interact with each other to self-assemble into skeleton-like superstructures in the nanocomposites [30]. Moreover, fillers with a smaller size give rise to a smaller interfiller distance, which would more easily trigger the far-field effect between neighboring nanoparticles. However, a very fine filler usually gives rise to a huge internal surface area in the composites, increasing the loss tangent approximately linearly [58–61]. Therefore, a careful selection of the sizes of the nanoparticles is crucial because large particles would suppress loss while small ones would benefit advantages from the nanocomposite concept. The realization of this proposed multicore model makes a conversion of bulk polymer into interfacial one, and holds the key to the dielectric polymer nanocomposites with high performances. Tailoring the nanoparticle surfaces as well as tuning the interfacial polymer shell structures are recognized as crucial approaches to meet this target. Successful examples of the application of core–shell nanostructure in FNDMs will be elaborated in detail in the next sections.

3. Inorganic particle/polymer nanodielectric materials

3.1. *Ceramic particle/polymer nanodielectric materials*

3.1.1. *Challenges for improving energy-storage density*

In general, the maximum energy density in a nanodielectric material is proportional to the product of the dielectric permittivity and the square of the dielectric breakdown strength, the maximum applied field it can withstand without short-circuiting. Therefore, the main approach to increase the energy density is to increase the dielectric permittivity. The currently available dielectric polymers, such as polypropylene or epoxy resins, usually possess dielectric permittivities of only 2–5. Much higher permittivity values are achieved from highly polarizable materials such as barium titanate ($BaTiO_3$) and polyvinylidene fluoride (PVDF) because of the polarization arising from ion displacements and permanent dipole alignment, respectively [62]. A material with a larger dielectric permittivity is recognized to be able to store proportionally more electrostatic energy. For instance, the PVDF-based terpolymer poly(vinylidene fluoride-trifluoroethylene-chlorofluoroethylene) [P(VDF-TrFE-CFE)] exhibits a dielectric permittivity as high as 65, and can be polarized with electric fields of 500 MV/m or more [63–66]. This produces a purely electrostatic capacitor with quite impressive recoverable energy density of 13 J/cm^3, whereas biaxially oriented polypropylene (BOPP) has similar electric field strength, but the maximum energy density is only 2 J/cm^3, which is limited by the low dielectric permittivity of BOPP (ca. 2.2) [10, 67, 68]. On the other hand, ion displacive dielectrics such as ferroelectric $BaTiO_3$ or lead manganate niobate (PMN) relaxors would seem much better choices for high-U_e capacitors because of their high dielectric permittivities ranging from hundreds to thousands [4, 7]. Unfortunately, they possess much lower dielectric strengths than polymers and, consequently, cannot take full advantage of their higher polarizability to store more energy in the material.

An appealing approach is to combine the polymers of high dielectric strength with the metal oxides of large dielectric permittivity by homogenously dispersing oxide nanoparticles into a polymer matrix [69]. In such a way, the nanoparticle fillers would contribute to the increment in the effective dielectric permittivity of the composite in terms of effective medium theory, and the polymer matrices would protect against the dielectric breakdown. Research has in past years concentrated on the inorganic/polymer nanodielectric materials, and succeeded in improving their overall dielectric performances using ferroelectric metal oxides such as TiO_2 [31, 51, 70–80], ZrO_2 [59, 81–83], $BaTiO_3$ [32, 33, 52, 59, 82, 84–91], $CaCu_3Ti_4O_{12}$ (CCTO) [92–95], $Pb(Zr,Ti)O_3$ (PZT) [96–99]. However, from the energy-storage point of view, several persistent issues still exist that need to be addressed. The primary challenging issue is associated with the physical dispersion of the oxide nanoparticles. The high surface energy of dielectric nanoparticles usually leads to agglomeration and phase separation from the polymer matrix, especially if they are forced into an incompatible matrix, yielding poor quality films and weakened dielectric properties, such as high dielectric loss and low dielectric strength [33, 52].

Another challenge lies in overcoming a more fundamental weakness, electric field intensification in the polymer matrix [62]. The field intensification associated with the highly inhomogeneous electric fields is as a result of the large difference in permittivity between the two phases, and generally gives rise to significant reduced effective breakdown strength in the composites [17, 100, 101]. In detail, nanoparticles of high permittivity act as electrical defect centers in filled polymers. Such defect centers effectively distort the distribution of electric field, making the local electrical field in the matrix much higher than the average electric field. This field expulsion lowers the overall energy storage because it results in the breakdown of the polymer at an electric field much lower than the intrinsic breakdown strength of the polymer. The extent of the field distortion or the degree of the field intensification is largely influenced by the particle size as well as the difference in permittivity or conductivity between the particle and polymer. We numerically simulate the electric field distortion in various inorganic/polymer nanodielectric materials using software Comsol Multiphysics. The results are shown in Fig. 6. It turns out that for composites with the same shape, size and distribution of the filler, the larger permittivity contrast in the two phases creates a larger local

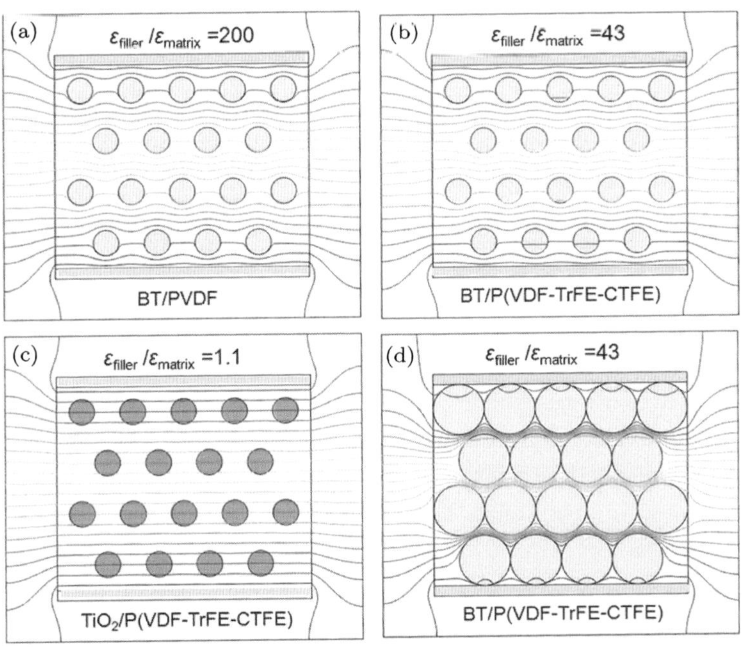

Figure 6. The outline contours of electrical potential in different nanodielectrics: (a) $BaTiO_3$/PVDF, (b) $BaTiO_3$/P(VDF-TrFE-CTFE), (c) TiO_2/P(VDF-TrFE-CTFE), and (d) $BaTiO_3$/P(VDF-TrFE-CTFE) nanocomposite with $BaTiO_3$ nanoparticles of larger size. Note that the permittivities for TiO_2, $BaTiO_3$, PVDF, and P(VDF-TrFE-CTFE) are 47, 2000, 10, and 42, respectively. This simulation is done in a capacitor with two parallel platelike electrodes (gray squares) and a dielectric polymer nanocomposite filled with spherical ceramic nanoparticles (the circles between the two electrodes).

field distortion. Moreover, as compared with the small-particle-filled polymer composite, a larger size of filler would be prone to inducing a higher degree of field intensification (see Fig. 6(b),(d)). These results imply that homogenous inclusion of small nanoparticles with comparable permittivities to the polymer matrix can effectively create dielectric homogeneity in the nanocomposites.

The last, but not least, challenge for the ceramic particle/polymer nanodielectric materials is to rationally optimize the particle concentration to maximize the energy-storage density. To achieve high effective permittivity and energy density, the fillers usually have a much greater permittivity than the surrounding polymer matrix. In this case, most of the increase in effective dielectric permittivity is contributed by an increase in the average field in the polymer matrix with very little of the energy being stored in the high permittivity filler phase [7, 102]. It takes quite a lot of high permittivity nanoparticles to afford a high permittivity in the composite. The dielectric strength, on the other hand, drops off precipitously with increasing particle concentration due to an increased defect density, as well as percolation with the low dielectric strength ceramic fillers [99]. The high proportion of inorganic fillers increases the particle packing density, intriguing a breakdown path through aggregated fillers, thus increasing the leakage current and lowering the electric breakdown strength and energy density [103]. Despite the enhancement in the dielectric permittivity of the polymer composites, the decrease of the breakdown strength leads to very minor enhancement of energy density as compared with the pure polymer matrix. Therefore, it is essential to determine an optimal volume fraction of inorganic nanoparticles in the nanocomposite to reach a high energy density at high electric breakdown strength, which ensures that capacitors fabricated from the films of this material could be operated at high voltages.

3.1.2. *Effects of surface modification of nanoparticles*

Remarkable progress in the field of FNDMs on improving the energy-storage capacity has been achieved by attacking at least one of the aforementioned problems. For instance, in order to improve the compatibility between the ceramic particles and polymer phases and thus avoid agglomerated nanoparticles, voids, and phase-separated mixture over large-length scales, surfactants such as phosphate esters and oligomers are usually employed, which allows for an increase in dispersion of the nanoparticles in the matrix, ultimately improving the overall film quality [48, 49, 104]. Unfortunately, residual free surfactant usually causes unexpectedly deteriorated dielectric properties at the applied fields, i.e., high leakage current and high loss tangent. An alternative approach is thus to strongly bind surface modifiers to nanoparticles via robust chemical bonds. It seems highly desirable to improve the compatibility between the filler and polymer matrix and consequently the energy-storage performance of the nanocomposites [27, 52]. Ramesh *et al.* [27] have used trialkoxysilane surface modifiers for the dispersion of $BaTiO_3$ nanoparticles in an epoxy polymer host. The resulting composites exhibited a reasonably high permittivity value up to 45. Unfortunately, the loss tangent increased compared with the unmodified $BaTiO_3$ nanocomposites. Kim and co-workers [52] have also demonstrated a strong surface binding of phosphonic acid ligands onto $BaTiO_3$ nanoparticles and examined the use of such specific surface functionalities to control $BaTiO_3$-polymer interfacial interaction and, thus, facilitate the formation of homogenously dispersed, high energy density nanocomposites. As shown in Fig. 7, the nanocomposites with unmodified $BaTiO_3$ nanoparticles exhibited large scale aggregates and pinhole defects. In contrast, the ones with modified $BaTiO_3$nanoparticles yielded comparatively uniform films with homogenous nanoparticle dispersions. The nanocomposites with 50 vol% modified $BaTiO_3$ particles afforded a large permittivity ($\approx$37), reasonably low loss (<0.007) and a high dielectric strength (over 200 MV/m). Thus, the modified $BaTiO_3$ nanocomposites featured a large calculated energy density of up to 6.1 J/cm^3. Being different from Kim's surface modification, Wang *et al.* [32] successfully produced nanosized, ethylene diamine surface modified $BaTiO_3$ in situ. This surface modification afforded an enhancement in the dispersion of $BaTiO_3$ nanoparticles into organic solvents and subsequently led to uniform polymer nanocomposite films. The energy density of the composites with 23 vol%

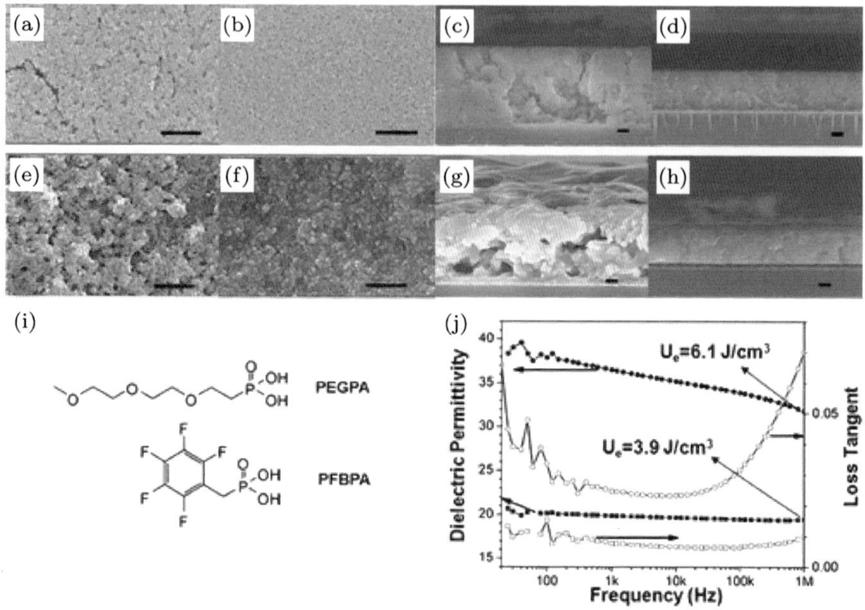

Figure 7. SEM images of surface and fractured cross-section of nanocomposite thin films. (a, c) $BaTiO_3$/PC, (b, d) PEGPA-$BaTiO_3$/PC, (e, g) $BaTiO_3$/P(VDF-HFP), and (f, h) PFBPA-$BaTiO_3$/P(VDF-HFP). All the scale bars are 1 μm. (l) Structures of the phosphonic acid ligands used to modify $BaTiO_3$ for dispersion in polycarbonate (PC) and poly(vinylidene- fluoride-co-hexafluoropropylene) (P(VDF-HFP)) host materials: {2-[-(2-methoxyethoxy)ethoxy]thyl} phosphonic acid (PEGPA) and pentafluorobenzyl phosphonic acid (PFBPA). (j) Frequency dependence of dielectric permittivity and loss tangent of PEGPA-$BaTiO_3$/PC (squares) and PFBPA-$BaTiO_3$/P(VDF-HFP) (circles). Reproduced with permission [52].

modified $BaTiO_3$ was nearly doubled from 1.9 J/cm^3 (neat polymer) to 3.7 J/cm^3.

It has been demonstrated that the introduction of ligands with long (C_6–C_{18}) hydrocarbon chains would adversely affect the bulk dielectric properties of nanocomposites, since the alkyl chain with low dielectric permittivity results in a disparity in the electric fields. This will induce discontinuous variations in the local electric field, and significantly reducing the intrinsic high dielectric strength of the polymers [17, 82]. Thus, approaches to directly coupling the host polymer with oxide nanoparticles to yield the covalent-bonded nanocomposites are highly desirable. Covalent assembly of polymer matrix and inorganic filler could be realized either by the functionalization of host polymer chains [82] or through the surface treatment of nanofillers [51, 103, 105]. Wang *et al.* [82] have proposed an approach including the synthesis of the ferroelectric polymer with phosphomic acid end-groups and successful utilization of the reactive terminal groups for direct coupling with ZrO_2 oxide nanofillers. The covalent bond is created based on the Zr–OH/P–OH condensation, allowing for a bonded layer around the nanofillers according to the multicore model proposed by Tanaka [56]. This layer thereby offers greater homogeneity in the composites compared with the composites with a pristine polymer matrix, as shown in Fig. 8. Of particular importance is that the covalent-bonded nanocomposite affords an energy density as large as $11.2\,J/cm^3$ with a considerably high dielectric strength of 270 MV/m, which is comparable to that of the pure polymer matrix. This result is most likely caused by the diminished defects, balanced permittivities, increased crystallinity in the matrix and interfacial coupling effects as a result of formation of a Gouy–Chapman diffusion layer in the "interaction zone". Wang's work has given an important hint for improving the polarization response and thus the energy density, by creating an effective bonded layer around the nanofillers in the polymer matrix. This is definitely the prerequisite to forming a multicore "interaction zone" between the nanoparticles and the polymer matrix. It should be noted that apart from polymer chain functionalization, the bonded layer can also be realized by a facile surface hydroxylation treatment of the nanoparticles. Zhou *et al.* [51] used hydrogen peroxide to prepare surface hydroxylated $BaTiO_3$ and subsequently mixed them with PVDF to create hydrogen bonded nanocomposites. Hydrogen bonds are formed between the electrophilic atoms on the PVDF and the -OH groups on the surface of the hydroxylated $BaTiO_3$ nanoparticles, as schematically illustrated in Fig. 9a. As compared with the unmodified $BaTiO_3$ nanocomposite, the hydroxylated $BaTiO_3$ nanocomposite exhibited a relatively lower permittivity yet much higher dielectric strength.

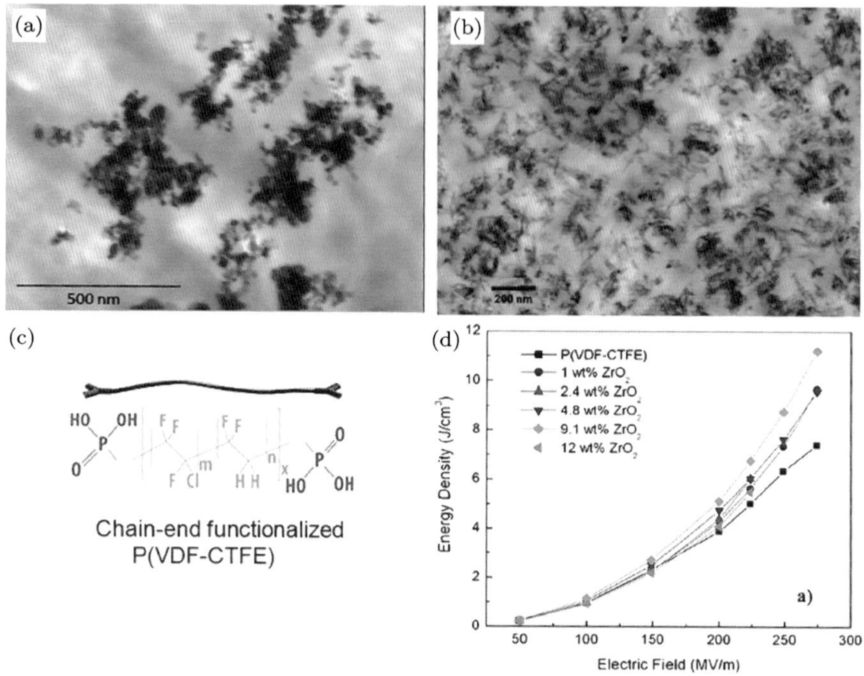

Figure 8. TEM images of the nanocomposites with 12 wt% ZrO_2: (a) covalent-bonded poly(vinylidene fluoride-chlorotrifluoroethylene) P(VDF-CTFE) nanocomposite and (b) pristine P(VDF-CTFE) nanocomposite. (c) Molecular structure of P(VDF-CTFE) polymers with phosphonic acid end-groups. (d) Energy-storage density of the polymer matrix and the nanocomposites versus the applied electric field for covalent-bonded P(VDF-CTFE) nanocomposite with various mass fractions of ZrO_2 nanofiller. Reproduced with permission [82]. Copyright 2010, American Chemical Society.

By balancing both, the maximum energy density can be calculated as 3.7 J/cm^3 with a loss tangent of 0.02 in the composite with 20 vol% of hydroxylated $BaTiO_3$ particles.

Another approach to directly binding the polymer matrix to the inorganic particles is that of host polymer growth on the surfaces of nanofillers via an in situ synthesis processes. This process has been widely applied to a variety of inorganic nanofillers, such as silica [107–109], nanotubes [110–113] and graphene oxide [114–116], and is demonstrated to be effective to realize multicore interfaces

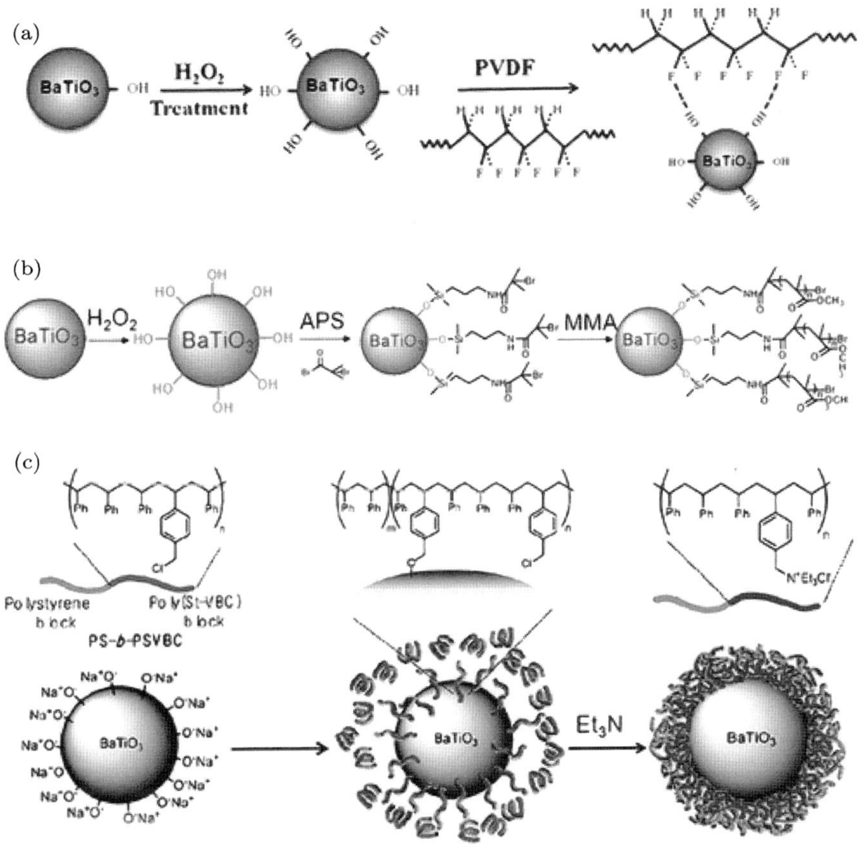

Figure 9. (a) Schematic illustration of the hydroxylation of $BaTiO_3$ nanoparticles and formation of hydrogen bond in h-$BaTiO_3$/PVDF nanocomposites. Reproduced with permission [51]. Copyright 2011, American Chemical Society. (b) Diagram schematically illustrating the process of in situ atom transfer radical polymerization of MMA monomer from the surface of $BaTiO_3$ nanoparticles. APS is the abbreviation for γ-Aminopropyl triethoxysilane. Reproduced with permission [106]. Copyright 2011, Royal Society of Chemistry. (c) Synthesis route to $BaTiO_3$ particle with the polymer shell. Reproduced with permission [86]. Copyright 2010, American Chemical Society.

in the dielectric nanocomposites [106, 117, 118]. For example, Huang and Jiang [106] successfully prepared core–shell structured $BaTiO_3$/poly(methyl methacrylate) (PMMA) nanocomposites via in situ atom transfer radical polymerization (ATRP) of methyl methacrylate (MMA) from the surface of $BaTiO_3$ nanoparticles. This process is schematically illustrated in Fig. 9(b). The $BaTiO_3$ core affords an increased permittivity (≈16 at 1 kHz), while the flexible PMMA shell endows the material with good dispersability and inherent low loss of the base polymer (≈0.037 at 1 kHz), as summarized in Fig. 10. The insulating polymer shell around the nanoparticles is found to be capable of serving as an effective barrier, retarding the movement of charge carriers in the composites. This should solve the aforementioned problem of nanoparticle percolation at high filler contents, and thereby leads to a suppressed loss

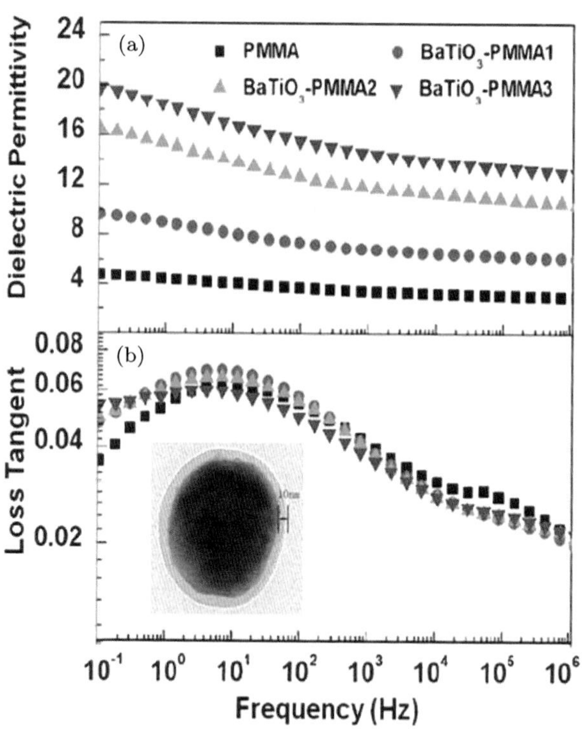

Figure 10. Room temperature (a) dielectric permittivity and (b) loss tangent as a function of frequency for $BaTiO_3$/PMMA nanocomposites prepared by ATRP. The inset in (b) is a typical TEM image of the as prepared core–shell $BaTiO_3$ nanoparticle in the PMMA matrix. Reproduced with permission [106]. Copyright 2011, Royal Society of Chemistry.

compared with conventional ceramic particle/polymer nanocomposites [26, 87]. It is a pity that this work did not report the high-field dielectric properties, such as dielectric strength or leakage current near the dielectric breakdown. The energy density of this $BaTiO_3$/PMMA nanocomposite has not yet been ascertained, and needs to be further investigated. Recently, Jung *et al.* [86] investigated the effect of the robust polymer shell on the dielectric insulation and energy-storage density in a polystyrene (PS) nanocomposite comprising core–shell $BaTiO_3$ particles with diblock copolymer shielding layers. The adoption of the diblock copolymer resulted in a well-defined layered polymer shell through two-step processes, as illustrated in Fig. 9c. A weakly adsorbed layer was first formed by reacting the chloromethyl group in polystyrene-*block*-poly(styrene-*co*-vinylbenzylchloride) (PS-*b*-PSVBC) with the oxy anion on the $BaTiO_3$ surface. This weakly adsorbed layer was subsequently performed through a triethylamine treatment to transform the nonpolar chloromethyl functional blocks to polar ammonium blocks, which formed a self-assembled stable and dense layer, and interacted strongly with the $BaTiO_3$ particle surfaces. As a result, the layered polymer shell was self-assembled with the nonpolar polystyrene block as the outer layer while the polar ammonium-containing block formed the inner one. This interesting feature not only benefits good dispersibility in a nonpolar medium, but also affords a high energy-storage capacity at a high applied electric field. In detail, the surface of particles can be largely charged with ammonium groups in the inner layer, which contributes to a high polarization level in the composites. Additionally, this charge-rich layer is directly buried and shielded by the 3–5 nm thick insulating outer layer so as to minimize the development of the breakdown path and leakage current. Thus, this shielding effect enables a large incorporation of $BaTiO_3$ particles, to reach a high enough permittivity while maintaining a high dielectric strength in the composites. As shown in Fig. 11, the diblock copolymer coated $BaTiO_3$ nanocomposite afforded a high dielectric permittivity ($\approx$44 at 10 kHz) with a significantly low leakage current density of 0.76 nA/cm^2; even the fraction of $BaTiO_3$ particle is as large as 75 wt%. The extremely large dielectric breakdown strength

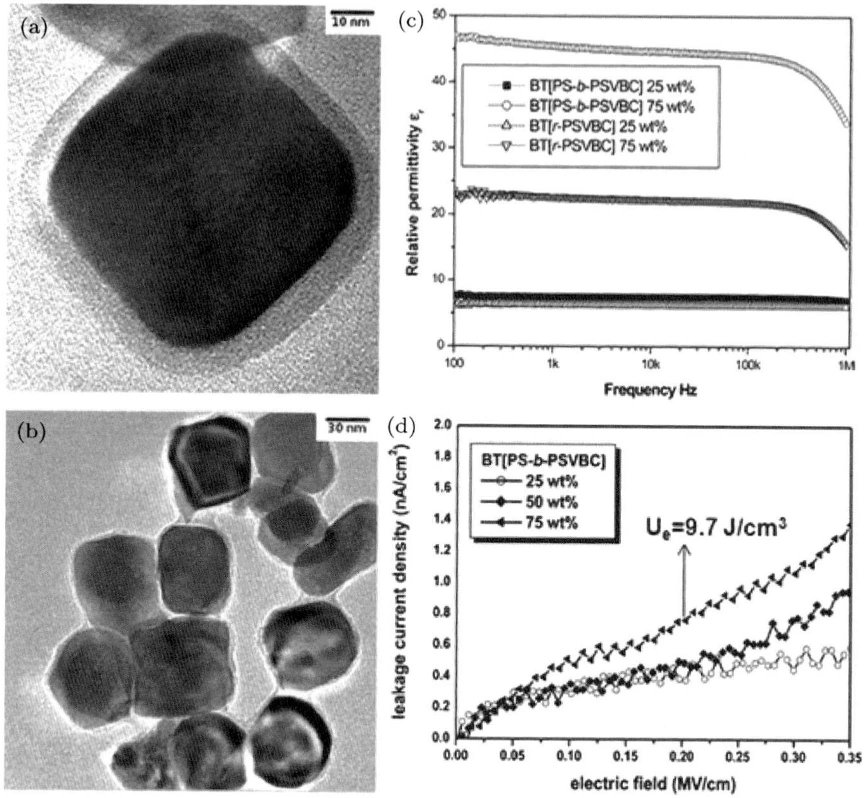

Figure 11. (a, b) TEM images of the core–shell $BaTiO_3$ nanoparticles with a dense PS-*b*-PSVBC shell (after amine treatment). The thickness of the polymer shell is in the range of 5–7 nm. (c) Frequency dependence of the dielectric permittivity of $BaTiO_3$[PS-*b*-PSVBC]/PS and $BaTiO_3$ [r-PSVBC]/PS nanocomposites at different particle fractions. (d) Leakage current measurements for the $BaTiO_3$ [r-PSVBC] nanocomposites. Reproduced with permission [86]. Copyright 2010, American Chemical Society.

(up to 222 MV/m) at high permittivity consequently gave rise to a high calculated energy density of 9.7 J/cm^3.

The achieved results imply that robust and stable insulating layers on inorganic particles can afford maximized dielectric properties of nanocomposites. Along this line, a very thin ceramic layer has been also realized on metal oxide nanoparticles for the purpose of facilitating dispersibility, preventing particle percolation, and most importantly, decreasing the permittivity contrast between filler and matrix. By establishing a buffer ceramic layer of medium permittivity between the filler and matrix, the aforementioned problem of reduced dielectric strength is anticipated to be addressed from the field intensification point of view. For the directly dispersed high permittivity ceramic nanoparticles in a polymer matrix (Fig. 12a), severe field intensification in the polymer matrix would prevent simultaneous realization of the maximum energy densities for both constituents. To minimize this effect, we proposed that an ideal composite should comprise nanoparticles with grading permittivities from the center to the border (Fig. 12(c)). Inclusion of core–shell nanoparticles (Fig. 12(b)) having a shell permittivity between that of the core and polymer matrix affords a realistic model to diminish the adverse effects of permittivity contrast on the dielectric properties of nanocomposites. In this way, Lanagan and Marks [37, 59, 83] conducted a series of synthesis of Al_2O_3 shell $\varepsilon_{\text{eff}} \approx 10$ coated metal oxide nanoparticles and investigated their effect on improving the energy-storage capacity in polymer nanocomposites as filler. They initially attacked the problems of nanoparticle aggregation and phase separation in the polymer nanocomposites. High-energy-density $BaTiO_3$-and TiO_2-isotactic

(a) (b) (c)

Figure 12. Dielectric nanocomposites filled with high permittivity spherdical nanoparticles. (a) High permittivity sphere in a low permittivity matrix, exhibiting a large permittivity contrast. (b) High permittivity sphere in a low permittivity matrix with a layer of medium permittivity between them. (c) A sphere with a grading permittivity from center to border. Reproduced with permission [59]. Copyright 2010, American Chemical Society.

polypropylene (^{iso}PP) nanocomposites were prepared via in situ-supported metallocene propylene polymerization. This polymerization was mediated by anchoring/alkylating/activating metallocene catalyst [rac-ethylenebisindenyl] zirconium dichloride ($EBIZrCl_2$) on the methylaluminoxane (MAO)-treated oxide nanoparticles (see Fig. 13(a)). The resulting nanocomposites exhibited relatively uniform nanoparticle dispersions and a remarkably high projected energy-storage density of 9.4 J/cm^3, as estimated from the permittivity ($\approx$6.1) and dielectric breakdown strength ($\approx$400 MV/m) [37]. It was speculated that numerous interfaces, which were formed between the ceramic nanoparticles and the polymer phases in a well-dispersed 0–3 nanocomposite, served as effective electron scatters and trapping centers to reduce the breakdown probability. Furthermore, well-dispersed ceramic nanoparticles could block degradation tree growth and thus increase the long-term breakdown strength. This approach proven promising, leading to effective dispersion and enhanced dielectric properties. The effect of the shape of the nanoparticles on the dielectric properties was further investigated and simulated. As illustrated in Fig. 13(b), a layer-by-layer MAO coating process was employed to prepare Al_2O_3 encapsulated metal oxide core-ell nanoparticles, the Al_2O_3 shell thickness of which was variable and dependent on the cycles of MAO coating. These core–shell nanoparticles subsequently served as metallocene catalyst support for the followed in situ propylene polymerization. Similarly, the in situ polymerization yielded homogeneously dispersed nanoparticles in an ^{iso}PP matrix. The effective permittivity of the composites increased with the particle concentration, affording a value as high as 6.2. It agreed well with Maxwell–Garnett (MG) effective medium approximation description for the volume fraction of nanoparticles below 0.06. MG theory was applied to model the permittivity variation with the shell thickness of Al_2O_3. As seen from Fig. 14, the thickness of the Al_2O_3 coating had little effect on the effective nanocomposite permittivity, only showing a marginal increase upon increasing the shell thickness. Nevertheless, the presence of Al_2O_3 shell dramatically suppressed the leakage

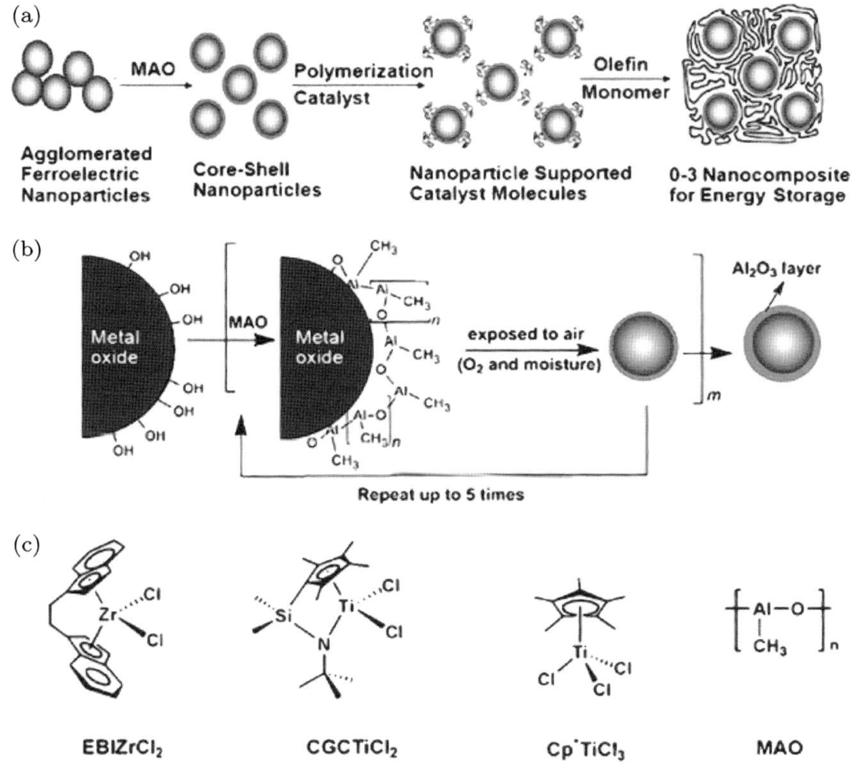

Figure 13. (a) A route to catalytically synthesized 0–3 ceramic/polymer nanocomposites via *in situ* supported metallocene olefin polymerization process. Reproduced with permission [37]. Copyright 2007, American Chemical Society. (b) Synthesis of aluminum oxide encapsulated high permittivity metal oxide core–shell nanoparticles via a layer-by-layer MAO coating process. Reproduced with permission [59]. Copyright 2010, American Chemical Society. (c) Various metallocene polymerization catalysts and MAO. (d) The dielectric permittivity of ^{iso}PP nanocomposite film as a function of nanoparticle volume fraction and composition. (e) Dependence of leakage current density on applied field, E, for different nanoparticle/^{iso}PP nanocomposites at a nanoparticle volume fraction of ≈ 0.07. Leakage current for neat ^{iso}PP is also plotted for comparison. The superscripted prefixes "s" and "r" in (d) and (e) denote sphere- and rod-shaped TiO_2 nanoparticles, respectively. Reproduced with permission [83]. Copyright 2010, American Chemical Society.

permittivity was observed in the rod-shaped TiO_2 nanocomposites, not in the composites with larger-permittivity $BaTiO_3$ nanoparticles. The authors claimed that the reason should be the greater degree of electric field exclusion associated with the rod shape of the inclusion [83]. Moreover, the rod-shaped TiO_2 nanocomposites had significantly lower breakdown fields than the sphere-shaped TiO_2-, $BaTiO_3$-, and ZrO_2-based composites because of the following two facts: (i) enhanced localized electric field around the TiO_2 nanorods facilitates increased charge injection, (ii) the high aspect ratio of the TiO_2 nanorods allows for percolation at lower volume fractions, thus creating enhanced conduction pathways in the composites so as to favor the movement of charge carriers. Improved charge mobility usually results in a reduced breakdown fields. The energy density of resulted composites varied from 0.3 to 9.4 J/cm^3. The maximum energy density was observed in the $BaTiO_3$-Al_2O_3/^{iso}PP system. At that point, the nanocomposite showed large permittivity ($\approx$6.1) and breakdown strength as high as 600 MV/m.

3.1.3. *Effect of permittivity contrast between filler and matrix*

As compared with establishing a buffer ceramic layer of medium permittivities on the nanofillers, a more straightforward strategy to mitigate the field intensification effect is balancing permittivity by a rational selection of filler and polymer matrix with comparable permittivity values. In this regard, pure oxide nanoparticles with relatively low average permittivities, such as TiO_2 ($\varepsilon_{\text{eff}} = 47$) [31], ZrO_2 ($\varepsilon_{\text{eff}} = 23$) [81], are preferred choices as dopants to improve the energy density of polymers. For example, Wang *et al.* [31] prepared a high energy density nanocomposites composed of ferroelectric polymer P(VDF-TrFE-CTFE) and surface functionalized TiO_2 nanoparticles with homogenous particle dispersions. Since the P(VDF-TrFE-CTFE) matrix possesses a dielectric permittivity ($\approx$42) comparable to TiO_2 filler, the low-field dielectric properties were not changed dramatically with increasing the filler content (Fig. 15), not as in the case of conventional nanocomposites with much higher permittivity fillers [35, 87, 93, 119, 120]. The effective

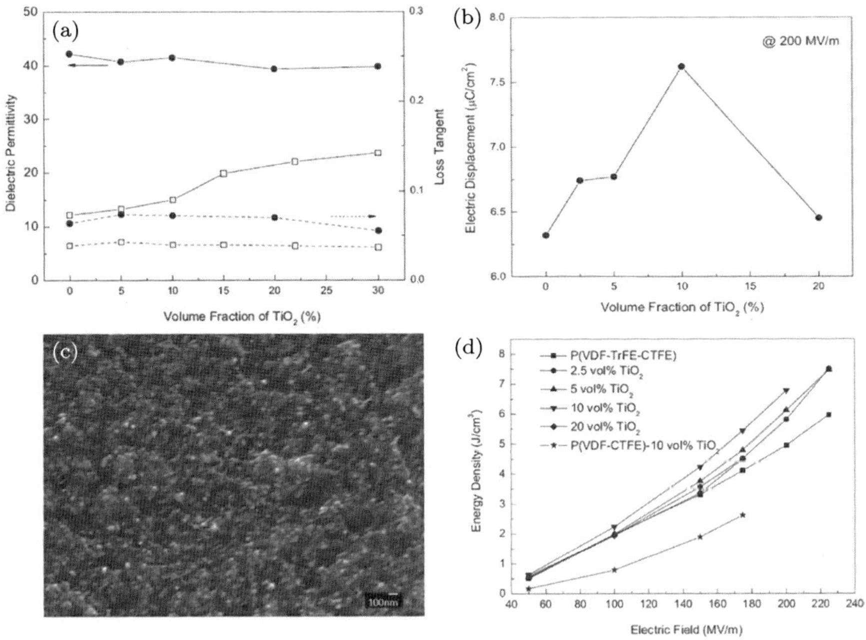

Figure 15. (a) Low-field dielectric permittivity and loss tangent of the TiO_2/P(VDF-TrFE-CTFE) (circle) and TiO_2/P(VDF-CTFE) (square) nanocomposites as a function of the volume fraction of TiO_2 nanoparticles, measured at 1 kHz and room temperature with a 1 V bias. (b) Dependence of maximum electric displacement on the TiO_2 content in the P(VDF-TrFE-CTFE) nanocomposites at 200 MV/m. (c) Cross-sectional SEM image of the P(V P(VDF-TrFE-CTFE) nanocomposite loaded with 30 vol% TiO_2, showing that the nanoparticles are homogenously dispersed in the polymer matrix with an average size about 50–70 nm. (d) Stored energy density of the polymer and nanocomposites at various electric fields. Reproduced with permission [31].

dielectric permittivity and loss tangent were similar to the neat polymer matrix. However, the incorporation of the TiO_2 nanoparticles into the polymer induced a substantial improvement in electric displacement and thus high energy density at high electric field due to the formation of a large-area so-called "interaction zone" as shown in the multicore model. The maximum energy density observed was 6.9 J/cm^3 at 200 MV/m in the composite with 10 vol% TiO_2. Notably, the dielectric strength was considerably larger than that of conventional dielectric polymer nanocomposites (below 100 MV/m) due to permittivity balancing between two phases. Overall, Wang's work successfully validates a principle that inorganic/polymer composites with high energy density can be achieved by judiciously selecting a combination of nanoparticle and polymer matrix with balanced dielectric properties.

3.1.4. *Effect of nanoparticle concentration*

Previously demonstrated results indicate that the energy density of inorganic/polymer composites depends on the volume fraction of inclusions non-monotonously. Nevertheless, the largely improved effective permittivity of a composite at high filler contents (>20 vol%) is rarely directly transferred into high energy density due to the significantly reduced dielectric strength relative to the neat polymer matrix [103, 121]. This trade-off between effective permittivity and breakdown field with increasing nanoparticle loading confounds efforts to improve the theoretical energy density, even leads to materials with lower energy density than the virgin polymer in extreme cases. Although measurable improvement in dielectric strength has been reported, this situation is only for rare nanocomposites specifically with slight volume fraction (<3 vol%) of randomly dispersed nanoparticles [91, 122–128]. Thereby, it is necessary to rationally optimize the filler volume fraction in nanocomposites to maximize the stored energy density. According to this viewpoint, Kim *et al.* [33] prepared nanocomposite comprising P(VDF-HFP) and phosphonic acid surface-modified $BaTiO_3$ nanoparticles, and systematically investigated the role of the nanoparticle volume fraction in a wide range (0–80 vol%) on the dielectric properties (effective permittivity, dielectric loss, and breakdown strength) of nanocomposites, aiming to determine the optimum volume fractions. It was found that the effective permittivity of nanocomposites increased with nanoparticle volume fraction to a maximum of 35 at around 50 vol%. However, as the fraction increased further up to 80 vol%, the permittivity followed a rapid decrease as a result of the formation of porosity in the spin-coated composite films (Fig. 16(a)). Also, the dielectric strength decreased upon addition of $BaTiO_3$ nanoparticles to the host polymer, as expected. Of particular note

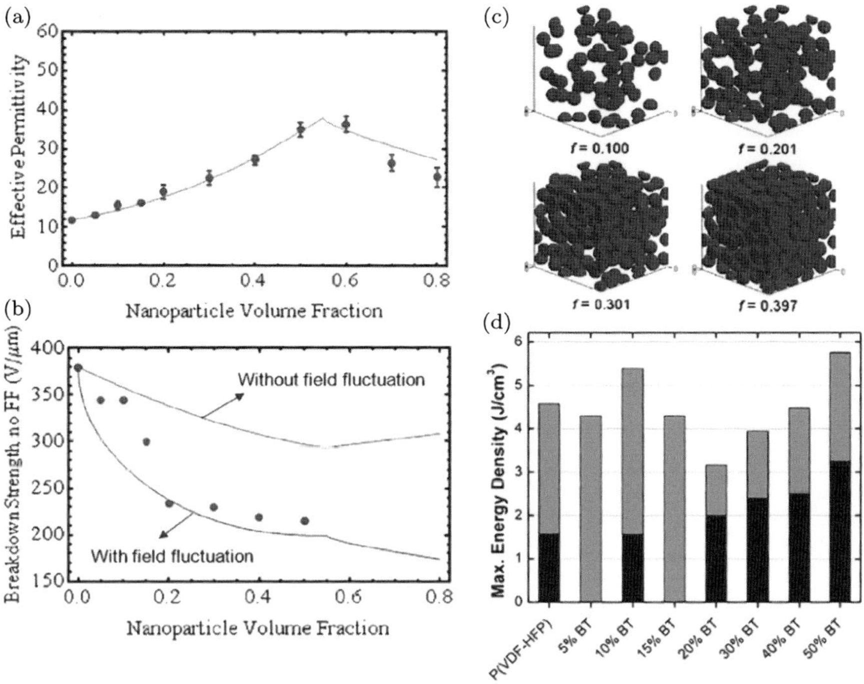

Figure 16. Comparison of: (a) effective permittivities (at 1 kHz) and (b) dielectric strengths predicted by the combination of self-consistent effective medium approximation and finite difference simulation (lines) with experimental values (dots). The prediction based on adjusting the onset volume fraction of air void formation from 39.7% to 55 vol% on the basis of SEM observations. (c) Particle packing examples in finite difference simulations at different volume fractions. Note that $f = 0.397$ is the maximum nanoparticle volume fraction, below which no air void is formed inside this model place. The intended increase of the particle volume fraction over $f = 0.397$ can only be made by decreasing the volume fraction of polymer host, inevitably creating air voids. (d) Measured energy density (black) and calculated maximum energy density (gray) of $BaTiO_3$/P(VDF-HFP) nanocomposites in a wide range of nanoparticle fractions. (a–d) Reproduced with permission [33]. Copyright 2009, American Chemical Society.

is that the dielectric strength exhibited an abrupt decrease when the volume fraction varied from 10 to 20 vol% (Fig. 16(b)), which should be derived from the occurrence of nanoparticle percolation as discussed above. The breakdown behavior of high fraction (>50 vol%) nanocomposites is highly related to the created air voids, which will significantly lower the dielectric strength of the nanocomposites due to the very low breakdown strength of air (≈3 MV/m) [129]. Apart from experimental observations, Kim *et al.* also used combined self-consistent effective medium theory (SC-EMT) [17] and finite difference simulations [130] to calculate the effective permittivity and dielectric breakdown strength. Being different from the commonly used composite model including the Clausius–Mossotti treatment [131], Maxwell–Garnet approximation [132–134], and Lichtenecker rule [135], the SC-EMT takes account of the effects of interactions between neighboring particles, the presence of possible air voids, and the field fluctuation in the host. Thus, it seems more realistic to model the dielectric properties of nanocomposites at high nanoparticle volume fractions. The effective permittivity, ε_{eff}, and dielectric breakdown strength of the nanocomposite can be expressed as:

$$\varepsilon_{\text{eff}} = \varepsilon_1 + f_2(\varepsilon_2 - \varepsilon_1)a_2 + f_3(\varepsilon_3 - \varepsilon_1)a_3 \tag{5}$$

$$E_0\left(a_1 + \sqrt{\frac{1}{f_1}\frac{\partial\varepsilon_{\text{eff}}}{\partial\varepsilon_1} - a_1^2}\right) \geq E_b, \tag{6}$$

where, phases 1, 2, and 3 represent the host matrix, filler and voids respectively. a_{r} denotes the electric field concentration factor in each phase r and relates to the ratio of average electric filed $\langle E_{\text{r}}\rangle$ in each phase to the applied electric field $\langle E_0\rangle$ and can be calculated as:

$$a_r = \frac{\langle E_r\rangle}{E_0} = 1 - \frac{1}{3}\left[(\varepsilon_r - \varepsilon_{\text{rff}})^{-1}\varepsilon_{\text{eff}} + \frac{1}{3}\right]^{-1}, \quad r = 2,3 \tag{7}$$

$$a_1 = 1 - f_2a_2 - f_3a_3 \tag{8}$$

The total volume fraction of air voids for a given particle volume fraction was overestimated using a statistical cellular packing model along with finite difference calculations (Fig. 16(c)). Using the obtained void volume fraction, the actual volume fractions of nanoparticles, polymer host and air void can be calculated and then used in the SC-EMT with electric field concentration factor together to obtain the theoretical permittivity and dielectric breakdown strength. The calculated results show a good agreement with experimental data when the onset volume fraction of air void formation was adjusted from 39.7% to 55 vol% on the basis of SEM observations. This indicates the important roles of nanoparticle percolation and porosity of the nanocomposites on the dielectric properties. As a result of the trade-off between the effective permittivity and the breakdown field, a maximum extractable energy density as high as 7–8 J/cm^3 was obtained at distinct values of $BaTiO_3$ volume fraction (i.e., 10 vol% and 50 vol%) (Fig. 16(d)). These results suggest the optimal nanoparticle volume fraction for the maximum energy density should be well below the onset of nanoparticle percolation (high dielectric strength yet low permittivity) or the onset of air void formation (high permittivity but low dielectric strength). The former filler inclusion is usually more desirable because of low cost, well-retaining flexibility and processability of the composites. More importantly, capacitors fabricated using films of the former composites can be operated at higher field conditions than their counterparts. Additionally, due to the field's quadratic relationship with electrostatic energy storage, a small variation in dielectric strength generally induces a dramatic change in the energy-storage density. Thus, approaches to inhibiting electric breakdown in the inorganic/polymer nanocomposite with high permittivities should be paid much more attention to improve the energy densities of dielectric polymer nanocomposites.

3.1.5. *Effect of anisotropy in nanocomposites*

Recent theoretical and experimental investigations indicate that a tailored architecture of oriented and spatially distributed constituents can improve the high field properties of the polymer nanocomposite [68, 127, 136–140]. The dielectric breakdown strengths and non-linear conduction at high fields can be controlled by employing an engineered anisotropy in heterogeneous systems. Therefore, manipulation of anisotropy in high-field dielectric properties can be exploited for the development of high energy density ceramic/polymer nanocomposites. Tomer and co-workers [136, 137] created an anisotropic composite by aligning $BaTiO_3$ nanoparticles into pearl chainlike structures inside a polydimethylsiloxane (PDMS) matrix using dielectrophoretic assembly. These field-structured composites

energy densities that increase significantly with field due to the corresponding relatively high permittivities. Such enhancement in the energy density for x–y aligned composites is far larger as compared with pure polymer, z-aligned, and randomly dispersed composites, as summarized in Fig. 18. Tomer's work suggests that a well-designed anisotropic nanocomposite provides a densely packed morphology capable of retarding breakdown and carrier mobility, thus substantially improving the energy-storage density of polymer nanocomposites.

Compared with the spherical particulates, aligned two-dimensional plates with large aspect ratios have proven to be more effective impediments to internal charge transport and allow for much more desirable high-field dielectric properties, and consequently, larger maximum energy-storage densities [127, 146]. This concept was first experimentally illustrated by Brandstetter [122] and Imai [128] with highly aligned organically modified montmorillonite (MMT) in thermoplastics and thermosets, respectively. Very recently, Tomer

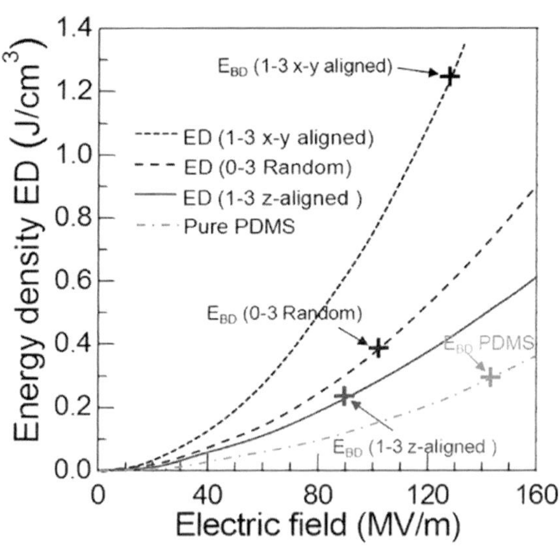

Figure 18. Recoverable energy density versus applied electric field for different systems. The x–y aligned composites show higher energy density than their counterparts. Reproduced with permission [137]. Copyright 2008, American Institute of Physics.

et al. [127] have exploited this concept to create an anisotropic polyethylene (PE) nanocomposite by controlling the spatial arrangement of high aspect ratio MMT nanofillers. The impact of filler orientation on the dielectric performances of the resulting nanocomposites has been fully discussed. It is found that at low field, the aligned nanocomposites were characterized by essentially the same dielectric properties with the randomly dispersed ones. At high electric fields however, the filler orientation significantly affected their dielectric behaviors, showing remarkable difference between composites with oriented and randomly fillers. Specifically, the oriented MMT composites demonstrated an increased dielectric strength when compared with that of the neat PE matrix, which is seldom observed for spherical $BaTiO_3$ composites, even when the $BaTiO_3$ nanoparticles are forced to form an alignment in the composites [136, 137]. This increased dielectric strength in turn led to superior energy-storage performance, whereas random fillers would result again in a deteriorated energy-storage performance due to the dielectric strength of their composites below that of PE matrix. The authors argued that the improved high field performance in oriented MMT nanocomposites was attributed to improved charge trapping by a synergistic effect of nanofillers and polar maleic anhydride (MAH) groups, introduced via polyethylene-MAH copolymers, as templated by the inorganic nanofillers [127]. Based on these findings, substantially greater number density of barriers for charge trapping can be anticipated by developing nanolaminate dielectric films, where the inclusions are large aspect ratio plates with a decreased thickness of several nanometers. Fillery and co-workers [68] have created uniform one-dimensional nanocomposites of highly aligned and dispersed MMT in polyvinyl butyral (PVB) matrix, as illustrated in Fig. 19(a). The resultant free-standing nanolaminates exhibited improvement in dielectric strength up to 30 vol% inorganic nanofillers, which largely extends the upper limit of volume fractions observed previously [91, 122–128] for the increased strength relative to the pure polymer matrix. Above this high upper limit however, the dielectric strength of nanolaminates decreased also with increasing of the filler content, as is the case of random

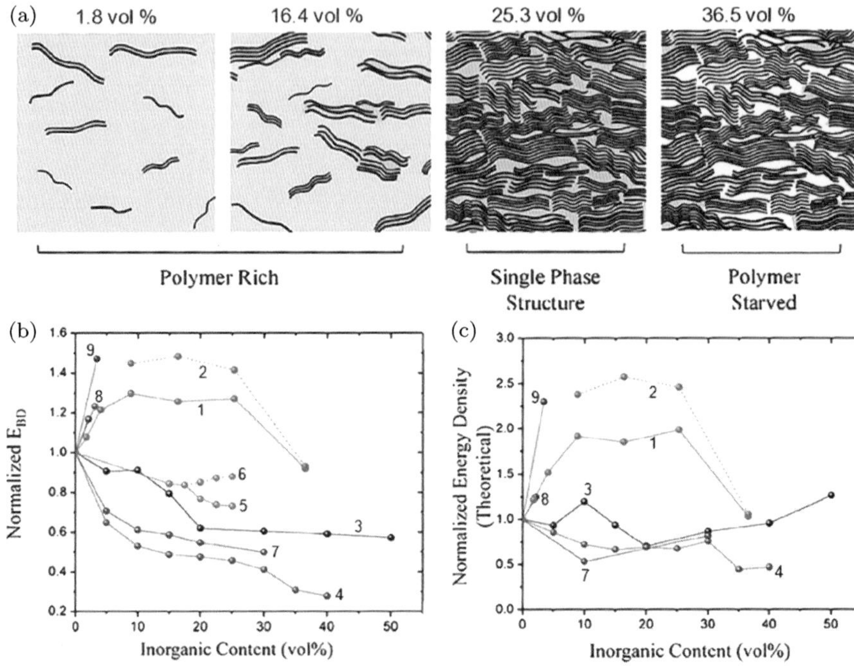

Figure 19. (a) Schematic illustration of the MMT/PVB nanolaminates. The black lines denote the MMT layers, while the gray parts represent PVB matrix and the white part signifies regions bereft of polymer at extremely high concentration of nanofillers. (b) Normalized breakdown field strengths (E_{BD} composite/E_{BD} polymer) and (c) normalized theoretical energy-storage density (U_e composite/U_e polymer) as a function of volume fraction of nanofillers in following systems. (a–c) Reproduced with permission [68]. (Copyright 2012, American Chemical Society): 1. MMT/PVB nanolaminate system, as-fabricated; 2. MMT/PVB nanolaminate system, compressed; 3. spherical $BaTiO_3$/P(VDF-HFP) nanocomposite [33]; 4. spherical $BaTiO_3$/polyimide nanocomposite [121]; 5. spherical $BaTiO_3$/epoxy nanocomposite, random dispersion [20]; 6. spherical $BaTiO_3$/epoxy nanocomposite, x–y aligned [20]; 7. spherical SiO_2/low-density polyethylene (LDPE) nanocomposite [124]; 8. MMT/LDPE nanocomposite [127]; 9. Kaolinite/P(VDF-HFP) nanocomposite [146].

nanoparticle composites. This is attributed to a trade-off between increased path tortuosity and polymer-deficient structural defects. Figures 19(b) and (c) provide a reasonable comparison of the normalized breakdown field strengths (E_{BD} composite/E_{BD} polymer) and theoretical energy-storage density (U_e composite/U_e polymer) for

MMT/PVB nanocomposites and prior inorganic/polymer systems. Most previous nanocomposites filled with spherical high permittivity nanoparticles suffered a steep decline in breakdown strength starting at very low vol% of fillers [33, 51, 121]. The subsequent increase in permittivity could not make up this decrease and yielded a lower energy density relative to the polymer matrix. However, lower permittivity nanoparticles, such as SiO_2 [124], layered silicate [127] or Kaolinite [146], provided the minimum change in the effective permittivity of composites, yet indeed allowed for enhancement in dielectric strength at low filler loading. Such a combination gave rise to a higher energy density compared with that of neat polymer. Therefore, the advantage of these MMT/PVB nanolaminates is that this positive effect on the energy-storage capacity could be largely tuned in a much wider range of nanofiller volume fraction, as shown in Fig. 19(c).

The introduction of anisotropy into the nanocomposites has been demonstrated to dramatically affect the dielectric properties and thus energy-storage capacity of composites. In order to obtain high dielectric permittivity and breakdown strength required to achieve materials with high energy density, the nanofiller orientation state or alignment degree should be well controlled. In general, the nanowire alignment can be realized in polymer composites by a variety of methods, including dielectrophoresis [136, 137, 140], melt extrusion [127, 147], microfluidics [148, 149], uniaxial strain assembly [99, 150], contact printing [151], blown bubble films [152, 153], strain-release assembly [154], and Langmuir–Blodgett troughs [155]. For instance, Sodano *et al.* [99] prepared nanocomposites with aligned PZT nanowires by uniaxial strain assembly. This method has been proved effective to align nanowires in bulk nanocomposites even at high filler concentrations, in which the dielectrophoresis usually failed. As shown in Fig. 20, the incorporation of PZT nanowires notably raised the electric displacement, especially in the case of 3-direction (z-aligned) nanocomposites. The larger electric displacement can usually generate higher stored energy density according to the integral expressed in Eq. (1). As a result, the 3-direction aligned nanocomposites had the largest energy density. It should

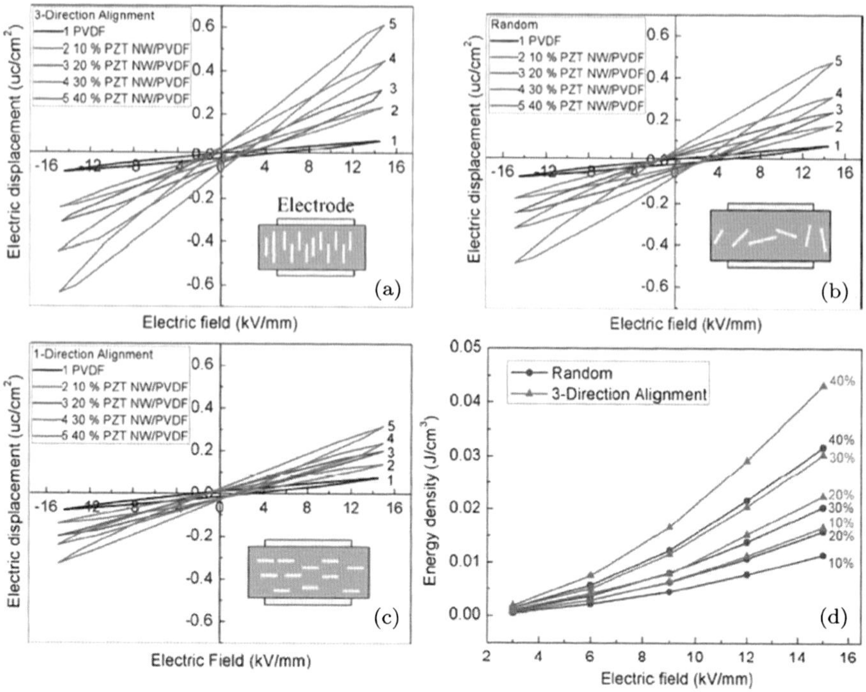

Figure 20. Electric displacement-electric field (D–E) loops for the polymer and (a) 3-direction (z-aligned), (b) Randomly dispersed and (c) 1-direction (x–y aligned) PZT/PVDF nanocomposites, measured under different applied fields at room temperature and 100 Hz. The insets in (a–c) show the schematic images of the orientation state of PZT nanowires in different nanocomposites. (d) Dependence of energy density on the PZT volume fraction in the PVDF matrix under 15 MV/m. Reproduced with permission [99].

be also noted that the energy density increased with the PZT volume concentration. This demonstrates that the energy density of nanocomposites is strongly dependent on the orientation state of the PZT nanowires, specifically, at any nanowire concentration, the 3-direction aligned nanocomposites represented higher energy density than the random nanowire composites. Here, we only cite the method of uniaxial strain assembly as an example to illustrate the impact of processing approaches on the energy-storage performances of nanocomposites. Principally, any other methods verified effective to align the nanoparticles inside composites can be employed to engineer the anisotropy and in turn, increase the energy density of the nanocomposites.

It is widely accepted that the effective permittivity of the ceramic/polymer composites are generally in the order of several tens, even high permittivity nanoparticles are involved as fillers. This limited permittivity enhancement could not offset the significant decrease in dielectric strength to afford an increased theoretical maximum energy density relative to the virgin polymer, if the local field exclusion or the interfacial morphologies are not properly controlled. In this case, the most important trend or potential research opportunity for further development of high-energy-density inorganic/polymer nanocomposites may lie in exploring approaches to increasing the dielectric strength of polymers rather than over focusing on increasing the permittivity at the expense of the dielectric strength. These approaches mainly rely on tailoring the interfacial morphology of composites or controlling the spatial arrangement of relatively low permittivity nanofillers, or combining both.

3.2. *Conductive particles/polymer nanodielectric materials*

3.2.1. *Percolation theory*

Despite great progress having been made on the improving energy density of ceramic nanoparticle/polymer nanodielectric materials, the common drawback of limited permittivity enhancement (several times in most cases, Fig. 21(a)) could not be fundamentally addressed, as the effective permittivity in these composites comes from the increase in the average field of the polymer matrix, with very little of the energy being stored in the high permittivity ceramic phase. This nature largely confounds the ability of increasing the energy densities mainly through significantly elevating the dielectric permittivity of composites. By replacing ceramic nanoparticles with conductive ones in the polymer composites, the percolative polymer composites can be made with the dielectric permittivity dramatically increased (several tens times higher than the polymer matrix, Fig. 21(a)) in the vicinity of the percolation threshold. Such

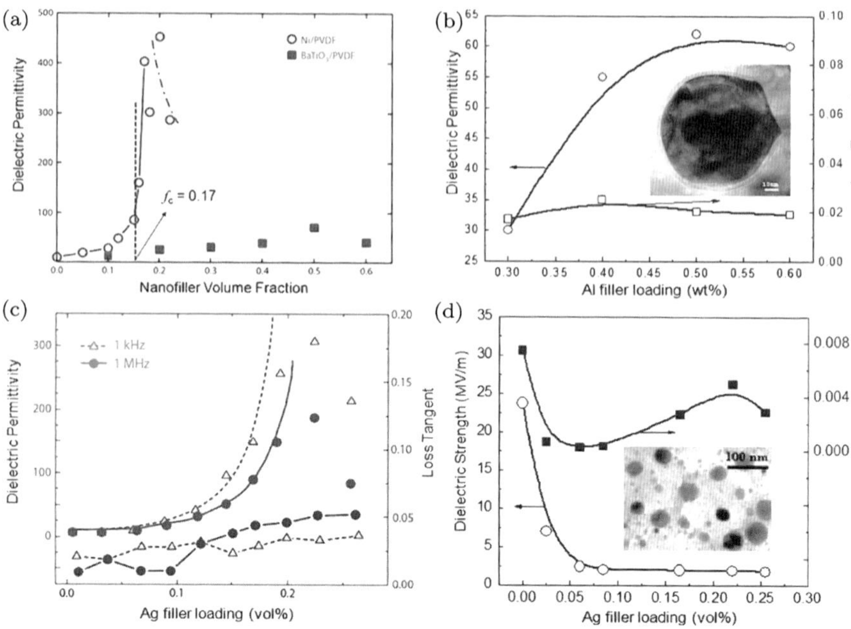

Figure 21. (a) Effective permittivity of Ni/PVDF and $BaTiO_3$/PVDF composites as a function of the volume fraction of particles. The data are taken from Refs. [158, 172]. (b) Dielectric permittivity and loss tangent of Al-filled epoxy composites as a function of Al filler loading. The 100 nm Al filler morphology can be clearly seen in the high resolution TEM image, as shown in the inset of (b). The thickness of self-passivation alumina shell is about 2.8 nm. Reproduced with permission from Ref. [217] Copyright 2005, American Institute of Physics. (c) Dependence of dielectric permittivity and loss tangent of MSA-modulated Ag/epoxy composites on the volume fraction of fillers, measured at different frequencies. The Ag nanoparticles used in these composites have a diameter of 40 nm. Reproduced with permission from Ref. [212]. (d) Dielectric breakdown strength and maximum calculated energy density of MAS-modulated Ag/epoxy composites as a function of the Ag volume fractions. The maximum energy density was calculated by using the permittivity data at 1 kHz. Inset in (d) shows the morphology of a single surfactant modified silver nanoparticle. Reproduced with permission from Ref. [212].

percolative systems offer a potential alternative to ceramic/polymer nanocomposites as nanodielectric material for high-U_e capacitors because of their extremely high dielectric permittivity as well as very low filler loading required, if they also have high dielectric breakdown strength and low loss dissipation factor.

In a percolative system, conducting particles are isolated by thin dielectric polymer layers to form a heterogeneous system, which could readily serve as a dielectric with excellent characteristics for energy storage [156]. Instead of following the linear rule of mixtures, the effective dielectric permittivity, ε_{eff}, of the percolative composites can be described by a classic explicit law [4, 157]:

$$\varepsilon_{\text{eff}}/\varepsilon_{\text{m}} = \left|\frac{f_{\text{c}} - f_{\text{filler}}}{f_{\text{c}}}\right|^{-q} \quad \text{for } f_{\text{filler}} < f_{\text{c}}, \tag{9}$$

where ε_{m} is the permittivity of the matrix, f_{c} the percolation threshold, and q dielectric critical exponent of approximately 1. This critical behavior of the dielectric permittivity near percolation can be explained by employing the microcapacitor model [158]. Namely, each two neighboring conductive particles are treated as a local capacitor with the fillers as two electrodes and a very thin host polymer layer in between as dielectric, and the network of local capacitors expands between two virtual electrodes with increasing filler content. Each microcapacitor contributes an abnormally large capacitance. The large capacitance contributed by each of these microcapacitors can then be correlated with a significant increase in the intensity of local electric field when the conductive filler are very close to each other near f_{c}. The significant increase in the intensity of the local electric field promotes the migration and accumulation of the charge carriers at the filler-matrix interfaces. These charges are generated either by surface plasma resonance or by charge injection from the external electrodes, depending on the relaxation times of the two phases. This interfacial polarization, also known as the Maxwell–Wagner–Sillars (MWS) effect, is responsible for the significant increase in permittivity at low frequency [158]. The critical exponent q depends on the spatial dimension of the composites [159], To some extent, it also determines the maximum value of the dielectric permittivity achieved at the percolation threshold, as pointed out by Efros and Shklovski [160]. Because percolation theory was initially established based on classic lattice percolation, the critical exponent values would deviate from its universal values (0.8–1) as interactions between the particles in composites occur, such as quantum tunneling effect and so on [161–164]. Meanwhile, the percolation threshold depends upon the size and shape of the conducting fillers. A theoretical value of f_c is approximately 0.16 for homogeneous composite containing randomly oriented fillers of spherical shape with similar dimensions [165]. Given its non-universality and strong dependency on the microstructure of composites, this value may not be observed in some practical continuum systems. The percolation threshold of different systems will fluctuate by one order of magnitude with respect to the theoretical value, 0.16. Specifically, for most spherical conductive fillers, the experimental percolation threshold values are distributed in a broad range, from 0.013 to 0.17 [166–172], most of which are far lower than the theoretical value of 0.16. A commonly accepted explanation for this is that the real existent form of filler in the polymer matrix is clusters formed by the randomly distributed conductive fillers with a complex shape and larger aspect ratio than each filler itself, due to the strong van der Waals forces between nanoparticles [173]. Thus, the dimension of the clusters plays a crucial role on determining the percolation threshold in percolative composites of spherical conductive fillers. Monte Carlo simulations demonstrated the influence of aspect ratio of ellipsoidal fillers on percolation threshold as:

$$f_c \propto \frac{1}{p}, \tag{10}$$

where, p is the ratio of length of long axis to short axis. This expression has been well verified by experimental results [174–177]. For instance, nanocomposites filled with $p \approx 12$ carbon fiber [178], $p \approx 200$ exfoliated graphite nanoplates [179], and $p \approx 1000$ graphene sheets [180], presented rapidly decreasing f_c values of 0.074, 0.010, and 0.001, respectively, as a result of the increase in aspect ratio. Of particular note is that a rather high dielectric permittivity of 2700 at 100 Hz could be observed at such a low filler concentrations [179]. A schematic description and detailed elaboration on the percolation theory can be found in Dang's and Nan's previous review papers [4, 157, 158]. Distinguished from approaches to realize high permittivities by ceramic inclusions in ceramic/polymer composites, the low percolation threshold in percolative composites allows for a superior mechanical flexibility as well as reduced volume and weight, which is thus highly desirable for satisfying the requirements of high-performance flexible energy-storage devices.

3.2.2. *Energy density of percolative nanodielectrics*

As discussed above, the effective dielectric permittivity of a percolative composite is inversely proportional to the difference between the percolation threshold f_c and the actual volume fraction of fillers (f_{filler}). Therefore, a high dielectric permittivity can be reached only as the filler loading is very close to but does not exceed the percolation threshold. By properly adjusting the filler loading, percolative composites can show very high dielectric permittivities. High permittivity values ranging from several hundreds to thousands have been observed in a number of percolative polymer composites with various conductive and semiconducting particles, including metal particles [168, 171, 172, 181–184], carbon nanofibers [185–188], CNTs [189–200], grapheme [201–204], and ZnO [205, 206]. Dang *et al.* [172] first proposed and demonstrated a percolative Ni/PVDF composite as dielectric for the energy-storage application. The dielectric permittivity of the composites can reach 400 with the volume fraction of Ni approaching the percolation threshold of 17 vol% from below. As compared with the spherical conductive particle, one dimensional carbon materials with large aspect ratios can lead to a much lower percolation level. A higher dielectric permittivity was reported as 600 near the percolation threshold of 8 vol% in multiwall carbon nanotube (MWNT)/PVDF nanocomposites, where the MWNT fillers were chemically modified with 3,4,5-trifluorobromobenzene (TFBB) [191]. The very high dielectric permittivity originated from the MWS effect at the percolation threshold, and was attributed to a thin insulating layer of PVDF that was part of the functionalized MWNT. The nomadic charge carriers were blocked at the internal interfaces between the functionalized MWNT and the PVDF, and the large π-orbital of the MWNT provided the nomadic electrons with large domains. Strongly electrophilic F groups on the functionalized MWNTs further reinforced the MWS effect. Another promising candidate for creating high permittivity percolative composites is two dimensional conductive filler with a disk or plate shape. He and coworkers [201] employed such a design and demonstrated a graphite nanoplate/polymer nanocomposite. The graphite nanoplates were aligned with their faces parallel to each other but still separated by a thin host polymer layer, forming numerous microcapacitors. As a result, a large dielectric permittivity of 2700 was observed near percolation threshold as low as 1.0 vol%. The large deviations of the dielectric permittivity observed in different systems can be largely attributed to size, shape, and spatial arrangement of nanoparticles as well as the intrinsic properties of polymer matrices and nanofillers [150, 207, 208]. Geometric factors and/or the conductivity of the fillers play key roles in determining the dielectric properties of the nanocomposites mainly through significantly affecting the value of percolation threshold [190, 209, 210]. Yao *et al.* [190] studied the effect of aspect ratio of carbon nanotube on the percolation threshold, and found that the percolation threshold first increased with increasing aspect ratio, reached a maximum and subsequently followed a decrease with further increasing aspect ratio. This result can be explained by using the concept of nanotube entanglement. Dang *et al.* [209] have also attempted to control the dielectric behavior of Li-doped NiO (LNO)/PVDF nanocomposites near the percolation threshold by fine tuning the conductivity of nanofiller itself. The percolation threshold was observed to increase with decreasing the conductivity of LNO nanofillers. Compared with the properties of the additives, the polymer matrix determines the dielectric response of composites directly as described by Eq. (9). In this regard, polar matrices with high permittivities are thus the preferred choices for the percolative composites formation.

Although desirable dielectric permittivity and low percolation threshold in conductive particle/polymer nanodielectric materials have readily been achieved, there are still some obstacles, such as high dielectric loss, low dielectric strength and narrow processing window, which largely inhibit their practical applications in high-U_e capacitors as dielectrics. The inclusion of conductive fillers not only leads to high even giant dielectric permittivities in composite due to the MWS polarization or microcapacitor effect, but also results in high dielectric loss and leakage current through the conductive particle network formed near f_c, which gives rise to an abrupt reduction in breakdown strength [167, 211, 212]. A general explanation for this is that the formation of microcapacitors leads to an abnormally large local electric field in the pure polymer region as the neighboring nanoparticles are very close to each other. As the polymer layer between neighboring particles (interparticle distance) is too thin to withstand the increasing local electric field, a short circuit occurs and induces large leakage current [167, 213, 214]. Tunneling is recognized as the dominant mechanism for electric conduction in percolative composites. According to the tunnelling model, the inter-particle tunnelling conductivity σ_{tun} is given by [215]:

$$\sigma_{\mathrm{tun}} \propto \exp\left(-\frac{l-2b}{d}\right) \tag{11}$$

where l is the distance between the centers of the particles, b and d are the radius of the particles and the typical tunnelling range, respectively. As described by this model, the conductivity of the nanocomposites increases exponentially when the loading level approaches the percolation threshold, since l decreases rapidly with increasing loading based on geometric calculation. The absence of insulation nature makes these materials useless for the application in capacitor dielectrics, since the significantly deceased dielectric strength is detrimental to the energy density of composites due to its quadratic relationship with electrostatic energy storage. Stoyanov *et al.* [167] compared the energy density of pure poly(styrene-*co*-ethylene-*co*-butylene-*co*-styrene) (SEBS) and carbon black(CB)/SEBS composites, and found that the increase in permittivity of CB/SEBS composite cannot counter the adverse impact of reduction in dielectric strength in relation to the stored electrical energy. The energy density of CB/SEBS composite continuously decreased with increasing CB concentration. However, the corresponding low field dielectric spectroscopy displayed an insulating behavior up to 3.5 vol%. Thus, they emphasized that the low field dielectric spectra could not provide adequate information to judge the usefulness of percolative composites at high field for the energy storage, because the tunnelling, which might guarantee a good insulating nature of composites at low fields, would occur at high fields well below the breakdown strength of the percolative composites. In addition, both of low field and high field dielectric properties in these percolative systems are extremely sensitive to the concentration of fillers, especially as approaching the percolation threshold. It has been demonstrated that, in some cases, a small deviation of 0.05% from the percolation threshold can decrease the dielectric permittivity by 100%. Such a narrow processing window indeed imposes a serious challenge and risk in controlling the percolative composites with the threshold composition in order to obtain reproducible composite for practical energy-storage application [216]. Thus, the optimal way is to adopt concentrations a little far below percolation threshold for security consideration, yet at which the high dielectric permittivity no longer exists.

During past years, several strategies have been proposed to prevent the direct contact between conductive particles near the percolation threshold, with the aim of suppressing the high dielectric loss, maximizing the dielectric strength and also expanding the compositional windows. One straightforward approach is to introduce an interparticle barrier layer into the percolative nanocomposites. As such, the tunneling current between neighboring conductive nanoparticles can be suppressed and hence leads to far lower dielectric loss and higher breakdown strength. Additionally, the interfacial layers introduced can control over the detailed and sensitive microcapacitor structures near f_c, which generally gives rise to a considerably expanded compositional window for percolative composites. By fine tuning the properties of the barrier layers, a desirable energy density can be expected in percolative composites due to the largely increased permittivity and slightly reduced breakdown strength. The introduction of interfacial layer into percolative composites was first demonstrated by Wong and co-workers in metal/polymer composite [217, 218]. The surfaces of 100 nm Al fillers were firstself-passivated by a very thin layer (2.8 nm) of Al_2O_3, then compounded with epoxy to form percolative composites. As shown in Fig. 21(b), the inclusion of such core–shell particles into an epoxy polymer matrix generated a significantly increased dielectric permittivity, while not affecting the inherent loss tangent of the matrix very much. A high dielectric permittivity up to 60 was achieved at filler content of 50 wt%, at which the loss tangent was as low as approximately 0.02. The Al_2O_3 interfacial layer effectively restricts the electron transfer between neighboring Al nanoparticles and ensures the loss tangent at very low level [217]. The slight decrease in dielectric permittivity at loading exceeding 50 wt% was attributed to the void formation at high filler content as a result of the presence of fractal clusters. The resultant nancomposites showed excellent low field dielectric properties; nevertheless, due to lacking of the characterization of the high field dielectric responses, it is hard to ascertain the energy density of such composites. A comprehensive picture of how the self-passivated insulating layer influences the energy-storage performance of percolative composites remains elusive.

As compared with the self-passivated inorganic shell, coating organic interfacial layers on the conductive nanoparticles has been demonstrated to be more versatile and allows for larger tunability in dielectric properties of percolative composites. Qi *et al.* [212] modulated 40 nm Ag nanoparticles with a thin layer of mercapto-succinic acid (MSA) molecules and then mixed them with epoxy by in situ polymerization process. The dielectric permittivity of the composites could increase up to 300 while the loss tangent still remained less than 0.05 (Fig. 21(c)). The presence of the organic coating makes the formation of conductive filler network very difficult, which thus significantly lowers the dielectric loss and affords a desirable high concentration tolerance for high permittivity. Such high permittivity and low loss tangent are promising for the energy-storage applications. However, the breakdown strength decreased substantially with increasing Ag volume (Fig. 21(d)), indicating the surfactant organic coating fails to prevent tunnelling current at high fields due to the loose morphology and possible structural defects. Because of the significant decline in the dielectric strength, the energy density of the composites largely decreased relative to the virgin polymer matrix, despite there was a small augmentation as approaching the percolation threshold. Therefore, a denser organic layer with sound structural integrity is needed for improving both of low field and high field dielectric properties, thus, in turn, increase the energy density of percolative composites. Shen and Nan [219, 220] recently made such an effort in Ag/epoxy percolative composites. Ag nanoparticles were first coated with carbonaceous organic shells via a one pot hydrothermal method, and then the as-prepared Ag@C core–shell nanoparticles were compounded with epoxy matrix to create percolative nanocomposites. The morphologies of the Ag@C core–shell nanoparticles are clearly shown in Fig. 22(a) and (b). The thickness of shells can be fine tuned to tailor the dielectric properties in terms of the specific requirements of practical applications. The core–shell nanoparticles form a network near the percolation threshold to connect the top and bottom electrodes on the sample. Each two neighboring particles can be treated as nanocapacitors, preventing the Ag cores from connecting with each other directly even at high filler loading (30 vol%). As shown in Fig. 22, the Ag@C/epoxy nanocomposites show high permittivity (450) and very low loss tangent ($\tan\delta < 0.05$), and the permittivity does not change with the applied high electric fields. However, the breakdown voltage drops with increasing filler concentration as expected, for example, from ca. 30 MV/m to 0.6–1 MV/m. Due to the dense and integral morphology of the organic shell layer, the decline in dielectric strength is not as sharp as in the case of MAS-Ag/epoxy systems. The energy density of the polymer matrix can be improved at certain optimal filler concentrations, and such enhancement becomes larger as increasing the thickness of the shells. It should be noted that these optimal concentrations are far below the percolation threshold, suggesting that the intensively studied high permittivity near f_c in most percolative systems may not be transferred into a high energy density at high fields, though the low field characteristics of composites could be insightful. For pulsed power application, the high-field behavior (applied electric voltage > 10 kV) is more important. It is plausible that the key for further developments of high energy density percolative composites lies in

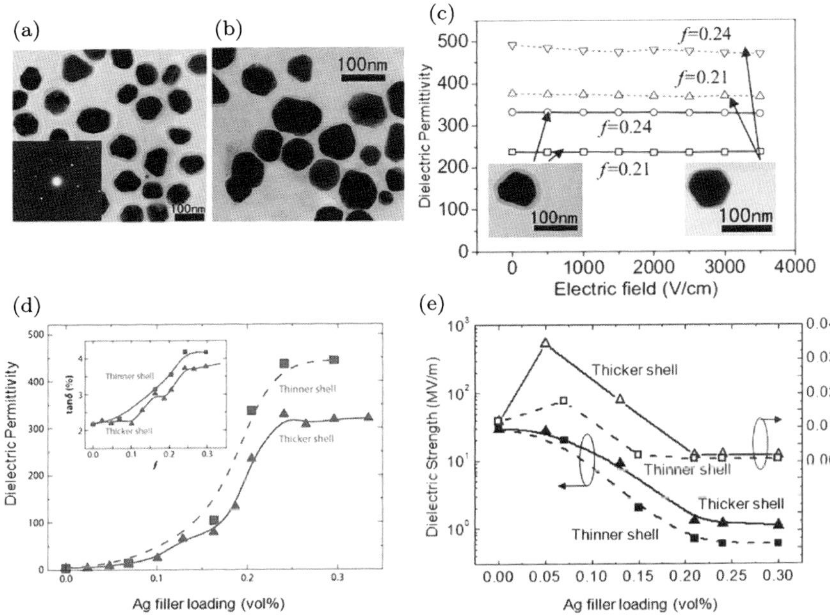

Figure 22. TEM image of the Ag@C core–shell nanoparticles with: (a) thick shells and (b) thinner shells. The single Ag crystals can be readily seen from the selected area electron diffraction pattern of the particle. (c) High field dielectric permittivity for the composites with different volume fraction of thinner and thicker Ag@C core–shell nanoparticles. (d) Low field dielectric permittivity as a function of the Ag volume fraction in the Ag@C filled composites. (e) Dielectric breakdown strength and calculated energy density of the composites as a function of the filler content. The energy density was calculated by using the permittivity data at 1 kHz from work by Shen *et al.* [219].

the improvement of the dielectric strength rather than the highly improved permittivity near percolation threshold. Nevertheless, how to improve the high-field dielectric responses of polymers by using conductive nanofillers is still an open question and deserves much more attention in the future.

It has been demonstrated that the dielectric behavior of the core–shell nanoparticle filled nanodielectrics was very sensitive to the thickness and electrical properties of the shells [221]. The dense and integral organic shells would anticipate to minimize the decrease in the dielectric strength and possibly give rise to an increased energy density relative to the neat polymer matrix. Very recently, Yang *et al.* [222] successfully coated the MWNTs with organic shells polypyrrole (PPy) by an inverse microemulsion polymerization. TEM images of the MWNT gave the thickness of the PPy layer to be on the order of 10 nm (Fig. 23). Such PPy-coated MWNT/polystyrene (PS) composites presented a high dielectric permittivity (44), low loss (<0.07), and a high energy density (4.95 J/cm^3). This value is about two times higher than that of pure PS and also larger than that of state-of-the-art commercial BOPP capacitor dielectrics. The significantly enhanced dielectric performance originated from the organic shell PPy, which not only enables good dispersion of MWNTs in the polymer matrix but also screens charge movement to shut off leakage current as a barrier.

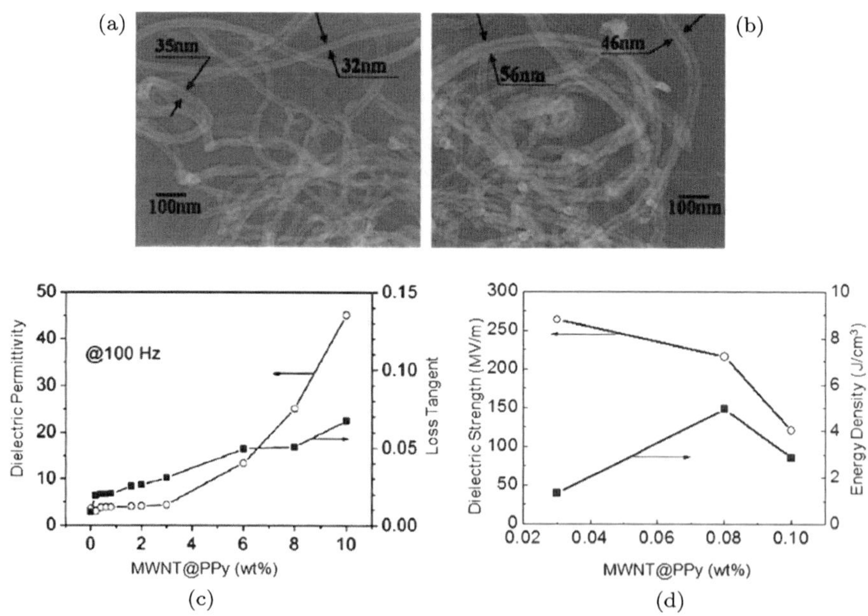

Figure 23. TEM micrographs of: (a) uncoated MWNTs and (b) MWNTs@PPy. (c) Dielectric permittivity and loss tangent of MWNTs@PPy composites as a function of the filler concentration, measured at 100 Hz and room temperature. Reproduced with permission from Ref. [222]. Copyright 2009, Elsevier. (d) Dielectric strength and energy density of MWNTs@PPy/PS nanocomposites as a function of filler content. Data taken from work by Yang *et al.* [222].

Very recently, Liu *et al.* [223] modified MWNTs with jeffamine T403 and established a strong interaction between the T403-functionalized MWNTs (MWNT-T403) and the polysulfone (PSF) polymer matrix due to the ester and ether groups on the surface of MWNT-T403. Afterwards, they used a simple electrospinning method (Fig. 24(a)) rather than directly added the MWNT-T403 into PSF to prepare MWNT-T403/PSF composite nanofibers with MWNT-T403 coated by PSF (Figure 24(b)). These electrospun composite nanofibers were then collected in the form of a sheet on

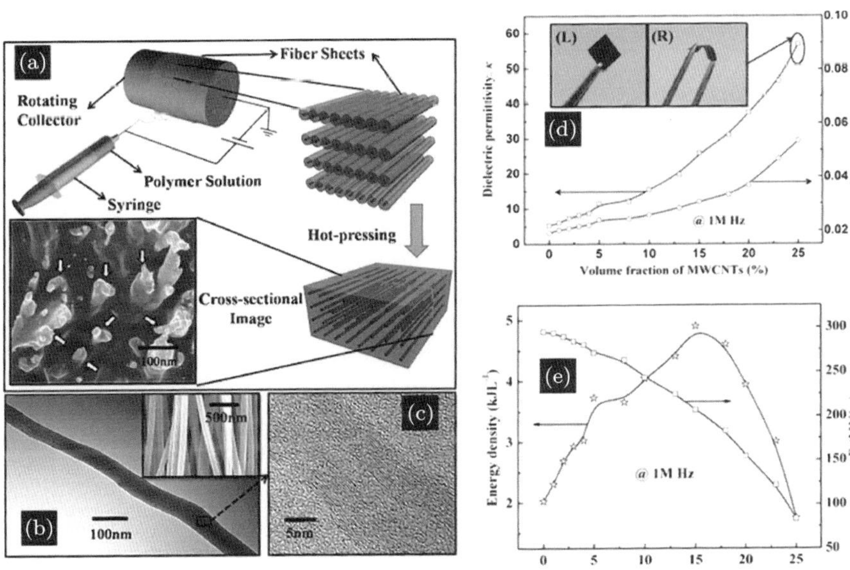

Figure 24. (a) Schematic for preparation of MWNT-T403/PSF composites. The inset shows the cross-sectional SEM image of the composites perpendicular to the fiber direction, MWNTs are indicated by arrows. (b) TEM images for MWNT-T403/PSF core–shell nanostructure. The SEM image in the inset shows the MWNT-T403/PSF nanofiberous sheet prepared by electrospinning. (c) High-resolution TEM image of a MWNT. (d) Dependence of the dielectric permittivity and loss of the composites on the MWNT-T403 loading, measured at room temperature and 1 MHz. The flexibility of the composite is demonstrated in the photograph in the inset. (e) Energy density and breakdown strength of the composites with different MWNT-T403 loading. Reproduced with permission from Ref. [223].

a rotating drum, and several plies of the sheets were stacked along the same fiber direction. These sheets with nanofibers lying down in the plane were finally molded by the hot-pressing near the softening temperature of PSF perpendicular to the fiber direction to form the MWNTs array (Fig. 24(a)). The organic PSF dielectric coating on MWNT not only acted as a barrier to prevent MWNTs from direct connection, but also as a polymer matrix after fusion by hot-pressing at 175°C. The resultant MWNT-T403/PSF composites showed a high permittivity (ca. 58) when the content of MWNT-T403 reached 25 vol%, while there was only a slight change in loss ($0.02 \rightarrow 0.05$ at 1 MHz) (Fig. 24(d)). In addition, a maximum energy density of 4.98 kJ/L was observed when MWNT-T403 was up to 15 vol%, as shown in Fig. 24(e).

Shells on the conductive fillers with good integrity are desirable in the pursuit of high dielectric permittivity and low loss tangent in that they provide a better insulating barrier for electron tunneling and suppress the leakage current resulting from quantum tunneling. Although the low-field characteristics of percolation nanocomposites are insightful, the high-field dielectric behaviors are much more important for the power-energy-storage applications. The introduction of thick and integral organic shells with a high breakdown voltage may effectively optimize the high-field tolerance and benefit in maximizing the energy density of the percolative composites.

3.3. *Three-phase inorganic particle/polymer nanodielectric materials*

Besides the effect of geometric or physical characteristics of fillers, the feature of matrix like viscosity, polarity, and degree of crystallinity could also influence the percolation behavior of the composite [166, 171, 224, 225]. The formed inorganic/polymer nanocomposite can be employed as a high-permittivity polymer matrix to host various conductive fillers, achieving both high permittivity and low dielectric loss by means of synergistic effect via combining the benefits of two fillers. Compared with single conductive filler/polymer nanocomposite, the dielectric permittivity values of three-phase inorganic particle/polymer nanodielectric materials usually increase remarkably at the similar concentration of conductive filler, accompanied with a slight variation in dielectric loss. A number of researchers have actively investigated the synergistic effect of two fillers on the dielectric properties of three-phase percolative nanocomposites [84, 226–236], but few focused on their influences on breakdown strength or energy-storage capacity [237]. The usefulness of this approach for high-energy-density capacitors requires further study.

4. Nanostructured all-polymer dielectric materials

Future pulsed-power and power electronic capacitors imposes strict requirements on the dielectric materials, such as ultimate energy densities of $>30\,\mathrm{J/cm^3}$, operating voltages of $>10\,\mathrm{kV}$, and charge/discharge times of milliseconds–microseconds with reliable operation near the dielectric breakdown limit [83]. An energy-storage performance closest to this target has been found in nanostructured polymeric material with a state-of-the-art energy density of $27.3\,\mathrm{J/cm^3}$ [21]. This shows another promising avenue for improving energy-storage capacity of polymers, that is, fine tuning molecular structure and/or microstructure of polymers themselves.

As indicated in the formula $U_e = \int_{D_{\max}}^{0} E dD$, the key to achieving high energy densities in polymers is to have high electric displacement and large dielectric breakdown strength simultaneously. However, the level of displacement is very low in dielectric polymers that are currently used. For example, the displacement in BOPP is below $0.012\,\mathrm{C/m^2}$ under a field of $600\,\mathrm{MV/m}$, thus resulting in a low energy density of only $2.5\,\mathrm{J/cm^3}$ [238]. Conversely, polar-polymers, such as PVDF and its copolymer P(VDF-TrFE), exhibit an electric displacement higher than $0.1\,\mathrm{C/m^2}$ owing to the large dipole moment of the C–F bond. This would potentially offer an order of magnitude increase in the energy density [19, 63, 239–241]. However, the discharged electric energy densities are actually less than $5\,\mathrm{J/cm^3}$, which is largely limited by the early electric displacement saturation at electric field far below E_b, as illustrated in the top shaded area in Fig. 25(a). In this regard, the electric displacement saturates and the charge density cannot be further increased with increasing the electric field. Even worse, only a very small portion of stored

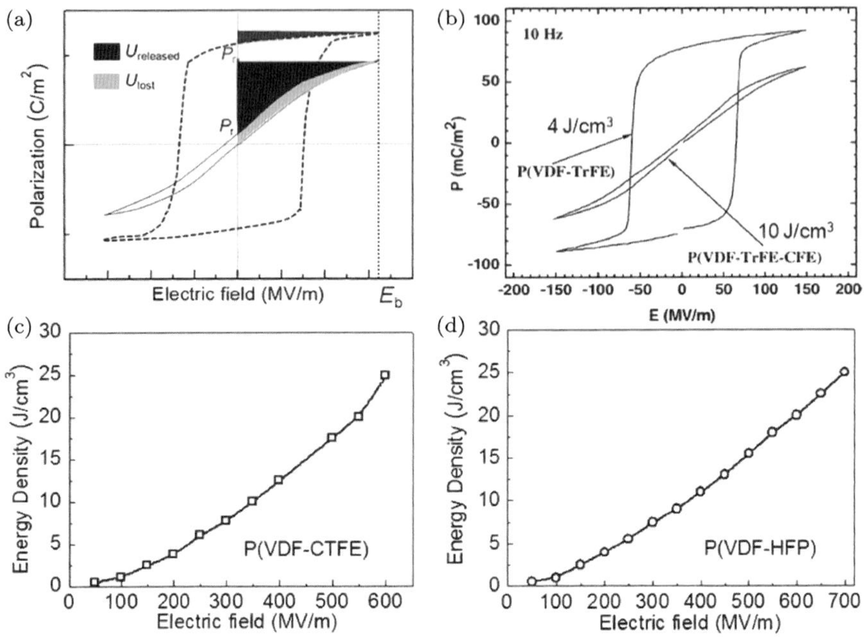

Figure 25. (a) Dipolar D–E hysteresis loops of typical polar fluoropolymer (dash line) and an ideal polar polymer (solid line) for energy storage, graphically representing high-field dielectric loss and the breakdown field that is previously used to calculate the maximum energy-storage density. (b) The polarization hysteresis loop of the terpolymer P(VDF-TrFE-CFE) 63/37/7.5 mol% and P(VDF-TrFE) copolymer measured at 10 Hz. Reproduced with permission from Ref. [242]. Copyright 2006, Springer. (c) Discharged energy density as a function of applied field deduced from the D–E loops for P(VDF-CTFE) 91/9 mol% measured at 10 Hz and room temperature. The data plotted are taken from Ref. [243]. (d) Discharged energy density of P(VDF-HFP) versus applied electric field at room temperature. Reproduced with permission from Ref. [244]. Copyright 2009, American Institute of Physics.

energy density can be effectively discharged due to the large remanent polarization (P_r). This irreversible polarization reduces their voltage reversal tolerance and also decreases the energy-storage efficiency (η), which can be quantitatively expressed as:

$$\eta = \frac{U_{\text{released}}}{U_{\text{released}} + U_{\text{lost}}} = 1 - \frac{U_{\text{lost}}}{U_{\text{released}} + U_{\text{lost}}}. \tag{12}$$

Overall, besides a high electric displacement, an ideal polar-polymer should also have a large dielectric breakdown strength, a very small P_r ($D \approx 0$ at $E = 0$) so as to afford a high discharged (released) energy density and a low electrical loss energy density, which correspond to the black and gray areas respectively in Fig. 25(a).

4.1. *Effects of molecular structure*

During the last few years, molecular structure as determined by monomer or comonomer nature has been demonstrated to play a crucial role on determining the dielectric performances at high electric fields. A few strategies, including chemical methods (copolymerization) [19, 64–66], mechanical approaches (uniaxial stretching) [244, 245], and high-energy radiation [246] have been developed to increase the energy density of PVDF and its copolymer P(VDF-TrFE) through modifying the molecular structure of polymers. It has been demonstrated that normal ferroelectric P(VDF-TrFE) polymers can be converted to relaxor ferroelectric polymers with very small P_r by the introduction of defect molecular structures. Herein, defect is defined as a different molecular conformation (nonpolar) from the all-trans one (polar) in the ferroelectric phase of PVDF-based polymers. These defects are introduced either by high-energy electron irradiation [246] or through chemical copolymerization of a third bulky monomer [247]. It was found that the electron irradiation could break up the long-range polarization domain (all-trans chains) into nanopolar regions with nanometer size by interrupting the all-trans chains with trans and gauche bonds (defects), thereby stabilizing the nonpolar phase and significantly reducing or even eliminating the transformation barrier at room temperature between the polar and nonpolar phases shown in the normal P(VDF-TrFE) polymer. Consequently, a continuous transformation from nonpolar to polar can be stimulated by an external electric field and shows a very small P_r [246]. The reducedP_r allows for a large variation in electric displacement with increasing the electric field and thus leads to a higher discharged energy density.

More recently, it has been shown that a similar modification of properties of P(VDF-TrFE) can also be realized by adding a third bulky monomer to form terpolymers. A terpolymer P(VDF-TrFE-CFE) has been developed via a suspension polymerization process [242]. The introduction of certain content of CFE into ferroelectric P(VDF-TrFE) copolymer could convert the crystal phase from high polar β-phase (all-trans TTTT conformation) into γ-phase (trans–trans–trans–gauche TTTG conformation) with medium polarity. Meanwhile, the crystallinity of copolymer is reduced, which provides essential lattice space for the poling and reversing of dipoles under high electric field. By careful molecular structure design, the terpolymer can exhibit a slim dipolar D–E loop, as shown in Fig. 25(b). The energy density of a specific terpolymer can reach up to $10\,\mathrm{J/cm^3}$, which is much larger than that ($4\,\mathrm{J/cm^3}$) of P(VDF-TrFE) copolymer.

On the other hand, copolymer P(VDF-CTFE) [19] and P(VDF-HFP) [244] have been demonstrated by importing bulky monomers of CTFE and HFP as defect sources into PVDF homopolymer. The combination of nonpolar (defect structure) and polar (PVDF) molecular structural changes of the polymers provides tunability in dielectric permittivity. By properly adjusting the permittivity of copolymers, the early electric displacement saturation can be effectively avoided and thus a high energy density can be observed as high as $17\,\mathrm{J/cm^3}$ in the early investigation of P(VDF-CTFE) copolymer [19]. By further improving the film processing conditions and quality of copolymer film, a dielectric breakdown strength more than 600 V/m has been realized in uniaxially stretched P(VDF-CTFE) [243]and P(VDF-HFP) [244] films, and consequently a discharged energy density of approximately $25\,\mathrm{J/cm^3}$ was achieved, as shown in Fig. 25(c) and (d). These improvements were attributed to the presence of Cl atom in defect molecular structures, which could stabilize the nonpolar α-phase of the copolymer and avoid displacement saturation or phase transition at high electric fields [238].

4.2. *Effect of crystal structure*

The energy-storage performance of polar polymers has been proven to be largely determined by P_r, E_b, and early D saturation. Apart from the defect modification and uniaxial mechanical stretching, simple thermal treatment process is also a promising approach to tune these parameters by changing crystal domain size and crystallinity, and consequently the energy-storage properties of polymers [238, 239]. Zhang and Li [238] have investigated the effect of crystal domain on the energy-storage performance of PVDF. α-, β-, and γ-PVDF films were first obtained via different thermal treatment processes and then suffered the charge–discharge procedure of electric energy storage. It was found that the β-PVDF could withstand a maximum electric field of only 200 MV/m due to the early D saturation, whereas α- and γ-PVDF endured an electric field as high as 500 MV/m. Unfortunately, near 300 MV/m, a phase transition occurred in α-PVDF, resulting in a high energy loss. Due to the absence of early D saturation and the phase transition, the γ-PVDF could be applied a high electric field (500 MV/m) and afforded a discharged energy density as high as 14 J/cm^3.

From previous cases, it can be easily noted that the dielectric strength relies highly on the crystal structure in PVDF-based polymers. According to Fröhlich theory [248], the dielectric strength of polymers is not only related to its molecular structure but also largely affected by the crystal size and crystallinity of polymers. The relationship between E_b and the diameter of the spherical crystal grain, D, is described as:

$$E_b = A\left[\sqrt{[(D-d)^2 + 1\times 10^{-10}]} - (D-d)\right] + E_0, \qquad (13)$$

where d is the thickness of sample film. E_0 represents the breakdown field as $D \geq d$, while $A$is a constant. For a polymer film with a given thickness, a reduced D could generally give rise to a higher E_b. Xia *et al.* [239] thermally quenched P(VDF-CTFE) films at various temperatures to vary the crystalline properties in an attempt to tailor their energy-storage properties. It was reported that the lower quenching temperature, the smaller the crystal grain and the lower the crystallinity in polymers. Therefore, quenching at very low temperature (liquid nitrogen:–195.8°) cannot only reduce or eliminate P_r induced by the large crystal grains, but also improve the dielectric strength from 230 MV/m to over 500 MV/m. The decrease in crystallinity can be another reason for the improvement in E_b. It has been postulated that an increase in the depth of shallow trap levels for electron was induced by the increase in crystal boundaries due to the decrease in crystallinity. Therefore, the author concluded that both the reduction of crystal grain size and crystallinity induced by quenching at low temperature favored the increment in crystal boundaries, thus the high E_{b}. As a result, the discharged energy density of corresponding films was improved from 3.7 to 15.3 J/cm^3.

4.3. *Effects of polymer blend structure*

In order to further optimize and/or balance the various characteristics of single polymer, polymer blend approaches have been proposed to exploit the merits of both the base polymer and the additive polymer. For instance, Chu and Zhang [249] investigated the dielectric properties of the blends of terpolymer P(VDF-TrFE-CFE) and copolymer P(VDF-CTFE). By the inclusion of a small amount (5 wt% and 10 wt%) of copolymer, the high-field polarization response and breakdown strength were largely improved as compared with the neat terpolymer. Consequently, the energy density was increased from 9 (for neat terpolymer) to 11.5 J/cm^3 in the polymer blends. The energy-storage performance of polymer blend is in the midst of both components. However, Rahimabady *et al.* [21] realized a higher energy-storage properties in polymer blends than either of the individual component by preparing a dense α-phase blend films of VDF oligomer and PVDF of various compositions from chemical solution deposition. The microstructure of the blend was schematically illustrated in Fig. 26(a). Despite the dielectric permittivity of the films being unexpectedly lower than either of the two components, the Young's modulus was higher than that of each component (Fig. 26(b)). According to the Stark–Garton model [250]:

$$E_b = 0.6\sqrt{\frac{Y}{\varepsilon_0 \varepsilon_r}}, \tag{14}$$

a low dielectric permittivity (ε_{r}) and a high Young's modulus (Y) may induce a large breakdown strength (E_{b}). It was observed that the dielectric breakdown strength was extremely high ($>$ 850 MV/m vs. 300–500 MV/m for typical PVDF films), leading to a maximum

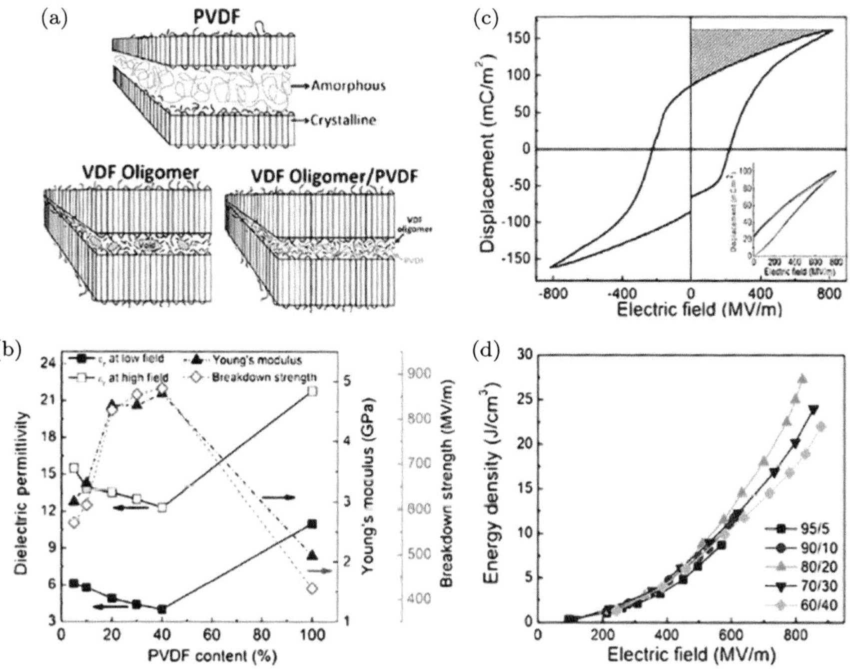

Figure 26. (a) Schematic illustrations of the phase structure of PVDF, VDF oligomer, and VDF oligomer/PVDF blend. (b) Low-field and high-field dielectric permittivity, Young's modulus and dielectric strength as a function of PVDF content in the VDF oligomer/PVDF blend films. (c) The dipolar *D–E* hysteresis loop of the VDF/PVDF (80/20). The shaded area represents a discharged energy density as high as $27.3\,J/cm^3$. The inset in (c) shows the hysteresis loops at unipolar electric fields, providing an energy loss of 33%. (d) Discharged energy density of VDF oligomer/PVDF blend as a function of applied electric field, measured at room temperature. (a–d) Reproduced with permission [21]. Copyright 2011, American Institute of Physics. (e) Unipolar *D–E* response under 10 Hz at room temperature of P(VDF-CTFE) 91/9 mol%, P(VDF-CTFE)/ECTFE 67/33 wt% blends, and crosslinked P(VDF-CTFE)/ECTFE 67/33 wt% blends. *D–E* response of a commercial BOPP film is also plotted for the sake of comparison. The lower and upper curve in each charging–discharging procedure is the charging and discharging curve, respectively. All the figures in (e) are reproduced with permission from Ref. [240]. Copyright 2011, American Institute of Physics.

(e)
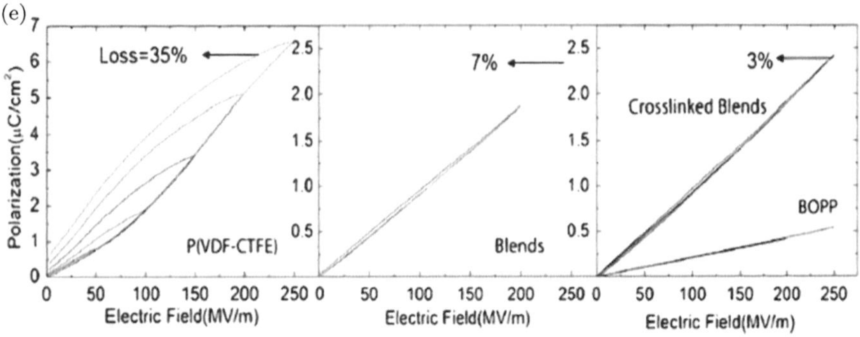

Figure 26. (*Continued*)

polarization of $0.162\,C/m^2$ and a large electric energy density up to $27.3\,J/cm^3$ (Fig. 26(c) and (d)). This value represents the state of the art in high-energy-density polymeric dielectrics. The properties were attributed to the unique blend structure with high crystallinity and densely packed rigid amorphous phase incorporating long and short chains (see Fig. 25(a)).

It has been demonstrated that the PVDF-based copolymer [243, 244] or polymer blends [21, 249] exhibit not only a high energy density but also a high loss at high electric field. Such a high field loss is usually caused by the formation of large polarization domains under high applied field and detrimental for their practical application in capacitors, since even moderate losses at high fields will cause an elevated temperature, thus severely limiting the energy-storage performance of the capacitor. A polymer blend approach can be also used to reduce the dielectric loss of PVDF-based polymers by introducing a low loss additive polymer. Wu and Zhang [240] observed a high energy density coupled with low loss in a transparent polar-fluoropolymer blend consisting of high energy density P(VDF-CTFE) copolymer and low loss poly(ethylene-chlorotrifluoroethylene) (ECTFE) polymer. At high fields, the polymer blends can significantly reduce the loss from 35% (for neat P(VDF-CTFE) copolymer) to 7%. Moreover, introducing crosslinking in the blends can further improve the mixing of two polymers at the nanoscale, thus leading to a further reduction of high-field losses to 3%, while maintaining a high energy density in blend films, as shown in Fig. 26(e). The author claimed that crosslinks had much stronger effect against the formation of large polarization domains as compared with the approach aforementioned of introducing isolated ECTFEs to serve as random defects to break up the large polarization domains in PVDF. The promise of this approach lies in that the reduction of high field loss does not affect the high level of energy density in capacitors. The good control of nanoscale mixing of the two polymers and the degree of their crosslinking density are the keys to yield a blend film with a high energy density and low loss at high electric fields.

5. Concluding remarks and prospects

FNDMs are receiving tremendous attention and research interest due to the increasing concern on power energy storage. To fulfill newly emerging applications of high-U_e capacitors, such as hybrid electric vehicles and electrical weapon systems, advanced dielectric materials with superior integrated performance including high energy and power density, good processability, environmentally friendly and flexible storage of energy are highly demanded. As described above, several general strategies for making advanced FNDMs have been developed, such as ceramic nanoparticle/polymer nanodielectrics, percolative nanodielectrics, and nanostructured all-polymer dielectrics.

Tremendous achievements have been made in recent years on improving the energy-storage capacity of flexible polymers through rationally selecting the identity and shape of the nanoadditives and engineering of the nanoparticle-polymer interfaces. The general idea is to accomplish universally desirable dielectric properties (large ε_{eff} and high E_b) necessary for a large energy density in dielectrics. It has been demonstrated that the high-ε_{eff} metal oxides generally lead to an improved dielectric permittivity of composites unfortunately at the expense of dielectric strength due to the significant local field exclusion. In most cases, such permittivity enhancement could not offset the significant decrease in dielectric strength to afford an increased theoretical maximum energy density relative to the virgin polymer. Therefore, the most significant progress in high-U_e ceramic/polymer FNDMs may rely on increasing the dielectric strength of polymers over a wide range of additive content rather than over focusing on increasing the permittivity. To meet this goal, two tactics should be preferentially considered. The first is to exploit manipulation of anisotropy in high-field dielectric properties. Designed anisotropy in dielectric properties can provide excellent paradigms for the development of high-U_e ceramic/polymer FNDMs and gain important insights into the mechanisms that control dielectric breakdown strengths and non-linear conduction at high fields. The low-ε_{eff} metal oxides with large aspect ratios should be given special attention for this approach, as the comparable permittivity to neat polymer could effectively minimize the severe field intensification in the polymer matrices; meanwhile the large aspect ratio is desirable to realize an anisotropic ceramic/polymer nanodielectric. The other approach is more versatile and associated with the multicore model proposed by Tanaka *et al.* [56]. It has been demonstrated a properly engineered multilayered nanoparticle-polymer interface is essential for achieving favorable high-field dielectric properties and thus high energy-storage capacity. The ideal multilayered interface would involve a polar inner layer and a robust, stable, and physically thick (3–5 nm) insulating nonpolar outer layer. The polar group of the inner layer could largely charge the surface of nanoparticles and substantially contribute to a high polarization level in the composites. While the charged surface and the connectivity of the charged layer is shielded and buried under a nonpolar shielding layer. The insulating nonpolar outer layer can not only effectively prevent the growth of a breakdown path and minimize the leakage current, but also increase the dispersibility of the hybrid particles by matching the solubility parameters of the outer surface of the particles with polymer matrix. Thus, these features enable a large incorporation of nanoparticles to reach a high enough permittivity while maintaining a high dielectric strength in the composites, which is highly desirable to realize a high energy density in ceramic/polymer nanodielectrics.

Although permittivity can be enhanced significantly in a percolative FNDM at much lower filler contents relative to the ceramic/polymer counterparts, the largely reduced electric strength generally leads to a lower energy density than the pure polymer matrix. The usefulness of percolative FNDM for high-U_e capacitors is widely questionable unless the conductive fillers are well coated with a sound and integral organic shell. This insulating barrier is capable of preventing electron tunneling and giving rise to a superior low field dielectric spectroscopy, but it fails to suppress the leakage current resulting from tunneling at high fields. Therefore, improved dielectric strength relative to the neat polymer matrix has not been reported so far despite a higher density having been attained in certain core–shell conductive filler/polymer percolative composites [219, 220, 223]. It is plausible that the key for further developments of percolative FNDM also lies in the improvement of the dielectric strength. Nevertheless, improving the high-field dielectric responses of polymers by using conductive nanofillers is very hard if not impossible. This open question deserves much more attention in the future.

Polymer nanocomposites consisting of nanoparticles allow for a large tunability of the energy density by varying the identity, shape, and content of additives, but they in fact suffer a limited energy density ($<10\,J/cm^3$) that is far lower than the required value ($30\,J/cm^3$) imposed by pulsed-power and power electronic capacitors. Energy-storage performance closest to this target has been found in a nanostructured polymeric material with a state-of-the-art energy density of $27.3\,J/cm^3$ [21]. This indicates the breakthrough of this target may be first realized in the nanostructured all-polymer FNDMs. New chemical methods (copolymerization), mechanical approaches, and thermal treatments are the key to optimizing and increasing the energy density of currently available polymeric dielectrics. As a newly emerging and highly interdisciplinary field, the progress in FNDMs for capacitive energy storage is critically dependent on collaborative efforts from various fields, such as nanoscience and nanotechnology, materials sciences and physical chemistry.

Acknowledgments

This work was financially supported by the NSF of China (Grant Nos. 51073015, 50977001), the NSF of Beijing City (Grant No. 2122040), the Ministry of Sciences and Technology of China through the China–Europe International Incorporation Project (Grant No. 2010DFA51490), and the State Key Laboratory of Power Transmission Equipment & System Security and New Technology (No. 2007DA10512712406).

References

[1] Landi, B. J., Ganter, M. J., Cress, C. D., DiLeo, R. A. and Raffaelle, R. P. (2009). Carbon nanotubes for lithium ion batteries, *Energy Environ. Sci.*, 2, pp. 638–654.

[2] Arico, A. S., Bruce, P., Scrosati, B., Tarascon, J. M. and Van Schalkwijk, W. (2005). Nanostructured materials for advanced energy conversion and storage devices, *Nat. Mater.*, 4, pp. 366–377.

[3] Simon, P. and Gogotsi, Y. (2008). Materials for electrochemical capacitors, *Nat. Mater.*, 7, pp. 845–854.

[4] Dang, Z.-M., Yuan, J.-K., Zha, J.-W., Zhou, T., Li, S.-T. and Hu, G.-H. (2012). Fundamentals, processes and applications of high-permittivity polymer matrix composites, *Prog. Mater. Sci.*, 57, pp. 660–723.

[5] Whittingham, M. S. (2008). Materials challenges facing electrical energy storage, *MRS Bull.*, 33, pp. 411–419.

[6] Sarjeant, W. J., Zirnheld, J. and MacDougall, F. W. (1998). Capacitors, *IEEE Trans. Plasma Sci.*, 26, pp. 1368–1392.

[7] Barber, P., Balasubramanian, S., Anguchamy, Y., Gong, S., Wibowo, A., Gao, H., Ploehn, H. J. and zur Loye, H.-C. (2009). Polymer composite and nanocomposite dielectric materials for pulse power energy storage, *Materials*, 2, pp. 1697–1733.

[8] Wang, Q. and Zhu, L. (2011). Polymer nanocomposites for electrical energy storage, *J. Phys. D: Appl. Phys.*, 49, pp. 1421–1429.

[9] Karanja, P. and Nath, R. (1993). Study of charge storage properties in biaxially oriented polypropylene, *IEEE Trans. Dielectr. Electr. Insul.*, 28, pp. 294–298.

[10] Rabuffi, M. and Picci, G. (2002). *Status quo* and future prospects for metallized polypropylene energy storage capacitors, *IEEE Trans. Plasma Sci.*, 30, pp. 1939–1942.

[11] Nash, J. L. (1988). Biaxially oriented polypropylene film in power capacitors, *Polym. Eng. Sci.*, 28, pp. 862–870.

[12] Zebouchi, N. and Malec, D. (1998). Combination of thermal and electromechanical breakdown mechanisms to analyze the dielectric breakdown in polyethylene terephthalate, *J. Appl. Phys.*, 83, pp. 6190–6192.

[13] Wang, Y., Zhou, X., Chen, Q., Chu, B. and Zhang, Q. (2010). Recent development of high energy density polymers for dielectric capacitors, *IEEE Trans. Dielectr. Electr. Insul.*, 17, pp. 1036–1042.
[14] Ihlefeld, J., Laughlin, B., Hunt-Lowery, A., Borland, W., Kingon, A. and Maria, J. P. (2005). Copper compatible barium titanate thin films for embedded passives, *J. Electroceram.*, 14, pp. 95–102.
[15] Yao, S.-H., Yuan, J.-K., Gonon, P., Bai, J., Pairis, S. and Sylvestre, A. (2012). Effect of oxygen vacancy on the dielectric relaxation of $BaTiO_3$ thin films in a quenched state, *J. Appl. Phys.*, 111, pp. 104–109.
[16] Wu, J. B., Nan, C. W., Lin, Y. H. and Deng, Y. (2002). Giant dielectric permittivity observed in Li and Ti doped NiO, *Phys. Rev. Lett.*, 89, pp. 217601.
[17] Li, J. Y., Zhang, L. and Ducharme, S. (2007). Electric energy density of dielectric nanocomposites, *Appl. Phys. Lett.*, 90, pp. 132901.
[18] Job, A. E., Alves, N., Zanin, M. and Ueki, M. M. (2003). Increasing the dielectric breakdown strength of poly(ethylene terephthalate) films using a coated polyaniline layer, *J. Phys. D: Appl. Phys.*, 36, pp. 1414–1417.
[19] Chu, B., Zhou, X., Ren, K., Neese, B., Lin, M., Wang, Q., Bauer, F. and Zhang, Q. M. (2006). A dielectric polymer with high electric energy density and fast discharge speed, *Science*, 313, pp. 334–336.
[20] Tuncer, E., Sauers, I., Randy James, D., Ellis, A. R., Parans Paranthaman, M., Goyal, A. and More, K. L. (2007). Enhancement of dielectric strength in nanocomposites, *Nanotechnology*, 18, pp. 325704.
[21] Rahimabady, M., Chen, S., Yao, K., Tay, F. E. H. and Lu, L. (2011). High electric breakdown strength and energy density in vinylidene fluoride oligomer/poly(vinylidene fluoride) blend thin films, *Appl. Phys. Lett.*, 99, pp. 142901.
[22] Yuan, J.-K., Dang, Z.-M., Yao, S.-H., Zha, J.-W., Zhou, T., Li, S.-T. and Bai, J. (2010). Fabrication and dielectric properties of advanced high permittivity polyaniline/poly(vinylidene fluoride) nanohybrid films with high energy storage density, *J. Mater. Chem.*, 20, pp. 2441–2447.
[23] Nelson, J. K. and Hu, Y. (2005). Nanocomposite dielectrics — properties and implications, *J. Phys. D: Appl. Phys.*, 38, pp. 213–222.
[24] Coppard, R. W., Bowman, J., Dissado, L. A., Rowland, S. M. and Rakowski, R. T. (1990). The effect of aluminum inclusions on the dielectric-breakdown of polyethylene, *J. Phys. D: Appl. Phys.*, 23, pp. 1554–1561.
[25] Sun, Y. Y., Zhang, Z. Q. and Wong, C. P. (2005). Influence of interphase and moisture on the dielectric spectroscopy of epoxy/silica composites, *Polymer*, 46, pp. 2297–2305.
[26] Dang, Z.-M., Xu, H.-P. and Wang, H.-Y. (2007). Significantly enhanced low-frequency dielectric permittivity in the $BaTiO_3$/poly(vinylidene fluoride) nanocomposite, *Appl. Phys. Lett.*, 90, pp. 012901.
[27] Ramesh, S., Shutzberg, B. A., Huang, C., Gao, J. and Giannelis, E. P. (2003). Dielectric nanocomposites for integral thin film capacitors: Materials design, fabrication, and integration issues, *IEEE Trans. Adv. Packaging*, 26, pp. 17–24.

[28] Manias, E. (2007). Nanocomposites — stiffer by design, *Nat. Mater.*, 6, pp. 9–11.
[29] Zhang, J., Manias, E. and Wilkie, C. A. (2008). Polymerically modified layered silicates: An effective route to nanocomposites, *J. Nanosci. Nanotechnol.*, 8, pp. 1597–1615.
[30] Tanaka, T., Montanari, G. C. and Mulhaupt, R. (2004). Polymer nanocomposites as dielectrics and electrical insulation-perspectives for processing technologies, material characterization and future applications, *IEEE Trans. Dielectr. Electr. Insul.*, 11, pp. 763–784.
[31] Li, J., Seok, S. I., Chu, B., Dogan, F., Zhang, Q. and Wang, Q. (2009). Nanocomposites of ferroelectric polymers with TiO_2 nanoparticles exhibiting significantly enhanced electrical energy density, *Adv. Mater.*, 21, pp. 217–219.
[32] Li, J., Claude, J., Norena-Franco, L. E., Seok, S. I. and Wang, Q. (2008). Electrical energy storage in ferroelectric polymer nanocomposites containing surface-functionalized $BaTiO_3$ nanoparticles, *Chem. Mater.*, 20, pp. 6304–6306.
[33] Kim, P., Doss, N. M., Tillotson, J. P., Hotchkiss, P. J., Pan, M.-J., Marder, S. R., Li, J., Calame, J. P. and Perry, J. W. (2009). High energy density nanocomposites based on surface-modified $BaTiO_3$ and a ferroelectric polymer, *ACS Nano*, 3, pp. 2581–2592.
[34] Lewis, T. J. (2004). Interfaces are the dominant feature of dielectrics at the nanometric level, *IEEE Trans. Dielectr. Electr. Insul.*, 11, pp. 739–753.
[35] Thomas, P., Varughese, K. T., Dwarakanath, K. and Varma, K. B. R. (2010). Dielectric properties of Poly(vinylidene fluoride)/$CaCu_3Ti_4O_{12}$ composites, *Compos. Sci. Technol.*, 70, pp. 539–545.
[36] Murugaraj, P., Mainwaring, D. and Mora-Huertas, N. (2005). Dielectric enhancement in polymer-nanoparticle composites through interphase polarizability, *J. Appl. Phys.*, 98, pp. 054304
[37] Guo, N., DiBenedetto, S. A., Kwon, D.-K., Wang, L., Russell, M. T., Lanagan, M. T., Facchetti, A. and Marks, T. J. (2007). Supported metallocene catalysis for *in situ* synthesis of high energy density metal oxide nanocomposites, *J. Am. Chem. Soc.*, 129, pp. 766–767.
[38] Li, J. Y. (2003). Exchange coupling in P(VDF-TrFE) copolymer based all-organic composites with giant electrostriction, *Phys. Rev. Lett.*, 90, pp. 217601.
[39] Lewis, T. J. (1994). Nanometric dielectrics, *IEEE Trans. Dielectr. Electr. Insul.*, 1, pp. 812–825.
[40] Hanemann, T. and Szabo, D. V. (2010). Polymer-nanoparticle composites: From synthesis to modern applications, *Materials*, 3, pp. 3468–3517.
[41] Chattopadhyay, S., Ayyub, P., Palkar, V. R., Gurjar, A. V., Wankar, R. M. and Multani, M. (1997). Finite-size effects in antiferroelectric $PbZrO_3$ nanoparticles, *J. Phys.: Condens. Matter*, 9, pp. 8135–8145.
[42] Yan, T., Shen, Z.-G., Zhang, W.-W. and Chen, J.-F. (2006). Size dependence on the ferroelectric transition of nanosized $BaTiO_3$ particles, *Mater. Chem. Phys.*, 98, pp. 450–455.

[43] Fan, B.-H., Zha, J.-W., Wang, D., Zhao, J. and Dang, Z.-M. (2012). Size-dependent low-frequency dielectric properties in the $BaTiO_3$/poly(vinylidene fluoride) nanocomposite films, *Appl. Phys. Lett.*, 100, pp. 012903.
[44] Arlt, G., Hennings, D. and Dewith, G. (1985). Dielectric-properties of fine-grained barium-titanate ceramics, *J. Appl. Phys.*, 58, pp. 1619–1625.
[45] Uchino, K., Sadanaga, E. and Hirose, T. (1989). Dependence of the crystal-structure on particle-size in barium-titanate, *J. Am. Ceram. Soc.*, 72, pp. 1555–1558.
[46] Hwang, U. Y., Park, H. S., Kim, Y. R. and Koo, K. K. (2004). Formation and mullitization mechanism of nano premullite particles prepared by the mixed solvents method, *Ind. Eng. Chem. Res.*, 43, pp. 3141–3150.
[47] Lewis, T. J. (2005). Interfaces: Nanometric dielectrics, *J. Phys. D: Appl. Phys.*, 38, pp. 202–212.
[48] Rao, Y. and Wong, C. P. (2004). Material characterization of a high-dielectric-constant polymer-ceramic composite for embedded capacitor for RF applications, *J. Appl. Polym. Sci.*, 92, pp. 2228–2231.
[49] Ogitani, S., Bidstrup-Allen, S. A. and Kohl, P. A. (2000). Factors influencing the permittivity of polymer/ceramic composites for embedded capacitors, *IEEE Trans. Adv. Packaging*, 23, pp. 313–322.
[50] Kim, S. W., Kim, T., Kim, Y. S., Choi, H. S., Lim, H. J., Yang, S. J. and Park, C. R. (2012). Surface modifications for the effective dispersion of carbon nanotubes in solvents and polymers, *Carbon*, 50, pp. 3–33.
[51] Zhou, T., Zha, J.-W., Cui, R.-Y., Fan, B.-H., Yuan, J.-K. and Dang, Z.-M. (2011). Improving dielectric properties of $BaTiO_3$/ferroelectric polymer composites by employing surface hydroxylated $BaTiO_3$ nanoparticles, *ACS Appl. Mater. Interfaces*, 3, pp. 2184–2188.
[52] Kim, P., Jones, S. C., Hotchkiss, P. J., Haddock, J. N., Kippelen, B., Marder, S. R. and Perry, J. W. (2007). Phosphonic acid-modiried barium titanate polymer nanocomposites with high permittivity and dielectric strength, *Adv. Mater.*, 19, pp. 1001–1004.
[53] Schandler, L. S., Brinson, L. C. and Sawyer, W. G. (2007). Polymer nanocomposites: A small part of the story, *Jom*, 59, pp. 53–60.
[54] Todd, M. G. and Shi, F. G. (2003). Characterizing the interphase dielectric constant of polymer composite materials: Effect of chemical coupling agents, *J. Appl. Phys.*, 94, pp. 4551–4557.
[55] Todd, M. G. and Shi, F. G. (2003). Molecular basis of the interphase dielectric properties of microelectronic and optoelectronic packaging materials, *IEEE Trans. Compon. Pack.* T., 26, pp. 667–672.
[56] Tanaka, T., Kozako, M., Fuse, N. and Ohki, Y. (2005). Proposal of a multi-core model for polymer nanocomposite dielectrics, *IEEE Trans. Dielectr. Electr. Insul.*, 12, pp. 669–681.
[57] Lewis, T. J. (2004). Interfaces and nanodielectrics are synonymous, *IEEE Int. Conf. solid Dielectrics*, 2, p. 792.

[86] Jung, H. M., Kang, J.-H., Yang, S. Y., Won, J. C. and Kim, Y. S. (2010). Barium titanate nanoparticles with diblock copolymer shielding layers for high-energy density nanocomposites, *Chem. Mater.*, 22, pp. 450–456.

[87] Dang, Z.-M., Wang, H.-Y. and Xu, H.-P. (2006). Influence of silane coupling agent on morphology and dielectric property in $BaTiO_3$/polyvinylidene fluoride composites, *Appl. Phys. Lett.*, 89, pp. 112902.

[88] Song, Y., Shen, Y., Liu, H., Lin, Y., Li, M. and Nan, C.-W. (2012). Enhanced dielectric and ferroelectric properties induced by dopamine-modified $BaTiO_3$ nanofibers in flexible poly(vinylidene fluoride-trifluoroethylene) nanocomposites, *J. Mater. Chem.*, 22, pp. 8063–8068.

[89] Song, Y., Shen, Y., Liu, H., Lin, Y., Li, M. and Nan, C.-W. (2012). Improving the dielectric constants and breakdown strength of polymer composites: Effects of the shape of the $BaTiO_3$ nanoinclusions, surface modification and polymer matrix, *J. Mater. Chem.*, 22, pp. 16491–16498.

[90] Tuncer, E., Sauers, I., James, D. R., Ellis, A. R. and Duckworth, R. C. (2008). Nanodielectric system for cryogenic applications: barium titanate filled polyvinyl alcohol, *IEEE Trans. Dielectr. Electr. Insul.*, 15, pp. 236–242.

[91] Dou, X., Liu, X., Zhang, Y., Feng, H., Chen, J.-F. and Du, S. (2009). Improved dielectric strength of barium titanate-polyvinylidene fluoride nanocomposite, *Appl. Phys. Lett.*, 95, 132904.

[92] Dang, Z.-M., Zhou, T., Yao, S.-H., Yuan, J.-K., Zha, J.-W., Song, H.-T., Li, J.-Y., Chen, Q., Yang, W.-T. and Bai, J. (2009). Advanced calcium copper titanate/polyimide functional hybrid films with high dielectric permittivity, *Adv. Mater.*, 21, pp. 2077–2082.

[93] Yang, W., Yu, S., Sun, R. and Du, R. (2011). Nano- and microsize effect of CCTO fillers on the dielectric behavior of CCTO/PVDF composites, *Acta Mater.*, 59, pp. 5593–5602.

[94] Prakash, B. S. and Varma, K. B. R. (2007). Dielectric behavior of CCTO/epoxy and Al-CCTO/epoxy composites, *Compos. Sci. Technol.*, 67, pp. 2363–2368.

[95] Amaral, F., Rubinger, C. P. L., Henry, F., Costa, L. C., Valente, M. A. and Barros-Timmons, A. (2008). Dielectric properties of polystyrene-CCTO composite, *J. Non-Cryst. Solids*, 354, pp. 5321–5322.

[96] Tang, H., Lin, Y., Andrews, C. and Sodano, H. A. (2011). Nanocomposites with increased energy density through high aspect ratio PZT nanowires, *Nanotechnology*, 22, pp. 015702.

[97] Yao, J., Xiong, C., Dong, L., Chen, C., Lei, Y., Chen, L., Li, R., Zhu, Q. and Liu, X. (2009). Enhancement of dielectric constant and piezoelectric coefficient of ceramic-polymer composites by interface chelation, *J. Mater. Chem.*, 19, pp. 2817–2821.

[98] Banerjee, S. and Cook-Chennault, K. A. (2011). Influence of Al particle size and lead zirconate titanate (PZT) volume fraction on the dielectric properties of PZT-epoxy-aluminum composites, *J. Eng. Mater. Technol.*, 133, pp. 041016.

[99] Tang, H., Lin, Y. and Sodano, H. A. (2012). Enhanced energy storage in nanocomposite capacitors through aligned PZT nanowires by uniaxial strain assembly, *Adv. Energy Mater.*, 2, pp. 469–476.

[100] Li, J. Y., Huang, C. and Zhang, Q. M. (2004). Enhanced electromechanical properties in all-polymer percolative composites, *Appl. Phys. Lett.*, 84, pp. 3124–3126.

[101] Cheng, Y., Chen, X., Wu, K., Wu, S., Chen, Y. and Meng, Y. (2008). Modeling and simulation for effective permittivity of two-phase disordered composites, *J. Appl. Phys.*, 103, pp. 034111.

[102] An, L., Boggs, S. A. and Callame, J. P. (2008). Energy storage in polymer films with high dielectric constant fillers, *IEEE Electr. Insul. Magn.*, 24, pp. 5–10.

[103] Almadhoun, M. N., Bhansali, U. S. and Alshareef, H. N. (2012). Nanocomposites of ferroelectric polymers with surface-hydroxylated $BaTiO_3$ nanoparticles for energy storage applications, *J. Mater. Chem.*, 22, pp. 11196–11200.

[104] Mikeska, K. R. and Cannon, W. R. (1988). Non-aqueous dispersion properties of pure barium-titanate for tape casting, *Colloid. Surf.*, 29, pp. 305–321.

[105] Chang, S.-J., Liao, W.-S., Ciou, C.-J., Lee, J.-T. and Li, C.-C. (2009). An efficient approach to derive hydroxyl groups on the surface of barium titanate nanoparticles to improve its chemical modification ability, *J. Colloid. Interface Sci.*, 329, pp. 300–305.

[106] Xie, L., Huang, X., Wu, C. and Jiang, P. (2011). Core–shell structured poly(methyl methacrylate)/$BaTiO_3$ nanocomposites prepared by in situ atom transfer radical polymerization: a route to high dielectric constant materials with the inherent low loss of the base polymer, *J. Mater. Chem.*, 21, pp. 5897–5906.

[107] Ramakrishnan, A., Dhamodharan, R. and Ruhe, J. (2006). Growth of poly(methyl methacrylate) brushes on silicon surfaces by atom transfer radical polymerization, *J. Polym. Sci. Part A.: Polym. Chem.*, 44, pp. 1758–1769.

[108] Ohno, K., Morinaga, T., Koh, K., Tsujii, Y. and Fukuda, T. (2005). Synthesis of monodisperse silica particles coated with well-defined, high-density polymer brushes by surface-initiated atom transfer radical polymerization, *Macromolecules*, 38, pp. 2137–2142.

[109] Boday, D. J., Keng, P. Y., Muriithi, B., Pyun, J. and Loy, D. A. (2010). Mechanically reinforced silica aerogel nanocomposites via surface initiated atom transfer radical polymerizations, *J. Mater. Chem.*, 20, pp. 6863–6865.

[110] Gao, Y., Gao, X., Zhou, Y. and Yan, D. (2008). Preparation of poly(methyl methacrylate) grafted titanate nanotubes by in situ atom transfer radical polymerization, *Nanotechnology*, 19, pp. 495604.

[111] Mulvihill, M. J., Rupert, B. L., He, R. R., Hochbaum, A., Arnold, J. and Yang, P. D. (2005). Synthesis of bifunctional polymer nanotubes from silicon nanowire templates via atom transfer radical polymerization, *J. Am. Chem. Soc.*, 127, pp. 16040–16041.

[112] Kong, H., Gao, C. and Yan, D. Y. (2004). Constructing amphiphilic polymer brushes on the convex surfaces of multi-walled carbon nanotubes by in situ atom transfer radical polymerization, *J. Mater. Chem.*, 14, pp. 1401–1405.

[113] Kong, H., Gao, C. and Yan, D. Y. (2004). Controlled functionalization of multiwalled carbon nanotubes by in situ atom transfer radical polymerization, *J. Am. Chem. Soc.*, 126, pp. 412–413.

[114] Goncalves, G., Marques, P. A. A. P., Barros-Timmons, A., Bdkin, I., Singh, M. K., Emami, N. and Gracio, J. (2010). Graphene oxide modified with PMMA via ATRP as a reinforcement filler, *J. Mater. Chem.*, 20, pp. 9927–9934.

[115] Fang, M., Wang, K., Lu, H., Yang, Y. and Nutt, S. (2010). Single-layer graphene nanosheets with controlled grafting of polymer chains, *J. Mater. Chem.*, 20, pp. 1982–1992.

[116] Fang, M., Wang, K., Lu, H., Yang, Y. and Nutt, S. (2009). Covalent polymer functionalization of graphene nanosheets and mechanical properties of composites, *J. Mater. Chem.*, 19, pp. 7098–7105.

[117] Maliakal, A., Katz, H., Cotts, P. M., Subramoney, S. and Mirau, P. (2005). Inorganic oxide core, polymer shell nanocomposite as a high K gate dielectric for flexible electronics applications, *J. Am. Chem. Soc.*, 127, pp. 14655–14662.

[118] Yang, K., Huang, X., Xie, L., Wu, C., Jiang, P. and Tanaka, T. (2012). Core–shell structured polystyrene/$BaTiO_3$ hybrid nanodielectrics prepared by in situ RAFT polymerization: A route to high dielectric constant and low loss materials with weak frequency dependence, *Macromol. Rapid Comm.*, 33, pp. 1921–1926.

[119] Fan, B.-H., Zha, J.-W., Wang, D.-R., Zhao, J. and Dang, Z.-M. (2012). Experimental study and theoretical prediction of dielectric permittivity in $BaTiO_3$/polyimide nanocomposite films, *Appl. Phys. Lett.*, 100, pp. 092903.

[120] Thomas, P., Satapathy, S., Dwarakanath, K. and Varma, K. B. R. (2010). Dielectric properties of poly(vinylidene fluoride)/$CaCu_3Ti_4O_{12}$ nanocrystal composite thick films, *Express Polym. Lett.*, 4, pp. 632–643.

[121] Dang, Z.-M., Lin, Y.-Q., Xu, H.-P., Shi, C.-Y., Li, S.-T. and Bai, J. (2008). Fabrication and dielectric characterization of advanced $BaTiO_3$/polyimide nanocomposite films with high thermal stability, *Adv. Funct. Mater.*, 18, pp. 1509–1517.

[122] Brandstetter, S. S., Drummy, L. F., Horwath, J. C., Schweickart, D. L. and Vaia, R. A. (2008). Breakdown voltage of thermoplastics with clay nanometer-sized fillers, *Proc. IEEE Int. Modulators and High Voltage Conf.*, p. 287.

[123] Ma, D. L., Siegel, R. W., Hong, J. I., Schadler, L. S., Martensson, E. and Onneby, C. (2004). Influence of nanoparticle surfaces on the electrical breakdown strength of nanoparticle-filled low-density polyethylene, *J. Mater. Res.*, 19, pp. 857–863.

[124] Roy, M., Nelson, J. K., MacCrone, R. K. and Schadler, L. S. (2007). Candidate mechanisms controlling the electrical characteristics of silica/XLPE nanodielectrics, *J. Mater. Sci.*, 42, pp. 3789–3799.

[125] Tuncer, E., Sauers, I., James, D. R., Ellis, A. R., Paranthaman, M. P., Aytug, T., Sathyamurthy, S., More, K. L., Li, J. and Goyal, A. (2007). Electrical properties of epoxy resin based nano-composites, *Nanotechnology*, 18, pp. 025703.

[126] Tuncer, E., Rondinone, A. J., Woodward, J., Sauers, I., James, D. R. and Ellis, A. R. (2009). Cobalt iron-oxide nanoparticle modified poly(methyl methacrylate) nanodielectrics, *Appl. Phys. A: Mater. Sci. Process.*, 94, pp. 843–852.

[127] Tomer, V., Polizos, G., Randall, C. A. and Manias, E. (2011). Polyethylene nanocomposite dielectrics: Implications of nanofiller orientation on high field properties and energy storage, *J. Appl. Phys.*, 109, pp. 074113.

[128] Imai, T., Sawa, F., Nakano, T., Ozaki, T., Shimizu, T., Kozako, M. and Tanaka, T. (2006). Effects of nano- and micro-filler mixture on electrical insulation properties of epoxy based composites, *IEEE Trans. Dielectr. Electr. Insul.*, 13, pp. 319–326.

[129] Davies, W. E. V. J., Dutton, J., Harris, F. M. and Jones, F. L. (1965). Electrical breakdown of air at high voltages, *Nature*, 205, pp. 1092–1093.

[130] Calame, J. P. (2006). Finite difference simulations of permittivity and electric field statistics in ceramic-polymer composites for capacitor applications, *J. Appl. Phys.*, 99, pp. 084101.

[131] Kittel, C. (1996) *Introduction to Solid State Physics* (*John Wiley and Sons Inc*, New York).

[132] Garnett, M. (1904) Color in metal glasses and in metallic films, *Philos. Trans. R. Soc. London*, 203, pp. 385–420.

[133] Smith, G. B. (1977). Dielectric-constants for mixed media, *J. Phys. D: Appl. Phys.*, 10, pp. L39–L42.

[134] Tuncer, E., Gubanski, S. M. and Nettelblad, B. (2001). Dielectric relaxation in dielectric mixtures: A pplication of the finite element method and its comparison with dielectric mixture formulas, *J. Appl. Phys.*, 89, pp. 8092–8100.

[135] Zakri, T., Laurent, J. P. and Vauclin, M. (1998). Theoretical evidence for 'Lichtenecker's mixture formulae' based on the effective medium theory, *J. Phys. D: Appl. Phys.*, 31, pp. 1589–1594.

[136] Tomer, V., Randall, C. A., Polizos, G., Kostelnick, J. and Manias, E. (2008). High- and low-field dielectric characteristics of dielectrophoretically aligned ceramic/polymer nanocomposites, *J. Appl. Phys.*, 103, pp. 034115.

[137] Tomer, V. and Randall, C. A. (2008). High field dielectric properties of anisotropic polymer-ceramic composites, *J. Appl. Phys.*, 104, pp.074106.

[138] Danikas, M. G. and Tanaka, T. (2009). Nanocomposites-a review of electrical treeing and breakdown, *IEEE Electr. Insul. Magn.*, 25, pp. 19–25.

[139] Vogelsang, R., Farr, T. and Frohlich, K. (2006). The effect of barriers on electrical tree propagation in composite insulation materials, *IEEE Trans. Dielectr. Electr. Insul.*, 13, pp. 373–382.
[140] Bowen, C. P., Newnham, R. E. and Randall, C. A. (1998). Dielectric properties of dielectrophoretically assembled particulate-polymer composites, *J. Mater. Res.*, 13, pp. 205–210.
[141] Wolak, M. A., Wan, A. S., Shirk, J. S., Mackey, M., Hiltner, A. and Baer, E. (2012). Imaging the effect of dielectric breakdown in a multilayered polymer film, *J. Appl. Polym. Sci.*, 123, pp. 2548–2557.
[142] Wolak, M. A., Pan, M.-J., Wan, A., Shirk, J. S., Mackey, M., Hiltner, A., Baer, E. and Flandin, L. (2008). Dielectric response of structured multilayered polymer films fabricated by forced assembly, *Appl. Phys. Lett.*, 92, pp. 113301.
[143] Mackey, M., Schuele, D. E., Zhu, L., Flandin, L., Wolak, M. A., Shirk, J. S., Hiltner, A. and Baer, E. (2012). Reduction of dielectric hysteresis in multi layered films via nanoconfinement, *Macromolecules*, 45, pp. 1954–1962.
[144] Mackey, M., Flandin, L., Hiltner, A. and Baer, E. (2011). Confined crystallization of PVDF and a PVDF-TFE copolymer in nanolayered films, *J. Polym. Sci., Part B: Polym. Phys.*, 49, pp. 1750–1761.
[145] Mackey, M., Hiltner, A., Baer, E., Flandin, L., Wolak, M. A. and Shirk, J. S. (2009). Enhanced breakdown strength of multilayered films fabricated by forced assembly microlayer coextrusion, *J. Phys. D: Appl. Phys.*, 42, pp. 175304.
[146] Tomer, V., Manias, E. and Randall, C. A. (2011). High field properties and energy storage in nanocomposite dielectrics of poly(vinylidene fluoride-hexafluoropropylene), *J. Appl. Phys.*, 110, pp. 044107.
[147] Chen, L. F., Hong, Y. P., Chen, X. J., Wu, Q. L., Huang, Q. J. and Luo, X. T. (2004). Preparation and properties of polymer matrix piezoelectric composites containing aligned $BaTiO_3$ whiskers, *J. Mater. Sci.*, 39, pp. 2997–3001.
[148] Huang, Y., Duan, X. F., Wei, Q. Q. and Lieber, C. M. (2001). Directed assembly of one-dimensional nanostructures into functional networks, *Science*, 291, pp. 630–633.
[149] Messer, B., Song, J. H. and Yang, P. D. (2000). Microchannel networks for nanowire patterning, *J. Am. Chem. Soc.*, 122, pp. 10232–10233.
[150] Yao, S.-H., Yuan, J.-K., Zhou, T., Dang, Z.-M. and Bai, J. (2011). Stretch-modulated carbon nanotube alignment in ferroelectric polymer composites: Characterization of the orientation state and its influence on the dielectric properties, *J. Phys. Chem. C*, 115, pp. 20011–20017.
[151] Yerushalmi, R., Jacobson, Z. A., Ho, J. C., Fan, Z. and Javey, A. (2007). Large scale, highly ordered assembly of nanowire parallel arrays by differential roll printing, *Appl. Phys. Lett.*, 91, pp. 203104.
[152] Yu, G., Li, X., Lieber, C. M. and Cao, A. (2008). Nanomaterial-incorporated blown bubble films for large-area, aligned nanostructures, *J. Mater. Chem.*, 18, pp. 728–734.

[153] Yu, G., Cao, A. and Lieber, C. M. (2007). Large-area blown bubble films of aligned nanowires and carbon nanotubes, *Nat. Nanotechnol.*, 2, pp. 372–377.

[154] Xu, F., Durham, J. W., III, Wiley, B. J. and Zhu, Y. (2011). Strain-release assembly of nanowires on stretchable substrates, *ACS Nano*, 5, pp. 1556–1563.

[155] Tao, A., Kim, F., Hess, C., Goldberger, J., He, R. R., Sun, Y. G., Xia, Y. N. and Yang, P. D. (2003). Langmuir–Blodgett silver nanowire monolayers for molecular sensing using surface-enhanced Raman spectroscopy, *Nano Lett.*, 3, pp. 1229–1233.

[156] Pecharroman, C., Esteban-Betegon, F., Bartolome, J. F., Lopez-Esteban, S. and Moya, J. S. (2001). New percolative $BaTiO_3$-Ni composites with a high and frequency-independent dielectric constant ($\varepsilon_r \approx$80,000), *Adv. Mater.*, 13, p. 1541.

[157] Nan, C. W. (1993). Physics of inhomogeneous inorganic materials, *Mater. Sci.*, 37, pp. 1–116.

[158] Nan, C. W., Shen, Y. and Ma, J., Physical properties of composites near percolation, *Annu. Rev. Mater. Res.*, 40, pp. 131–151.

[159] Stauffer, D. (1992). *Introduction to Percolation Theory* (Taylor & Francis, London).

[160] Efros, A. L. and Shklovskii, B. I. (1976). Critical behavior of conductivity and dielectric-constant near metal-non-metal transition threshold, *Phys. Status Solidi B*, 76, pp. 475–485.

[161] Toker, D., Azulay, D., Shimoni, N., Balberg, I. and Millo, O. (2003). Tunneling and percolation in metal-insulator composite materials, *Phys. Rev. B*, 68, pp. 041403.

[162] Nan, C. W., Birringer, R., Clarke, D. R. and Gleiter, H. (1997). Effective thermal conductivity of particulate composites with interfacial thermal resistance, *J. Appl. Phys.*, 81, pp. 6692–6699.

[163] Del Coso, R., Requejo-Isidro, J., Solis, J., Gonzalo, J. and Afonso, C. N. (2004). Third order nonlinear optical susceptibility of Cu: Al_2O_3 nanocomposites: From spherical nanoparticles to the percolation threshold, *J. Appl. Phys.*, 95, pp. 2755–2762.

[164] Chiteme, C. and McLachlan, D. S. (2000). Measurements of universal and non-universal percolation exponents in macroscopically similar systems, *Physica B*, 279, pp. 69–71.

[165] Zallen, R., (1983) *The Physics of Amorphous Solids* (*Wiley*, New York).

[166] Chen, Q., Du, P., Jin, L., Weng, W. and Han, G. (2007). Percolative conductor/polymer composite films with significant dielectric properties, *Appl. Phys. Lett.*, 91, pp. 022912.

[167] Stoyanov, H., Mc Carthy, D., Kollosche, M. and Kofod, G. (2009). Dielectric properties and electric breakdown strength of a subpercolative composite of carbon black in thermoplastic copolymer, *Appl. Phys. Lett.*, 94, pp. 232905.

[168] Huang, X. Y., Jiang, P. K. and Kim, C. U. (2007). Electrical properties of polyethylene/aluminum nanocomposites, *J. Appl. Phys.*, 102, pp. 124103.

[169] Macutkevic, J., Seliuta, D., Valusis, G., Banys, J., Kuznetsov, V., Moseenkov, S. and Shenderova, O. (2009). High dielectric permittivity of percolative composites based on onion-like carbon, *Appl. Phys. Lett.*, 95, pp. 112901.
[170] Dang, Z.-M., Peng, B., Xie, D., Yao, S.-H., Jiang, M.-J. and Bai, J. (2008). High dielectric permittivity silver/polyimide composite films with excellent thermal stability, *Appl. Phys. Lett.*, 92, pp. 112910.
[171] Panda, M., Srinivas, V. and Thakur, A. K. (2011). Role of polymer matrix in large enhancement of dielectric constant in polymer-metal composites, *Appl. Phys. Lett.*, 99, pp. 042905.
[172] Dang, Z. M., Lin, Y. H. and Nan, C. W. (2003). Novel ferroelectric polymer composites with high dielectric constants, *Adv. Mater.*, 15, p. 1625.
[173] Hakim, L. F., Portman, J. L., Casper, M. D. and Weimer, A. W. (2005). Aggregation behavior of nanoparticles in fluidized beds, *Powder Technol.*, 160, pp. 149–160.
[174] Neda, Z., Florian, R. and Brechet, Y. (1999). Reconsideration of continuum percolation of isotropically oriented sticks in three dimensions, *Phys. Rev. E*, 59, pp. 3717–3719.
[175] Bug, A. L. R., Safran, S. A. and Webman, I. (1986). Continuum percolation of permeable objects, *Phys. Rev. B*, 33, pp. 4716–4724.
[176] Berhan, L. and Sastry, A. M. (2007). Modeling percolation in high-aspect-ratio fiber systems. I. Soft-core versus hard-core models, *Phys. Rev. E*, 75, pp. 041120.
[177] Garboczi, E. J., Snyder, K. A., Douglas, J. F. and Thorpe, M. F. (1995). Geometrical percolation-threshold of overlapping ellipsoids, *Phys. Rev. E*, 52, pp. 819–828.
[178] Dang, Z.-M., Wu, J.-P., Xu, H.-P., Yao, S.-H., Jiang, M.-J. and Bai, J. (2007). Dielectric properties of upright carbon fiber filled poly(vinylidene fluoride) composite with low percolation threshold and weak temperature dependence, *Appl. Phys. Lett.*, 91, pp. 072912.
[179] He, F., Lau, S., Chan, H. L. and Fan, J. (2009). High dielectric permittivity and low percolation threshold in nanocomposites based on poly(vinylidene fluoride) and exfoliated graphite nanoplates, *Adv. Mater.*, 21, p. 710.
[180] Stankovich, S., Dikin, D. A., Dommett, G. H. B., Kohlhaas, K. M., Zimney, E. J., Stach, E. A., Piner, R. D., Nguyen, S. T. and Ruoff, R. S. (2006). Graphene-based composite materials, *Nature*, 442, pp. 282–286.
[181] Panda, M., Srinivas, V. and Thakur, A. K. (2008). On the question of percolation threshold in polyvinylidene fluoride/nanocrystalline nickel composites, *Appl. Phys. Lett.*, 92, pp. 132905.
[182] Huang, X., Jiang, P. and Xie, L. (2009). Ferroelectric polymer/silver nanocomposites with high dielectric constant and high thermal conductivity, *Appl. Phys. Lett.*, 95, pp. 242091.
[183] Panda, M., Srinivas, V. and Thakur, A. K. (2008). Surface and interfacial effect of filler particle on electrical properties of polyvinyledene fluoride/nickel composites, *Appl. Phys. Lett.*, 93, pp. 242908.

[184] Kofod, G., Risse, S., Stoyanov, H., McCarthy, D. N., Sokolov, S. and Kraehnert, R. (2011). Broad-spectrum enhancement of polymer composite dielectric constant at ultra low volume fractions of silica-supported copper nanoparticles, *ACS Nano*, 5, pp. 1623–1629.

[185] Barick, A. K. and Tripathy, D. K. (2012). Preparation and characterization of carbon nanofiber reinforced thermoplastic polyurethane nanocomposites, *J. Appl. Polym. Sci.*, 124, pp. 765–780.

[186] Sun, L.-L., Zhao, Y. and Zhong, W.-H. (2011). Dependence of dielectric properties and percolative behavior on phase separation structure induced by heterogeneous carbon nanofiber distribution in polymer blend nanocomposites, *Macromol. Mater. Eng.*, 296, pp. 992–1001.

[187] Sun, L.-L., Zhang, Z.-G. and Zhong, W.-H. (2011). Fluorination deposition on carbon nanofibers by PTFE decomposition as a facile method to enhance dispersion and interaction in PVDF composites, *J. Mater. Chem.*, 21, pp. 944–950.

[188] Sun, L. L., Li, B., Zhao, Y., Mitchell, G. and Zhong, W. H. (2010). Structure-induced high dielectric constant and low loss of CNF/PVDF composites with heterogeneous CNF distribution, *Nanotechnology*, 21, pp. 305702.

[189] Yao, S.-H., Dang, Z.-M., Xu, H.-P., Jiang, M.-J. and Bai, J. (2008). Exploration of dielectric constant dependence on evolution of microstructure in nanotube/ferroelectric polymer nanocomposites, *Appl. Phys. Lett.*, 92, pp. 082902.

[190] Yao, S.-H., Dang, Z.-M., Jiang, M.-J., Xu, H.-P. and Bai, J. (2007). Influence of aspect ratio of carbon nanotube on percolation threshold in ferroelectric polymer nanocomposite, *Appl. Phys. Lett.*, 91, pp. 212901.

[191] Dang, Z.-M., Wang, L., Yin, Y., Zhang, Q. and Lei, Q.-Q. (2007). Giant dielectric permittivities in functionalized carbon-nanotube/electroactive-polymer nanocomposites, *Adv. Mater.*, 19, p. 852.

[192] Wang, L. and Dang, Z. M. (2005). Carbon nanotube composites with high dielectric constant at low percolation threshold, *Appl. Phys. Lett.*, 87, pp. 042903.

[193] Li, Q., Xue, Q., Hao, L., Gao, X. and Zheng, Q. (2008). Large dielectric constant of the chemically functionalized carbon nanotube/polymer composites, *Compos. Sci. Technol.*, 68, pp. 2290–2296.

[194] Yuan, J.-K., Yao, S.-H., Sylvestre, A. and Bai, J. (2012). Biphasic polymer blends containing carbon nanotubes: Heterogeneous nanotube distribution and its influence on the dielectric properties, *J. Phys. Chem. C*, 116, pp. 2051–2058.

[195] Yuan, J.-K., Yao, S.-H., Dang, Z.-M., Sylvestre, A., Genestoux, M. and Bai, J. (2011). Giant dielectric permittivity nanocomposites: Realizing true potential of pristine carbon nanotubes in polyvinylidene fluoride matrix through an enhanced interfacial interaction, *J. Phys. Chem. C*, 115, pp. 5515–5521.

[196] Yuan, J.-K., Li, W.-L., Yao, S.-H., Lin, Y.-Q., Sylvestre, A. and Bai, J. (2011). High dielectric permittivity and low percolation threshold in polymer composites based on SiC-carbon nanotubes micro/nano hybrid, *Appl. Phys. Lett.*, 98, pp. 032901.

[197] Zhang, S., Wang, H., Wang, G. and Jiang, Z. (2012). Material with high dielectric constant, low dielectric loss, and good mechanical and thermal properties produced using multi-wall carbon nanotubes wrapped with poly(ether sulphone) in a poly(ether ether ketone) matrix, *Appl. Phys. Lett.*, 101, pp. 012904.

[198] Simoes, R., Silva, J., Vaia, R., Sencadas, V., Costa, P., Gomes, J. and Lanceros-Mendez, S. (2009). Low percolation transitions in carbon nanotube networks dispersed in a polymer matrix: Dielectric properties, simulations and experiments, *Nanotechnology*, 20, pp. 035703.

[199] Wu, C., Huang, X., Wu, X., Yu, J., Xie, L. and Jiang, P. (2012). TiO_2-nanorod decorated carbon nanotubes for high-permittivity and low-dielectric-loss polystyrene composites, *Compos. Sci. Technol.*, 72, pp. 521–527.

[200] Chang, J., Liang, G., Gu, A., Cai, S. and Yuan, L. (2012). The production of carbon nanotube/epoxy composites with a very high dielectric constant and low dielectric loss by microwave curing, *Carbon*, 50, pp. 689–698.

[201] He, L., Lau, S., Chan, H. L. and Fan, J. T. (2009). High dielectric permittivity and low percolation threshold in nanocomposites based on poly(vinylidene fluoride) and exfoliated graphite nanoplates, *Adv. Mater.*, 21, pp. 710–715.

[202] Fan, P., Wang, L., Yang, J. Chen, F. and Zhong, M. (2012). Graphene/poly(vinylidene fluoride) composites with high dielectric constant and low percolation threshold, *Nanotechnology*, 23, pp. 365702.

[203] Wu, C., Huang, X., Wang, G., Wu, X., Yang, K., Li, S. and Jiang, P. (2012). Hyperbranched-polymer functionalization of graphene sheets for enhanced mechanical and dielectric properties of polyurethane composites, *J. Mater. Chem.*, 22, pp. 7010–7019.

[204] Yu, J., Huang, X., Wu, C. and Jiang, P. (2011). Permittivity, thermal conductivity and thermal stability of poly(vinylidene fluoride)/graphene nanocomposites, *IEEE Trans. Dielectr. Electr. Insul.*, 18, pp. 478–484.

[205] Wang, G., Deng, Y. and Guo, L. (2010). Single-crystalline ZnO nanowire bundles: Synthesis, mechanism and their application in dielectric composites, *Chem. Eur. J.*, 16, pp. 10220–10225.

[206] Wang, G., Deng, Y., Xiang, Y. and Guo, L. (2008). Fabrication of radial ZnO nanowire clusters and radial ZnO/PVDF composites with enhanced dielectric properties, *Adv. Funct. Mater.*, 18, pp. 2584–2592.

[207] Kerr, C. J., Huang, Y. Y., Marshall, J. E. and Terentjev, E. M. (2011). Effect of filament aspect ratio on the dielectric response of multiwalled carbon nanotube composites, *J. Appl. Phys.*, 109, pp. 094109.

[208] Dubnikova, I., Kuvardina, E., Krasheninnikov, V., Lomakin, S., Tchmutin, I. and Kuznetsov, S. (2010). The effect of multiwalled carbon nanotube dimensions on the morphology, mechanical, and electrical properties of melt

mixed polypropylene-based composites, *J. Appl. Polym. Sci.*, 117, pp. 259–272.

[209] Dang, Z. M., Nan, C. W., Xie, D., Zhang, Y. H. and Tjong, S. C. (2004). Dielectric behavior and dependence of percolation threshold on the conductivity of fillers in polymer-semiconductor composites, *Appl. Phys. Lett.*, 85, pp. 97–99.

[210] Deepa, K. S., Kumari Nisha, S., Parameswaran, P., Sebastian, M. T. and James, J. (2009). Effect of conductivity of filler on the percolation threshold of composites, *Appl. Phys. Lett.*, 94, pp. 142902.

[211] Huang, C. and Zhang, Q. M. (2004). Enhanced dielectric and electromechanical responses in high dielectric constant all-polymer percolative composites, *Adv. Funct. Mater.*, 14, pp. 501–506.

[212] Qi, L., Lee, B. I., Chen, S. H., Samuels, W. D. and Exarhos, G. J. (2005). High-dielectric-constant silver-epoxy composites as embedded dielectrics, *Adv. Mater.*, 17, p. 1777.

[213] Gyure, M. F. and Beale, P. D. (1992). Dielectric-breakdown in continuous models of metal-loaded dielectrics, *Phys. Rev. B*, 46, pp. 3736–3746.

[214] Flandin, L., Prasse, T., Schueler, R., Schulte, K., Bauhofer, W. and Cavaille, J. Y. (1999). Anomalous percolation transition in carbon-black-epoxy composite materials, *Phys. Rev. B*, 59, pp. 14349–14355.

[215] Balberg, I., Azulay, D., Toker, D. and Millo, O. (2004). Percolation and tunneling in composite materials, *Int. J. Mod. Phys. B*, 18, pp. 2091–2121.

[216] Pecharroman, C. and Moya, J. S. (2000). Experimental evidence of a giant capacitance in insulator-conductor composites at the percolation threshold, *Adv. Mater.*, 12, pp. 294–297.

[217] Xu, J. W. and Wong, C. P. (2005). Low-loss percolative dielectric composite, *Appl. Phys. Lett.*, 87, pp. 082907.

[218] Xu, J. W., Moon, K. S., Tison, C. and Wong, C. P. (2006). A novel aluminum-filled composite dielectric for embedded passive applications, *IEEE Trans. Adv. Packaging*, 29, pp. 295–306.

[219] Shen, Y., Lin, Y., Li, M. and Nan, C.-W. (2007). High dielectric performance of polymer composite films induced by a percolating interparticle barrier layer, *Adv. Mater.*, 19, p. 1418.

[220] Shen, Y., Lin, Y. and Nan, C.-W. (2007). Interfacial effect on dielectric properties of polymer nanocomposites filled with core/shell-structured particles, *Adv. Funct. Mater.*, 17, pp. 2405–2410.

[221] Sun, X. M. and Li, Y. D. (2005). Ag@C core/shell structured nanoparticles: Controlled synthesis, characterization, and assembly, *Langmuir*, 21, pp. 6019–6024.

[222] Yang, C., Lin, Y. and Nan, C. W. (2009). Modified carbon nanotube composites with high dielectric constant, low dielectric loss and large energy density, *Carbon*, 47, pp. 1096–1101.

[223] Liu, H., Shen, Y., Song, Y., Nan, C.-W., Lin, Y. and Yang, X. (2011). Carbon nanotube array/polymer core/shell structured composites with high dielectric permittivity, low dielectric loss, and large energy density, *Adv. Mater.*, 23, pp. 5104–5108.

[224] Grossiord, N., Loos, J. and Koning, C. E. (2005). Strategies for dispersing carbon nanotubes in highly viscous polymers, *J. Mater. Chem.* 15, pp. 2349–2352.

[225] Sun, L. L., Li, B., Zhang, Z. G. and Zhong, W. H. (2010). Achieving very high fraction of beta-crystal PVDF and PVDF/CNF composites and their effect on AC conductivity and microstructure through a stretching process, *Eur. Polym. J.*, 46, pp. 2112–2119.

[226] Dang, Z. M., Shen, Y. and Nan, C. W. (2002). Dielectric behavior of three-phase percolative Ni-$BaTiO_3$/polyvinylidene fluoride composites, *Appl. Phys. Lett.*, 81, pp. 4814–4816.

[227] Dang, Z. M., Fan, L. Z., Shen, Y. and Nan, C. W. (2003). Dielectric behavior of novel three-phase MWNTs/$BaTiO_3$/PVDF composites, *Mater. Sci. Eng. B*, 103, pp. 140–144.

[228] Dang, Z. M., Fan, L. Z., Shen, Y. and Nan, C. W. (2003). Study on dielectric behavior of a three-phase CF/(PVDF+$BaTiO_3$) composite, *Chem. Phys. Lett.*, 369, pp. 95–100.

[229] Yao, S. Y., Dang, Z. (2009). Application of dielectric materials, Proc. 9th Int. Conf. Properties, *Applications of Dielectric Materials*, p. 765.

[230] Dang, Z.-M., Yao, S.-H., Yuan, J.-K. and Bai, J. (2010). Tailored dielectric properties based on microstructure change in $BaTiO_3$-carbon nanotube/polyvinylidene fluoride three-phase nanocomposites, *J. Phys. Chem. C*, 114, pp. 13204–13209.

[231] Yao, S.-H., Yuan, J.-K., Dang, Z.-M. and Bai, J. (2010). High dielectric performance of three-component nanocomposites induced by a synergetic effect, *Mater. Lett.*, 64, pp. 2682–2684.

[232] Li, Y., Huang, X., Hu, Z., Jiang, P., Li, S. and Tanaka, T. (2011). Large dielectric constant and high thermal conductivity in poly(vinylidene fluoride)/barium titanate/silicon carbide three-phase nanocomposites, *ACS Appl. Mater. Interfaces*, 3, pp. 4396–4403.

[233] Wang, G. (2010). Enhanced dielectric properties of three-phase-percolative composites based on thermoplastic-ceramic matrix ($BaTiO_3$+PVDF) and ZnO radial nanostructures, *ACS Appl. Mater. Interfaces*, 2, pp. 1290–1293.

[234] Dang, Z.-M., Xia, B., Yao, S.-H., Jiang, M.-J., Song, H.-T., Zhang, L.-Q. and Xie, D. (2009). High-dielectric-permittivity high-elasticity three-component nanocomposites with low percolation threshold and low dielectric loss, *Appl. Phys. Lett.*, 94, pp. 042902.

[235] Lu, J., Moon, K.-S. and Wong, C. P. (2008). Silver/polymer nanocomposite as a high-K polymer matrix for dielectric composites with improved dielectric performance, *J. Mater. Chem.*, 18, pp. 4821–4826.

[236] Lu, J. X., Moon, K. S., Xu, J. W. and Wong, C. P. (2006). Synthesis and dielectric properties of novel high-K polymer composites containing in-situ formed silver nanoparticles for embedded capacitor applications, *J. Mater. Chem.*, 16, pp. 1543–1548.

[237] Yang, W., Yu, S., Sun, R., Ke, S., Huang, H. and Du, R. (2011). Electrical modulus analysis on the Ni/CCTO/PVDF system near the percolation threshold, *J. Phys. D: Appl. Phys.*, 44, pp. 475305.

[238] Li, W., Meng, Q., Zheng, Y., Zhang, Z., Xia, W. and Xu, Z. (2010). Electric energy storage properties of poly(vinylidene fluoride), *Appl. Phys. Lett.*, 96, pp. 192905.

[239] Xia, W., Xu, Z., Wen, F., Li, W. and Zhang, Z. (2010). Crystalline properties dependence of dielectric and energy storage properties of poly(vinylidene fluoride-chlorotrifluoroethylene), *Appl. Phys. Lett.*, 97, pp. 222905.

[240] Wu, S., Lin, M., Lu, S. G., Zhu, L. and Zhang, Q. M. (2011). Polar-fluoropolymer blends with tailored nanostructures for high energy density low loss capacitor applications, *Appl. Phys. Lett.*, 99, pp. 132901.

[241] Chu, B., Zhou, X., Neese, B., Zhang, Q. M. and Bauer, F. (2006). Relaxor ferroelectric poly(vinylidene fluoride-trifluoroethylene-chlorofluoroethylene) terpolymer for high energy density storage capacitors, *IEEE Trans. Dielectr. Electr. Insul.*, 13, pp. 1162–1169.

[242] Zhang, S. H., Klein, R. J., Ren, K. L., Chu, B. J., Zhang, X., Runt, J. and Zhang, Q. M. (2006). Normal ferroelectric to ferroelectric relaxor conversion in fluorinated polymers and the relaxor dynamics, *J. Mater. Sci.*, 41, pp. 271–280.

[243] Zhou, X., Chu, B., Neese, B., Lin, M. and Zhang, Q. M. (2007). Electrical energy density and discharge characteristics of a poly(vinylidene fluoride-chlorotrifluoroethylene) copolymer, *IEEE Trans. Dielectr. Electr. Insul.*, 14, pp. 1133–1138.

[244] Zhou, X., Zhao, X., Suo, Z., Zou, C., Runt, J., Liu, S., Zhang, S. and Zhang, Q. M. (2009). Electrical breakdown and ultrahigh electrical energy density in poly(vinylidene fluoride-hexafluoropropylene) copolymer, *Appl. Phys. Lett.*, 94, pp. 162901.

[245] Yamada, K., Oie, M. and Takayanagi, M. (1983). Relationship between piezoelectricity and superstructure in highly oriented poly(vinylidene fluoride) film prepared by zone drawing, *J. Polym. Sci. Part B: Polym. Phys.*, 21, pp. 1063–1077.

[246] Zhang, Q. M., Bharti, V. and Zhao, X. (1998). Giant electrostriction and relaxor ferroelectric behavior in electron-irradiated poly(vinylidene fluoride-trifluoroethylene) copolymer, *Science*, 280, pp. 2101–2104.

[247] Xia, F., Cheng, Z. Y., Xu, H. S., Li, H. F., Zhang, Q. M., Kavarnos, G. J., Ting, R. Y., Abdul-Sedat, G. and Belfield, K. D. (2002). High electromechanical responses in a poly(vinylidene fluoride-trifluoroethylene-chlorofluoroethylene) terpolymer, *Adv. Mater.*, 14, p. 1574.

[248] Fröhlich, H. (1947) *Theory of Dielectrics* (Oxford university press, London).

[249] Chu, B., Neese, B., Lin, M., Lu, S.-G. and Zhang, Q. M. (2008). Enhancement of dielectric energy density in the poly(vinylidene fluoride)-based terpolymer/copolymer blends, *Appl. Phys. Lett.*, 93, pp. 152903.

[250] Stark, K. H. and Garton, C. G. (1955). Electric strength of irradiated polythene, *Nature*, 176, pp. 1225–1226.